Concrete
In The Service
Of Mankind

Radical
Concrete Technology

Related books from E & FN Spon

Alternative Materials for the Reinforcement and Prestressing of Concrete
Edited by J.L. Clarke

Cement-based Composites: Materials, Mechanical Properties and Performance
A.M. Brandt

Continuous and Integral Bridges
Edited by B. Pritchard

Design Aids for Eurocode 2: Design of Concrete Structures
The Concrete Societies of The UK, Netherlands and Germany

Durability Design of Concrete Structures
Edited by A. Sarja and E. Vesikari

Fracture Mechanics of Cementitious Materials
Y.W. Mai and B. Cotterell

HAPM Component Life Manual
HAPM Publications Ltd

High Performance Fiber Reinforced Cementitious Composites 2
Edited by A.E. Naaman and H.W. Reinhardt

Introduction to Eurocode 2: Design of Concrete Structures
D. Beckett and A. Alexandrou

Non-metallic (FRP) Reinforcement for Concrete Structures
Edited by L. Taerwe

Size-Scale Effects in the Failure Mechanisms of Materials and Structures
Edited by A. Carpinteri

Strip Method Design Handbook
A. Hillerborg

Structural Design of Polymer Composites: Eurocomp Design Code and Handbook
Edited by J.L. Clarke

For details, contact the Promotions Department, E & FN Spon, 2-6 Boundary Row, London SE1 8HN, UK. Tel: 0171-865 0066, Fax: 0171-522-9623.

Radical
Concrete Technology

Proceedings of the International Conference
held at the University of Dundee, Scotland, UK
on 27-28 June 1996

Edited by

Ravindra K. Dhir
Director, Concrete Technology Unit
University of Dundee

and

Peter C. Hewlett
Director, British Board of Agrément

E & FN SPON
An Imprint of Chapman & Hall

London · Weinheim · New York · Tokyo · Melbourne · Madras

Published by E & FN Spon, an imprint of
Chapman & Hall, 2—6 Boundary Row, London SE1 8HN, UK

Chapman & Hall, 2—6 Boundary Row, London SE1 8HN, UK

Chapman & Hall, GmbH, Pappelallee 3, 69469 Weinheim, Germany

Chapman & Hall USA, 115 Fifth Avenue, New York, NY 10003, USA

Chapman & Hall Japan, ITP-Japan, Kyowa Building, 3F, 2-2-1 Hirakawacho, Chiyoda-ku, Tokyo 102, Japan

Chapman & Hall Australia, 102 Dodds Street, South Melbourne, Victoria 3205, Australia

Chapman & Hall India, R. Seshadri, 32 Second Main Road, CIT East, Madras 600 035

First edition 1996

© 1996 E & FN Spon

Printed in Great Britain by St Edmundsbury Press, Bury St Edmunds, Suffolk

ISBN   0 419 21480 1

        0 419 21500 X (5 volume set)

A catalogue record for this book is available from the British Library

Publisher's Note This book has been prepared from camera ready copy provided by the individual contributors in order to make the book available for the Conference.

∞ Printed on acid-free paper, manufactured in accordance with
    ANSI/NISO Z39.48-1992 (Permanence of Paper)

The cover illustration shows the construction of the Colchester Eastern Approach road embankment using ultra-low density foamed concrete. Photograph courtesy Ready Mixed Concrete (UK) Ltd.

PREFACE

Concrete is ubiquitous and unique. Indeed, there are no alternatives to concrete as a volume construction material. This raises important questions of how concrete should be designed and constructed for cost effective use in the short and long-term, and yet encourage further radical development. Equally, it must also be environmentally-friendly during manufacture and in its aesthetic presentation in structures.

The Concrete Technology Unit (CTU) of the University of Dundee has organised this major 5 day International Congress, following the conferences, Protection of Concrete in 1990 and Concrete 2000: Economic and Durable Construction Though Excellence in 1993, as part of its continuing commitment to the development of excellence in concrete construction.

The central theme of the Congress was Concrete in the Service of Mankind, under which 5 self-contained conferences were organised; *(i) Concrete for Environment Enhancement, (ii) Concrete for Infrastructure and Utilities, (iii) Appropriate Concrete Technology, (iv) Radical Concrete Technology* and *(v) Concrete Repair, Rehabilitation and Protection.* In total 350 papers were presented by authors from 70 countries worldwide.

The Congress Opening Addresses were given by the Lord James Douglas-Hamilton MP, Minister of State for the Construction Industry, Scotland and by Dr Ian J. Graham-Bryce, Principal and Vice-Chancellor of the University of Dundee. The Opening Papers were presented by Emeritus Professors P. Kumar Metha and Ben C. Gerwick, University of California, Berkeley, USA and Professor John Morris, University of Witwatersrand and Mr Spencer S. Sephton, PPC Cement (pty), South Africa. The closing address was given by Professor Peter C. Hewlett, Director of the British Board of Agrément, UK and Visiting Industrial Professor, Department of Civil Engineering, University of Dundee.

The Congress was supported by 14 major International Institutions together with 23 Sponsors and 50 Exhibitors, highlighting the importance of concrete and the close cooperation between the CTU and industry.

A Congress of this size and scope was a major undertaking. The immense efforts of the Organising, International Advisory and National Technical Committees, who advised on the selection and review of papers is gratefully noted. The efforts of all the Authors and Chairmen of the various Technical Sessions and, in particular, those who travelled from afar to come to Dundee are greatly appreciated as are all the CTU staff and research students for their sterling efforts in ensuring the smooth running of the Congress. Particular thanks must be given to the two Congress joint Secretaries, Mr Neil A. Henderson and Dr Michael J. McCarthy and the Unit Secretaries Mr Steven Scott and Miss Diane Sherriff.

All the Proceedings have been prepared directly from the manuscripts provided by the authors and, therefore, there may be some errors or inaccuracies that have been inadvertently overlooked.

Dundee   Ravindra K Dhir
February 1996

INTRODUCTION

Concrete has been regarded as a mature technology which has become tainted by its very familiarity. However, by unlocking its potential, cement-based composite materials can offer a performance as yet unexplored. There is, of course, the 'Holy Grail' of the development of high strength, ductile concrete which can rival steel directly. Indeed, it is now possible to produce compressive strengths, on a commercial basis, in excess of 250MPa but equal tensile stress capacity remains as elusive as ever.

Many countries now regularly consider the use of high performance concrete, both for strength and/or durability. Such materials are already being proven in tall structures, bridges and tunnels. However, as these new concretes become widely available, it will be necessary to develop design tools that can exploit these, both safely and economically.

Furthermore, it would appear that there must be an improved interaction between architects and engineers in order to advance the so-called ultra structures, utilising radical concrete technologies. Equally, specifications must be developed which can be used to provide confidence in the use of many new materials that are becoming available. Perhaps the most pressing need is the explicit design for durability.

The Proceeding of this Conference; *'Radical Concrete Technology'* dealt with all of these subject areas and the issues thus raised, under six clearly identified themes: (i) High Performance Concrete, (ii) Ultra Structures, (iii) Maximising Strength/Durability, (iv) Performance Specifications, (v) Construction Techniques and (vi) New Materials Technology. Each theme started with a Leader Paper presented by the foremost exponents in their respective fields. There were a total of 71 papers presented during the 2 day International Conference and compiled into these Proceedings.

Dundee   Ravindra K Dhir
February 1996   Peter C Hewlett

CONTENTS

OPENING ADDRESSES

Chairman   Professor R K Dhir, University of Dundee, United Kingdom

Opening of the Congress
Lord James Douglas Hamilton MP, Minister of State for the Construction Industry, Scotland

Welcoming the Delegates to the University
Dr I J Graham-Bryce, Principal and Vice Chancellor, University of Dundee

THEME 1   HIGH PERFORMANCE CONCRETE

Chairmen   Dr Habib Bein-Al-Abideen, Ministry of Public Works & Housing, Saudi Arabia
Dr J B Menzies, Engineering Consultant, United Kingdom
Professor H R Sasse, University of Technologie Aachen, Germany
Professor S P Shah, Northwestern University, USA

THEME 2   ULTRA STRUCTURES

Chairmen   Professor C Andradé, Institute Educardo Tonoja of Construction
Sciences, Spain
Professor G Somerville, British Cement Association, United
Kingdom

Leader Paper

THEME 3   MAXIMISING STRENGTH/DURABILITY

Chairmen   Mr L H McCurrich, Fosroc International Limited, United Kingdom
Professor S Nagataki, Tokyo Institute of Technology, Japan
Professor A E Sarja, Technical Research Centre of Finland, Finland

THEME 4 PERFORMANCE SPECIFICATIONS

Chairmen   Mr J A Bickley, Concrete Canada, Canada
Professor S Y N Chan, Hong Kong Polytechnic University, Hong Kong
Mr P Livesey, Castle Cement Limited, United Kingdom

Leader Paper

THEME 5 CONSTRUCTION TECHNIQUES

Chairmen   Professor S Besari, Institut Teknologi Bandung, Indonesia
             Professor J H Bungey, University of Liverpool, United Kingdom

Leader Paper

THEME 6   NEW MATERIALS TECHNOLOGY

Chairmen   Dr R J Collins, Building Research Establishment, United Kingdom
                Professor F P Glasser, University of Aberdeen, United Kingdom

Late Papers

CLOSING ADDRESS

Chairman	Dr T A Harrison
	British Ready Mixed Concrete Association, United Kingdom
Presented by	*Professor P C Hewlett, Director*
	British Board of Agrément, United Kingdom

OPENING ADDRESS

Lord James Douglas Hamilton MP
Minister of State for the Construction Industry,
Scotland

WELCOMING ADDRESS

Dr I J Graham-Bryce
Principal and Vice-Chancellor
University of Dundee

Chairman

Professor R K Dhir
University of Dundee
United Kingdom

HIGH PERFORMANCE CONCRETES

Chairmen **Dr Habib Bein-Al-Abideen**
Ministry of Public Works & Housing
Saudi Arabia

Dr J B Menzies
Engineering Consultant
United Kingdom

Professor H R Sasse
University of Technologie Aachen
Germany

Professor S P Shah
Northwestern University
USA

Leader Paper

The Concrete Canada Programme - From 1 MPa to 1,000 MPa

Mr J A Bickley
Concrete Canada
Canada

THE CONCRETE CANADA PROGRAMME - FROM 1 MPa TO 1,000 MPa

J A Bickley

Concrete Canada

Canada

ABSTRACT. Concrete Canada is a Network of Centres of Excellence on High Performance Concrete. It comprises twelve principal researchers from ten universities plus three industrial partners. In the current phase of its eight year programme the emphasis is on technology transfer to the construction industry. One of the most effective technology transfer mediums is demonstration projects. Concrete Canada has funded a large number of these. The paper reports on this programme and results to date.

Keywords: Demonstration projects, high performance concrete.

John A. Bickley, P. Eng. is Implementation Manager for Concrete Canada, the Canadian Network of Centres of Excellence on High Performance Concrete. He is a Fellow of the Institution of Civil Engineers and ACI, and a member of CSA, ACI and ASTM committees.

Radical Concrete Technology. Edited by R K Dhir and P C Hewlett. Published in 1996 by E & FN Spon, 2–6 Boundary Row, London SE1 8HN, UK. ISBN 0 419 21480 1.

INTRODUCTION

In 1990 the Canadian Government established the Networks of Centres of Excellence Programme and, in Phase 1, allocated 240 million dollars to fifteen research networks. These networks, chosen from 158 applicants, were in diverse, generally high-tech, disciplines such as genetics, telecommunications, robotics, microelectronics and neuroscience.

One group was chosen in the field of civil engineering: The Network of Centres of Excellence on High Performance Concrete. In 1994, after a detailed review process, ten of the original fifteen networks were re-funded for Phase 2 of the programme. The concrete network, now called Concrete Canada, was again successful.

In Phase 1 the Network comprised eleven Principal Investigators, nine from seven universities across Canada, and two from industry. In Phase 2 four Principal Investigators were added, one each from three additional universities and one more from industry.

Most of the funding supports research projects at the ten universities, but there is a much stronger emphasis in Phase 2 on technology transfer to the construction industry. For this purpose the position of implementation manager has been created, and significant funding has been applied to demonstration projects. These are current construction and industrial projects to which high performance concrete technology is being transferred.

NETWORKING

One of the fundamental concepts embodied in the Centres of Excellence programme is that of "Networking". The principals of Concrete Canada come from the three solitudes of the construction industry: design, materials and construction. In the early stages of Phase I effective and sustained networking was not easily achieved. Each principal tends to be focused on a specific and limited number of subjects. Historically there has been little interaction between the three main sectors of the construction industry. It was probably year three of Phase 1 before the level of networking between principals reached the level intended by the government. The interaction, once started, achieved a healthy ongoing momentum, and has proved to be very fruitful. The principals of the network have a wide range of specialties within their chosen field, and combining several specialties on a research project has proven to be synergistic and stimulating.

Management

Figure 1 shows the management structure of the Phase 2 network.

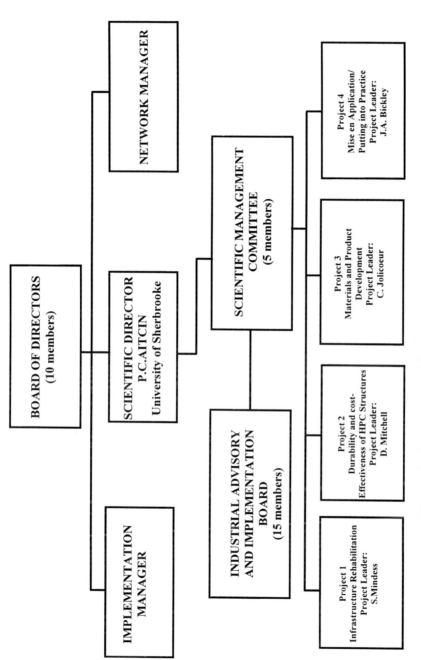

Figure 1. Structure and Organisation of the Network

4 Bickley

Code Committees

The general adoption of high performance concrete will only come about when standards for design and construction are adopted and published by the relevant code committees. Active membership on technical committees is thus an important role for Concrete Canada principals. Changes to national codes are prime deliverables.

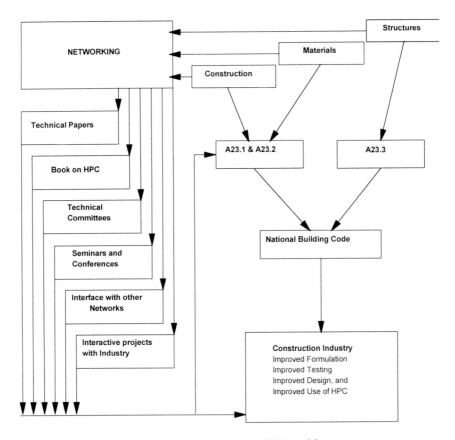

Figure 2   Network Focus and Deliverables

Figure 2 summarises the network focus and deliverables. The essential step in getting high performance concrete into general use is the adoption of changes in national codes and standards. In Canada the most important national standards for concrete are Canadian Standards CSA Committees A23.1 and A23.2 - Concrete Materials and Methods of Construction and Methods of Test for Concrete, and CSA Committee A23.3 - Code for the Design of Concrete Structures.

In 1994 new editions of these three standards were published, and significant input resulting from Concrete Canada research was incorporated into both documents.

In A23.1 a new Section 29 was added entitled "Special Requirements for High Strength Concrete". The definition of high-strength concrete was changed from concrete having a specified strength of 40 MPa or higher to concrete having a specified strength of 70 MPa or higher.

The new clauses in full plus a commentary are detailed in papers available from Concrete Canada. [1,2]

Principals of Concrete Canada sit on approximately one hundred CSA, ACI, ASTM, RILEM, FIP, ASCE and CEB technical committees.

The High Performance Concrete Market

Initially the main use of high performance concrete in North America was in structures requiring high strength. It was soon realised that while most high performance concrete was high-strength, due to the invariable use of low and very low water-cement ratios, other beneficial properties resulted. High-strength is usually easy to achieve, given aggregates of good quality, but the pre-occupation today is with durability. The deterioration of structures due to de-icing salt induced corrosion, as well as freeze-thaw damage and chemical attack, has been of epidemic proportions in North America. In Canada alone corrosion damage to parking structures has been estimated to be as high as two billion dollars.[3]

The beneficial properties of high performance concrete can be summarised as follows:

> High strength, including very high strength and high strength
> at very early ages.
> Abrasion resistance.
> Low permeability to water and chemical ions.
> Low absorption.
> Low diffusion coefficient.
> High resistivity.
> Good resistance to chemical attack.
> High modulus of elasticity.
> High resistance to freezing and thawing
> damage.
> Volume stability.
> Inhibition of bacterial or mould growth.

It can thus be seen that high performance concrete can have many applications, other than in outstanding major engineering structures, that would not normally be considered. Residential basements, agricultural structures and mine backfilling use very significant quantities of concrete and traditionally use low quality concretes and low-tech practices. Unexpected benefits can be obtained by taking a high-tech approach using the more expensive high performance concretes. In major engineering structures requirements for durability have in recent years been partially met by the use of epoxy-coated steel. There are now some reservations about the use of coated steel in all situations. On projects such as Hibernia, The Prince Edward Island Link and precast tunnel segments for new subways in Toronto, long service life requirements under conditions of severe exposure are being met by reliance on high performance cover concrete protecting uncoated steel.

There is a trend to turnkey projects where private investment finances, designs, builds, operates and maintains infrastructure. In these projects a long service life and low maintenance costs become significant factors. The assurance of long term durability by the use of high performance concrete is one way in which these objectives can be met.

While high performance concrete has a price delivered to site that is higher than normal structural concrete it can be less expensive to place. Additionally in some structures the superior properties of the high performance concrete can be used to effect savings at the design stage. In the construction of a bridge in Quebec a first cost saving of 5% was reported by the Ministry of Transportation.4

Further there is generally a higher level of QA/QC on these projects since the achievement of high quality is in the interests of the builder. It is thus more likely that well established good construction practices will be conscientiously followed. The concrete even gets cured!

The potential use of high performance concrete should be evaluated for all projects, since as reviewed above, and as illustrated by the network's demonstration projects, it is beneficial in more applications than may at first seem obvious.

Demonstration Projects

Possibly the most effective method of technology transfer is the demonstration project. High performance concrete is introduced into a project under controlled conditions, with adequate pre-testing, indoctrination, and hands-on site supervision. Nearly all such projects require special testing, additional to normal requirements, and, typically, large scale trials to provide field practice for the construction team. In some cases the demonstration project may provide testing and monitoring on a high performance concrete structure in order to obtain data that would not otherwise be obtained by the construction team. Additionally some demonstration projects are in highly specialised fields. A summary of the demonstration projects, complete or in progress, is given below.

Project	Project Description
Bridges The Portneuf Bridge [A]	First demonstration bridge using 70 MPa air-entrained concrete.
Montreal Autoroute [B]	Design changes from standard resulted in a 5% saving in first cost.
Yamaska Bridge	Laboratory and field trials of pumping planned to determine a methodology to assure a high quality air void system in place.
Bridge 95-39, Highway 20	Replacement bridge, all concrete 60 MPa air-entrained. Uncoated steel.
Prince Edward Island Link	12.9 km long precast structure. Main units weigh 7000 tonnes. Temperature and corrosion monitoring.
Jacques Cartier Bridge [B]	Rehabilitation of underwater piers using superfine cement grouts.
Highways Highway 407	10^6 m^3 of concrete, 600 lane kilometres of exposed concrete pavement, 128 concrete bridges.
Canals Rideau Canal	Rehabilitation of masonry walls using superfine cement grouts. Evaluation of quality of grouting using cross-hole tomography.
Residential Experimental Basements, Sherbrooke	Project to investigate practical aspects including placing and labour costs.
National Basement Project	National Research Council study which will end with the construction of demonstration basements.

	Durability of HPC in Pulp	First project with Pulp and Paper .
	Mill Effluents and Marine Exposure	Industry to evaluate in-service durability of HPC in an aggressive environment.
	1 MPa Mine Backfill	Use of superplasticiser technology produces cheaper backfill material.
	Pressure Pipe	Improved mixtures and state-of-the-art shotcrete technology improve quality of corrosion protective coating.
	HPC Moulds for Metal Casting and Metal Deposition	Trials with three metal casting and metal deposition companies are demonstrating the suitability and economies in using HPC for moulds.
Tunnels	Precast Concrete Segments New Toronto Subways	HPC to ensure 100 year service life in an aggressive environment
Agricultural	Pig Pens Constructed [B] using HPC	Impermeable concrete inhospitable to disease bearing bacteria. Increase in market weight of pigs.

More details of some of the above projects can be obtained as follows:

A   Canadian Portland Cement Association, 5500 Royalmount, Suite 250, Ville Mont-Royal, Quebec H4P 1M7

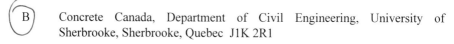

B   Concrete Canada, Department of Civil Engineering, University of Sherbrooke, Sherbrooke, Quebec J1K 2R1

Results of all demonstration projects will be published by Concrete Canada. There is not enough space to describe all the above projects in detail. The following are selected to illustrate the range of the network involvement.

Bridge 95-39, Highway 20

This is a small replacement bridge, under the jurisdiction of the Ontario Ministry of Transportation. Starting in 1994, a joint committee of the Ministry and Concrete Canada held a number of meetings to discuss and agree design changes, and specification and special provision requirements.   The cast-in-place deck is supported by CPCI 1400 precast girders. All the site cast concrete, abutments, deck and barrier walls will be of 60 MPa air-entrained concrete. Standard Ministry designs use 30 MPa or 35 MPa air-entrained concrete, epoxy coated bars, and the decks are waterproofed and then paved with asphaltic concrete. So that the success, or otherwise, of the use of high performance concrete can be measured, the demonstration bridge has uncoated reinforcement and a bare deck. Graphite probes will be installed in the deck and barrier walls to determine if any corrosion occurs. The contract documents contain a number of non-standard requirements.

Pre-bid meeting - Contractors were not allowed to bid unless they attended this meeting. A presentation was made outlining the objectives of the project and all the non-standard items, so that the bidders were clear on what was wanted.

Trials   1   A full scale trial batch is required at an early stage in the contract.

         2   A trial deck slab, 13m wide by 10m x 225mm thick, is to be constructed before the actual deck. It is to be placed, compacted, finished and textured using the same equipment and procedures as will be used for the deck.

The contract is very specific about curing requirements, including fog curing of the slab immediately after finishing, but prior to texturing, and the provision of insulation to allow temperature gradients to be controlled to minimise cracking.

An extensive programme of testing before and during construction will check strength, air void systems and salt scaling resistance. Additional tests to complete the documentation of the concrete may include modulus determinations, stress-strain curve, permeability, tensile strength and diffusivity.

Concrete temperature, limited to a maximum of 25°C on delivery, will be monitored until it has cooled to ambient as part of the crack control programme.

Type 10 SF cement, that is type 10 blended with silica fume, is specified with a minimum cementitious content of 450 kg/m^3. Certification of compatibility between the chosen superplasticiser and the cement is required, and a retarder is to be used. The concrete will be supplied by ready-mixed concrete and all loads must be at least 1 m^3 less than the truck capacity.

The bridge will be monitored visually, by tests on cores, 1/2 cell tests and graphite probe measurements for an extended period in order to evaluate its performance.

Prince Edward Island Link

This 12.9 km bridge, now under construction, will join Prince Edward Island to the Canadian mainland at Jourmain Island, New Brunswick. The bridge is being built by Strait Crossing Inc. This group has financed and designed the bridge and, after building it, will operate and maintain it for 35 years, until 2032. It will then revert to the Federal Government. In addition to tolls the builder will receive an annual subsidy less than that currently provided to the ferry service.

The structure is almost entirely comprised of precast units. The main spans are 250 m and the main precast girders weigh 7000 tonnes! The concrete is 60 MPa air-entrained high performance concrete. Reinforcing steel is uncoated, so that protection in the severe marine environment of Canada's East Coast is provided entirely by the cover concrete. This concrete is designed to have very low permeability, and the curing controls used during production are designed to produce cover concrete that is free of any significant cracks.

Principals of Concrete Canada have had ongoing input into a number of areas of concrete technology. In this demonstration project two Principal Investigators, Professor Walter Dilger of Calgary University and Professor Carolyn Hansson of Queen's University have combined to take advantage of this outstanding construction in high performance concrete. The two demonstration projects are as follows. Both have construction and post-construction phases.

1.   **Monitoring of Thermal Stresses: Professor Dilger**

During construction the maximum temperature reached in an element, and the maximum temperature differences developed within a member and between adjacent members are the prime interest.

After construction diurnal and seasonal changes in temperature, and the stresses induced by them, will be monitored for a long period.

During construction, data will be collected on site using a 200 channel data logger. After construction all data will be transmitted to Calgary University by modem.

In parallel with the site measurements a laboratory test programme is being carried out. Short term creep tests will be made. Stresses will be determined in axially restrained prisms subjected to different rates of temperature changes. These tests will be repeated at intervals as long as the temperature monitoring programme continues.

2.   **Corrosion Monitoring: Professor Hansson**

The builder has an interest in minimising maintenance costs during the initial 35 year tenure, and is obliged to hand over the bridge in close to new condition. Since the corrosion protection to the reinforcement is the cover concrete it is important to monitor how well it is performing.

Clearly, if any significant corrosion is detected during the early life of the structure, the chosen protection concept has failed. It has been assumed that this will not happen, and that any corrosion activity during the early and ongoing life of the bridge will be at a level at which damage does not occur to the reinforcement or the cover concrete. A very sensitive system has therefore been developed to monitor any signals from the steel.

The instrumentation will provide a method of evaluating the durability performance of the bridge throughout its service life.

HPC as Slab Floors in Swine Nurseries (from reference 5)

"After a few years of use, a number of the floors built in swine nurseries in the seventies evidenced real performance problems. In response, owners gradually turned to rubber or plastic flooring to protect the concrete surface. In addition to poor resistance to the acidic environment, the concrete generally used at the time was relatively porous, providing ideal living conditions for *Escherichia colibacillus* (commonly referred to as *E. coli*), which can be present in pig manure and can cause diarrhea. *E. coli* contamination of the nursery requires complete disinfection of the barn. When this is necessary, the operator suffers loss of income related to the cost of the disinfection as well as costs associated with growth disruption of the piglets.

Installing a heated floor in the nursery provides the characteristics listed below:

- greater comfort (nonabrasive);
- nonskid;
- good thermal conductivity, resulting in more uniform distribution of heat;
- high thermal inertia to prevent abrupt heat fluctuations;
- sealed surface to prevent bacterial and viral contamination;
- easy to clean;
- durable; and
- inexpensive.

Piglets normally weigh 32 kg at age 8 weeks. An eight-week old piglet suffering from diarrhea, however, weighs 1.4 kg less on average and needs 3.5 days of recuperation to get back up to normal weight. In monetary terms, a bout of diarrhea increases production costs by around $0.50 a piglet. Considering that a major operator can produce up to 11,000 hogs per year in a single nursery, annual losses can exceed $5000.

A major producer in St-Gregoire, Quebec, was selected for the construction of a heated 70-MPa HPC slab for a new comfort pen for piglets. This concrete was chosen for its very low permeability and porosity and its high resistance to acids, the combined effect of which should greatly attenuate the risks of contamination and surface erosion.

The resulting installation has resulted in diarrhea free piglets."

CONCLUDING REMARKS

The fundamental properties of HPC make it eminently suitable for a wide range of applications, beyond those requiring high strength. The concrete has a higher first cost but may be cheaper in placing costs in some applications. Service life costs, particularly with structures subjected to aggressive environments, will be lower than with the use of traditional concrete mixtures. In some instances where the properties of HPC permit design modifications, it may be possible to reduce first cost as well as service life costs.

ACKNOWLEDGEMENTS

The work of Concrete Canada is funded by the Centres of Excellence programme of the National Sciences and Engineering Research Council. Material for this paper was provided by principal and collaborating investigators of Concrete Canada.

REFERENCES

1. Bickley, J.A., "Requirements for High Strength Concrete" in the 1994 CSA Standard A23.1.

2. Mitchell, D., "Requirements for High Strength Concrete" in the 1994 CSA Standard A23.3.

3. Litvan, G.G., "Deterioration of Parking Structures Report", National Research Council Canada, Report CR 5493/5517/5518/5519.9 of Nov 2nd, 1992.

4. Coulombe, L-G., and Ouellet, C., "Construction of Two Experimental Bridges Using High-Performance Air-Entrained Concrete." Concrete Canada publication.

5. Gagne, R., Chagnon, D., and Parizeau, R., "Utilization of High Strength Concrete in the Agricultural Industry", Concrete Canada publication.

HIGH PERFORMANCE METAKAOLIN CONCRETE: RESULTS OF LARGE SCALE TRIALS IN AGGRESSIVE ENVIRONMENTS

A H Asbridge

T R Jones

ECC International Ltd

G J Osborne

Building Research Establishment

UK

ABSTRACT: Extensive research has been carried out in recent years into the beneficial effects of metakaolin when used as a partial replacement for Portland cement in concrete. Benefits include increased resistance to acids and sulphates, reduced chloride ion diffusivity, reduced porosity, and prevention of alkali-silica reaction. These have been demonstrated in the laboratory by a number of researchers. As a result, it is anticipated that structures built of metakaolin concrete would prove to be highly durable in aggressive environments.

The purpose of this paper is to review the results of construction projects in which metakaolin concrete was used. The constructions were exposed to environments such as tidal sea water, high sulphate soils and acidic waters.

Practical aspects of using metakaolin concrete, such as handling, mixing, rheology, placing and finishing are discussed. Performance of the concrete constructions was assessed after exposure to the environment for three years. The effect of exposure on permeability, chloride penetration and general integrity are discussed.

Keywords: Chloride, Concrete, Durability, Metakaolin, Ordinary Portland Cement (OPC), Permeability, Pozzolan.

Anthony H. Asbridge has worked in the Research and Development Department of ECC International since 1984 and is currently concerned with the development of metakaolins for use in the construction industry.

Dr. Thomas R. Jones has worked in the Research and Development Department of ECC International since 1972 and is presently Manager of the New Materials Group specialising in the chemistry and processing of industrial minerals.

Geoffrey J. Osborne is a Principal Scientific Officer and Head of the Concrete Durability Section at the Building Research Establishment, where he has worked for 40 years. His main research interests include the properties of Portland and calcium aluminate cements and the long term durability of concrete.

Radical Concrete Technology. Edited by R K Dhir and P C Hewlett. Published in 1996 by E & FN Spon, 2–6 Boundary Row, London SE1 8HN, UK. ISBN 0 419 21480 1.

INTRODUCTION

Concrete is one of mankind's most versatile and useful building materials. However concrete sometimes displays two undesirable features: poor durability in hostile environments and poor aesthetic properties - ie. poor visual appearance. There is growing evidence that these disadvantages can be overcome by using certain pozzolanic materials which are able to modify the chemistry and micro-structure of concrete.

Maintenance costs of concrete can be high, so the concept of life-time cost is increasingly being taken into account. By using the correct pozzolanic materials, the overall economics of a project can be improved, and an aesthetic problem can be converted into a desirable feature.

Metakaolin is a reactive pozzolan, produced by the thermal activation of the mineral kaolin. It is available in a high state of purity (greater than 90%) and can react with more than its own weight of calcium hydroxide to give new cementitious compounds [1]. By replacing part (typically 10 to 20%) of the Portland cement content of concrete, metakaolin reduces calcium hydroxide levels in the cured concrete. The pozzolanic reaction is rapid, i.e. within 28 days at ambient temperatures, and at the higher replacement level virtually all the calcium hydroxide is removed [1]. Although the calcium hydroxide levels are significantly reduced, the pH of the pore solution is maintained above 12.5 [2].

Metakaolin improves the micro-structure of concrete:

- average pore size is reduced [3][4]
- calcium hydroxide levels are reduced [5][6][7]
- bonding between cement paste and aggregate is improved [8]

In turn, these micro-structural changes have a large beneficial effect on key properties which determine durability and visual appearance:

- penetration of water, salts and acids is reduced [1][4][7]
- diffusion of salts within the structure is reduced [2]
- resistance to sulphate attack improves [9]
- ASR can be prevented [6][10]
- compressive and tensile strengths are improved [8][11][12][13]

The use of "burnt clay" as a pozzolanic material dates back as far as Roman times. Despite this long history, there are few reports of modern constructions which have utilised metakaolin as a partial cement replacement to improve the durability of the concrete used. Examples include a series of dams constructed in Brazil in the mid 1960's [11][14]. In these cases the local aggregate was known to be alkali reactive and metakaolin was used as a partial cement replacement for economic reasons. To date no problems or failures relating to these major constructions have been reported.

SCOPE OF THE PRESENT WORK

Wide-ranging laboratory research has been carried out into the use of metakaolin as a partial cement substitute in concrete. However, concrete is essentially a practical material. In real situations it must be mixed, placed and finished with as little trouble as possible and with minimum of variation in the final product quality. With this in mind, and to evaluate metakaolin containing concrete at a practical level, a number of projects have been undertaken. Up to 800m^3 of metakaolin concrete were involved in each case. In all cases the metakaolin concrete was placed in an aggressive environment. The metakaolin concrete placement locations varied for each project. Consequently, durability requirements differed, ie. chemical (acid resistance) or physical (erosion by wave action).

This report will focus on three of the projects undertaken. The adverse exposure conditions and concrete mixing and placement will be discussed. The results of measurements (compressive strength, oxygen permeability and chloride penetration) carried out on the hardened concrete up to 2½ years after placement will be presented and discussed.

MATERIALS USED

The bulk materials used for each of the trials were those available at the ready-mix batching plants used. These were aggregates in accordance with BS882, Ordinary Portland Cement in accordance with BS12:1989 and plasticisers (Fosroc P509 for projects 1 and 2 and Cormix P7 for project 3). The metakaolin used was a commercially available product of high purity and pozzolanic reactivity (MetaStar 500 supplied by ECC International Europe). Chemical (by X-ray fluoresence) and physical (by sedigraph for the particle size distribution, BET gas adsorption for the surface area and Chapelle test [15] for the pozzolanic reactivity) analyses are shown in Tables 1 and 2. MetaStar is a pozzolan which complies with ASTM C618.

Table 1   Chemical analysis of metakaolin

Oxide	wt. %
SiO_2	51.6
Al_2O_3	41.3
Fe_2O_3	4.64
TiO_2	0.83
CaO	0.09
MgO	0.16
K_2O	0.62
Na_2O	0.01

Table 2   Physical analysis of metakaolin

<10μm (wt.%)	88
<2μm (wt.%)	37
Surface area (m²/g)	15
Pozzolanic reactivity (mg Ca(OH)$_2$/g)	1060

CONCRETE ENVIRONMENT AND MIX DESIGN

In each of the three projects to be discussed, concrete incorporating metakaolin was placed in environments which could be considered as aggressive. Details of the construction sites and the prevailing ground and exposure conditions are given below.

Project 1

The concrete was used for the foundations of an industrial processing plant. It was required to give high durability in sulphate-containing soil (Class II sulphate resistance specified). Additionally, the concrete was at the edge of a concrete jetty and consequently exposed to chloride ions from sea spray and cyclic wetting/drying processes. The concrete was placed in December 1992.

The minimum compressive strength requirement was 30 MPa at 28 days. Coarse and fine granite aggregates, and crushed granite fines sand, were used with the maximum aggregate size being 20mm. A plasticiser was used to obtain the desired workability. The mix design was as follows:

Ordinary Portland Cement	335 kg/m^3
Metakaolin (15 wt.% substitution for OPC)	60 kg/m^3
Water/binder ratio	0.5
Aggregate/binder ratio	4.5
Plasticiser	1.4 l/m^3

binder = OPC + metakaolin

Test cubes for compressive strength and density measurements were made from concrete taken on-site, slump measurements being carried out at the same time. Samples (c.350mm×150mm×100mm) intended for subsequent oxygen permeability and chloride penetration measurements were cast with the test cubes but were left exposed to the same conditions as the mass concrete.

Project 2

The concrete was used for a marine slipway. Exposed to the combined chemical and physical attrition of the sea, the concrete not only had to be physically strong but also impermeable, particularly with regard to chloride ions, to protect the steel reinforcement from corrosion. Aggregates were as for project 1. A plasticiser was used to obtain a slump of c.50mm. The compressive strength requirement was 45 MPa at 28 days. To give the required chemical and physical resistance a relatively high binder content and low water/binder ratio were used. The concrete was placed in March 1994.

Ordinary Portland Cement	375 kg/m³
Metakaolin (16 wt.% substitution for OPC)	70 kg/m³
Water/binder ratio	0.45
Aggregate/binder ratio	3.9
Plasticiser	1.56 l/m³

As with Project 1, slump measurements and the preparation of test cubes for compressive strength and density measurements were made using concrete taken on-site. Samples intended for subsequent oxygen permeability and chloride penetration measurements were 150mm test cubes. These were cast at the same time as the other test cubes and cured under water for 24 hours prior to being clamped down immediately adjacent to the marine slipway at a point mid-way between the high and low tide marks.

Figure 1   Marine slipway constructed in project 2

Project 3

The concrete was used in a process water diversion scheme. Culverts, sumps and pipe sections were cast in-situ. Resistance to sulphate (present at concentrations up to c.100 ppm) and low pH (c.3) was required. The scheme was designed to carry large volumes of water which contained suspended solids, so good resistance to abrasion was important. It was necessary for the concrete to have a slump of c.125mm to facilitate the placement of 3-4m high wall sections. A plasticiser was used to help give the desired workability. The minimum strength requirement was 45 MPa at 28 days. Coarse and fine limestone aggregates, and a crushed limestone sand, were used to ensure even wear of the aggregate and cement paste in the presence of effluent containing a mixture of dilute mineral and organic acids. The concrete was placed during August to December, 1994.

Ordinary Portland Cement	340 kg/m^3
Metakaolin (15 wt. % substitution for OPC)	60 kg/m^3
Water/binder ratio	0.5
Aggregate/binder ratio	4.7
Plasticiser	1.2 l/m^3

As with the previous two projects, slump measurements and preparation of test cubes for compressive strength and density measurements were made from concrete taken on-site. A cylinder (c. 500mmx300mm) was bored from a section of cast concrete approximately 7 days after placement and left exposed to the flowing liquids in a section of culvert, ie. the same conditions as the in-situ concrete. Samples for oxygen permeability measurements were cut from this cylinder immediately prior to the tests being carried out.

Figure 2   Culvert constructed in project 3
(This Figure shows only a small part of the project,
which used a total of 800m^3 of metakaolin concrete)

MIXING, PLACEMENT, FINISHING AND CURING

For all three projects the metakaolin was added to the mixer lorries at the same time as the Ordinary Portland Cement.   This was achieved either manually by splitting 25 kg bags, or by using an automated bulk silo.

Different compaction and finishing techniques were adopted during placement; vibrating poker and steel float for projects 1 and 3, and vibrating poker and beam for project 2. Plywood formwork was used for the three cases discussed.

Different curing conditions were used for each of the three projects as follows;

Project 1:   Formwork removed after 24 hours.   The concrete was then covered with a plastic sheet for a further period of approximately 48 hours.

Project 2:   The fresh concrete was loosely covered with a plastic sheet to minimise the possibility of wash-out of cement paste as the tide came in over the slipway. The sheeting and formwork was removed after 24 hours and the concrete left exposed to air/sea.

Project 3:   The formwork was removed after 24 hours. The concrete was left uncovered to obtain a carbonated surface layer which would give additional sulphate resistance [9][16].

CONCRETE DURABILITY STUDIES

The chloride penetration and/or oxygen permeability of metakaolin concrete samples were evaluated using the methods detailed below. The samples' ages at the time of testing were 2½ years for the plant foundation (project 1), 14 months for the marine slipway (project 2) and 10 months for the water culvert (project 3).

Method for Oxygen permeability measurements

100mm cores were taken from concrete specimens from each location and 50mm slices were cut from these cores. These were then stored in air at 20°C, 65% relative humidity for at least 28 days for conditioning.

Oxygen permeability measurements were carried out using a gas permeability cell developed by Lawrence [17]. The cell allows a gas pressure to be applied to one of the flat surfaces of the 50mm concrete slice of 100mm diameter while providing a seal to the curved surface.   The flow rate from the opposite flat surface is then measured using a bubble flow meter. For each sample, the flow rate of oxygen was recorded at one bar intervals up to 5 bars. The permeability coefficient was calculated using a combination of Darcy's law and the Poiseuille equation for flow through a regular capillary.

Method for Chloride Penetration measurements

The concrete samples, at 20°C and 65% relative humidity, from projects 1 and 2 were used for chloride ingress measurements by drilling powder samples for analysis. A 13mm masonry drill was used to obtain representative samples at a range of depths, discarding the top 1mm surface.   The powder samples were analysed for total chloride (Cl⁻) in accordance with procedures given in BS1881:Part 124:1988.

RESULTS

Slump, Compressive Strength and Density data

Slump values were obtained on the fresh metakaolin concrete in accordance with BS1881:Part 102:1983.   Compressive strength and density measurements were carried out on concrete test cubes made at the time of placement and cured and tested in accordance with BS1881:Parts 111,114,116:1983.   The number of individual measurements made in each case are given in parentheses next to the mean values .

Project 1 *(Plant foundation)*

	Range	Mean
Slump (mm)	30-50	40 (4)
Compressive strength: 7 days (MPa)	40.0-43.0	41.7 (4)
" " : 28 days (MPa)	60.5 - 62.0	61.6 (4)
Density : 7 days (kg/m^3)	2360-2370	2368 (4)
: 28 days (kg/m^3)	2370	2370 (4)

Project 2 *(Marine slipway)*

	Range	Mean
Slump (mm)	40-65	55 (2)
Compressive strength: 7 days (MPa)	48.5-49.0	48.8 (2)
" " : 28 days (MPa)	63.5-66.0	64.5 (2)
Density: 7 days (kg/m^3)	2370-2390	2380 (2)
: 28 days (kg/m^3)	2360-2390	2377 (2)

Project 3 *(Water culvert)*

	Range	Mean
Slump (mm)	90-135	110 (5)
Compressive strength: 7 days (MPa)	50.7-59.3	54.8 (16)
" " : 28 days (MPa)	60.9-69.4	64.5 (16)
: 56 days (MPa)	62.8-70.7	66.4 (12)
: 180 days (MPa)	70.3-71.4	70.9 (3)
Density: 7 days (kg/m^3)	2393-2429	2407 (16)
: 28 days (kg/m^3)	2394-2424	2408 (16)
: 180 days (kg/m^3)	2403-2404	2404 (3)

Oxygen Permeability and Chloride Penetration data

Oxygen permeability and chloride penetration measurements were carried out on specimens exposed to the same environments as the constructions themselves.

Table 3   Oxygen Permeability data

Position in sample	O_2 Permeability (x 10^{-18} m²)		
	Project 1	Project 2	Project 3
Top	2.26	2.07	1.59
Middle	-	2.25	1.56
Bottom	-	2.56	1.74

Table 4   Chloride Penetration data

Penetration depth (mm)	Chloride concentration (% Cl^- by wt. of binder)	
	Project 1	Project 2
1-6	1.08	1.07
6-11	0.86	0.82
11-16	0.68	0.56
16-21	0.06	0.06
21-26	0.06	0.04
26-31	0.05	0.04
31-36	0.05	0.04
36-41	0.06	0.04
41-46	0.06	0.04

DISCUSSION

The concretes produced for the three trials incorporated a partial replacement of Ordinary Portland Cement by metakaolin of 15 to 16 weight.%. However, they differed with regard to their workabilities (indicated by the range of slump values), aggregate types, aggregate to binder ratios and water to binder ratios.

Being hydrophilic, the metakaolin was rapidly incorporated when added to the ready-mix lorries.   The metakaolin concretes showed no signs of either inhomogeneities or agglomerates of the metakaolin itself.   The metakaolin concretes could be discharged on-site without difficulty.

For the three concrete mixes, the incorporation of metakaolin as a partial replacement for OPC tended to make the concrete slightly thixotropic relative to a conventional Portland Cement concrete of comparable mix design.   However the metakaolin concrete readily became fluid when a shearing force was applied, eg. vibrating poker, and could be compacted well.   It was noted that the metakaolin concrete appeared less susceptible to segregation due to over-vibration than conventional concrete. It is noteworthy that the slight thixotropy of the metakaolin concrete was a positive benefit during project 2 (marine slipway) where it was necessary to place the concrete at a slope of approximately 20°.

The metakaolin concretes could be finished without difficulty. The metakaolin concrete for project 3 (water culvert) was noted to be slightly "sticky" but this effect was mainly attributed to the use of all limestone aggregates (and crushed limestone sand). The combination of plywood formwork and steel float finishing gave a good quality surface effect to the metakaolin concrete.

It was noted that the concrete used for the marine slipway constructed in project 2 did not exhibit any signs of "wash-out" of the cement paste when viewed approximately 24 hours after placement. This was despite the fresh concrete being immersed by sea-water, which could get under the loose plastic covering sheet, less than 4 hours after placement.

The minimum compressive strength requirements for the concrete for each of the projects were exceeded by the metakaolin concrete. No control (Ordinary Portland Cement) concrete was placed to provide a reference for each of the three projects. However work at the Building Research Establishment has shown metakaolin to have no detrimental effect on the rate of development of or ultimate compressive strength of concrete at substitution levels of 10-20 wt.% [2]. Additionally, other large-scale projects and further laboratory studies that we have undertaken have indicated that improvements in compressive strength can be achieved as a result of the incorporation of metakaolin as a partial cement substitute. For example, at a partial replacement level of 15 wt.% compressive strength relative to a control concrete (same mix design and total cementitious content but incorporating only Ordinary Portland Cement) can be improved by 5% after 3 days curing, 21% after 7 days and 33% after 28 days. This trend has been reported by other researchers [13].

Oxygen permeability measurements gave values ranging from 2.07×10^{-18} to $2.56 \times 10^{-18} m^2$ for concrete from projects 1 and 2. Slightly lower values ranging from 1.56×10^{-18} to $1.74 \times 10^{-18} m^2$ were observed for concrete from project 3. Regardless of the slight differences noted, all of the concretes tested had oxygen permeabilities of the order of $10^{-18} m^2$ indicative of average, tending to low, permeability concrete [18]. It would be expected for the use of metakaolin as a partial cement replacement to produce relatively impermeable concrete due to pore size refinement and densification of the aggregate-cement paste interfacial zone.

Chloride penetration beyond the surface 16mm of the samples evaluated was very low ($<0.07\%$ Cl⁻ by weight of binder) indicating the metakaolin concretes to be resistant to chloride ingress. These chloride levels are below the threshold level for corrosion activation (0.4% by weight of binder) [19].

Observation of the structures built using metakaolin concrete has indicated that over periods of up to 3 years, their integrity has been maintained. No evidence of erosion or deterioration is apparent.

CONCLUSIONS

In recent years, many investigations at laboratory and pilot scales have shown that metakaolin, as a partial cement replacement, significantly improves the properties of concrete. Metakaolin reduces calcium hydroxide concentrations, reduces the volume of coarse pores, and improves the paste-aggregate interface zone. We predict that metakaolin will increase durability and improve service life.

We have now demonstrated that metakaolin concrete (where metakaolin replaces up to 16% of the OPC) can be conveniently prepared in standard ready-mix batching plants. Placing and finishing the metakaolin concrete does not present any unexpected difficulties, compared with conventional concrete.

Tests carried out on the metakaolin concrete after 10-30 months demonstrated excellent strength development, average-to-low oxygen permeability and low chloride penetration. We will continue to monitor the structures and assess their resistance to abrasive fluids, acid and sulphate attack, and freeze-thaw damage.

The extra cost of using metakaolin concrete is modest, compared with the total project cost (detailed discussion on costs will be given in a separate paper). We expect that, over the life-time of the constructions reported here, we will demonstrate improved durability and reduced maintenance costs. Overall, on the basis of previously published laboratory data, we expect to see a significant reduction in life-time costs for these projects.

Metakaolin concrete is rapidly developing a track record in specialist applications such as water resistant mortars and renders, pre-cast products and glass fibre reinforced mouldings. We have now demonstrated the potential for metakaolin concrete in large engineering projects.

It is expected that metakaolin concrete will be particularly desirable where the environment is aggressive and durability is an important issue.

KNOWLEDGEMENTS

The authors would like to thank Mr B. Singh and Mrs J. Hardcastle (BRE) and Mr J. Kostuch and Mr G.V. Walters (ECCI) for their contributions to this work.

REFERENCES

1. KOSTUCH, J.A., WALTERS, G.V. AND JONES T.R., High performance concretes containing metakaolin - A review, Concrete 2000, Ed. Dhir R.K and Jones M.R., E&FN Spon, Vol. 2, 1993, pp 1799-1811.

2. HALLIWELL, M.A., Private communication, BRE Report TCR 48/92.

3. BREDY, P. CHABANNET, M. AND PERA J. Microstructure and porosity of metakaolin blended cements : Mat. Res. Soc. Symp. Proc., Boston, 1989, Vol. 137, pp 431-436.

4. LARBI, J.A. AND BIJEN, J.M. Ph.D. Thesis, Delft University, Ch. 8, 1991.

5. LARBI, J.A. AND BIJEN, J.M. ibid, Ch.6.

6. JONES, T.R., WALTERS, G.V. AND KOSTUCH, J.A., Role of metakaolin in suppressing ASR in concrete containing reactive aggregate and exposed to saturated NaCl solution, 9th Int. Conf. AAR in Concrete, Vol. 1, 1992, pp 485-496.

7. COLLIN-FEVRE, I., Use of metakaolinite in the manufacture of concrete products, CIB, Montreal, 1992, Poster 479.

8. LARBI, J.A. AND BIJEN, J.M. ibid, Ch.9.

9. SINGH, B. AND OSBORNE, G.J., Private communication, BRE Report CR 291/94.

10. WALTERS, G.V. AND JONES, T.R., Effect of metakaolin on alkali-silica reaction in concrete manufactured with reactive aggregate, 2nd Int. Conf. Durability of Concrete, Canada, Ed. V.M. Malhotra, 1991, pp 941-953.

11. SAAD, M.N.A. de ANDRADE, W.P. AND PAVLON, V.A., Properties of mass concrete containing an active pozzolan made from clay, Concrete International, July 1982, pp 59.

12. AMBROISE, J. MAXIMILIEN, S. AND PERA, J., Properties of metakaolin blended cements, Advn. Cem. Bas. Mat., 1994, 1, pp 161-168.

13. ZHANG, M.H. AND MALHOTRA, V.M., Characteristics of a thermally activated alumino-silicate pozzolanic material and its use in concrete. 2nd CANMET/ACI Int. Symp. Adv. in Concrete Technology, LasVegas, USA, June 1995.

14. ANDRIOLO, F.R. AND SGARABOZA, B.C., Proc. 7th Int. Conf. AAR, Ed. P.E. Gratton-Bellow, 1985, pp 66-70.

15. LARGENT, R. Bull. Liasons Lab. Pont Chausses, v.93, 1978, pp.63.

16. OSBORNE, G.J., The effectiveness of a carbonated outer layer to concrete in the prevention of sulphate attack, Proc. Int. Conf. The Protection of Concrete, Dundee,1990, pp.74-79.

17. LAWRENCE, C.D., Measurements of permeability. Proc. 8th Int. Congress on the chemistry of cements, Rio de Janeiro, FINEP, 1986,Vol. 5, pp 29-34.

18. THE CONCRETE SOCIETY REPORT. Permeability of concrete and its control. London, December 1985.

19. BAMFORTH, P.B., Concrete classifications for R.C. structures exposed to marine and other salt-laden environments, Presented at Structural Faults and Repair, Edinburgh, 1994.

HOLISTIC APPROACH TO DURABILITY OF STEEL REINFORCED CONCRETE

N S Berke

M P Dallaire

M C Hicks

A C MacDonald

W R Grace & Co

USA

ABSTRACT. Steel reinforced concrete is one of the most widely used construction materials in the world. The concrete matrix provides a protective alkaline environment for the steel and the steel provides ductility to the concrete. Because the steel and concrete composite has good durability and strength at an attractive cost, reinforced concrete is being used in corrosive marine and deicing salt environments. These concretes usually have compressive strengths exceeding 35 MPa. Unfortunately, there is an ever increasing catalog of structures that are showing early demise to chloride induced corrosion of steel even though the concrete had high strength, and was defined as being "high performance" concrete.

In this paper, the authors provide data showing that a holistic approach to concrete design is necessary for high durability. Extensive long-term corrosion testing shows that not all high strength concretes have the same durability. However, concrete designs that take durability into consideration typically will produce higher strength concretes. The more durable concrete designs have water-to-cementitious ratios below 0.4, concrete covers in excess of 40 mm (65 mm in marine environments), and corrosion inhibitors. Chloride ingress is significantly reduced compared to concretes having higher water-to-cementitious contents and lower covers, and tolerance to chloride is increased by the addition of the corrosion inhibitor. If freezing and thawing conditions exist these concretes will be air entrained. Such concretes typically have compressive strength values above 50 MPa and even above 90 MPa. Thus, proper consideration of durability will result in truly high performance concrete, that is, concrete with good durability as well as high strength.

Keywords: Durability, Concrete, Steel Reinforcing, Chloride, Corrosion, Corrosion Inhibitors, Calcium nitrite, Long-Term Testing, Shrinkage, Shrinkage Reducing Admixtures

Dr Neal Berke is a Research Manager in the Cement and Concrete Products Research Section, Construction Products Division, W. R. Grace & Co. Conn., Cambridge, Massachusetts, USA. He is a member of ACI, ASTM, TRB, NACE International and ASM and serves on numerous committees within these organisations.

Mr Michael Dallaire is a Research Associate in the Cement and Concrete Products Research Section, Construction Products Division, W. R. Grace & Co. Conn., Cambridge, Massachusetts, USA. His area of expertise are high performance concrete.

Mrs Maria C Hicks is a Senior Research Engineer in the Cement and Concrete Products Research Section, Construction Products Division, W. R. Grace & Co. Conn., Cambridge, Massachusetts, USA. Most of her work involves concrete durability.

Mr Alistair Macdonald is the Engineer Services Group representative for the UK. Ireland and Scandinavia. He has been with W.R. Grace & Co. Conn. Since 1988. He graduated in 1986 from the University of Abertay, Dundee, UK.

Radical Concrete Technology. Edited by R K Dhir and P C Hewlett. Published in 1996 by E & FN Spon, 2–6 Boundary Row, London SE1 8HN, UK. ISBN 0 419 21480 1.

INTRODUCTION

Due to its excellent structural properties and relatively low cost reinforced concrete is one of the most widely used construction materials in the world. Concrete provides a protective alkaline environment for the steel, and as a result, reinforced concrete is generally durable in all but the more severe environments. Some of the most severe exposures involve marine or deicing salt environments, and the chloride ingress disrupts the passivity of the steel. The resulting corrosion causes cracking and spalling of the concrete due to the increase in volume of the corrosion products. All too often the corrosion-induced damage results in major repairs or loss of functionality before reaching the expected service life.

Improving the durability of steel reinforced concrete is a major challenge facing researchers and engineers. This can result in a significant savings in life-cycle costs, and be beneficial to the environment by extending the useful life of materials that require energy to produce. For example, a bridge that lasts for 120 years is less expensive than building two bridges that last for 60 years each, and requires only half of the cement, aggregates and steel.

Improvements in concrete mix designs and increased concrete cover can improve the durability of reinforced concrete. This is recognized, for example, in ACI 318, ACI 357, CSA S413-94, and BS6349, which specify maximum water-to-cement levels and minimum concrete covers over the steel for severe chloride environments [1-4]. In this paper it will be shown that improved mix designs alone might not be adequate for long-term corrosion protection, and that supplemental corrosion protection measures such as adding corrosion inhibitors, such as calcium nitrite, are needed, even when mineral additives are present.

Other factors that can affect corrosion performance are the formation of cracks. Though various codes discuss the use of steel placement to minimize the size of cracks, e.g., ACI 244R [5] and BS8110 [6], reducing the tendency for cracking is not as well defined. The choice of mix design and curing procedures can have a significant impact on thermal gradients and drying shrinkage, and therefore, on the development of thermal cracks and restrained drying shrinkage. These cracks can increase the ingress of chloride, therefore reducing their size and number is beneficial. In addition to showing the effects of different mix designs and curing on these properties, it is shown that a new class of shrinkage reducing admixtures can significantly decrease drying shrinkage.   The combination of low water-to-cementitious ratios with corrosion inhibitors and shrinkage reducing admixtures provides an holistic concrete approach to improved long-term durability.

EXPERIMENTAL PROCEDURES

Numerous experimental techniques are needed to evaluate the long-term durability of concrete beyond the typical measurements of plastic properties and compressive strength measurements. The procedures used to develop the data in this paper are described in this section.

Concrete Production

Concrete was produced according to procedures described in an ASTM standard [7]. All materials complied with ASTM specifications for cement, aggregates, silica fume, and for cement and slag [8-12]. Commercial admixtures used met the requirements of one of the classifications in ASTM C 494 [13].

Concrete properties of slump, plastic air, unit weight, setting times, and compressive strength as a function of time were determined for all of the concretes studied using ASTM procedures [14-18]. After demolding (at 24 hours) specimens were cured in a fog room meeting the specifications of C 192 [7].

Reinforcing steel met the requirements of ASTM A 615 [19]. Dimensions of the steel and reinforced concrete specimens were varied.

Lollipop specimens were 76 mm diameter x 152 mm tall cylinders with an embedded reinforcing bar 9.5 mm in diameter. The reinforcing bar was positioned 38 mm off the bottom of the cylinder and had 2720 mm^2 of exposed area.

Minibeam specimens had dimensions as specified in ASTM G 109 with varying depths of cover used and occasionally only one cathodic bar in the lower mat [20].

In general, lollipops are a more severe environment from the point of view of chloride attack due to the multidirectional ingress of chloride. However, calculation of the diffusion coefficients is easier in the case of one-dimensional minibeam geometry, since the solution to Fick's second law can be used, as described in reference [21].

Chloride Analysis and Diffusion Coefficients

The ability of the concretes to resist the ingress of chloride was determined by analyzing for chloride as a function of depth and time. Chloride was also determined at the reinforcing bar level at corrosion sites. The analysis was for total acid soluble chloride according to ASTM C 1152 [22].

When chloride was determined on minibeams the ingress was one-dimensional and Fick's Second Law was used to determine the effective diffusion coefficient, D_{eff}. The validity of this approach for exposure times over 1 year in the laboratory and several years in the field is quite good [23-27].

Another method used to estimate chloride ingress include ASTM C1202 which is known as the "Rapid Chloride Technique" [28]. Data from this test were compared to D_{eff} values when available.

The concrete resistivity was also determined using electrochemical impedance spectroscopy, EIS, at a frequency of 20,000 Hz on lollipop specimens as described in references 29 and 30. The EIS technique is nondestructive and described in further detail below.

Corrosion Measurements

Corrosion measurements were performed for various steel reinforced concrete specimens that were subjected to chloride ingress. Brief descriptions of the tests methods used are given below and more detail can be found in other publications [31-33].

Polarization Resistance

This is a nondestructive test method that gives an indication of the corrosion rate of steel embedded in concrete. The basic technique is described in ASTM G 59, but modified for this application [34]. A potentiostat is used to displace the equilibrium of the embedded reinforcement by about 10 to 20 mV as measured relative to a reference electrode, and the current between the reinforcing steel and a counter electrode is measured.

The polarization resistance is defined as the slope at zero current:

$$R_p = \Delta E/\Delta i (ohm\text{-}cm^2)$$

where R_p is the polarization resistance, E is potential and i is the current density. The corrosion rate is estimated as:

$$i = B/R_p \ (\mu A/cm^2)$$

where B is a constant with a typical value of 26 mV for steel in concrete.

R_p values less than 50 kohm-cm^2 indicate the onset of corrosion. Values below 20 kohm-cm^2 are associated with severe corrosion [35].

One problem with this technique is that if not corrected for, the high resistivity of the concrete will add to the measured R_p and thus cause a significant underestimation in the corrosion rate. In this paper a current interruption technique [36] was used to correct for this error.

Electrochemical Impedance Spectroscopy (EIS)

The EIS technique utilizes an alternating current over a range of frequencies to determine impedance as a function of frequency. At high frequencies (10 to 40 kHz) the resistivity of the concrete can be determined and at very low frequencies (less than 1 mHz) the impedance is inversely proportional to the corrosion rate.

Macrocell Corrosion

This technique involves the creation of a macrocell between steel in the top portion of a concrete specimen and that in a lower mat. This is accomplished by ponding with chloride to provide a higher concentration of chloride at the upper steel mat. The current is measured as the voltage drop across a resistor between the two reinforcing bar levels. Specimens in this paper were similar to those described in ASTM G109 [20], however, mixture proportions and concrete covers were modified in some cases to provide a better indication of longer-term performance.

Shrinkage

Shrinkage measurements were conducted according to ASTM C157 at various combinations of moist and dry curing. [37]. A ring test method as described by Shah et al [38] was also employed. These specimens were cured for one day and then exposed to 50% RH at 22 °C.

Reduction in Concrete Permeability

Since the passivity of steel in concrete is lost in the presence of chloride, preventing the ingress of chloride to the steel is an effective method to improve corrosion resistance. Increased concrete cover over the steel is desirable, but structural considerations often limit the maximum cover from under 50 to 75 mm. Thus reduced concrete permeability is needed to further restrict chloride ingress. In this section the changes in concrete mix designs that result in reduced permeability are discussed and examples highlighting improved corrosion performance are given.

Reducing the water-to-cement ratio (w/c) of concrete results in a decrease in the diffusion coefficient for chloride ingress [25,39]. This is illustrated in Figure 1. An example of the improved corrosion resistance is given in Figure 2 which shows the long-term corrosion performance as a function of w/c for steel reinforced minibeams. At the same cement factor the reduction in w/c via the use of superplasticizers resulted in a significant reduction in chloride ingress and improvement in corrosion resistance. However, it should be noted that over time chloride diffuses into even the lower w/c concretes.

Supplemental pozzolans such as silica fume and ground blast furnace slag (GGBFS) are used to reduce chloride ingress. The improved performance in long-term corrosion resistance is illustrated for silica fume in Figure 3.

Figure 4 shows that fly ash provides some benefit in corrosion performance. Further benefits are obtainable by combining fly ash and silica fume [21] in reducing chloride ingress.

Figures 2 and 3 show significant benefits in corrosion performance as w/c is lowered or silica fume is added. However, once chloride reaches the steel corrosion does initiate. As will be shown later, chloride levels above the 0.9 kg/m^3 which are sufficient to initiate corrosion can arise in long-term exposures in severe environments, even for low permeability concretes with concrete covers of 75 mm.

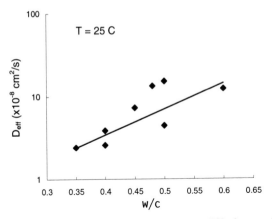

Figure 1   Effect of concrete quality on the diffusion coefficient

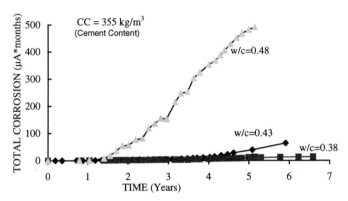

Figure 2 Effect of w/c on the corrosion of minibeams with 35 mm cover

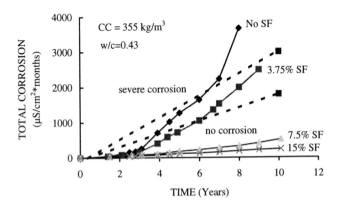

Figure 3   Effect of silica fume on the corrosion of lollipops

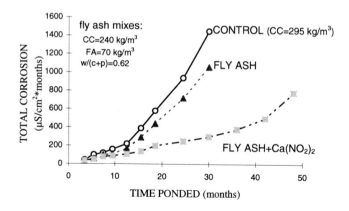

Figure 4 Effect of pozzolan addition on the corrosion of lollipops
10 L/m^3 of a 30% sol. of calcium nitrite

Corrosion Inhibitors

Corrosion inhibiting admixtures protect the embedded steel in the presence of chloride. The most widely used corrosion inhibitor for steel in concrete is calcium nitrite. It is beneficial to the hardened concrete properties [40,41] and meets the requirements of an ASTM C 494 Type C admixture [13]. Numerous articles demonstrating its corrosion performance improvements have been published [39-46].

Examples of the improvement in corrosion performance with calcium nitrite are given in Figures 4-6 which show that even low permeability concretes need a corrosion inhibitor to attain long term durability.

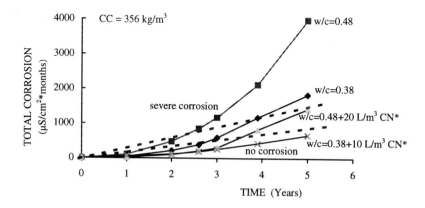

Figure 5 Effect of w/c and calcium nitrite on the corrosion of lollipops
* CN=30% solution of calcium nitrite

Increased calcium nitrite dosages are necessary to protect to higher chloride contents at the reinforcing level [40]. Thus, reducing chloride permeability will require less calcium nitrite to protect for a given time. Conversely, reducing permeability for a given calcium nitrite content will result in significantly increased times to corrosion.

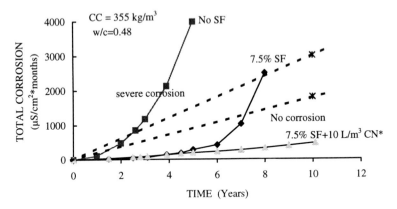

Figure 6 Effect of silica fume and a corrosion inhibitor on the corrosion of lollipops.
*CN=30% solution of calcium nitrite

Shrinkage reducing admixtures

Drying shrinkage can result in cracking due to restraints in the structure [47]. Thus, experiments to determine the effects of mix designs on drying shrinkage were performed. Figure 7 shows the effect of cement content and w/c on drying shrinkage. If increased cement is used to achieve a lower w/c, drying shrinkage will increase, thus the use of superplasticizers to reduce w/c is preferred. Drying shrinkage tests showed that curing times play a significant role in the magnitude of drying shrinkage for ordinary Portland cement concrete, Figure 8.

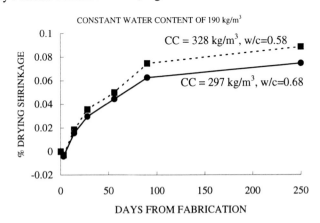

Figure 7 Effect of cement content on drying shrinkage performance of concrete (3 days curing)

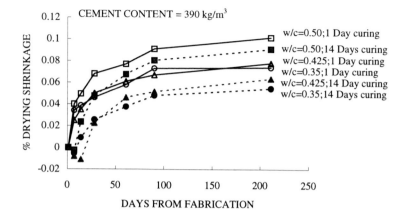

Figure 8 Influence of w/c and curing on long term drying shrinkage performance
(ASTM C 157)

This is even more pronounced if silica fume is used, Figure 9. In general, reducing permeability results in reduced drying shrinkage, if the concrete is moist cured for several days and the improvement in the water to binder ratio (w/b), is obtained by reducing water versus increasing cement or binder, especially at early ages.

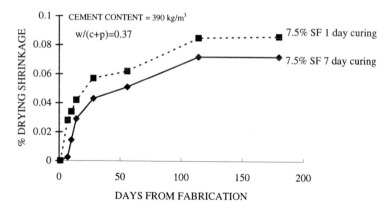

Figure 9 Influence of curing on the long term drying shrinkage performance
of silica fume concrete (ASTM C 157)

Shrinkage reducing admixtures (SRA's), can produce significant additional reductions in drying shrinkage. Figure 10 shows that significant reductions in long-term drying shrinkage are obtained with a glycol based SRA and that the shrinkage reduction is enhanced at lower w/c ratios. Figure 11 demonstrates that even though longer cure times are beneficial, that there is still a significant benefit of using SRA's at shorter curing times, especially for low w/c concretes.

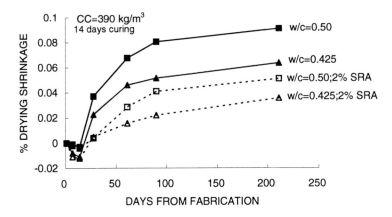

Figure 10 Effect of SRA and w/c on the long term drying shrinkage
(ASTM C 157)

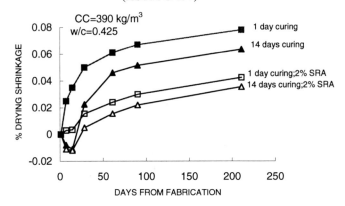

Figure 11 Effect of SRA and curing times on the long term drying shrinkage
performance of concrete (ASTM C 157)

Ring tests can be used to show the beneficial reduction in drying shrinkage on
reducing cracking in restrained concrete [37]. An example is given in Figure 12,
which demonstrates reduced cracking by using a glycol based SRA.

Table 1 shows a substantial reduction in drying shrinkage using a glycol based SRA
with silica fume or silica fume and calcium nitrite. In these tests the lowering of
permeability with silica fume alone could result in a greater tendency to drying
shrinkage cracking due to the higher modulus of the concrete and increased drying
shrinkage [47]. The addition of the SRA reduced shrinkage to levels less than that of
a reference, thus providing improved corrosion protection without increasing the risk
of cracking which could offset permeability reductions.

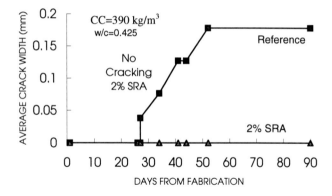

Figure 12 SRA evaluation-restrained shrinkage rings: through cracking

Table 1 Effect of SRA on concrete properties and drying shrinkage
(7 days curing)

MIX	HRWR (mL/m³)	w/(c+p)	Slump (mm)	Plastic air (%)	Compressive Strength (MPa)			ASTM C157 free shrinkage length change at 56 days (%)
					1 day	7 days	28 days	
Reference	1370	0.40	140	2.5	20.0	35.0	44.0	0.049
7.5% SF	1990	0.37	165	2.6	27.4	49.7	62.3	0.051
7.5% SF + 2% SRA	1860	0.37	133	2.0	23.8	39.6	54.4	0.035
5% SF 853	1263	0.38	171	3.6	21.0	31.2	51.0	0.058
5% SF + 1% CN*+2% SRA	825	0.38	159	3.0	14.7	36.3	51.0	0.043

*CN = calcium nitrite

Due to the relatively recent introduction of cost effect SRA's, long term corrosion data are not yet available. Work in the authors' laboratory to date indicate SRA containing concrete has equal or better corrosion resistance relative to untreated and corrosion uninhibited reference concrete. However, exposure times are as yet too short to quantify the benefit of SRA addition (Table 2).

Other Considerations

If the concrete is to be subjected to freezing and thawing environments then it is necessary to properly entrain air at the correct spacing factor. Excellent freezing and thawing resistance is obtainable with air entrained concrete containing either silica fume or calcium nitrite or combinations of the two [41].

Table 2 Effect of SRA addition on the corrosion of lollipops

CC = 390 kg/m³ w/c = 0.44	Corrosion rate (µS/cm²)after one year of ponding with 3% NaCl*
Reference	39.8
2% SRA1	4.7
2% SRA 2	21.4
2% SRA 3	23.6

Corrosion rate ----------> above 25 µS/cm² indicates severe corrosion
at 15 µS/cm² indicates corrosion initiation
below 15 µS/cm² indicates no corrosion

Compressive strengths in excess of 40 MPa are often specified to produce durable concrete. As noted in reference 40 compressive strengths in excess of 40 MPa are easily reached in durable concretes, however, not all concrete with compressive strengths above 40 MPa are necessarily durable. Thus, compressive strength alone is a poor measure of durability.

Modeling Durability

Based upon the data presented or reviewed above an holistic approach to durability design would start with a low w/c concrete to reduce chloride permeability. However, as will be shown below, even if further reductions in permeability are achieved using silica fume or other pozzolans, long-term corrosion free performance in severe chloride exposures will not be obtained. Thus the use of a corrosion inhibitor is recommended. Calcium nitrite corrosion inhibitor is effective in raising the chloride levels necessary to initiate corrosion as shown in Table 3, which is based upon extensive testing in concrete and verified by autopsies of the concrete [40]. Several examples as to how to estimate the service life as a function of environment, mix design, and calcium nitrite content are given below.

Though many of the examples given below use silica fume, one can substitute other materials if their effect on the diffusivity of chloride in concrete is known. Ground granulated blast furnace slag (GGBFS) and fly ash are not as efficient as silica fume in reducing chloride ingress. However, in the UK, where GGBFS and high quality fly ash are readily available, the industry is well accustomed to their properties.

Bridge and car park decks are subjected to drying conditions that could cause drying shrinkage to occur. In these structures stresses can develop and produce drying shrinkage cracking. The addition of SRA's could minimize these cracks and prevent

Table 3. Calcium nitrite dosage rates vs. chloride protection

Calcium Nitrite (30% solution) (L/m^3)	Chloride Concentration (kg/m^3)
10	3.6
15	5.9
20	7.7
25	8.9
30	9.5

a more rapid ingress of chloride than predicted from the diffusion equations.   In addition, since the main cause of chloride ingress in these structures is often deicing salts, air entrainment to prevent freezing and thawing and salt scaling is recommended.

Square Piles in a Marine Exposure

Figure 13 gives the estimated chloride concentration, in the splash-tidal zone, at 75 mm of cover off the diagonal of a square pile in a severe marine environment with an average yearly temperature of 10 C, as a function of time. Diffusion coefficients were chosen to represent concrete with a w/c=0.5 (D_{eff}=5.3x10^{-8} cm^2/s), w/c=0.4 (D_{eff}=1.3x10^{-8} cm^2/s) and w/b=0.38 with silica fume or GGBFS (D_{eff}=0.67x10^{-8}cm^2/s), equivalent to concrete with 750 Coulombs. A two-dimensional diffusion model is used to account for the fact that chloride enters from both sides. The results show that even with a low permeability concrete considerable chloride will be at the reinforcement level in 50 years. The addition of 25 L/m^3 of a 30% calcium nitrite solution would provide an elevated corrosion threshold and the necessary protection for the anticipated chloride levels.

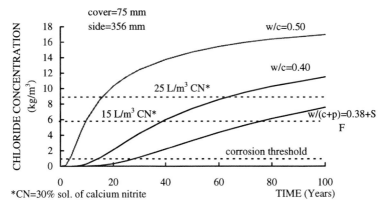

Figure 13 Estimated chloride concentration along the diagonal of a square pile in a marine environment (10 C).

The dotted lines represent the chloride concentrations at which corrosion will initiate for reference, 15 L/m^3 and 25 L/m^3 concretes. By determining the time at which the curves representing the various types of concrete intersect the dotted lines, service life can be predicted.

Marine Wall Splash-Tidal Zone

A typical chloride profile as a function of depth is given in Figure 14 for a marine wall at the splash-tidal zone after 75 years of exposure at an average yearly temperature of 10 C. Depending on the quality of the concrete and concrete cover used different quantities of calcium nitrite corrosion inhibitor could be used to extend the service life. For example, 20 L/m^3 of a 30% calcium nitrite solution could provide protection for concrete covers above 75 mm for the w/c=0.40 concrete, whereas at a lower permeability either less cover would be needed or a lower inhibitor content could be used. Use of high quality 0.38+SF or GGBFS is seen not to be adequate to provide the 75 year design life at any reasonable cover. Calcium nitrite provides the needed protection.

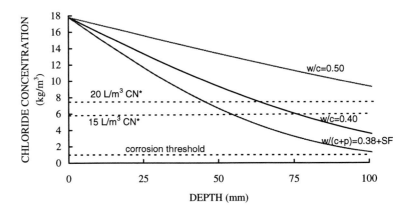

Figure 14 Estimated chloride profiles of a marine wall in the splash/tidal zone after 75 years of exposure (T = 10 C). *CN=30% sol. of calcium nitrite.

Bridge Decks and Car Parks

Horizontal decks, such as bridges and car parks, that are subjected to chloride deicing salts will see an increase in the chloride surface concentration over time [49]. Thus, the generation of a chloride profile is obtained by solving Ficks' second law

$$\partial C/\partial t = D \, \partial^2 C/\partial x^2$$

numerically using a finite difference scheme and reducing the partial differential equation to a series of ordinary differential equations in time. The resulting ordinary differential equations are integrated numerically using the Gear method (LSODE subroutine). The initial conditions are at t=0, C(0,x)=0, at x=0, and, after a number of years the surface concentration reaches a maximum and remains constant. At the bottom of the slab $\partial c/\partial x=0$.

An example of a bridge deck in which the chloride content at the surface increases over time to a constant value is given in Figure 15. The average diffusion coefficient used was reported by Weyers and Hoffman for bridges in New York, USA [50]. A surface concentration of 7.4 kg/m^3 and a chloride buildup of 0.6 kg/m^3 per year was assumed. The increase in time to corrosion initiation can be seen to be from 10 to 55 years with the addition of 10 L/m^3 of calcium nitrite corrosion inhibitor. A dosage of 15 L/m3 would protect the structure beyond 100 years.

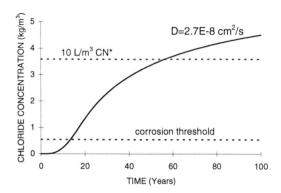

Figure 15   Estimated range of chloride concentrations of a New York bridge deck at 64 mm depth.. *CN=30% sol. of calcium nitrite.

Car park decks are a particularly severe exposure due to the fact that chloride is usually not washed off the covered decks. Chloride builds up much faster on the surface than occurs for a bridge deck and after 15 years of exposure can be at 19 kg/m^3 [49]. Typical chloride concentrations at 50 mm depth as a function of time are shown in Figure 16 for concrete of w/c=0.4 (D=1.3x10^{-8} cm^2/s) or w/(c+p)=0.4 with 5% silica fume (D=0.88x10^{-8} cm^2/s). Clearly, for a 40 year service life, additional corrosion protection would be required. This protection could be provided with calcium nitrite. Increased concrete cover or less permeable concrete is not necessarily an option due to increased dead loads and ceiling height requirements, and because of difficulties in placing the concrete. Furthermore, the combination systems are often less expensive.

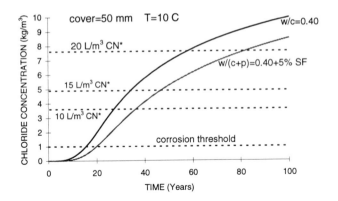

Figure 16 Estimated chloride concentrations of parking garage
decks at 50 mm depth.*CN=30% sol. of calcium nitrite

CONCLUSIONS

1. Extensive work shows that excellent long-term durability for steel reinforced concrete exposed to chloride environments is achievable by modifying the concrete mixture. A mix design that produces low permeability concrete with reduced tendency for drying shrinkage can be readily produced using ordinary Portland cement, superplasticizers, and shrinkage reducing admixtures. However, these measures are not sufficient to provide long-term protection in severe chloride environments.   To provide the long-term service lives corrosion inhibitors are needed with or without silica fume, fly ash or ground blast furnace slag.

2. The improvements in durability are easily modeled using standard theories for diffusion of chlorides into good quality concrete, and understanding of the mechanisms by which calcium nitrite protects steel in concrete.

3. Thus, it is possible to produce steel reinforced concrete structures with long service lives by optimizing the concrete mixture components to provide a balance between permeability, concrete cover, corrosion inhibiting capability, and resistance to drying shrinkage induced cracking.

4. The extended service life obtained by the use of concrete admixtures benefits the owner by reducing the long-term cost of his structure. Mankind also benefits in that the new raw materials will not need to be processed and consumed due to the early demise of the structure.

REFERENCES

1. ACI 318 "Building Code Requirements for Reinforced Concrete", ACI Manual of Concrete Practice, Part 3, Use of Concrete in Buildings - Design, Specifications, and Related Topics, American Concrete Institute, Detroit, MI, USA.

2. ACI-357, "Guide for the Design and Construction of Fixed Offshore Concrete Structures", ACI Manual of Concrete Practice, Part 4, Bridges, Substructures, Sanitary, and Other Special Structures. Structural Properties, American Concrete Institute, Detroit, MI, USA.

3. CSA Standard S413-94, *Parking Structures,* Canadian Standard, pp. 115, 1994.

4. BS6349, "Maritime Structures", British Standards Institution, Part 1, 1984.

5. ACI 224R "Control of Cracking in Concrete Structures", ACI Manual of Concrete Practice, Part 3, Use of Concrete in Buildings - Design, Specifications, and Related Topics, American Concrete Institute, Detroit, MI, USA.

6. BS8110, "Structural Use of Concrete", British Standards Institution, Part 2, 1985.

7. ASTM C192-90a, "Practice for Making and Curing Concrete Test Specimens in the Laboratory", American Society for Testing and Materials, Philadelphia, USA.

8. ASTM C150-92, "Standard Specification for Portland Cement", American Society for Testing and Materials, Philadelphia, USA.

9. ASTM C1240, "Standard Specification for Silica Fume for Use in Hydraulic-Cement Concrete and Mortar", American Society for Testing and Materials, Philadelphia, USA.

10. ASTM C 33-90, Standard Specification for Concrete Aggregates, American Society for Testing and Materials, Philadelphia, USA.

11. ASTM C 618-92a, "Standard Specification for Fly Ash and Raw or Calcined Natural Pozzolan for use as a Mineral Admixture in Portland Cement Concrete", American Society for Testing and Materials, Philadelphia, USA.

12. ASTM C 989-89, "Specification for Ground Granulated Blast-Furnace Slag for Use in Concrete and Mortars", American Society for Testing and Materials, Philadelphia, USA.

13. ASTM C 494-92, "Standard Specification for Chemical Admixtures for Concrete", American Society for Testing and Materials, Philadelphia, USA.

14.   ASTM C 143-90, "Test Method for Slump of Hydraulic Cement Concrete", American Society for Testing and Materials, Philadelphia, USA.

15.   ASTM C 231-91b, "Test Method for Air Content of Freshly Mixed Concrete by the Pressure Method", American Society for Testing and Materials, Philadelphia, USA.

16.   ASTM C 39-86, "Test Method for Compressive Strength of Cylindrical Concrete Specimens", American Society for Testing and Materials, Philadelphia, USA.

17.   ASTM C 403-90, "Standard Test Method for Time of Setting of Concrete Mixtures by Penetration Resistance", American Society for Testing and Materials, Philadelphia, USA.

18.   ASTM C 138-81, "Test Method for Unit Weight, Yield, and Air Content (Gravimetric) of Concrete", American Society for Testing and Materials, Philadelphia, USA.

19.   ASTM A 615, "Specification for Bars, Deformed and Plain, Billet-Steel, for Concrete Reinforcement", American Society for Testing and Materials, Philadelphia, USA.

20.   ASTM G109-92, "Standard Test Method for Determining the Effects of Chemical Admixtures on the Corrosion of Embedded Steel Reinforcement in Concrete Exposed to Chloride Environments", American Society for Testing and Materials, Philadelphia, USA.

21.   BERKE, N. S. AND HICKS, M. C, "Predicting Chloride Profiles in Concrete", Corrosion, pp. 234-239, March 1994

22.   ASTM C 1152-90, "Test Methods for Acid-Soluble Chloride in Mortar and Concrete", American Society for Testing and Materials, Philadelphia, USA.

23.   WEST, R. R. AND HIME, W. G., "Chloride Profiles in Salty Concrete", Corrosion/85, Paper No. 256, National Association of Corrosion Engineers, Houston, 1985. 24

24.   WEYERS, R. E. AND CADY, P. D., "Deterioration of Concrete Bridge Decks from Corrosion of Reinforcing Steel", Concrete International, pp. 15-20, January, 1987.

25.   BERKE, N. S. AND HICKS, M. C., "Estimating the Life Cycle of Reinforced Concrete Decks and Marine Piles using Laboratory Diffusion and Corrosion Data", STP 1137, Corrosion Forms and Control of Infrastructure, Victor Chaker Editor, ASTM, Philadelphia, pp. 207-231, 1992.

26.   BAMFORTH, P. B., "Specification and Design of Concrete for the Protection of Reinforcement in Chloride Contaminated Environments", UK Corrosion & Eurocorr 94, Bournemouth, UK, 31 October-3 November 1994.

27.   BROWNE, R.D., "Design Prediction of the Life for Reinforced Concrete in Marine and Other Chloride Environments", Durability of Building Materials, V. 1, pp. 113-125 (1982).

28.   ASTM C1202-94, "Test Method for Electrical Indication of Concrete's Ability to Resist Chloride Ion Penetration", American Society for Testing and Materials, Philadelphia, USA.

29.   BERKE, N.S., "The Effects of Calcium Nitrite and Mix Design on the Corrosion Resistance of Steel in Concrete (Part 2, Long-Term Results)". Proceedings of the Corrosion-87 Symposium on Corrosion of Metals in Concrete, NACE, Houston (1987), p. 134-144.

30.   SCALI, M. J., CHIN, D. AND BERKE, N. S., "Effect of Microsilica and Fly Ash upon the Microstructure and Permeability of Concrete", Proceedings of the Ninth International Conference on Cement Microscopy, April 5-9, 1987, Reno, Nevada, U.S.A., 375-397, International Cement Microscopy Association, Dallas, Texas, USA.

31.   BERKE, N.S. AND HICKS, M.C., "Electrochemical Methods of Determining the Corrosivity of Steel in concrete", ASTM 25th Anniversary Symposium, STP 1000, R. Baboian and S. Dean Eds., ASTM, Philadelphia, pp. 425-440, 1990.

32.   TOURNEY, P. G. AND BERKE, N. S., "Put to the Test", Civil Engineering, pp. 62-63, December 1992

33.   BERKE, N.S. AND HICKS, M.C., HOOPES, R. J. AND TOURNEY, P. G., "Use of Laboratory Techniques to Evaluate Long-Term Durability of Steel Reinforced Concrete Exposed to Chloride Ingress", Durability of Concrete, Third International Conference, Nice, France, ACI SP 145, pp. 299-329, 1994.

34.   ASTM G 59-91, "Standard Practice for Conducting Potentiodynamic Polarization Resistance Measurements", American Society for Testing and Materials, Philadelphia, USA.

35.   BERKE, N. S., "The Effects of Calcium Nitrite and Mix Design on the Corrosion Resistance of Steel in Concrete (Part 1)", NACE Corrosion-85, Paper No. 273, Houston, 1985

36.   BERKE, N.S., SHEN, D.F., AND SUNDBERG, K.M., "Comparison of Current Interruption and Electrochemical Impedance Techniques in the Determination of the Corrosion Rates of Steel in Concrete", Presented at the ASTM Symposium on Ohmic Electrolyte Resistance Measurement and Compensation, Baltimore, MD, May 17, 1988.

37. ASTM C 157-91, "Standard Test Method for Length Change of Hardened Hydraulic-Cement Mortar and Concrete", American Society for Testing and Materials, Philadelphia, USA.

38. SHAH, S. P., KARAGULER, M. E. AND SARIGAPHUTI, M. "Effects of Shrinkage-Reducing Admixtures on Restrained Shrinkage Cracking of Concrete", ACI Materials Journal, V. 89, No. 3, May-June 1992, pp. 289-295.

39. PAGE, C. L., SHORT, N. R. AND TARRAS, A El, "Diffusion of Chloride Ions in Hardened Cement Pastes", Cement and Concrete Research, Vol. 11, pp. 395-406, 1981.

40. BERKE, N. S. AND ROSENBERG, A. M., "Technical Review of Calcium Nitrite Corrosion Inhibitor in Concrete", Transportation Research Record 1211, Transportation Research Board, Washington , DC (1989), pp. 18-27.

41. BERKE, N.S. AND ROBERTS, L.R., "Use of Concrete Admixtures to Provide Long-Term Durability from Steel Corrosion", Proceedings, Third CANMET/ACI International Conference on Superplasticizers and Other Chemical Admixtures in Concrete, Ottawa, Canada, October 4-6, 1989, ACI SP 119, 1989 (Ed. V.M. Malhotra) p. 383

42. BERKE, N.S. AND EL-JAZAIRI, B, "The Use of Calcium Nitrite as a Corrosion Inhibiting Admixture to Steel Reinforcement in Concrete", Corrosion of Reinforcement in Concrete, C.L. Page, K.W.J. Treadaway, and P.B. Bamforth, eds., Elsevier Applied Science, London (1990), pp. 571-585.

43. BERKE, N. S., "Corrosion Inhibitors in Concrete", Concrete International, Vol. 13, No. 7, July 1991, pp. 24-27.

44. TOMOSAWA, F., MASUDA, Y., TANAKA, H., FUKUSHI, I., TAKAKURA, M., HORI, T. AND HIGASHI, S., "An Experimental Study on the Effectiveness of Corrosion Inhibitor in Reinforced Concrete Under High Chloride Content Conditions". Niho Architecture Society, October 1987.

45. HOPE, B. B. AND IP, A. K. C., "Corrosion Inhibitors for Use in Concrete", ACI Materials Journal, pp. 602-608, November-December 1989.

46. ANDRADE, C., ALONSO, C. AND GONZALEZ, J.A., "Some Laboratory Experiments on the Inhibitor Effect of Sodium Nitrite on Reinforcement Corrosion", Cement, Concrete and Aggregates, CCAGDP, Vol. 8, No. 2, Winter 1986 pp. 110-116.

47. CARINO, N.J. AND CLIFTON, J.R., "Prediction of Cracking in Reinforced Concrete Structures", NISTIR 5634, Building and Fire Research Laboratory, National Institute of Standards and Technology, Gaithersburg, MD, USA.

48. BERKE, N. S., HICKS M.C. AND HOOPES R. J., "Condition Assessment of Field Structures with Calcium Nitrite", Philip D. Cady International Symposium, Concrete Bridges in Aggressive Environments, Richard E. Weyers Ed., SP-151, American Concrete Institute, Detroit, MI, pp. 43-72, 1994.

49. BERKE, N.S., PFEIFER, D.E. AND WEIL, T.G. "Protection Against Chloride Induced Corrosion - A Review of Data and Economics on Microsilica and Calcium Nitrite", Concrete International, Vol. 10, No. 12, pp. 45-55. December 1988

50. WEYERS, R.E. AND HOFFMAN, P.C., "Estimating Service Life and Remaining Service Life of Concrete Bridges in Chloride-Laden Environments", to be published.

STABILITY OF MECHANICAL PROPERTIES AND INTERFACE DENSITY OF HIGH PERFORMANCE FLY ASH CONCRETE

M S Besari

D R Munaf

Hanafiah

M M Iqbal

Institut Teknologi Bandung

Indonesia

ABSTRACT. Relationships between various mechanical properties of High Strength Concrete made with Fly Ash with strengths of up to 85 MPa are discussed, as well as the strengthening behaviour of interface between matrix and the coarse aggregates as observed from Scanning Electron Microscope. Two types of fly ash which come from different Power Plants were used, with their chemical composition differing slightly. The results show that equal effects independent of fly ash source could be reached by modifying the percentages of fly ash replacement and their specific surface areas. The result will be useful for concrete manufactures as well as for scientists working in the field of Concrete Technology.

Keywords: Carbonation, Chemical composition, Density, High Performance Concrete (HPC), Fly Ash Suralaya (FAS), Fly Ash Paiton (FAP), Specific Surface Area (SSA), Scanning Electron Microscope (SEM), Lubricating effect.

Professor Dr M Sahari Besari is dean of the Faculty Civil Engineering and Planning, Institut Teknologi Bandung (ITB), Bandung, Indonesia. He specialises in the Numerical Modelling and Structural Repair in Steel and Concrete Structures.

Dr Dicky R Munaf is Head of Structures and Materials Laboratory, Dept Civil Engineering, ITB, Indonesia. His interests include the Strengthening and Crack Density Analysis in Interface Zone of High Performance Fly Ash Concrete.

Hanafiah is a PhD candidate of Department of Civil Engineering, Institut Teknologi Bandung (ITB). His main research is Constitutive Model for High Strength Fly Ash Concrete. He is lecturer in University of Sriwijaya, Palembang, Indonesia.

Maulid M Iqbal is a PhD candidate of Department of Civil Engineering, ITB. His main research is Confinement Effect of High Performance Fly Concrete. He is lecturer in University of Sriwijaya, Palembang, Indonesia.

Radical Concrete Technology. Edited by R K Dhir and P C Hewlett. Published in 1996 by E & FN Spon, 2–6 Boundary Row, London SE1 8HN, UK. ISBN 0 419 21480 1.

INTRODUCTION

Fly ash is currently accepted as suitable cement supplementary material for the production of High Performance Concrete mixes. This material is commonly the byproduct of coal firing in the industry and possesses cementitious and pozzolanic properties and can be used to strengthen the interface zone beetwen the matrix and coarse aggregates, also to increase the impermeability of concrete to environment influence.

Selection of a combination of suitable cementitious materials for a concrete mix is normally made by the concrete producer based on given concrete specifications and also on his experience. Use of fly ash provide significant cost saving to the concrete producer. If the composition of a concrete mix is selected only on the basis of its compressive strength, performance of concrete in service may not be guaranteed without the necessary knowledge on the role of cement supplementary materials on other engineering properties.

This paper presents the results of an experimental study of the effects of partially replacing cement with two kinds of fly ash, which come from different types of coal, on the properties of high strength concrete. The binder materials used in this investigation were Ordinary Portland Cement and fly ash. Concrete properties studied included the density, compressive and tension strength and the interface zone density. Especially for the observation of interface zone density, various enlargements using Scanning Electron Microscope was applied.

EXPERIMENTAL DETAILS

Fly ash from two Power Plants in Indonesia (Suralaya and Paiton) and Ordinary Portland Cement (ASTM C 150-92) were used in the preparation of the concrete mixes. The mix design adopted to produce concrete with compressive strengths of 85 MPa and optimum mix designs obtained from previous results [1]. Table 1 shows the details of the adopted mix design.

Table 1. Mix Proportioning

Material	Weight (kg/m^3 concrete)	
Cement	560	476
Water/Cement	0.28	0.28
Fly Ash	0	84 (15 %)
Fine Aggregate	751.53 (0 - 4.8 mm)	741.53 (0 - 4.8 mm)
Coarse Aggregate	978.82 (4.75-9.5 mm = 45%)	978.82 (4.75-9.5 mm= 45%)
	(9.5 - 19.5 mm = 55%)	(9.5-19.5 mm=55%)
Superplasticizer	2.4 %	2.4 %

Water to bider ratio was kept constant at 0.28 by weight for all the mixes.

The levels of fly ash replacing cement was varied between 10 to 25 %. The control datum contains fly ash 15 % of cement weight. The coarse aggregate was crushed andesit consistinng of size gradations as shown in table 1. The fine aggregates consisted of fine volcanic sand. Especially for the coarse aggregate, the petrography result as shown in figure 1 indicated that 25 % of the volume consists of surface and non structural cracks which is typical with common coarse aggregate in Indonesia [2].

Figure 1. Result of Petrography Test of Coarse Aggregate

The mixing process was carried out in a revolving type the mixer at 30 RPM and the fresh concrete was tested for its slump and density accordance with ASTM procedures. The temperature of the concrete mixes was about 24.5°C (± 1.5°C).

A number of standard cylindrical specimens were cast in steel moulds using a vibrating table for compaction purposes. 150 x 300 mm Cylinder Specimens of several ages, up to 90 days, were tested for compressive strength. A new type of specimens, as shown in figure 2 were used for direct tension tests [3].

Figure 2. New Type Specimen for Tension Test

Upon demoulding, after 24 hours, the test specimens were subjected to water curing at room temperature of about 21°C (± 2°C).

FLY ASH PARAMETER

Three types of fly ash parameters were cosidered, the chemical composition, specific surface area and morphology of the fly ash. Chemical composition tests were performed based on ASTM C 114-88 Standard and the specific surface area was found by Blaine tests. The results are shown in table 2. [2],[4].

Table 2. Chemical Composition and Specific Surface Area Test Result

NO	TYPE OF TEST		TYPE OF FLY ASH		CEMENT
			SURALAYA	PAITON	
I	**Chemical Composition**				
	• Si O_2	(%)	59.4	41.1	21.0
	• Al_2 O_3	(%)	24.7	22.0	4.0
	• Fe_2 O_3	(%)	4.6	15.0	5.9
	• Ca O	(%)	3.1	8.8	64.3
	• Mg O	(%)	1.7	3.0	3.8
	• Na_2 O	(%)	2.5	1.5	NA
	• K_2 O	(%)	0.5	0.9	NA
	• Ti O_2	(%)	0.8	0.6	NA
	• P_2 O_5	(%)	0.4	0.5	NA
	• SO_3	(%)	NA	6.6	NA
	• Mn_3 O_4	(%)	NA	NA	NA
II	**Specific Surface Area (cm^2/gram)**		4682	3817	3175

Scanning Electron Microscope (SEM) type JEOL 5310 LV was used to carry out morphological investigation of fly ash and cement. It is capable of providing informations on the size, the shape and texture of grains down to milimeter or even micrometer scale. The result are shown in figure 3 and table 3, with enlargement of 4000X.

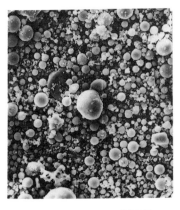

a) SEM Result of FAS   **b) SEM Result of CEMENT**

Figure 3.

Table 3. Morphology Investigation of Cement, FAS and FAP

Material Type of Observation	Cement	Fly Ash Suralaya (FAS)	Fly Ash Paiton (FAP)
Size of Grains	114-370 μm	50-210 μm	933-291 μm
Shape of Grains	Sharp	Spherical	Spherical
Texture	Rough	Smooth	Smooth

RESULT AND DISCUSSION

Workability of Concrete

Figure 4 shows the effect of partially replacing cement with fly ash. When fly ash was used, concrete slump increased noticeably (for FAS and FAP). Slump of the concrete mix with 25 % fly ash (FAP) content was 124 mm compared to 52 mm of that with 15 % fly ash (FAS) concrete (control mix). This improvement in workability of concrete is due to the combined effect of :

a. Increased paste volume content in the concrete as the result of lower particle density of fly ash compare to that of cement, and

b. Lubricating effect ("Ball Bearing Effect") due to the spherical shape of fly ash particles.

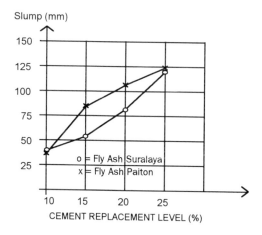

Figure 4. Slump Measurement Result of 25% FAP and 15% FAS

Comparing the slump test result of FAS and FAP, it is coincides with the Owen [5] prediction that the workability of fly ash concrete is mainly affected by the proportion of coarse material in the fly ash in addition to the level of cement replacement.

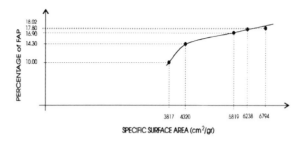

**Figure 6. Effect of SSA of FAP in the Optimum Volume
of Cement Replacing by FAP**

Tensile Strength of Concrete

Table 4 summarizes the compressive and direct tensile strengths at 28 days for concretes, having different combination of fly ash types. Similar to the compressive strengths, tensile strengths of concrete is lower compared to those of the control mix at early ages.

**Table 4. Result of Tension and Compression Strength
of HPC - FAP and HPC - FAS**

Type of Fly Ash	Percentage of Cement Replacement	Concrete Strength (Mpa), 28 days	
		Tensile	Compression
FAS	10 %	162.4	804
	15 %	163.1	848
	20 %	151.8	803
	25 %	120.6	754
FAP	10 %	111.2	794
	15 %	127.7	743
	20 %	118.7	781
	25 %	109.4	701

Figure 7 shows the existence of a linear relationship between tensile and compressive strength of the concrete. It shows that the tensile strength of fly ash concrete is about 17.32 % of its compressive strength for FAS, while 15.48 % for FAP ad is relatively higher compared to that of normal concrete (10 %) [6].

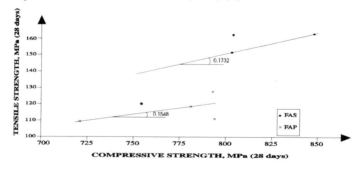

Figure 7. Relationship Between Tensile and Compressive Strength

Density of Fresh Concrete

Since the specific gravities of FAS and FAP are about 2.3 and 2.26 respectively compared to 3.14 of that of cement, the density of concrete is expected to decrease when cement is replaced by fly ash. At cement replacement of 25 %, the densities of fresh concrete with fly ash were 2368 kg/m^3 for fresh FAS-concrete and 2341 kg/m^3 for FAP-concrete respectively, compared to 2469 kg/m^3 for the control mix.

Compressive Strength of Concrete

Figure 5 shows the effect of partially replacing cement with FAS and FAP on the development of Compressive Strength with age for water-cured concrete. The optimum use of FAS was found to be 15 % (coinciding with previous investigation results [1]), but the optimum use of FAP was found to be 10 % replacement of cement.

The effect of fly ash content on the strength was noted to depend on the age of concrete as pozzolanic reaction is slower than hydration of cement. Specifically, the early age strength of fly ash concrete is governed by the water to cement ratio, where as the final strength it is influenced by the water to the total cementitious material ratio.

Furthermore, comparing optimum percentages of FAS and FAP for replacing cement to produce High Performance Concrete, it was found that the optimum percentage of FAP replacing cement is about 10 %. It is not significant number compared to the production of Fly Ash in Paiton Power Plant and to the demand of concrete in the Construction industry. The investigation shows that the optimum percentage could be increased to 17.8 % by increasing the specific surface area of fly ash, keeping the strength of concrete constant to 85 MPa. Figure 6 shows a process to get the optimum percentage of FAP replacing cement weight to produce 85 MPa Concrete.

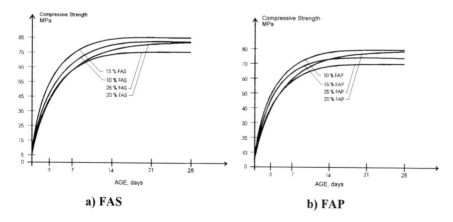

a) FAS   **b) FAP**

Figure 5. Strength Development for Various Percentage of Fly Ash

Interface Density Behavior

Fracture surfaces at interface zones were examined using SEM for each percentage and type of fly ash in concrete. The examination includes the densities of microcracks and microporous areas. Table 5 gives the results of quantification on microcracks and microporosities using dye penetration technique.

Table 5. Quantification of Microcrack and Microporosities

Type of Fly Ash	Percentage of Cement Replacement	Interface Behavior (cm^{-1})		
		Microcrack	Paste/Aggregate Interface	Microporous Zone
FAS	10 %	0.14	2.41	1.01
	15 %	< 0.10	1.21	0.81
	20 %	0.19	2.42	1.08
	25 %	0.21	2.42	1.11
FAP	10 %	<0.10	1.38	0.74
	15 %	0.17	2.78	1.34
	17.8 %	0.11	1.41	0.84
	20 %	0.21	2.94	1.42
	25 %	0.14	2.80	1.38

The densities of microcracks in 15 %- FAS concrete and 17.8 % - FAP concrete are low as shown in figure 8. Conversely, that of the 10 % - FAS concrete and 10 % FAP concrete are relatively larger and are also of large sizes. These results indicate the influence of carbonation and reduction of Ca(OH)$_2$ and also the effect of specific surface area of the fly ash.

Based on the above results, it was plannned also to abserve the influence of the curing method, such as drying method and hot weather curing method to retard the crack due to the hydration process.

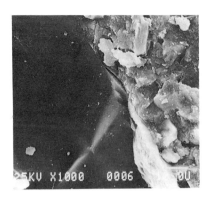

a) Interface Density of 15 % FAS   b) Interface Density of 17.8 % FAP

Figure 8

CONCLUSIONS

From the results reported in the previous sections, the following conclusions can be made :

1. Slump of concrete increases with increasing content of fly ash.
2. Compressive and tensile strength of concrete decrease with the increase or decrease in fly ash content. However, there is an optimum percentage of fly ash to produce the "best" concrete, which is 15 % for FAS (59,4% SiO_2) and 17.8 % for FAP (41.1 % SiO_2).
3. It was found that the "ball bearing effect" was strongly influenced by the specific surface area.
4. Decrease of SiO_2 percentages inn fly ash will require a greater value of Specific Surface Area to reach the same compressive strength of 85 MPa.

ACKNOWLEDGEMENTS

The authors would like to express their appreciation for the research grants made by the Directorate General of Higher Education, Department of Culture and Education, Government of Indonesia through contract no : 040/P4M/DPPM/94/PHB II/2/1994 to carry out the reported work, which forms part of the major research program on High Performance Concrete and Cement Based Materials at the Institut Teknologi Bandung.

REFERENCE

1. Besari M.S., Munaf D.R. & Hanafiah, **"The Effect of Fly Ash and Strength of Aggregate to The Mechanical Properties of High Strength Concrete",** International Conference of Concrete and Structures, Singapore, August 25-27, 1992.

2. Munaf D.R. et al., **"The Investigation of Based Material for High Strength Concrete",** Research report through funding for Directorate General of Higher Education-World Bank, Contract No. 9/P4M/DPPM/BD XXI/ID&R, 1993 (in Indonesian).

3. Munaf D.R. et al., **"The Investigation of Behavior of Structural Components Made By High Performance Concrete-Fly Ash",** Research Report through funding from Competitive Grant Project of Directorate General of Higher Education, Contract No. 010/P4M/DPPM/PHB III/1994. (in Indonesian).

4. Mangunwijaya A., **"The Use of Coal Fly Ash and its Related for Environmental Protection",** 1995 (in Indonesian).

5. Owen P.L., **"Fly Ash and its Usage in Concrete",** Journal of Concrete Scociety, vol. 13, 1976, pp. 21-26.

6.   CEB/FIP., **"Recommendations International Pour le Calcul et l'execution des ouvrages en Beton"**, 1990.

SHEAR-FATIGUE BEHAVIOUR OF ULTRA-HIGH STRENGTH REINFORCED CONCRETE BEAMS

K-H Kwak

K-W Jang

K-H Kwak

W T Kim

Won Kwang University

Korea

ABSTRACT. This paper examines the characteristics of the shear-fatigue behaviour of ultra-high strength reinforced concrete beams. The mix proportion for the concrete was determined from a series of trial mixes using different cement contents and silica fume replacement levels. A constant $W/(C+SF)$ ratio was used throughout, and the Powercon-100 superplasticizer was added. The mix type with more cement and silica fume contents results in higher compressive strength and was used to produce the tested beams. Shear reinforcement ratios of 0, 50 and 100% according to the ACI 318-89 code were chosen for the study. Both static flexural tests and fatigue tests were carried out. The shear failure mode of the ultra-high strength concrete beams was found to be similar to that of the normal RC beams, except that brittle characteristics had been clearly shown. The results also indicated that the ultimate shear strength value of the ultra-high strength reinforced concrete beam was about twice the value calculated from the formula of ACI 318-89 code. A repair method with epoxy injection to the flexural and the diagonal tensile cracks was also investigated. No further propagation of the flexural cracks in the repaired specimens was found when reloading.

Keywords: Ordinary Portland cement, Silica fume, Superplasticizer, Ultra-high strength concrete, Shear reinforcements, Ultimate shear strength, Fatigue strength, Flexural cracks, Diagonal tensile cracks, Epoxy injection.

Professor & Dr Kae-Hwan Kwak is a Professor in the Department of Civil Engineering, WonKwang University.

Ki-Woong Jang is a PhD Candidate in the Department of Civil Engineering, WonKwang University.

Kyung-Hern Kwak and Won-Tae Kim are graduate students in the Department of Civil Engineering, WonKwang University.

Radical Concrete Technology. Edited by R K Dhir and P C Hewlett. Published in 1996 by E & FN Spon, 2–6 Boundary Row, London SE1 8HN, UK. ISBN 0 419 21480 1.

INTRODUCTION

It is a recent trend that concrete structures are becoming bigger, higher, longer, and more specialized. Owing to this trend, high strength concrete need to be developed urgently. Although various researches on utilizing high strength reinforced concrete have been undertaken in Korea, their results are not satisfactory yet. This paper, therefore, tries to clarify the characteristics of shear-fatigue behaviour of high strength reinforced concrete beams through the analysis on the production and the tests of materials of ultra-high strength reinforced concrete for which silica fume is added.

PRODUCTION OF ULTRA-HIGH STRENGTH CONCRETE (UHSC)

Ordinary Portland cement was used and silica fume was added as an admixture in this experiment. Powercon-100 was also used as a superplasticizer. To make them high strength, W/(C+SF) was fixed at 19 percent and the amount of cement and silica fume was changed. Several mixture proportions were studied in this paper. Three mixture proportions of UHSC were produced with different cement content such as $500kg/m^3$(A-type), $600kg/m^3$(B-type) and $700kg/m^3$(C-type). Each type was further divided into four types according to the content of silica fume, that is 0% (SF0), 10% (SF10), 20% (SF20) and 30% (SF30) by cement weight. In this paper, hence, 12 types of mix proportions were presented and tested.

In this experiment, C-SF30 series showed the highest compressive strength. The mixture proportion was given in Table 1.

EXPERIMENTAL PROGRAM

Among the 12 mixture proportions, the mixture of C-SF30 series which showed the highest compressive strength was used to make ultra-high strength reinforced concrete beams. Deformed bars were used as tensile steel bars and the longitudinal reinforcement ratio (ρ_v) was constant at 0.033. Deformed bars were used as stirrups and shear reinforcement ratio ($\rho_v/\rho_{v(ACI)}$) was 0%, 50% and 100% following ACI 318-89 (11-17) equation. Hence, the first character in the specimen name gives the shear reinforcement ratio (A, without stirrup; B, 6 stirrups; C, 12 stirrups) and the

Table 1   Mixture proportion for C-SF30 series

W/C+SF (%)	Slump (cm)	S/A (%)	SF (%)	SP (%)	Unit Weight (kg/m^3)					
					Cement	Water	SF	Fine Aggre.	Coarse Aggre.	SP
19	10	35	30	1.9	700	173	210	481	889	15.4

SF = silica fume, SP = superplasticizer

second character gives the test method (S, static test; F, fatigue test). Also, the stirrup of CS(CF) series was designed by ACI 318-89 code and BS(BF) series was designed to induce shear fracture prior to the failure of CS(CF) series. The details of the specimens were shown in Figure 1.

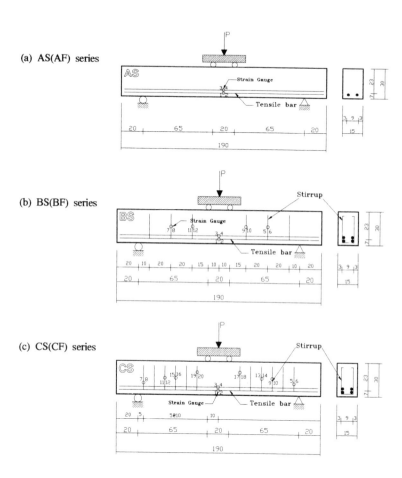

(a) AS(AF) series

(b) BS(BF) series

(c) CS(CF) series

Figure 1 Details of specimens tested

RESULTS OF EXPERIMENTS

Tests of Materials

100 x 200mm cylinders were made according to the variance of the amount of cement and admixture and measured for the compressive strength on 3, 7, 28 and 90 days. The tensile strength was also measured at 28-age. Each of the tests was conducted by the universal testing machine and Table 2 shows the compressive strength and tensile

Table 2   Variation of strength with time

Mix Series	Compressive Strength (kg/cm^2)				Split Tensile Strength (kg/cm^2)
	3 days	7 days	28 days	90 days	28 days
A-SF0	364	426	525	577	45.4
A-SF10	533	630	749	808	55.4
A-SF20	534	684	865	908	58.1
A-SF30	507	697	790	845	47.0
B-SF0	395	486	608	655	39.5
B-SF10	446	552	708	778	46.6
B-SF20	621	689	851	927	43.3
B-SF30	566	720	809	880	38.2
C-SF0	497	610	700	780	41.7
C-SF10	638	800	940	990	52.0
C-SF20	582	815	995	1,040	47.5
C-SF30	560	990	1,100	1,180	53.6

strength of each series. As we can observe in Table 2, the more amount of cement was used in the experiment, the higher the strength was. It was also noticeable that the optimum content of admixtures for strength increase was present when we use the same amount of cement. From this result, we can observe that there is an optimum content of admixtures for strength increase and that the amount of admixture increases as the amount of cement increases.

Static Tests

For this experiment, the universal testing machine was used of 200 ton capacity. As the tests proceeded, we increased the load by 0.5 ton each time and the midspan deflection, strain and crack propagation were checked at each stage.

Mode of failure

Initial flexural cracks developed in the pure bending region of beams as the load increases. When the load was increased more, flexural cracks in the pure bending region propagated more obvious and flexural cracks were observed up to the shear region. Afterwards when the load is added, either a singular diagonal tensile cracks occurs in the middle of the web on the shear span of beams or flexural cracks on the shear span was developed into diagonal tensile cracks at 45 deg. from the neutral axis of the beams. The diagonal tensile cracks developed simultaneously toward both the loading points and the supports.

The beams without shear reinforcement showed initial diagonal tensile cracks first. After a minor additional load, the width of the cracks widened rapidly, and then there came shear-tension failure. But as shear reinforcement became increased, the reserve shear strength was increased even after the initial diagonal tensile cracks. Shear reinforcing bars was able to control the diagonal tensile cracks from the additional loads. By this mechanism, the redistribution of resistance reduced the increase of tensile crack region.

Finally, shear-compressive failure and shear-tension failure occurred. Figure 2 and Figure 3 below illustrate crack growth patterns and failure modes of the beams without shear reinforcement.

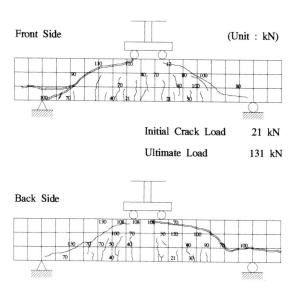

Initial Crack Load   21 kN

Ultimate Load   131 kN

Figure 2 Relation of load and crack growth for AS1

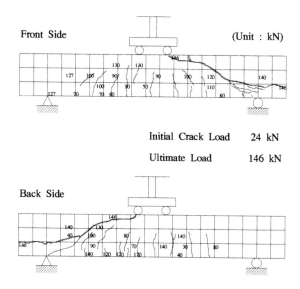

Initial Crack Load   24 kN

Ultimate Load   146 kN

Figure 3 Relation of load and crack growth for AS2

Load and midspan deflection

The deflection was obtained by installing a dial gauge at the centre of the specimen. As we can observe in Figure 4, the midspan deflection increased linearly as the load increased, but the beams without shear reinforcing bars were failed immediately after the cracks were occurred, showing brittle behaviour. As shear reinforcing bars are increased, however, we could observe ductile behaviour.

Load and strain

The strain on tensile steel bars was recorded by the steel strain gauges. The tensile steel bars showed the strain increase with a constant angle when the load increased.

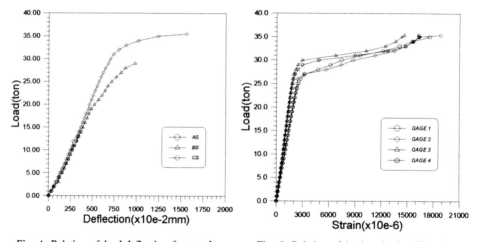

Fig. 4. Relation of load-deflection for test beams   Fig. 5. Relation of load-strain for CS series

Shear strength

Since our country, South Korea, uses ACI 318-89 code, we would like to show the test value of shear strength for the experimental and the calculated value in ACI 318-89 as in Table 3 below. The ACI formula for shear strength can be obtained by adding the shear crack strength of the concrete and the shear strength of shear reinforcing bars. Since a/d of the experimental beams is greater than 2.5, the nominal shear strength of the concrete V_{cr} and that of the steel bars are as follows:

$$V_{cr} = 0.53\sqrt{\sigma_{ck}} + 176 \; \rho_w \; (\frac{d}{a}) \; \leq \; 0.93\sqrt{\sigma_{ck}}$$

$$V_s = \frac{A_v \; \sigma_y \; d}{s}$$

Table 3 Comparison of shear strength

Specimen ID	Test Results			ACI
	P_u (t)	P_{dc} (t)	P_{fc} (t)	P_u (t)
AS1	13.1	13.0	2.1	6.06
AS2	14.6	14.0	2.4	6.06
AS3	13.4	13.1	1.8	6.06
BS1	28.8	11.5	1.5	10.98
BS2	29.3	12.8	2.1	10.98
BS3	29.4	13.1	1.9	10.98
CS1	35.4	11.7	1.3	15.90
CS2	35.7	12.6	1.8	15.90
CS3	36.3	12.9	2.0	15.90

P_u = shear ultimate load ; P_{dc} = initial diagonal tensile cracking load ;
P_{fc} = initial flexural cracking load

Table 3 shows that the shear crack strength remain constant at the same shear span ratio no matter how large the shear reinforcement ratio was, and that shear reinforcing bars were not affected by the shear cracks were not occurred. When we compare these with our test results, we can notice that ACI formula for beams of shear span ratio 2.5 - 3.0 underestimates the shear strength of the concrete and the shear reinforcing steel bars.

Repair effect to cracks

In this paper, an epoxy injection was used to repair flexural cracks and diagonal tensile cracks, and the repaired specimens were tested under reloading. Figure 6 shows repair job to flexural cracks and diagonal cracks by epoxy injection method. According to the result, flexural cracks were no longer propagated and it was proved that epoxy repair to flexural cracks is very effective.

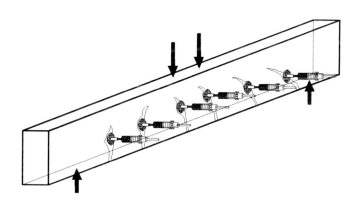

Figure 6 Epoxy injection repair to cracks

64   Kwak et al

Dynamic Tests

Since the amplitude of load is understood to be the most important factor among the factors effective to the fatigue strength, we carried out the experiments by making the minimum value of cyclic load constant with the flexural crack load for the members from 2 to 3 tons and by making the maximum value of cyclic load 50% and 60% of the ultimate static load. The three kinds of members utilized for these are the same as those utilized for the static tests.

Mode of failure

Under fatigue load most of the members showed initial flexural cracks when a cyclic load was effective. In the shear region, the initial flexural cracks were increased as the cycle increased first, but at some stage they developed into diagonal tensile cracks at the mid point. Figure 7 and Figure 8 below indicate the crack growth pattern and the failure mode of the experimental beams without shear reinforcement.

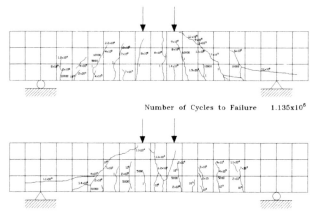

Number of Cycles to Failure 1.135×10^6

Figure 7   Relation of crack growth and number of cycles for AF series

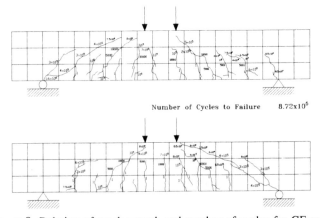

Number of Cycles to Failure 8.72×10^5

Figure 8   Relation of crack growth and number of cycles for CF series

Table 4 Results of fatigue test

Specimen	P_{min} (t)	P_{max}(t)	R*	P_u (t)	N_u (cycle)	P_{dc} (t)	P_{fc} (t)
AF1	2.1	6.9	0.50	13.7	1,135,620	2.1	13.4
AF2	2.1	8.2	0.60		964,000		
BF1	1.8	14.6	0.50	29.2	594,800	1.8	12.5
BF2	1.8	17.5	0.60		1,073,000		
CF1	1.7	17.9	0.50	35.8	872,400	1.7	12.4
CF2	1.7	21.5	0.60		393,800		

Relationship between stress level and the number of cycles

Table 4 illustrates the number of loading cycles to failure for the members according to the maximum stress level. According to ACI Committee 215 Report, it is preferable that the fatigue limit of concrete is obtained from the fatigue strength at such a high repetition as 1×10^6, that the flexural strength is about 55% of the ultimate static strength. The results of this research, however, showed that the failure of the experimental beams without shear reinforcement occurred with the repetition little over 1×10^6 when the maximum stress level was 50% of the ultimate static strength. It was also noticeable that the repetition dropped drastically when the maximum stress level was below 60%. As the shear reinforcement increased, the ultimate maximum stress level was below 60%. As the shear reinforcement increased, the ultimate static strength increased, and the increase in the ultimate static strength in turn increased the maximum stress level, which caused the repetition at failure drop rapidly. These results clearly indicate that the fatigue limit of the high strength concrete is lower than that of normal concrete and that it is so because of the higher brittle characteristics of the high strength concrete.

CONCLUSIONS

1. The compressive strength at 28-day could be increased up to maximum 1,100 kg/cm^2, the shear fracture happens as soon as diagonal tension cracks appear and the width of diagonal tension cracks increase in the definite scope and the shear failure mode of the ultra-high strength concrete beams was similar to that of the normal RC beams.

2. The ultra-high strength reinforced concrete beams showed the initial shear cracks when the load was about seven times of the initial flexural crack whereas the normal RC beams showed the cracks when the load was about twice of the initial flexural crack.

3. When we compare the shear strength of this experiment with that of ACI 318-89 formula, the value of experiment was about twice of the ACI 318-89 value. Therefore, ACI 318-89 formula may not applicable to the high strength reinforced concrete.

4. In this paper, an epoxy injection was used to repair flexural cracks and diagonal tensile cracks, and the repaired specimens were tested under reloading. According to the result, flexural cracks were no longer propagated and it was proved that epoxy repair to flexural cracks is very effective.

ACKNOWLEDGEMENTS

The financial support of the WonKwang University is gratefully acknowledged.

REFERENCES

1. COLLINS, M P. Shear design of high-strength concrete structures, Utilization of High Strength Concrete, Symposium in Lillehammer, Norway.

2. ACI COMMITTEE 318, Building code requirements for reinforced concrete (ACI 318-89) and Commentary-ACI 318R-89, American Concrete Institution, Detroit, 1989.

3. ACI COMMITTEE 363, State-of-the-Art Report on high-strength concrete, ACI Manual of Concrete Practice 1992, Part 1, Detroit, 1992.

USE OF BLASTFURNACE SLAG CEMENT WITH HIGH SLAG CONTENT FOR HIGH-PERFORMANCE CONCRETE

E Lang

J F Geiseler

Forschungsgemeinschaft

Germany

ABSTRACT. In this paper the effect of blastfurnace slag cement in high-performance concrete (HPC) is presented. The results show that with all types of German blastfurnace slag cements can be produced with and without silica fume.

The HPC's with high slag blastfurnace cement show a very low capillary porosity and a high impermeability to organic liquids. The freeze-thaw and deicing agent resistance of these concretes without air-entraining agents is improved. Additionally the depth of carbonation after 180 days in HPC is reduced to about 25% compared to that in ordinary concrete.

The drying shinkage of HPC is decreased and with a reduction of the cement content decreased the shrinkage additionally.

Keywords: Blastfurnace slag cement, High-performance concrete, Durability, Freeze-thaw-resistance, Impermeability, Carbonation

Dr Eberhard Lang is the head of the of the research department for building materials of Forschungsgemeinschaft Eisenhüttenschlacken, Duisburg-Rheinhausen, Germany. His main research interests include blastfurnace slag cement, durability of concrete with particular reference to freeze-thaw resistance and multiaggressive attack on concrete with blastfurnace slag cement.

Professor Dr Jürgen Geiseler is manager of the Forschungsgemeinschaft Eisenhüttenschlacken and a professor of industrial byproducts at Technical University Clausthal, Germany. His special research interests are use of metallurgical byproducts for cement, concrete, road and waterway construction and the environmental impact of metallurgical byproducts.

Radical Concrete Technology. Edited by R K Dhir and P C Hewlett. Published in 1996 by E & FN Spon, 2–6 Boundary Row, London SE1 8HN, UK. ISBN 0 419 21480 1.

INTRODUCTION

Results of investigations in laboratory scale as well as long time experience in practice indicate the special properties of blastfurnace slag cements. This is besides others their comparatively low capillary porosity, their high resistance against sulfate attack, sea-water or other aggressive attack, against alkali-aggregate reaction and against the diffusion of chlorides into the concrete. Recent research has also shown a high binding capacity of chlorides in blastfurnace slag cement pastes. A further aspect of importance for the corrosion protection of reinforcement is the comparatively low electrical conductivity of concrete made with blastfurnace slag cement.

Due to their low heat of hydration, blastfurnace slag cement helps to prevent cracks in concrete structures resulting from temperature stresses at early ages. Contrary to Portland cement blastfurnace slag cements without artificial air voids has a high freeze-thaw-resistance.

This paper described researches to use these special properties in high performance concretes.

EXPERIMENTAL DETAILS

Materials

Blastfurnace slag cement with different slag content and two different strength classes were used. Table 1 shows the chemical composition and Table 2 the most important physical properties. Silica fume was used in form of a slurry with a water content of 50%. The coarse and fine aggregates consisted of natural gravel and sand from the Rhine area and crushed air-cooled blastfurnace slag (bfs) with a particle size distribution of 2/8 mm and 8/16 mm.

Using a sulfoneted melamine formaldehyde-based superplasticizer or a Naphthalensulfonic acid based admixture and in some cases a calcium phosphate based retarder.

Table 1 Chemical composition of the blastfurnace slag cements

Constituents %	CEM III/A 42,5	CEM III/A 32,5	CEM III/B 32,5 NW/HS/NA*
SiO_2	25,99	27,61	31,91
Al_2O_3	7,45	8,80	10,12
FeO	1,50	1,46	1,62
TiO_2	0,54	0,35	0,37
MnO	0,14	0,27	0,32
CaO	52,50	50,13	45,11
MgO	3,64	5,21	6,46
Na_2O	0,29	0,30	0,29
K_2O	0,75	0,79	0,86
Slag content	48,3	63,0	77,8

* NW - low heat, HS - high sulfat resistance, NA - low effective alkali content

Table 2 Physical properties of blast furnace slag cements

Properties		CEM III/A 42,5	CEM III/A 32,5	CEM III/B 32,5 NW/HS/NA
specific surface, cm^2/g		3960	3860	4050
setting time				
initial set	h:min	3 : 00	3 : 55	4 : 20
final set	h:min	3 : 30	4 : 45	5 : 00
Compressive strength, MPa				
2 d		19,8	10,4	7,9
7 d		38,7	27,0	28,5
28 d		59,0	55,0	50,5
91 d		70,4	69,7	58,5
dyn. Elastic Modulus, Mpa				
2 d		27 368	22 810	19 510
28 d		38 252	37 210	39 162
91 d		41 946	39 883	41 264
Heat of hydration after 7 days, J/g		285	220	199

Mix Proportions

The mix proportions used are summarised in Table 3.

Table 3 Mix proportions of the HPC

		1	2	3	4	5	6	7
CEM III/A 42,5	kg/m³	455	455	435	415	-	-	-
CEM III/A 32,5		-	-	-	-	455	-	-
CEM III/B 32,5 NW/HS/NA		-	-	-	-	-	455	455
Sand 0/2 mm	kg/m³	618	618	541	549	667	666	667
Gravel 2/8 mm	kg/m³	360	180	438	445	389	389	389
Gravel 8/16 mm		738	369	881	899	797	797	797
bfs 2/8		-	180	-	-	-	-	-
bfs 8/16		-	369	-	-	-	-	-
Silika suspension	kg/m³	60	60	30	27	30	-	30
Plasticizer	l/m³	10	10	13,9	13,3	-	16	-
Plasticizer	kg/m³	-	-	-	-	6,8	-	6,8
Retarder	l/m³	1,6	1,6	-	-	-	-	-
w/c-ratio		0,34	0,37	0,33	0,33	0,29	0,28	0,29

Test Methods

The fresh and hardened concrete properties were tested in accordance with the German standard DIN 1048 [1]. The dynamic modulus of elasticity was tested with a Grindo-Sonic MK 4x-instrument [2].

The pore size distribution was measured by a Mercury pressure porosimeter series 4000 by Carlo Erba in the range 1.85 to 7500 nm radius. The results obtained are cumulative pore size distribution and the total porosity [3].

The freeze-thaw and deicing salt resistance of concrete was determinded according to Rilem draft recommendation for test method for the freeze-thaw-resistance - tests with sodium chloride solution - CDF (Capillary suction of Deicing solution and Freeze thaw test) [4].

The equipment used for measuring the penetration of organic liquid is shown in Fig. 1.

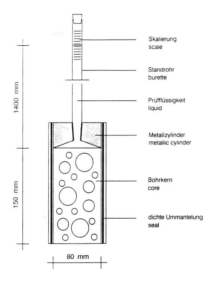

Figure 1 Equipment used for penetration tests (principle drawn)

The samples were drill cores with hights of 150 mm and diameters of 80 mm which were obtained at the 7-day moist storage. The cores were stored for 56 days at 20° C and 65% relative humidity until tested. In order to achieve a one-dimensonal transport process the cores were sealed at their circumferential surfaces. For this purpose they were coated with epoxy resin adhesive and at the same time wrapped in special steel foil. This surrounded also a metal cylinder served to hold a burette with graduations and also made it possible to distribute the test liquid uniformly over the sample surface.

After sealing, the core and metal cylinder were also wrapped with rubber tape to ensure that the fresh and the hardened epoxy resin were constantly pressed against the circumferential surface of the core. This ensured that the circumferential surface was impermeable even when the resin was dissolved by the test liquid.

During the test the burette was filled rapidly with the test liquid to a level of 1400 mm above the ample, and the exact initial level recorded from the graduations. During the subsequent 72-hours test the quantity of liquid which had penetrated could be red at any time from the graduations, which made it possible to follow the course of the penetration. After completion of the test period of 72 hours the cores were unwrapped and split parallel to the longitudinal axis, and the visible penetration front of the liquid was recorded [5, 6].

RESULTS AND DISCUSSION

Properties of Fresh Concrete

The consistence of the fresh concretes is shown in Table 4. All concretes are classified according to the flow classes F3 and F4. The decrease of the cement content from 455 kg/m³ to 435 kg/m³ reduce the flow class for one grade, but the decrease to 415 kg/m³ gives nearly the same result as with 435 kg/m³. The consistence of the concrete 1 with the retarder belongs after 45 min to the same flow class as after 10 min. Without retarder the consistence class decreases for one class in this period. The consistence of the concrete 2 with the absorbent air cooled blastfurnace slag decreases for one class, too.

Table 4 Consistence and air void content of the fresh concrete

		1	2	3	4	5	6	7
Consistence a_{10}	cm	60	54	48	47	52	48	44
Consistence a_{45}	cm	57	47	39	37	43	40	34
Air void content	%	1,2	1,7	0,6	0,6	1,3	1,9	0,9

* Consistence 10 or 45 min after mixing

Strength and Dynamic Modulus of Elasticity

The development of strength and of dynamic modulus of elasticity is given in Table 5.

Table 5 Compressive strength and dynamic modulus of elasticity

	1	2	3	4	5	6	7
strength				MPa			
1 d	39	39	39	36	44	13	17
2 d	-	-	-	-	-	37	40
7 d	72	77	74	71	85	58	67
28 d	100	102	87	81	97	91	105
180 d	-	-	93	85	104	101	114
E-modul.				N/mm²			
1 d	-	-	38 320	39 770	36 800	34 480	34 750
7 d	-	-	45 600	45 660	49 300	47 540	47 090
28 d	-	-	46 640	48 830	50 240	52 660	51 260

The guidline for high strength concrete in Germany includes the strength classes from B 65 to B 115 [7]. In accordance with this guidline the criteria for compressive strength for initial testing depends on the strength class. For strength class B 75 the criteria is 87 Mpa and for B 85 it is 99 Mpa. These results show that it is possible to produce a high strength concrete with all types of blastfurnace slag cements, even with slag contents > 70%. Only the 1 day strength of this cement type is lower.
The dynamical modulus of elasticity increases with the compressive strength.

Porosity

Figure 2 gives an overview about the pore size distribution of a standard mortar as given in Table 2 (w/c-ratio 0,50) and three mortars without, with 4% and with 10% silica fume. The silica fume was used as replace of cement. These 3 mortars get the same consistence with the addition of superplasticizer. The best workability is given with mortar with 4% silica fume which also has the lowest w/c-ratio and the lowest volume of capillary pores. The pore size distribution of blastfurnace slag cement mortar with and without silica fume is very similar. The porosity in the hardened cement paste of the concretes is comparable with the results in Figure 2.

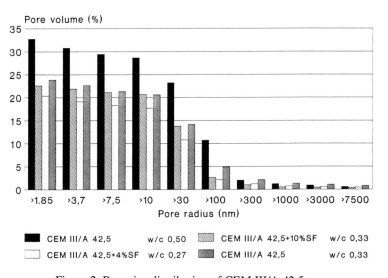

Figure 2 Pore size distribution of CEM III/A 42,5

As shown in Figure 3, the volume of capillary pores (30-100 nm) one day after producing of the concrete No. 7 is high (11.7%). Already after 7 days the volume of capillary pores is very low, but in the range from 10-30 nm the volume has increased. After 28 days the porosity has increased in this range of 10-30 nm. Between 28 and 180 days the change of porosity is insignificant.

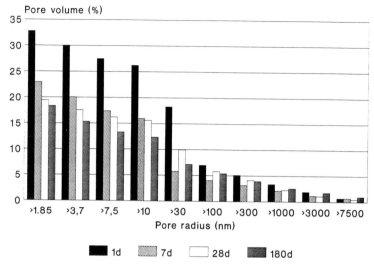

Figure 3 Development of pore size distribution

Shrinkage

The drying shrinkage of concrete depends upon the unit amount of cement and aggregate, the w/c-ratio, the types and fineness of cement, the types and unit amount of admixture and other. Figure 4 shows the shrinkage of the concretes No. 1 - 4.

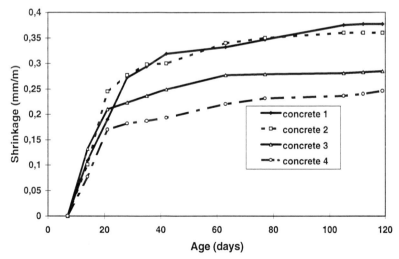

Figure 4 Shrinkage of high performance concretes with blastfurnace slag cement

The drying shrinkage of concrete prepared by demolding after 1 day , curing in water for 6 days, and drying in the atmosphere of RH of 65% at 20°C decreases with the decrease of the cement content. In comparison with the results of Mori [8] the concretes with blastfurnace slag cement show a lower drying shrinkage then the Portland cement at the same w/c-ratio.

Carbonation
In general in the atmosphere of RH of 65% at 20°C blastfurnace slag cements with high slag content have a higher carbonation depth than Portland cements. In the natural atmosphere this difference is not so high. In comparison with ordinary concrete is the carbonation depth in HPC very low, as shown in Table 6. The curing conditions are the same as for the measuring of shrinkage with a normal CO_2-content of 0.03%.

Table 6 Carbonation depth, curing in natural atmosphere

	carbonation depth (mm)	
	90 d	180 d
ordinary concrete with		
CEM III/A 42,5	3,5	4,0
CEM III/B 32,5 NW/HS/NA	4,5	5,5
HPC with		
CEM III/A 42,5	0,5	1,0
CEM III/B 32,5 NW/HS/NA	0,5	1,5

Freeze-thaw and deicing agent resistance
High slag blastfurnace cements have in laboratory tests not only with carbonation but with the freeze-thaw restistance greater differences to the practice as rapid hardening cements. Concretes with high slag blastfurnace cements have proved their durability by lasting for decades by high working stress par example in sewage clarification plants or in sea locks [8]. Even without artificial air voids, these concretes have proved a good resistance to freezing and thawing cycles, to seawater or deicing salt attack.
The results of measurement of the freeze-thaw and deicing agent resistance with a solution of water and 3% by weight of NaCl is shown in Figure 5. All the concretes were prepared without an AE agent and the HPC's without silica fume. The resistance to freezing and thawing is significantly improved by lowering the w/c-ratio. The resistance of the HPC's with silica fume is improved additionally. A sufficient resistance to freezing and thawing is given in practice already for concretes with blastfurnace slag cement and w/c-ratio ≤ 0,50. The decrease of the w/c-ratio to the range of 0.3 improved the resistance significantly.
In laboratory tests ordinary concretes with high slag blastfurnace cements show a higher scaling in the first freeze-thaw cycles than Portland cement. This scaling ranges between 0.1 to 0.3 mm cement grout from the surface. Afterwards the scaling rate decreases significantly. HPC with high slag blastfurnace cement does not have this effect. With these strong laboratory conditions the scaling rate is uniform and the freeze-thaw-resistance very high. The scaling of a cement with 50% blastfurnace slag is very low.

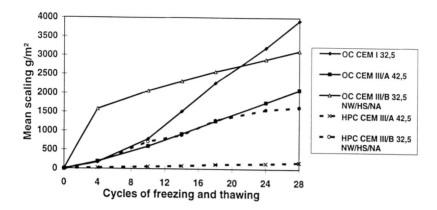

Figure 5 Freeze-thaw resistance of ordinary concrete (OC) and high performance concrete (HPC) without artificial air voids

Impermeability to Organic Liquids

Catchment tanks made of uncoated concrete in the chemical industry are secondary barriers for limited periods of time (in general ≥ 72 hours) if the storage tank has a damage. The capillary porosity of concrete with blastfurnace slag cement is lower than of concrete with Portland cement and the chemical resistance in general is higher. Therefore a lot of catchment tanks in Germany are carried out with blastfurnace slag cement. For this application the impermeability of HPC was tested. The research on this field is going on. The Table 7 gives a first information for the penetration depth of Diesel fuel.

Table 7 Penetration depth of diesel fuel in HPC

	Penetration depth mm
Concrete according [5] *	12
Concrete No.1	5
Concrete No. 4	6
Concrete No. 7	4

* cement content 320 kg/m³, w/c-ratio 0.50, 1930 kg/m³ sand and gravel.

CONCLUSIONS

Based on the results presented in this paper it can be stated thatHPC can be produced using all types of high slag blastfurnace cement (CEM III/A 42,5, CEM III/A 32,5, CEM III/B 32,5). These concretes have a good workability and the well known special properties of durability are additional improved.

Further research will be necessary, particularly to use the low heat of hydration, the electrical conductivity, the influence of an internal curing by use of porous air-cooled blastfurnace slag as aggregate.

Acknowledment

This projekt was supported by the Arbeitsgemeinschaft industrieller Forschungsvereinigungen „Otto von Guericke" e.V. (AiF).

REFERENCES

1.   DIN 1048. Prüfverfahren für Beton. Part 1-5, Issue June 1991

2.   Operating instructions for the Grindo-Sonic MK 4x instrument.
      J.W. Lemmens-Elektronika N.V., Leuven, Belgium (1988).

3.   PO 4000 Instruction manual. Carlo Erba Strumentazione, Rodano (Milan),
      Italy.

4.   Rilem draft recommendation for test method for the freeze-thaw resistance
      ofconcrete - Tests with water (CF) or with sodium chloride solution (CDF).
      Materials and Structures, 1995, 28, pp 175-182.

5.   Richtlinie für Betonbau beim Umgang mit wassergefährdenden Stoffen.
      Deutscher Ausschuß für Stahlbeton - DAfStb - im DIN Deutsches Institut
      für Normung e.V. 9/1992.

6.   Paschmann, H.; Grube, H. The effect of mineral and organic additions on
      the impermeability to organic liquids and an other properties of concrete.
      Beton, Vol. 44, No. 1, 1994, pp 86-91

7.   Richtlinie für hochfesten Beton.
      Deutscher Ausschuß für Stahlbeton - DAfStb - im DIN Deutsches Institut
      für Normung e.V. 5/1995.

7.   Mori, D. Study of drying shrinkage of High-strength concrete. ref. by
      Uchikawa, H. Durability of High-Strength Concrete with Superior
      Workability
      Estimated from the Composition and Structure. P.K.Mehta-Symposium on
      durability of Concrete. Nice 1994, pp 61-97

8.   Geiseler, J.; Lang, E. Long-term durability of non-air-entrained concrete
      structures to marine environments and freezing and thawing cycles.
      Third CANMET/ACI International Conferenze on Durabilty of Concrete.
      Nice, 1994, Supplementary Papers pp 715-737

HIGH STRENGTH CONCRETE

B Mezghiche

M Bouabaz

M Chebbah

Centre Universitaire

Algeria

ABSTRACT . This study shows how to obtain high strength concrete combined into hydraulic binding elaborated from granulated slag, base and aggregates. Basic binding presents a mixture of two components in which silico- aluminuous is presented by granulated slag finely crushed and the alcaline component composed of basic metals giving an alkaline reaction in water. Concrete made up from basic slag possesses high physico-mechanic properties, such as resistance to corrosive agents, is water-proof and capable of hardening at low and high temperatures. Binding and basic concrete harden at naturel conditions including negative temperatures under treatment conditions by heat into stream (drying) and autoclave; they can also be used for thermic treatment at low energy consumption, at a temperature of 30 to 50 °C.

Concrete from base of basic binding can reach a mechanical resistance to compression at 28 days equal to 90,5 MPa for natural hardening and 96,4 MPa for dried concrete.

Keywords: basic slag, basic cement, basic concrete, liquid glass, binding from basic slag, alcaline component, normal hardness, drying.

Dr Bouzidi Mezghiche is a lecturer, head of Hydraulic Department, and Director of the Concrete Technology Institue, University of Biskra (Algeria). He is specialist in the use of binders and durability of concrete.

Mr Mohamed Bouabaz is an Assistant lecturer, Member of Concrete Technology and Construction Management at University of Biskra (Algeria).

Mr Mohamed Chebbah is a Research Assistant in Concrete Technology, University of Biskra (Algeria).

Radical Concrete Technology. Edited by R K Dhir and P C Hewlett. Published in 1996 by E & FN Spon, 2–6 Boundary Row, London SE1 8HN, UK. ISBN 0 419 21480 1.

INTRODUCTION

Binding from basic slag presents a mixture of two components in which silico-aluminous is presented by granulated slag finely crushed, and the alcaline component composed of basic metals giving an alcaline reaction in water [1].

The use of binding from basic slag and concrete made up from their base opens a wide perspective of a sensitive extension from a scale of materials used in construction owing to the use of various substances originally natural and artificial.

The mechanical resistance of basic concrete depends on the type of binding , the basic module, the fineness of grinding, the nature of basic metals and the density of basic solutions and the relation solution-slag [2].

In the present paper, we study the possibility to obtain basic concrete of high mechanical resistance to corrosive agents and the hardening under different temperatures.

EXPERIMENTAL DETAILS

Characteristics of Materials Used

Granulated slag from high furnace

The present study made focuses on the granulated slag from high furnance derived from metallurgic overall [Annaba, Algeria]. Granulated slag is obtained by savage cooling during the realization of granulation by humide way, which leads to the formation of high quality vitreous phases and consequently a great deal of cristalline phases. In a state of non grinding the slag gives a rounded granule with high porosity.

The results of chemical analysis and granulometric of slag are shown on table 1 and 2.

Table 1 Chemical analysis of slag

	GRADE (TENOR) %						Mb	Ma	K
SiO	Al_2O_3	CaO	Mg	MnO	Fe_2O_3	S			
40.4	5.8	42.5	6.9	2.7	0.48	0.7	1.07	0.14	1.4

Table 2 Granulometric analysis of slag

REFUSAL %		STRAINER MESH , mm					RIDDLED BY MESH
	5.0	2.5	1.25	0.63	0.315	0.14	(0.14)
PARTIAL							
SUMMARY	-	2.3	39.58	29.23	22.21	4.19	2.49
TOTAL	-	2.3	41.88	71.11	93.32	97.51	100

The volumic density of granulated slag in a dry state is 1410 Kg/m^3 and the density is 2780 Kg/m^3.

According to the quality factor and the chemical composition the slag refers to a third class basic slag (Gost 37476-74).

The experimentation have been carried out, using grinded slag in a grinder with a density surface of 3000-4000 cm^2/g.

The fineness scale was obtained by Vicat apparatus.

Waterglass

We used for experimental work waterglass; its chemical composition of an aqueous solution of glass is shown on table 3.

Table 3 Chemical analysis of waterglass

GRADE (TENOR) %						DENSITY	Mc
SiO	$Al_2O_3 + Fe_2O_3$	CaO	SO	Na_2O	H_2O	Kg/m^3	
31.19	0.18	0.1	0.35	11.03	57.15	1400	3.16

Mc : module of liquid glass

For experimental work we used waterglass with silicical module of 3; 2.5; 2; 1.5; 1 and density of 1150, 1200, 1250, 1300, 1400 and 1500 kg/m^3 which was obtained by adding corresponding quantity of caustic soda and water into the first preparation.

Bulky Aggregats and Fine

Crushed Stones.

Instead of bulk aggregates, we used crushed granite into fraction of 5 to 10 mm and 10 to 20 mm. The stone density is 2700 Kg/m^3; the compacted volumic density is 1690 Kg/m^3; the apparent volumic density is 1380 Kg/m^3; the granulometric space is 0.39 and the specific surface is 0.3 m^2/Kg.

River Sand.

Instead of fine aggregates, we used river quartz sand with a density of 2650 Kg/m^3; and a compacted volumic density of 1870 Kg/m^3; the apparent volumic density is 1620 Kg/m^3; the granulometric space is 0.30 and the specific surface is 130 m^2/kg.

STUDY FOR A CEMENT BASIC

Binding Qualitative Rating as a Function of Silica Module of Waterglass

It was established that binding based on waterglass on silica module of 3 to 1 was on the type of hydraulic [3].
Table 4 shows the data which indicates the dependence between slag activity liquid glass module, the density and hardeness pattern. We can see that in equal conditions the binding activity depends on the module, the glass density and the crushed fineness of slag.
The treatment by heating up into steam intensifies the binding interaction with waterglass, which means that the slag activity at the age of 28 days increases by 10% and much more, according to the binding which hardens at naturel conditions.

The binding resistance depends considerably on the relation waterglass-slag. For the value of 0.3 the resistance is very high in all cases. With the rise of the relation glass-slag, the binding resistance decreases [3].

It was established experimentaly that in equal conditions, the optimal resistance was proved into binding based on waterglass at silica module of 1.5 (see table 4). Any increase or decrease of the module makes shadding off the binding mechanical properties.

Resistance into Agressive Environment

For resistance tests into agressive environment of basic cement we have prepared six cubic test (4x4x4 cm) by series, with pure pastry of basic cement.
The resistance of materials depends on the nature of agressive liquid and the hardness delay of test tubes before immersion into agressive environment. The results are shown in table 5.

Table 4 Basic cement resistance

WATER GLASS DENSITY Kg/m³	GRINDING FINENESS OF SLAG cm²/g	RESISTANCE LIMIT TO COMPRESSION (MPa)							
		Natural hardness Silica Module of Waterglass				Drying Silica Module of Waterglass			
		1.0	1.5	2.0	3.0	1.0	1.5	2.0	3.0
1150	3000	$\frac{47}{34}$	$\frac{51}{36}$	$\frac{50}{35}$	$\frac{39}{28}$	$\frac{50}{37}$	$\frac{55}{39}$	$\frac{52}{38}$	$\frac{39}{28}$
	3500	$\frac{52}{39}$	$\frac{57}{43}$	$\frac{55}{41}$	$\frac{46}{35}$	$\frac{54}{41}$	$\frac{61}{45}$	$\frac{59}{43}$	$\frac{43}{33}$
	4000	$\frac{63}{48}$	$\frac{66}{52}$	$\frac{64}{50}$	$\frac{57}{44}$	$\frac{65}{51}$	$\frac{70}{56}$	$\frac{68}{54}$	$\frac{59}{46}$
1200	3000	$\frac{62}{44}$	$\frac{72}{52}$	$\frac{69}{50}$	$\frac{53}{38}$	$\frac{65}{47}$	$\frac{76}{54}$	$\frac{72}{51}$	$\frac{52}{37}$
	3500	$\frac{67}{48}$	$\frac{77}{59}$	$\frac{74}{56}$	$\frac{60}{46}$	$\frac{69}{53}$	$\frac{81}{62}$	$\frac{78}{57}$	$\frac{58}{44}$
	4000	$\frac{73}{57}$	$\frac{81}{64}$	$\frac{79}{60}$	$\frac{66}{52}$	$\frac{75}{61}$	$\frac{85}{68}$	$\frac{83}{64}$	$\frac{69}{55}$
1300	3000	$\frac{87}{62}$	$\frac{103}{74}$	$\frac{93}{66}$	$\frac{66}{48}$	$\frac{91}{65}$	$\frac{108}{76}$	$\frac{96}{67}$	$\frac{68}{49}$
	3500	$\frac{91}{67}$	$\frac{110}{79}$	$\frac{98}{70}$	$\frac{72}{54}$	$\frac{94}{69}$	$\frac{115}{88}$	$\frac{103}{73}$	$\frac{75}{56}$
	4000	$\frac{98}{73}$	$\frac{117}{85}$	$\frac{104}{76}$	$\frac{80}{58}$	$\frac{101}{76}$	$\frac{121}{89}$	$\frac{107}{79}$	$\frac{84}{61}$

Note: - Numerator indicate the resistance cement to compression of sample on normal pastry

- Denominator corresponds to a same indice for samples on mortary 1:3

STUDY FOR A BASIC CONCRETE

Selection Method of Concrete Composition

Basic slag concrete is an artificial stone which is formed during the hardening of a mixture: cement and basic slag, water and aggregates.
The choice of concrete composition allows us to determine the proportion of materials in use which guarantees the resistance of concrete, the preparation techniques and the plasticity for a minimum binding.

Table 5 Resistance into agressive environment

WATER GLASS DENSITY Kg/m^3	DELAY OF HARDNESS BEFORE IMMERSION (days)	IMMERSION TIME OF TEST TUBE INTO AGRESSIVE ENVIRONMENT	COMPOSITION OF AGRESSIVE LIQUID			
			Water	0.1 % HCl	5 % Na$_2$SO$_4$	3 % MgSO$_4$
1200	7	3	56.5	$\frac{48}{0.85}$	$\frac{59.3}{1.05}$	$\frac{53.7}{0.95}$
		6	65.0	$\frac{53.3}{0.82}$	$\frac{65.0}{1.0}$	$\frac{63.7}{0.98}$
		12	74.0	$\frac{62.9}{0.85}$	$\frac{75.5}{1.02}$	$\frac{72.5}{0.98}$
	28	3	97.0	$\frac{85.4}{0.88}$	$\frac{106.7}{1.1}$	$\frac{95.1}{0.98}$
		6	112.0	$\frac{98.6}{0.88}$	$\frac{125.4}{1.12}$	$\frac{112.0}{1.0}$
		12	127.5	$\frac{114.8}{0.9}$	$\frac{142.8}{1.12}$	$\frac{133.9}{1.05}$

For the calculation of concrete components we used appropriate formula in which the mixture components of concrete represents, in the best case, a material for construction of a conglomerate type with securing compact of its components.

For the study of concrete we have prapared three cube series of 15 x 15 x 15 cm and prism of 10 x 10 x 40 cm each series containing 03 samples. As an alcaline component we used waterglass in module 1.5. The density of alcaline solution was established at 1150-1300 Kg/m^3. The handiness of concrete was established at 2 seconds.
One series of samples was tested after 28 days of hardening at normal conditions.

The second series of samples was dryed in accordance to pattern 2 + 6 + 2 hours at temperature of 90 + 5°C at one day ages. The samples were removed from the framework and placed inside a hardness room were they have been kept untill test.
The tests to compression have been done at 28 days with hydraulic press according to ASTM norm. The results are shown on table 6.

Table 6 Basic material resistance

SERIES N°	WATER GLASS DENSITY Kg/m³	MECANICAL RESISTANCE TO COMPRESSION (MPa)					
		Natural Hardness at 28 days			Drying		
		Basic Binder	Cubic	Prismatic	Basic Binder	Cubic	Prismatic
1	1150	$\frac{51}{36}$	43.0	35.0	$\frac{55}{39}$	46.5	38.0
2	1200	$\frac{72}{52}$	49.5	37.5	$\frac{76}{54}$	52.0	40.0
3	1250	$\frac{96}{68}$	90.5	72.0	$\frac{102}{74}$	96.5	77.5
4	1300	$\frac{103}{74}$	96.5	75.0	$\frac{108}{76}$	102.5	80.0

Note: - Numerator indicate the resistance cement to compression of sample on normal pastry
      - Denominator corresponds to a same indice for samples on mortary 1:3

Study of Thaw Basic Concrete

For study of thaw basic concrete we have prepared a series of cubes (15x15x15 cm), with six samples by series.
The test tube has frosting-defrosting cycles of temperatures (-20°C , -50°C). The loss in weight after frost-defrost test was in order of 4 - 6%. The results are shown in table 7.

Table 7 Resistance of concrete to the frost

SERIES N°	WATERGLASS DENSITY Kg/m³	CYCLE	MECHANICAL RESISTANCE TO COMPRESSION (MPa)	
			After Test into Freeze Resistance	Control Test
1	1150	456	38.0	42.5
2	1200	600	46.0	49.5
3	1250	1000	78.5	87.0
4	1300	1115	86.0	95.0

CONCLUSIONS

1. It has been established that it is possible to prepare heavy concrete with high mechanical resistance using liquid glass and basic granulated slag from high furnace.

2. It has been shown that the mechanical resistance of basic concrete depends on the fineness of slag grinding, on the density of silical module of liquid glass and the hardening mode of concrete.

REFERENCES

1. Pakhomov V. A. , Basic-Concrete Based Construction. Vichtcha Chkola, Kiev, 1984, 184 pp.

2. Gontchar V. I. , Mezghiche B. Bétons basiques à base de verre soluble à délais de prises prolongés, III conférence scientifique des Républiques Soviétiques, Kiev, Vol.2, 1989, pp 48-49.

3. Mezghiche B. Propriétés mécaniques du béton basique. Séminaire international. Ghardaia (Algérie) Mars, 1994, pp 323-335.

MOMENT REDISTRIBUTION EFFECTS IN HIGH STRENGTH REINFORCED BEAMS

R H Scott

University of Durham

UK

ABSTRACT. This paper describes a programme of tests designed to investigate moment distribution effects in three high strength reinforced concrete beams. Each beam was 5.2 m long and carried on three supports with each of the two spans being loaded with a central point load. Strain gauged reinforcement was used to obtain very detailed information regarding reinforcement strain distributions over the centre support and in one of the spans. Similar specimens using normal strength concrete had already been tested as part of a previous test programme. Early results indicate that the use of HSC can result in a change of failure mode from shear to bending with considerably higher reinforcement strains being developed as a consequence. Levels of moment redistribution appear to be similar with the two concrete types.

Keywords: Moment redistribution, high strength concrete, two-span beams, strain gauged reinforcement.

Dr R.H. Scott is a Reader in the School of Engineering at the University of Durham. He specialises in the measurement of reinforcement strain and bond stress distributions using reinforcing bars which have been internally strain gauged. To date these bars have been used in investigations into the behaviour of tension specimens, beam/column connections, single span and two-span beams, slab/column connections and curved anchorages. Before coming to Durham in 1978 he spent ten years in industry where he was involved in the design and construction of a wide range of structures, mainly in reinforced concrete and structural steelwork, but also including timber and loadbearing brickwork.

Radical Concrete Technology. Edited by R K Dhir and P C Hewlett. Published in 1996 by E & FN Spon, 2–6 Boundary Row, London SE1 8HN, UK. ISBN 0 419 21480 1.

INTRODUCTION

During the drafting of EC2, the new Eurocode for reinforced concrete (1), there were discussions concerning the related areas of moment redistribution, reinforcement ductility and allowable rotation of a reinforced concrete section, which initiated an examination of UK reinforcement to re-assess the consequences of ductility on the design and detailing of reinforced concrete beams and slabs.

As part of this work, the author undertook an SERC (now EPSRC) funded research programme to examine the associated problems of reinforcement ductility and moment redistribution by conducting laboratory tests on a series of two-span beams, the simplest specimens to exhibit moment redistribution. Specimens were 5.2 m long overall and 300 mm wide, with each span of 2.5 m being loaded with a central point load. The seventeen specimens in the series covered three beam depths - 400, 250 and 150 mm - and a range of reinforcement percentages, in order to study the effects of both shear and flexural failure. Internally strain gauged reinforcement was used to measure the strain distributions associated with various levels of moment redistribution and, with up to 200 electric resistance strain gauges per specimen, the volume of data generated was considerable. An overview of the technique, which has been used by the author in a number of major investigations, is given in reference 2

The test beams were all made with normal strength concrete having a cube strength at test in the order of 40 MPa. In view of the current interest in high strength concrete, three tests are being repeated using concrete having a characteristic compressive strength of 110 MPa. This additional work was prompted by suggestions made by Ove Arup and Partners and is funded as part of a Brite Euram Project administered by Arups and Taywood Engineering Ltd. This paper describes these additional tests and compares results with those for specimens made from normal strength concrete.

TEST SPECIMENS

Details of the three test specimens, designated B2T20BH, B5T12BH and B2T12DH are given in Table 1 (the normal strength concrete equivalents were designated B2T20B, B5T12B and B2T12D respectively). Specimens B2T20BH and B5T12BH had a similar percentage of reinforcement over the centre support, but using different bar combinations - two T20's (0.93%) for B2T20BH and five T12's (0.83%) in B5T12BH respectively. With normal strength concrete these two specimens exhibited shear failures. B2T12DH had two T12's over the centre support giving 0.58% tension steel for the reduced depth of 150 mm. The normal strength version of this specimen failed in bending. All three specimens were designed for 30% moment redistribution, which determined the area of tension steel provided in the bottom of each span. Two strain gauged bars were installed in each specimen, one as part of the top reinforcement over the centre support and one as part of the bottom reinforcement in the left hand span. Each bar contained 51 electric resistance strain gauges. Surface strains were measured using a grillage of Demec studs on one side face.

Concrete for each specimen used 10 mm granite aggregate and microsilica (in slurry form) in addition to OPC and sand. The overall moisture/binder ratio was around 0.3 with adequate workability being achieved by the addition of appropriate admixtures. The mix was supplied by Tarmac Topmix. Nine 100 mm cubes and four cylinders were cast along with each specimen. Thermocouples were installed in each beam and in three of the nine cubes so that temperature matched curing could be attempted. A maximum temperature of around 45° C was reached approximately eight hours after casting.

Specimens were tested by applying equal load increments to each span until failure occurred, with measurements of the two applied loads and three support reactions being recorded at every load stage in addition to a full set of Demec readings. Concrete surface strains were measured at selected load stages.

Table 1 : Specimen Details

SPECIMEN	OVERALL DEPTH	SUPPORT REINFORCEMENT		SPAN REINFORCEMENT	
	mm	Top	Bottom	Top	Bottom
B2T20BH	250	2T20 (0.93%)	3T12	3T12	3T20
B5T12BH	250	5T12 (0.83%)	3T12	3T12	3T20
B2T12DH	150	2T12 (0.58%)	3T10	3T10	3T12

RESULTS

At the time of preparing this paper results were only available for specimen B2T20BH. However, results for the other two tests will be ready very shortly and will be presented at the conference.

Strain Distributions

Specimen B2T20B (normal strength concrete) failed in shear at beam loads of 175 kN. B2T20BH (high strength concrete) exhibited a similar pattern of diagonal shear cracks but finally failed in bending at loads of 246 kN - an increase of 40% - when gross yield occurred in the reinforcement over both the centre support and in the two spans. Cracks were first observed at applied loads of 15 kN and 30 kN respectively in the two specimens, a reflection of the higher Young's modulus attained using high strength concrete.

Typical strain distributions in the reinforcement over the centre support of B2T20B are shown in Figure 1, the two small peaks corresponding to shear cracks either side of the support itself. The maximum strain was only 2390 microstrain, an indication of the brittle nature of the shear failure. By contrast, strains for B2T20BH are shown in Figure 2. The first set of points plotted shows strains for beam loads of 175 kN, equivalent to the last load stage in the test of B2T20B. Strain distributions up to this point were remarkably similar for the two specimens. Beyond 175 kN, Figure 2 indicates how gross yield developed at two crack positions, with strains peaking at around 25 000 to 30 000 microstrain. Strains of similar magnitude were recorded in the span reinforcement of this specimen.

Moment Redistribution

The percentage of moment redistribution was calculated at the support and span positions for each and every load stage using the following expression:-

$$\% \text{ Re distribution} = \frac{(\textbf{Experimental BM} - \textbf{Calculated BM})}{\textbf{Calculated BM}} \times 100$$

The experimental bending moment distributions were calculated using the applied loads and measured support reactions. The calculated bending moments were derived from an elastic analysis which assumed constant stiffness along the specimen, as would normally be the practice in design office calculations. The above expression gave negative values at the support, where bending moments were reduced numerically as a result of redistribution, and positive values for the spans, where moments were numerically increased. Plots of percentage redistribution against experimental bending moment at the centre support are shown in Figure 3 for both B2T20B (open symbols) and B2T20BH (solid symbols).

Figure 3 indicates that, once the beams had settled firmly onto their supports, the redistribution characteristics were remarkably similar for the two specimens over their common loading range. Typical values (numerically) were around 25-28% at the centre support and 15% in both spans. It is interesting to note that reinforcement strains were elastic over much of this load range in both specimens, the moment redistribution being essentially a consequence of changes in flexural stiffness resulting from cracking. Higher moments were generated in B2T20BH due to its greater failure load and led to gross yield of the reinforcement at the centre support and in both spans. This resulted in higher levels of percentage redistribution right at the end of the test with maximum values (numerically) of 37% and 22% being developed at the support and span positions respectively. A possible implication for design is that B2T20BH came much closer to developing a full plastic collapse mechanism than B2T20B as a consequence of the gross yield of the reinforcement. This will receive further consideration as more results become available.

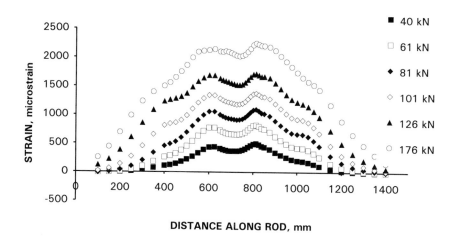

Figure 1: Strain Distributions for Specimen B2T20B

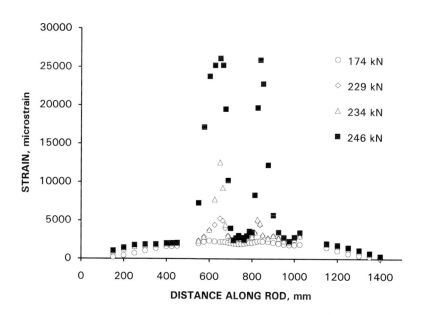

Figure 2: Strain Distributions for Specimen B2T20BH

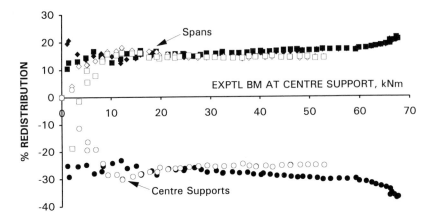

Solid symbols high strength concrete, open symbols normal strength concrete

Figure 3: Comparison of Percentage Redistributions

CONCLUSIONS

Comparing results for the first specimen in a test programme of three high strength concrete beams with those for a similar specimen made from normal strength concrete indicates that the use of HSC can result in a change of failure mode from shear to bending with considerably higher reinforcement strains being developed as a consequence. Levels of moment redistribution appear to be similar with the two concrete types.

ACKNOWLEDGEMENTS

The assistance of Ove Arup and Partners, Taywood Engineering Ltd, Tarmac Topmix, the Science and Engineering Research Council (now EPSRC) and the technical staff of the University of Durham is gratefully acknowledged.

REFERENCES

1.   EC2 : Eurocode No. 2, "Design of Concrete Structures", Part 1, General Rules and Rules for Buildings, Final Draft, December 1988

2.   Scott R.H. & Gill P.A.T., "Short Term Distributions of Strain and Bond Stress along Tension Reinforcement", The Structural Engineer, Vol. 65B, No. 2, June 1987, pp 39-43, 48

STRESS-STRAIN RELATIONSHIP OF CONCRETE: HOW IMPORTANT IS IT?

S Setunge

Monash University

P A Mendis

University of Melbourne

Australia

ABSTRACT. For normal strength concrete with compressive strengths between 20 and 50 MPa, there are several well established reliable models to predict the stress-strain relationship of concrete. Recent developments in concrete technology has facilitated production of high strength/ high performance concrete with compressive strengths over 100 MPa. Developing a unified stress-strain model which predicts the behaviour of reinforced concrete with a reasonable accuracy and valid for the full range of compressive strengths of concrete, is a challenging task. Most important consideration should be given to the degree of influence of the assumed stress-strain relationship of concrete on the prediction of the structural behaviour of reinforced concrete. This paper reports the early results of an investigation of the influence of the stress-strain curve assumed for concrete on the predicted behaviour of structural members. Structural elements studied include reinforced high strength concrete columns under axial compression, high strength concrete beams and slender high strength concrete walls.

Keywords: High Strength Concrete, Stress-Strain Relationship, Moment-Curvature Behaviour, Instability Analysis, Axial Compression, Lateral Confinement

Dr Sujeeva Setunge is a Lecturer in Structural Engineering, Monash University, Gippsland School of Engineering. Her main research interests include the structural behaviour of high performance concrete with particular reference to the triaxial behaviour and time dependent deformations.

Dr Priyan Mendis is a Senior Lecturer in Structural Engineering at the University of Melbourne. He worked as a Structural Engineer with Connell Wagner Ltd. before joining the University in 1991. His research interests include high strength/high performance concrete, Earthquake Engineering and multi-storey buildings.

Radical Concrete Technology. Edited by R K Dhir and P C Hewlett. Published in 1996 by E & FN Spon, 2–6 Boundary Row, London SE1 8HN, UK. ISBN 0 419 21480 1.

INTRODUCTION

In ultimate strength design of concrete structures, the stress-strain relationship is one of the main material properties required. Since part of the concrete compression zone at ultimate strength is subjected to strains beyond the strain at the peak stress, information on both the ascending and descending portions of the curve are of importance. For normal strength concrete (with compressive strengths up to 50 MPa) the shape of the curve in uniaxial compression is well established. There are a large number of analytical expressions proposed to predict the stress strain curves of normal strength concretes in uniaxial compression (1,2,3,4,5,6) and under lateral confinement (4,7,8,9,10).

In recent years, a large number of researchers have attempted to predict the stress-strain relationship of high strength/ high performance concrete with compressive strengths in the range of 50 to 130 MPa (11,12,13,14,15). These models were developed based on experimental results of one particular test condition: eg. columns under axial compression, cylinders under triaxial compression or flexural members. While conducting a comprehensive test program on the behaviour of different types of structural elements, authors noted that a particular stress-strain relationship applicable to one element may not be appropriate for another type of element. Therefore a study of the effect of stress-strain curve on predicting the structural behaviour is needed before attempting to develop a unified stress-strain model which is valid for concrete of a range of compressive strengths and structural members.

STRESS-STRAIN MODELS

Mathematical functions proposed for the stress-strain relationship of concrete can be divided into three main categories: one continuous function for the ascending and descending branches of the stress-strain curve (4,6), two separate curves for the ascending and the descending parts meeting at the peak stress (12,13) and a parabolic curve for the ascending branch together with a bi-linear descending curve (10,11).

In predicting the stress-strain relationship of laterally confined concrete, the relationship between the degree of confinement and the peak stress, and the strain at the peak stress should also be incorporated into the stress-strain models. Some stress-strain models take the lateral confinement into account by establishing a relationship between the effective lateral confining pressure and the parameters of the stress-strain curve(6,13). Others developed relationships between the configuration of the confining and longitudinal steel and the parameters of the stress-strain curve (10,11).

Four stress-strain models proposed for concrete were selected for the investigation reported here to represent different types of functions used in literature. A brief description of each model is given below:

Fafitis and Shah (13)

Equations given below define the stress-strain relationship proposed for confined/unconfined high strength concrete with uniaxial compressive strengths up to 80 MPa. Two separate expressions are proposed for the ascending and descending portions of the curve. Model was developed using experimental results of axially loaded laterally confined concrete columns as well as concrete cylinders tested under triaxial compression.

$$f = f_{cc} x (1 - (1 - \frac{\varepsilon}{\varepsilon_{cc}})^A) \quad \text{when } \varepsilon < \varepsilon_{cc}$$

$$f = f_{cc}(\exp(-k(\varepsilon - \varepsilon_{cc})^{1.15})) \quad \text{when } \varepsilon > \varepsilon_{cc}$$

where;

ε_{cc} is the strain at the peak stress, f_{cc} is the peak stress , A is a constant depending on the ratio of the initial elastic modulus to the secant modulus at the peak stress and k is a constant depending on the lateral confining pressure and the compressive strength of concrete under uniaxial compression.

Mander et al. (6)

Mander et al. proposed a stress-strain model for concrete with compressive strengths up to about 33 MPa. This was based on a mathematical expression proposed by Popovics (5) for the stress-strain relationship of NSC in uniaxial compression, and was found to give a very good fit to the results for laterally confined normal strength concrete columns. The main advantage of this relationship (Equations given below) is that, in addition to having one continuous function for both ascending and descending branches, the whole curve is only a function of the uniaxial peak stress (f_c) and strain (ε_c), the confining pressure (f_l) and the initial elastic modulus (E_c). Experimental results of laterally confined, axially loaded columns had been used in calibrating the model.

$$f = \frac{f_{cc} \, X \, r}{r - 1 + X^r}$$

where, $X = \dfrac{\varepsilon}{\varepsilon_{cc}}$, $r = \dfrac{E_c}{E_c - E_{sec}}$, $E_{sec} = \dfrac{f_{cc}}{\varepsilon_{cc}}$, ε_{cc} is a function of the confining pressure and the compressive strength of concrete. f_{cc}/f_c was determined using a constitutive model reported for concrete under triaxial compression (16).

Bjerkeli et al (11)

The model of Bjerkeli et al consists of an ascending curve, a linear descending branch and a residual. Equations below give the full range stress strain curve. The relationship is proposed for concrete with cylinder compressive strengths up to about 80 MPa (cube strengths up to 90 MPa). This model is also developed for laterally confined axially loaded high strength concrete columns.

$$f = \cfrac{E_c \varepsilon}{1 + (\cfrac{E_c}{E_{cc}} - 2)(\cfrac{\varepsilon}{\varepsilon_{cc}}) + (\cfrac{\varepsilon}{\varepsilon_{cc}})^2} \qquad \text{when } \varepsilon \le \varepsilon_{cc}$$

$$f = f_{cc} - Z(\varepsilon - \varepsilon_{cc}) \qquad \text{when } \varepsilon \ge \varepsilon_{cc} \text{ and } f \le f_{res}$$

where,

f_{res} = 4.87 f_1, Z is the initial slope of the descending branch given as a function of the strain corresponding to $0.85f_{cc}$ which again depends on the compressive strength of concrete, confining pressure and the reinforcement configuration, E_{cc} is the secant modulus at the peak stress and f_{cc}/f_c ratio is given as a linear function of the lateral confining pressure and the reinforcement configuration.

Modified Scott Model (14)

Based on the stress-strain model proposed by Scott et al (10), Setunge et al (14) proposed the following stress-strain model for flexural stress-block of reinforced concrete beams. The model was calibrated using experimental results on high strength concrete beams reported by Pendyala et al (16). Setunge et al (14) reported that the moment curvature curves of high strength concrete beams predicted using the existing stress-strain models are in good agreement with the experimental results only up to the peak moment. None of the available models were observed to give even a reasonable estimate of the full range moment curvature behaviour of high strength concrete beams. The proposed stress-strain model could easily be calibrated to accommodate the changes in the material behaviour.

$$f = Kf_c\left[\frac{2\varepsilon}{0.002K} - \left(\frac{\varepsilon}{0.002K}\right)^2\right] \qquad \text{for } \varepsilon < 0.002K$$

$$f = Kf_c\left[1 - Z_m(\varepsilon - 0.002K)\right] \ge f_{res} \quad \text{for } \varepsilon > 0.002K$$

where, $f_{res} = 0.2Kf_c \times R$, Zm and K are functions of lateral confining pressure and the compressive strength of concrete.

Figure 1 shows the stress-strain curve of an 80 MPa concrete predicted using the four different stress-strain models.

LATERALLY CONFINED COLUMNS UNDER AXIAL COMPRESSION

In order to investigate the influence of the stress-strain relationship on predicting the behaviour of laterally confined columns, the results reported by Martinez et al (17) of spirally confined high strength concrete columns were compared with those predicted by the four stress-strain models. In a column confined with spiral steel, the effective confining pressure can be calculated directly without using any approximations required in columns confined with rectangular ties. A typical comparison is given in Figure 2. It was observed that for concrete with compressive strengths below 50 MPa, prediction using Mander (6) model is very good whereas for high strength concrete, model of Bjerkeli et al (11) gives a reasonable prediction. Figure 2 shows the expected range of the predicted stress-strain behaviour using the four different models.

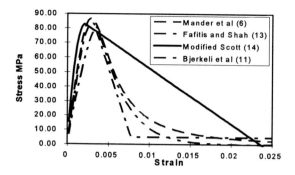

Figure 1: Stress-strain curves of an 80 MPa Concrete subjected to 1 MPa confining pressure

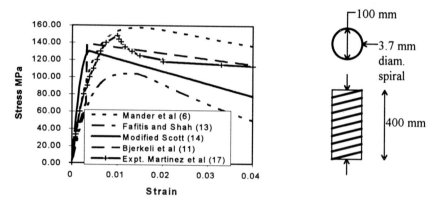

Figure 2: Stress-Strain Curve of a Spirally Reinforced High Strength Concrete Column Reported in (17)

MOMENT CURVATURE BEHAVIOUR OF BEAMS

The results of a comprehensive experimental study on high strength concrete beams tested at The University of Melbourne (16) was used to investigate the influence of the stress-strain relationship of concrete on the prediction of the moment-curvature behaviour of high strength concrete flexural members. Computer program FRMPHI (14) was used to predict the moment curvature curves of the results reported in reference 19. Figure 3 shows a typical comparison of the full range moment curvature behaviour of an experimental beam with those predicted using the four stress-strain models. It is evident that the stress-strain model proposed by Setunge et al (14) which had been calibrated using the results of beam tests gives a better indication of the moment-curvature behaviour compared to the relationships developed using results of column tests..

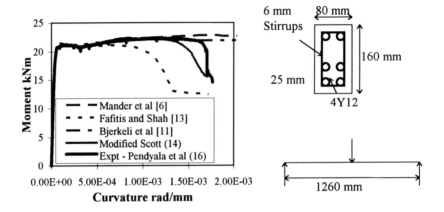

Figure 3: Moment Curvature Curve of Beam 9 (16)

STABILITY OF HIGH STRENGTH CONCRETE WALLS

Two typical walls from an experimental study (18) of slender high strength concrete walls were used to compare the effect of the stress-strain relationship of concrete on the prediction of the instability failure. The computer program WASTAB developed for stability analysis was modified to accommodate two stress-strain models proposed by Fafitis and Shah (13) and Scott et al (10). Since the wall panels were unconfined, the confinement was assumed to be zero in the stress-strain models. Above two stress-strain models were selected for the reason that they have been calibrated for the behaviour of unconfined concrete under uniaxial compression.

Table 1: Results of Instability Analysis

Wall Panel	Concrete Strength f'c (MPa)	Failure Expt. (kN)	Ult. Load WASTAB Faf. & Shah (kN)	Ult. Load WASTAB Scott et al (1982) (kN)
1a	40.7	162.0	139.3	160.4
1b	58.9	187.0	179.8	229.8

Wall panels 1a and 1b had the dimensions 1000x200x50mm reinforced with a wire mesh consisting of 4 mm diameter mild steel bars placed in vertical and horizontal directions with a reinforcement ratio of 0.0025 in each direction. Walls were loaded with an eccentricity of 8.33 mm (thickness/6).

The effect of the selected stress-strain relationship on the prediction of the instability failure load of slender high strength concrete walls was investigated using the computer program WASTAB (18). Table 1 shows the failure loads predicted using the two stress-strain models, Scott et al (10) and Fafitis and Shah (13). Also tabulated in Table 1 are the actual failure loads of the two walls tested by Fragomeni (18).

It was observed that although the stress-strain model of Fafitis and Shah (13) does not give a good indication of the behaviour of spirally reinforced high strength concrete columns or the moment curvature behaviour of flexural members, it gives a good prediction of the instability failure load of the walls tested by Fragomeni (18).

CONCLUSIONS

In design of reinforced concrete structural members, the rectangular stress block is universally accepted as having the advantages of great simplicity and predicting the ultimate moment capacity with dependability and accuracy. However, it is applicable only at the ultimate concrete strain. A more accurate stress-strain relationship is required to follow the full-range behaviour of a structural element.

The results of the above investigation clearly indicate the influence of the stress-strain relationship of concrete in predicting the structural behaviour. It is important to realise that a model developed for confined columns may not necessarily be applicable to beams. Main observations made in the study were that accurate prediction of the tail of the stress-strain curve is important in developing a flexural stress block whereas it may not be as important in columns under predominantly axial compression. Furthermore it was observed that stress-strain models which are calibrated for uniaxial compression would be more representative of the stress-strain relationship of unconfined members such as slender high strength concrete walls.

In proposing a function for the stress-strain relationship of high performance concrete, it is important to calibrate any proposed model for confined concrete under axial compression, reinforced concrete under flexural compression and also for unconfined concrete. The exercise should incorporate concrete members with compressive strength of concrete ranging from 20 to 100 MPa.

ACKNOWLEDGMENT

The authors would like to express their gratitude to Mr. Sam Fragomeni for providing details of the tests conducted on slender high strength concrete walls, which have been included in this paper.

REFERENCES

1. ROY H.E.H. AND SOZEN M.A., " A Model to Simulate the Response of Concrete to Multi-Axial Loading", Structural Research Series, No.268, Civil Engineering Studies, University of Illinois, June 1963.

2. STURMAN, GERALD M., SHAH S.P. AND WINTER G., "Effects of Flexural Strain Gradients on Microcracking and Stress Strain Behaviour of Concrete", ACI Journal Proceedings, Vol.62 (7), July 1965, pp 805-822.

3. DESAYI P. AND KRISHNAN S., "Equation for Stress Strain Curve of Concrete", ACI Journal, Proceedings , Vol.61 (3), March 1964, pp 345-350.

4. SARGIN M. "Stress Strain Relationship for Concrete and the Analysis of Structural Concrete Sections" Study No.4, Solid Mechanics Division, University of Waterloo, Ontario, 1971, pp 167.

5. POPOVICS S., "A review of the Stress-Strain Relationships for Concrete", ACI Journal, Vol.67 (3), March 1970, pp 243-248.

6. MANDER J.B., Priestly M.J.N and Park R., "Theoretical Stress Strain Model for Confined Concrete", Journal of Structural Division, ASCE, Vol.114 (8), Aug. 1988, pp 1804-1849.

7. VALLENAS J., BERTERO V.V. AND POPOV E.P., "Concrete Confined by Rectangular Hoops Subjected to Axial Loads", Report No. UCB/EERC, No.13, 1977, Earthquake Engineering Research Centre, University of California, Berkeley.

8. SHIEKH S.A. AND UZUMERI S.M., "Strength and Ductility of Tied Concrete Columns", Journal of the Structural Division, ASCE, Vol.106 (ST5), May 1980, pp 1079-1102.

9. AHMAD S.H. AND SHAH S.P., "Stress Strain Curves of Concrete Confined by Spiral Reinforcement", ACI Journal, Vol.79 (6), 1982, pp 484-490.

10. SCOTT B.D., PARK R. AND PRIESTLY M.J.N., "Stress-Strain Behaviour of Concrete Confined by Overlapping Hoops at Low and High Strain Rates", ACI Journal, Proceedings Vol.79, No.1, Jan-Feb. 1982 pp 13-27.

11. BJERKELI L., TOMASZEWICZ A. AND JENSEN J.J. "Deformation Properties and Ductility of Very High Strength Concrete", Second International Symposium on Utilisation of High Strength Concrete, May 20-23, 1990, Berkely, California.

12. YONG Y.K., NOUR M.G., NAWY E.G., "Behaviour of Laterally Confined High Strength Concrete Under Axial Loads", Journal of Structural Division, ASCE, V.114 (2), February 1988, pp.332-350.

13. FAFITIS A. AND SHAH S.P., "Lateral Reinforcements for High Strength Concrete Columns", ACI Special Publication, SP-87, American Concrete Institute, 1986 pp 213-232.

14. SETUNGE S., MENDIS P.A., DARVALL P.LeP., "Full Range Moment Curvature Behaviour of Reinforced Concrete Sections" Proceedings, Australasian Structural Engineering Conference, IEAust., 1994, pp 507-514.

15. ATTARD M.M., SETUNGE S., "The Stress-Strain Relationship of Confined and Unconfined Concrete", Paper accepted for publication by the ACI Structural Journal, 1995.

16. PENDYALA, R., MENDIS P., PATNAIKUNI I., "Full Range Behaviour of High Strength Concrete Flexural Members - Comparison of the Ductility Parameters", Paper accepted for publication by the ACI structural journal.

17. MARTINEZ S., NILSON A.H. AND SLATE F.O., "Spirally Reinforced High Strength Concrete Columns", ACI Journal, Vol.81 (5), September-October 1984, pp 431-442.

18. FRAGOMENI, S. "Design of Normal and High Strength Reinforced Concrete Walls", Ph.D. Thesis, The University of Melbourne, Australia, 1995.

THE DEVELOPMENT OF PHC PILES IN CHINA

Z L Yan

S J Huang

J C Zhang

Guangzhou Yangcheng Pile Co. Ltd

China

ABSTRACT. This paper briefly discusses the history of the development of pre-stressed high-strength concrete (PHC) piles in China. It also includes the manufac-turing technology and the important characteristics of the manufacturing process. Then we talk about the quality control and its level of raw material, concrete and the ultimate piles. The last part is the study and application of replacing part of the cement with ground ordinary building sand.

Keywords: PHC piles, Manufacturing technology, Quality control, Autoclave cur-ing, Centrifugal casting, Ground sand.

Mr Yan Zhilong is a vice professor in Concrete Technology, Southeast University, China. Now he is chief engineer and vice general manager in YangCheng Pile Co. Ltd. He specialises in the technology and quality control of concrete, prestressed concrete and its products, with particular reference to the research, production, application and development of PHC piles.

Mr Huang Shaojiang is an engineer in Concrete Technology. And he is general manager and director in YangCheng Pile Co. Ltd. He has much experience in com-modity concrete, design of concrete construction and plant management.

Mr Zhang Jiechun is a Master of Science in Concrete Technology and manager of the Technology Unit in YangCheng Pile Co. Ltd. He specialises in the technology and quality control of high-strength concrete and its products.

Radical Concrete Technology. Edited by R K Dhir and P C Hewlett. Published in 1996 by E & FN Spon, 2–6 Boundary Row, London SE1 8HN, UK. ISBN 0 419 21480 1.

INTRODUCTION

The development of prestressed high-strength concrete (PHC) piles in China was based on the technology used in the production of prestressed concrete (PC) piles. PC piles were first developed and manufactured by Fengtai Bridge Factory of the Railway Ministry in the 1970s. The concrete average compressive strength was in the area of $60N/mm^2$. The level of prestressing was fairly low, which was at around $4N/mm^2$. The prestressing steel bar mostly used homemade high-strength steel wires or building grade $I \sim IV$ ribbed steel. The manufacturing process was not advanced and every stage in manufacturing was controlled manually. From 1989, China began to introduce PHC piles manufacturing facilities and technology from Japan and other countries. Special grooved bar for prestressed concrete piles also started to be adopted partially. During the trial at this stage, most piles produced still belonged to PC piles due to various reasons[1]. It was not until 1990 when PHC piles were produced on a large scale with concrete strength keeping at $80N/mm^2$ and the level of prestressing exceeding $5N/mm^2$[2]. In 1994, the concrete piles produced in China reached 5000km all together, most of which are PHC piles. They are mostly used in eastern and southern China, Hongkong and Macao, where the layer of soft earth is thick.

MANUFACTURING TECHNOLOGY AND ITS CHARACTERISTICS

Currently, manufacturing base for PHC piles is concentrated in Guangdong province of Southern China. According to incomplete statistics, the number of companies in this area has exceeded 50. Among these manufacturers, the typical manufacturing process (taking YangCheng Pile Co. Ltd. for example) can be summarized as Figure 1. From Figure 1, we can conclude main characteristics of manufacturing technology in PHC pile producing process as follows:

(1) Prestressing steel bar mostly uses grooved high strength steel bar special for prestressed concrete. They are cut into certain length by machines special for cutting, and heads produced at both ends using special head-forming machines. Prestressing steel bar and circular steel wires are spot-welded to form cages, and all prestressing tension is produced by pretensioning technique.

(2) Some modern factories use computerised control in the production of concrete. Raw material is automatically weighed and moisture content of aggregates are automatically measured and corrected. These ensure adequate control over the actual water-cement ratio in the mix as well as slump.

(3) Centrifugal concrete casting technique is commonly adopted and this process is often automatically controlled. Curing uses a two-stage process, i. e. the initial low temperature steam curing and autoclave curing.

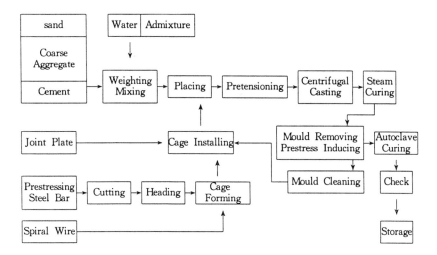

Figure 1   Manufacturing process of PHC piles

QUALITY CONTROL AND ITS LEVEL

We conclude the quality control and its level of raw material, concrete and PHC piles during the PHC pile production as follows:

Raw Material

Coarse Aggregate: Crushed granite, grain diameter 5-20mm, average grain diameter about 12. 5mm; Fine Aggregate: medium quartz sand, module of fineness 2. 5 ～ 3. 0; Admixture: superplasticising water-reducing agent; Cement: grade 525 Portland cement [5].

Characteristics of Pile Concrete and Steel Bar

The main characteristics of prestressing steel bar and concrete used by YangCheng Co. Ltd are sumarized below: The compressive strength and modulus of elasticity of concrete are above 80N/mm² and 3. 92×10⁴N/mm² respectively. For the prestressing steel bar, the tensile strength (σ_b) and yield strength ($\sigma_{0.2}$) are 1442. 5N/mm² and 1275. 3N/mm² separately. Its elongation (δ_8) is above 5% and relaxation is below 1. 5%. (Relaxation testing method complying with JISG3109).

Quality Control And Its Level of PHC Pile Concrete

Most PHC piles concrete produced has a cement content at about 540kg per cubic metre and around 1% of superplasticising admixture. The mix proportion in theory is 1 : 1. 34 : 2. 29 : 0. 01 (cement : sand : coarse aggregate : admixture) The water-cement ratio is about 0. 30. The slump is around 3 to 5cm. Taking YangCheng PHC piles for example, the actual control level of concrete quality is as follows: The monthly average compressive strength is between 95N/mm² and 98N/mm². The standard deviation "σ" is between 3. 5N/mm² and 4. 0N/mm².

Products Range and Main Engineering Properties

PHC piles have a diameter of 300 to 600mm. Their length tends to vary but is mainly between 7 and 12m. Due to various reasons, the percentage of prestressing steel bar is fairly low in PHC piles in China and the effective level of prestressing is low at around $5N/mm^2$ [4]. Table 1 is the range of PHC piles produced by YangCheng Pile Co. Ltd and their engineering properties[1][2].

Table 1 Range and engineering properties of PHC piles

Products Range			Concrete Compressive Strength	Effective Prestress	Allowable Bearing Capacity	Crack Bending Moment	Ultimate Moment	Shearing Capacity
Dia. mm	Thick mm	Length m	N/mm^2	N/mm^2	kN	kNm	kNm	kN
400	95	7,9,11	≥ 80	5.53	1570	78.0	135.4	250.2
500	100	7,9,11	≥ 80	5.72	2256	145.0	258.8	303.1
500	125	7,9,11	≥ 80	5.38	2700	147.0	261.4	359.0
600	110	7,9,11	≥ 80	5.12	3000	205.0	342.0	381.0

RESEARCH DEVELOPMENT

From above, we know that as the cement content of PHC piles ($\geq 540kg/m^3$ concrete) is high, their cost is high. To be able to use them extensively in China, it is important to reduce cement amounts and to maintain, if not improve the properties of PHC piles.

Every one knows, according to Interface Theory, it is of enormous significance for making high-strength concrete to improve concrete interface structure. As SiO_2 of ground sand can react with $Ca(OH)_2$ in concrete to produce Tobermorite when we use autoclave curing, it decreases weak link of interfaces of coarse and fine aggregates [3]. Owning to above-mentioned, we studied the corresponding properties of PHC piles in which part of the cement was replaced equally by ground ordinary building sand and got a succes. Now we have put the PHC piles with ground sand into wide practice.

The experimental details are listed as follows:

(1) The quality of raw material is the same as what has been mentioned, Ordinary building sand is used for ground sand, whose content of SiO_2 is above 92%.

(2) When 30% of the cement is equally replaced by ground sand, the mixing proportion of concrete is 0.3 : 0.7 : 1.34 : 2.29 : 0.01 (ground sand : cement : sand : coarse aggregate : admixture). The water-cement (including ground sand) ratio is about 0.30.

From Table 2,3,4 and 5, we find that the fineness and SiO_2 content of ground sand are important to improve concrete strength and PHC pile strength; distribution of ground sand, mixing method and curing system has great influence on concrete mechanical properties; when 30% cement is replaced equally by ground sand, the compressive, bending and tensile-splitting strength of the concrete increase after low temperature steam curing and autoclave curing. In addition, the PHC piles with ground sand increase in crack bending moment and ultimate moment. At the same time, we use PHC piles with ground sand in full scale hammer-driving condition. All the performance including lash property proves to be satisfactory.

Table 2 The concrete compressive strength influenced by amounts of ground sand

No.	Level of Replacement (% of cement)	Compressive Strength, N/mm²		Remarks
		Low Temp. Steam Curing	Low Temp. Steam Curing-Autoclaving	
1	0	51.8	81.4	manual
2	10	47.2	87.4	manual
3	15	48.0	(81.9)	manual
4	20	45.9	85.5	manual
5	25	41.2	87.1	manual
6	30	40.5	86.5	manual

Table 3 The concrete compressive strength influenced by sand fineness

No.	Level of Replacement (% of cement)	Mixing Method	The Specific Surface of Ground Sand, cm²/g	Compressive Strength after Low Temp. Steam Curing —Autoclaving N/mm²
1	0	manual		81.4
2	30	manual	3502	76.0
3	30	manual	4328	79.0
4	30	manual	4924	86.5

Table 4 The mechanical properties of high strength concrete with ground sand

No.	Level of Replacement (% of cement)	Mixing Method	Steam Curing-Autoclaving Concrete		
			Compressive Strength N/mm²	Bending Strength N/mm²	Tensile-Splitting Strength N/mm²
1	0	manual	83.2	5.39	5.28
2	0	machine	110.2	7.62	6.10
3	30	manual	80.4	5.66	5.25
4	30	machine	114.8	7.67	6.44

Table 5 Bending properties of PHC piles with and without ground sand

Pile Dimension, mm (Dia.-Thick-Length)	Crack Bending Moment, kNm	Ultimate Moment, kNm	Remarks
500-100-7000	159.5	284.7	30% with ground sand
500-100-7000	145.0	258.8	without ground sand

CONCLUSIONS

1. Choosing and quality control of raw materials (example sand, coarse aggregate, cement, admixture and etc) are strict. The compressive strength of piles concrete exceeds 80N/mm² and it exceeds 98N/mm² in some advanced companies.

2. PHC piles have many ranges. They mostly have a diameter of 300 to 600mm, their length tends to vary but mainly between 7 and 12m. But the percentage of prestressing steel bar is fairly low and the effective level of prestressing is lower. The piles quality in some advanced companies has come to Japan standards[1].

3. The prestressing steel bar mostly adopts PC steel bar which is similar to "UL-BON" made in Japan. They are handled with special machines. The control of concrete mixing is different. There is some computer control of mixing process in a few advanced companies. Raw material and moisture content of aggregates are automatically measured and determined. It is adopted that centrifugal concrete casting technique and low temperature steam curing-autoclave curing.

4. Using PHC piles with part of cement replaced by ground sand, the optimum amount of replacement is 25～30% using low temperature steam curing-autoclave curing. It is very important for concrete strength that ground sand contains a high amount of SiO_2 and fineness is unusually fine. When ground sand is added to concrete, mixing method and distribution of ground sand have great infulence on concrete mechanical properties.

5. Replacing some cement with ground sand to produce PHC piles has better mechanical properties, economical and social benefits.

REFERENCES

1. JAPANESE INDUSTRIAL STANDARDS. Pretensioned Spun High-strength Concrete piles, JISA5337-1993, 1993, pp 4-5.

2. P. R. CHINA STANDARDS. Pretensioned Spun Concrete Piles GB13476-92, 1992, pp6-7.

3. Wu Zhongwei. With and without Autoclave Curing Early High-strength Concrete Used For Piles. Precast Concrete Piles, No. 2, 1994, pp 1-3.

4. Yan Zhilong, Huang Shaojiang. Discussing PHC piles at present in China and difference compared with foreign countries. Precast Concrete piles, No. 1, 1994, pp 16-17.

5. Yan Zhilong, Huang Shaojiang, Zhang Jiechun, Li Lun. Discussing a few problems about quality of PHC piles in China, Precast concrete piles, No. 1, 1995, pp 20-22.

THE DEVELOPMENT OF VERY HIGH STRENGTH CONCRETE

S Y N Chan

M Anson

S L Koo

Hong Kong Polytechnic University

Hong Kong

ABSTRACT. Attempts have been made to maximize the strength of 'very high strength' concrete, that is concrete with a compressive strength at 28 days of over 100 N/mm^2. The measures taken for maximizing the strength include: a) using high strength aggregate; b) incorporating silica fume; c) employing steel fibres; d) lowering the water-cementitious ratio. Results show that the 28-day compressive strength of a 95 MPa concrete can be increased by about 70% to 160 MPa by adopting the above methods.

Keywords: Very high strength, High strength aggregate, Silica fume, Steel fibres.

Dr Sammy Yin Nin CHAN is an Assistant Professor at the Hong Kong Polytechnic University. He is a committee member of the Structural and Materials Divisions of the Hong Kong Institution of Engineers and a member of the Hong Kong Reinforced Concrete Design Code Review Committee. His research interests include high strength concrete and durability of concrete structures.

Professor Mike ANSON is the Dean of the Faculty of Construction and Land Use at the Hong Kong Polytechnic University. He was a researcher in concrete technology in earlier years and into the behaviour of structural concrete and the productivity of concrete supply and placing on site in later years. He is an experienced supervisor of research in concrete and in other fields. He is also a panel member and a Specialist Referee of the Research Grants Council.

Mr Siu Long KOO is a Civil Engineering Graduate in the Works Group of Departments, Hong Kong Government and a research student in the Concrete Technology Group, Hong Kong Polytechnic University. His research topic focuses on the development and study of very high strength concrete(VHSC).

Radical Concrete Technology. Edited by R K Dhir and P C Hewlett. Published in 1996 by E & FN Spon, 2–6 Boundary Row, London SE1 8HN, UK. ISBN 0 419 21480 1.

INTRODUCTION

In a major financial centre like Hong Kong, construction of high-rise buildings using in situ high strength concrete(HSC) is becoming popular owing to the attractive economic advantages[1]. In 1990, HSC of grade 60 was used for the first time in the construction of the Central Plaza. In 1995, grade 100 concrete was used for the re-development of City Plaza in Tai Koo Shing.

A 120-storey building has been recently scheduled for construction soon in Hong Kong. With this ever increasing demand for taller buildings, a research programme focusing on the development and study of VHSC and aiming to promote this concrete technology in the design of high-rise buildings was established at the Hong Kong Polytechnic University.

The programmme includes the development of mix designs and the study of a wide range of rheological, mechanical [2] and durability [3] properties, such as the workability retention [4] and performance under elevated temperatures [5]. The study reported in this paper forms part of the research programme and aims to maximize the strength of concrete using locally available materials, a conventional casting method and curing regime.

EXPERIMENTAL DETAILS

Methodology

According to FIB/CEB, concrete with cylinder compressive strength above $60N/mm^2$ is defined as high strength and concrete above $100N/mm^2$ as very high strength [6].

In this study, the following methods were attempted to produce in situ VHSC: a) using high strength aggregate; b) incorporating silica fume in the binder; c) employing steel fibres; d) lowering the effective water-cementitious ratio. The individual effects of each parameter on the compressive strength of concrete were studied.

As a result, the optimum values required for each of these four parameters to produce the maximum concrete compressive strength with the appropriate workability of about 150mm slump was determined. A concrete mix with maximized compressive strength was then formulated based on the optimum values obtained for each parameter [7]. The uniaxial compressive strength results of concretes subjected to a water curing regime are presented.

Materials

The following commercially available materials were used in this study:

a)   ordinary Portland cement of Green Island Brand to BS12;
b)   superplasticizer of sulphonated naphthalene formaldehyde condensates satisfying the requirements of ASTM C494, type A and F;
c)   condensed microsilica in powder form;

d)   coarse aggregates including normal strength granite(NSG) and high strength granite(HSG), with nominal sizes of 10 mm and 20 mm;
e)   fine aggregate of zone F river sand to BS882;
f)   hooked ends steel fibre of 25 mm long and 0.4 mm^2 cross sectional area;

RESULTS AND DISCUSSIONS

Effects of Type and Size of Coarse Aggregate

Concrete mixes with a water-cement ratio of 0.22 and incorporating NSG and HSG were cast. The compressive strength development of concrete is shown in Figure 1. The 28-day compressive strength of concrete using HSG was about 14% higher than that of concrete using NSG. For concrete with HSG, the compressive strength development of concrete with 10mm maximum size aggregate was similar to that of concrete with 20mm maximum size aggregate.

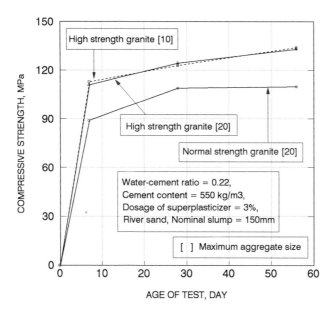

Figure 1 Effects of type and size of coarse aggregate

Effect of Silica Fume

Two series of concrete mixes have been cast. Series I consisted of silica fume concrete with a constant dosage of superplasticizer of 1.5% by weight of cement. Series II consisted of silica fume concrete with various dosages of superplasticizer to provide a constant slump of about 150 mm. The results are shown in Figure 2.

The results show that the optimum level of silica fume addition was 10% for series I and was 15% for series II, with the 28-day compressive strength increased by 15% and 25% respectively. For both series of mixes, concrete incorporating silica fume above the optimum level resulted in strength reduction.

The addition of silica fume to concrete with constant dosage of superplasticizer decreased the workability of the mixes. For concrete with constant dosage of superplasticizer, the low workability of the concrete mixes with silica fume content exceeding 10% eventually caused a reduction in the compressive strength of concrete due to poor compaction.

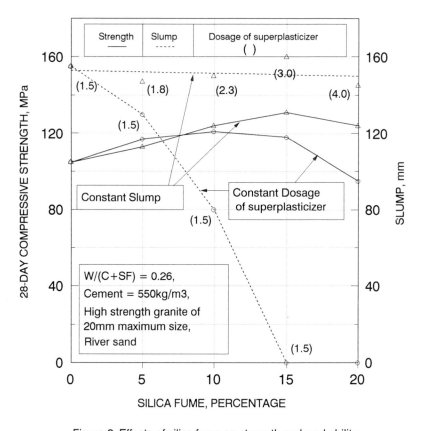

Figure 2 Effects of silica fume on strength and workability

Effect of Steel Fibre

A commercially available hooked ends steel fibre was used in this study. Concrete incorporating 1, 2 and 3% fibre by volume of mortar content and concrete without steel fibre were prepared. The 28-day compressive strength versus the percentage of fibre addition is plotted in Figure 3.

The compressive strength of concrete incoporating fibre was higher than that of concrete without steel fibre, regardless of the percentage of addition. At 1% steel fibre addition level, compressive strength of concrete was increased by about 6% and 9% for concrete with w/c ratio of 0.30 and 0.24 respectively. Concrete incorporating 2% steel fibre demonstrated a further increase in compressive strength of about 6% and 4% for concrete with w/c 0.30 and 0.24 respectively. Concrete incorporating 3% steel fibre showed very little improvement in the compressive strength.

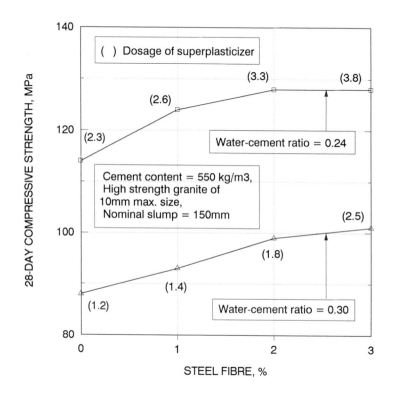

Figure 3 Effect of hooked ends steel fibre on 28-day strength

Effect of Water-cement Ratio

Superplasticized concretes with cement content 550kg/m³ and a nominal slump of 150mm were cast. The 28-day compressive strength of concrete with a range of w/c ratio is shown in Figure 4.

For w/c ratio below 0.20, the compressive strength tended to decrease with decreasing w/c ratio. It was probably caused by the high dosage of superplasticizer required for achieving adequate workability [8].

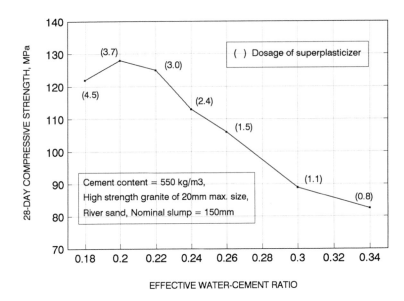

Figure 4 Relationship between 28-day strength and water-cement ratio

Maximizing the Strength of VHSC

A HSC mix, M1, with 28-day compressive strength of 95MPa and workability of about 150mm slump was prepared as the control (Figure 5 refers). The mix proportions of the control were then altered step by step based upon the optimum mix parameters determined from the previous series of studies.

Step 1 - Replacing normal strength granite by high strength granite
By replacing normal strength granite in mix M1 with high strength granite, the 28-day compressive strength was increased by about 9% (M2).

Step 2 - Addition of silica fume
Fifteen percent silica fume by weight of cement was added to the concrete mix M2 with the effective water-cementitious ratio kept constant. The compressive strength was further increased by 23% (M3).

Step 3 - Adding steel fibre
The addition of 1% steel fibre to the concrete mix M3 further increased compressive strength of concrete by 7% (M4).

Step 4 - Reducing water-cementitious ratio
The effective water-cementitious ratio of the mix M4 was reduced from 0.26 to 0.20 to form mix M5. Consequently, the compressive strength was further increased by about 18%. The total increase in compressive strength of M5 relative to the control M1 was about 68%.

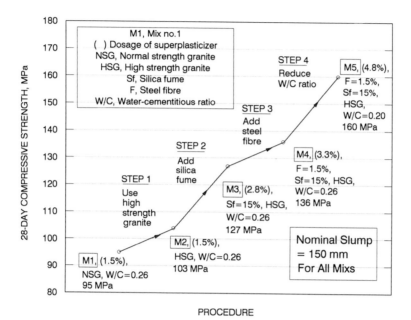

PROCEDURE

Figure 5 Maximizing strength of very high strength concrete

CONCLUSIONS

1. For concrete using NSG and with 28-day compressive strength over 95 MPa, the replacement of NSG by HSG increased the strength of concrete by 8% to 14%.

2. Concrete incorporating 15% silica fume (by addition) demonstrated the highest compressive strength provided that additional dosage of superplasticizer was used to maintain the workability of concrete mixes to achieve good compaction.

3. The compressive strength of concrete incoporating steel fibre was higher than that of concrete without steel fibre. The addition from 2% to 3% showed very little further improvement in the compressive strength of concrete.

4. The compressive strength of concrete and doasge of superplasticizer required for maintaining workability increased with decreasing water-cement ratio.

5. No improvement in compressive strength of concrete was observed when water-cement ratio fell below 0.2.

6. The 28-day compressive strength of a 95 MPa concrete can be increased by about 70% to 160 MPa by using the following methods: a) replacement of normal aggregate by high strength aggregate; b) addition of silica fume; c) incorporation of steel fibres; d) reduction of water-cementitious ratio.

REFERENCES

1.   Chan, Sammy Y.N. and Anson, Mike. High-strength concrete: the Hong Kong experience. Magazine of Concrete Research, Vol. 46, No. 169, December 1994, pp. 235-236.

2.   Chan, Sammy Y.N., Dhir, Ravindra K., Tham, K.W. and Chang, Da Yong. Superplasticized normal workability concrete: physical and mechanical properties. Transactions, Hong Kong Institution of Engineers, Vol. 1, No. 2, December 1994.

3.   Chan, Sammy Y.N., Dhir, Ravindra K., Tham, K.W. and Chang, Da Yong. Superplasticized normal workability concrete: permeation and durability properties. Transactions, Hong Kong Institution of Engineers, Vol. 1, No. 1, August 1994.

4.   Chan, Sammy Y.N., Feng, N.Q. and Tsang, Mathew K.C. Workability retention of high strength concrete. Magazine of Concrete Research (accepted for publication).

5.   Chan, Sammy Y.N. and Peng, G.F. High strength concrete subjected to elevated temperatures. RILEM Materials and Structures (accepted for publication).

6.   CEB/FIP Report. High strength concrete - state of the art report, SR90/1, Bulletin d' Information No.197, August 1990.

7.   Rougeron, P. and Aitcin, P.C. Optimization of the composition of a high performance concrete. Cement, Concrete and Aggregates, CCAGPD, Vol. 16, No. 2, Dec. 1994, pp. 115-124.

8.   Ping Gu, Ping Xie, Beaudoin, J.J. and Jolicoeur, C. Investigation of the retarding effect of superplasticizers on cement hydration by impedance spectroscopy and other methods. Cement and Concrete Research, Vo.24, No.3, 1994, pp. 433-442.

HIGH STRENGTH CONCRETE INCORPORATING CARRIER FLUIDIFYING AGENT

S Y N Chan

Hong Kong Polytechnic University

Hong Kong

N Q Feng

Tsinghua University

China

M K C Tsang

Hong Kong Polytechnic University

Hong Kong

ABSTRACT. The problem of rapid loss of workability of high strength concrete (HSC) must be resolved to facilitate the insitu use of HSC. This paper presents the effectiveness of an admixture, carrier fluidifying agent (CFA), in retaining the workability of HSC and the mechanical properties of the resulting concrete. The results showed that the workability of HSC mixes incorporating CFA can be maintained for a period of up to two hours. Laboratory trials using repeated dosing of superplasticizer were also carried out. Unlike the presently adopted repeated dosing method, incorporating CFA will not impose manpower, quality control and cost penalty as well as the risk of over-dosing of superplasticizer. Although the setting times were marginally extended and there was a slight air entrainment, the mechanical strength of HSC incorporating CFA showed no reduction comparing to that of the control.

Keywords: High strength concrete (HSC), Carrier fluidifying agent (CFA), Workability retention, Repeated dosing.

Dr Sammy Y.N. Chan is the Leader of the Concrete Technology Group at the Department of Civil and Structural Engineering of the Hong Kong Polytechnic University. His main research interests include high strength concrete, workability retention and concrete admixtures.

Professor N.Q. Feng is the Director of Research Laboratory of Building Materials at the Tsinghua University. He is a member of the RILEM TC 78MCA Committee. His research interests include high performance concrete, concrete admixtures and a variety of construction materials.

Mr Matthew K.C. Tsang is a PhD research student at the Hong Kong Polytechnic University. The theme of his research is the study of HSC incorporating CFA.

Radical Concrete Technology. Edited by R K Dhir and P C Hewlett. Published in 1996 by E & FN Spon, 2–6 Boundary Row, London SE1 8HN, UK. ISBN 0 419 21480 1.

INTRODUCTION

One of the main problems for insitu construction with superplasticized HSC is that its workability is almost solely provided by superplasticizer. The rapid consumption of superplasticizer by the hydrating paste (repulsive negative ions of the superplasticizer being neutralised by the metallic ions of the hydrating cement) results in a rapid loss of workability. For a given mix proportion, higher the dosage of superplasticizer, higher is the initial slump and longer is the time of the workability retention of the mix[1]. However, if the w/c ratio is very low and the dosage is increased beyond an operation limit as a means of retaining the workability of concrete for a sufficient period for insitu casting, adverse effects such as retardation in setting and hardening, air entrainment, segregation, and/or bleeding could result.

Besides the influence of the type and quality of cement, the slump loss of concrete with time is mainly due to the hydration reaction of cement, physical agglomeration of hydrating cement, loss of effectiveness of superplasticizer in the liquid by adsorption on the surface of C_3S phase and the drop of zeta potential in the mix[2]. In order to retain the workability, a sufficient quantity of active superplasticizer and hence the zeta potential in the concrete should be maintained within the required period. One of the methods commonly adopted to maintain the workability is to redose the concrete. The dosage of each redose being the same as the initial[3,4]. However, it was claimed that the addition of a third dosage was not recommended because the concrete, although maintaining its consistency, lost its workability[4]. Another method that has been used to retain the concrete workability is the addition of superplasticizer together with a set retarder.   The presence of retarding admixtures reduces the rate of slump loss although in the case of hot weather or concrete with low w/c ratio[5,6], the resulting effect is usually not satisfactory and there is the accompanying adverse effect of retardation in setting and hardening.

In recent years in China, an admixture named carrier fluidifying agent (CFA)[2,7] has been developed and patented. This admixture was developed by one of the authors Professor Feng. Research works reported in this paper investigated the effectiveness and mechanism of the workability retention of this new admixture, its effects on the properties of HSC and improvement works made to this admixture.

EXPERIMENTAL DETAILS

Materials

Ordinary Portland cement used complies with BS12: 1991[8]. The coarse aggregates used were crushed granite in 20mm and 10mm single sizes. The fine aggregate was river sand. Sulphonated naphthalene formaldehyde condensates was the superplasticizer used throughout the study.

Carrier fluidifying agent (CFA) was developed to retain the workability of fresh concrete[2]. It was produced in the form of pellets of up to 10mm in length and 5mm in diameter and it consisted of an inorganic strengthening agent, zeolitic mineral admixture (ZMA)[9], as the carrier that absorbed a definite proportion of superplasticizer and plasticizer. The basic proportion of the constituents was 70% ZMA, 20% superplasticizer, and 10% plasticizer. In the presence of mixing water, the absorbed superplasticizer and plasticizer were progressively released from the carrier to the fresh concrete. The workability of concrete can be retained by this continuous redosing of superplasticizer.

Concrete Mix Proportions

Design mixes given in Table 1 were used to study the effectiveness of CFA in the workability retention of concrete. Mixes PS1 to PS4 and Control were used to study the effect of the pellet size of CFA on the workability retention. Mixes RD1/4, RD1/2, RD1, CFA were used in the comparison between CFA and repeated dosing of superplasticizer.

Mixing Sequence, Specimen Preparation and Testing

Concrete mixes were prepared using a pan mixer with a nominal capacity of 50 litres. Air-dried coarse aggregates were first mixed with one third of mixing water followed by the addition of cement and sand. The remaining mixing water was then added followed by the addition of superplasticizer. In different parts of the study, CFA was added either at the beginning of the mixing with cement and sand or added following the addition of superplasticizer and measurement of the initial slump in comparison with repeated dosing of superplasticizer.

Table 1 : Design mixes for the study of the effectiveness of CFA

Mix	Constituent Materials (kg/m^3)						W/C Ratio	A/C$^\oplus$ Ratio	Dosage[#] (%)
	Cement	CFA	Water[+]	Aggregate[*]					
				20mm	10mm	Sand			
PS1	550	5.5	143	866	433	433	0.26	3.15	1.6
PS2	550	5.5	143	866	433	433	0.26	3.15	1.5
PS3	550	5.5	143	866	433	433	0.26	3.15	1.37
PS4	550	5.5	143	866	433	433	0.26	3.15	1.22
Control	550	0	143	866	433	433	0.26	3.15	1.42
RD1/4	550	0	143	866	433	433	0.26	3.15	0.57$^\otimes$
RD1/2	550	0	143	866	433	433	0.26	3.15	0.57$^\otimes$
RD1	550	0	143	866	433	433	0.26	3.15	0.57$^\otimes$
CFA	550	5.5	143	866	433	433	0.26	3.15	0.57$^\otimes$

1. A/C ratio - Aggregate/cement ratio
2. + Water content values shown do not include that arising from superplasticizer used.
3. * Saturated surface-dry.
4. # Dosage of superplasticizer was in terms of the solid content of superplasticizer by weight of cement.
5. $\otimes$ The dosage was the superplasticizer requirement for acquiring the initial slump.

The workability of fresh concrete was tested by means of a slump test in accordance with BS1881: Part 102: 1983[10]. Air content and the setting time of the concrete samples were tested in accordance with method B in BS1881: Part 106: 1983[11] and the method described in BS 5057: Part 1: 1982[12], respectively. All the specimens were cured in water at 27 °C immediately after demoulding at 24 hours after casting.

RESULTS AND DISCUSSION

Workability Retention By CFA

CFA is a new chemical admixture developed in China, and when tested in accordance with BS 5075: Part 1: 1982[12], CFA can be considered as a kind of plasticizer. However, CFA is not simply a plasticizer that increases workability, it also prolongs the workability of concrete. In this part of study, it was revealed that the mechanism of CFA in the retention of concrete workability is physical.

The physical effect is provided by the pellet size of CFA which affects its solubility and effectiveness on the workability retention. The admixtures (superplasticizer and plasticizer) absorbed in CFA are released gradually to deflocculate the cement conglomerates when the fresh concrete is being mixed mechanically. This continuous replenishment maintains sufficient quantity of active superplasticizer and plasticizer in the concrete mix and hence the workability of fresh concrete for a required period of time.

Physical effect (pellet size of CFA)

To study the effect of particle size on the workability retention, CFA was ground and sieved to obtain 4 sizes of pellets :
- retained by 5.00mm sieve,
- passed through 5.00mm sieve, but retained by 2.36mm sieve,
- passed through 2.36mm sieve, but retained by 600μm sieve,
- passed through 600μm sieve.

In this part of the study, CFA was added to the pan mixer together with cement and sand. Additional quantity of naphthalene based superplasticizer was used to acquire an initial slump of about 200mm. The concrete mixes PS1, PS2, PS3, and PS4 were used in this series of study and are given in Table 1.

The particle sizes of CFA for the design mixes PS1 through to PS4 were the ones which were retained by 5mm, 2.36mm, 600μm sieve, and those passed through 600μm sieve respectively. The slump values against elapsed time are plotted in Figure 1. It shows that the workability retention was better with the larger particle CFA. One point should be emphasized is that even with the finest particle, the workability retention by CFA was better than that by superplasticizer only.

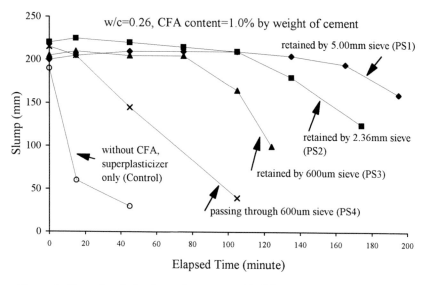

Figure 1: Slump loss behaviour of concrete with different pellet sizes of CFA

The specific surface area of CFA pellets increases with their fineness. The total surface area of CFA increases from that in mix PS1 to that in mix PS4. Such an increase led to a more rapid release of the absorbed admixtures. This can be illustrated in Table 1 that the finer the CFA pellet was used, the lesser the quantity of the superplasticizer required to acquire the initial slump.

Comparison Between CFA And Repeated Dosing

CFA retains the workability of fresh concrete by continuously redosing the concrete mix with the absorbed admixtures. It is similar to the redosing of superplasticizer manually at very short time interval to recover the concrete workability but with less additional cost and problem. This admixture is particularly useful in circumstances such as at the pumping stage, where it is not feasible to recover the workability of concrete in the pipe by redosing from outside. Table 2 and Figure 2 show the results and compare the slump retention of concrete by means of using CFA and repeated of superplasticizer. The concrete mixes used in this series of study are given in Table 1. The nominal slump of the concrete mixes was set at 200mm.

From Table 2, the total amount of naphthalene required to maintain the slump decreased as the time interval between the consecutive redosing increased. From the columns of recovery rate, it can be observed that, in a particular redosing scheme, the quantity of naphthalene required to recover the slump in each redosing process was similar. This phenomenon agrees with the results obtained by Hattori[3] the reason may be explained by referring to Figure 3.

Table 2: Result of slump retention by means of repeated dosing and using CFA.

Time	RD1/4				RD1/2			
(min.)	Slump before dosing (mm)	Cum. naph. (%)	Slump after dosing (mm)	Recovery Rate (mm / %)	Slump before dosing (mm)	Cum. naph. (%)	Slump after dosing (mm)	Recovery Rate (mm / %)
0		0.570	190			0.57	220	
15	60	0.63	180	1970				
30	85	0.69	210	2160	45	0.69	185	1150
45	90	0.75	200	1900				
60	90	0.80	195	2020	75	0.78	195	1330
75	100	0.86	200	1720				
90	110	0.90	195	1810	125	0.84	190	1180
105	110	0.99	205	1170				
120		0.99	190		125	0.88	180	1250
145		0.99	140			0.88	115	
	Total Naphthalene =0.99%				Total Naphthalene =0.88%			

Time	RD1				CFA	
(min.)	Slump before dosing (mm)	Cum. naph. (%)	Slump after dosing (mm)	Recovery Rate (mm / %)	Cum. naph. (%)	Slump (mm)
0		0.57	195		0.57	215
15					0.60	185
30					0.61	205
45					0.63	230
60	30	0.7	220	1520	0.64	210
75					0.65	210
90					0.66	195
105					0.67	135
120	75	0.79	220	1560		
145		0.79	170		0.67	50
	Total Naphthalene =0.79%				Total Naphthalene =0.67%	

Note:
1. Mixes RD1/4, RD1/2, and RD1 were used to study the repeated dosing of superplasticizer (naphthalene based) at 1/4 hr, 1/2 hr, and 1 hr intervals respectively.
2. Cum. naph. means the cumulative solid content of naphthalene by weight of cement in the mix.
3. In mix CFA, CFA was added following the addition of superplasticizer and measurement of the initial slump.
4. The cumulative naphthalene solid content in the mix incorporating CFA was calculated according to the solubility of CFA[2].

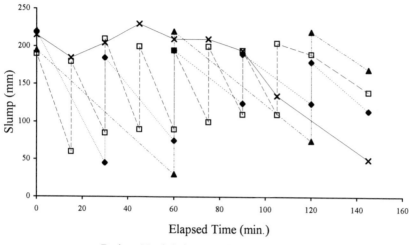

Figure 2: Comparison of slump retention by means of repeated dosing and using CFA

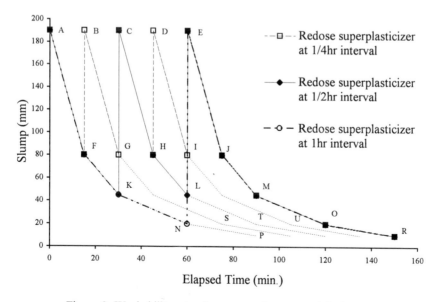

Figure 3: Workability retention process by repeated dosing

In Figure 3, curve AFKNP is a slump behaviour against time. If the concrete mix is redosed at F, the curve will shift to B, and the slump will lose again and drop to G and then S. If the concrete is redosed at every 15min., the slump will be pushed up from the points F, G, H, I to the points B, C, D, E respectively. In the same manner, the slump will be pushed up from the points K and L to points C and E, and from the point N to E for 30min. and 60min. redosing intervals respectively. With the total amount of slump value that needs to be recouped by the superplasticizer increased with decreasing redosing interval, it can be postulated that the total quantity of superplasticizer should also be increased.

Setting Times

Owing to the plasticizer, the setting time of the concrete incorporating CFA was extended. Each sample was mortar sieved from the concrete mixture after the last slump test was carried out. The time was measured starting from the completion of filling of mould with mortar. From Figure 4, the initial and final setting times were about 75 and 105 min. longer than those of manually redosed concrete.

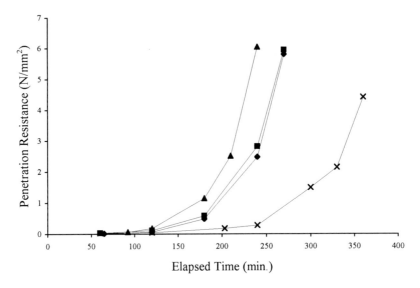

Figure 4: Setting times of concrete by repeated dosing and using CFA

Strength Development

Owing to ZMA, there is an air entraining effect of CFA [9,13]. The air contents of concretes with and without CFA were 1.70% and 0.95% respectively. However, ZMA which is a pozzolanic material made a compensation to the loss of strength caused by the prolonged setting and the air entrainment. As a result, the compressive strength of HSC with or without CFA was similar. The compressive strength development is shown in Figure 5.

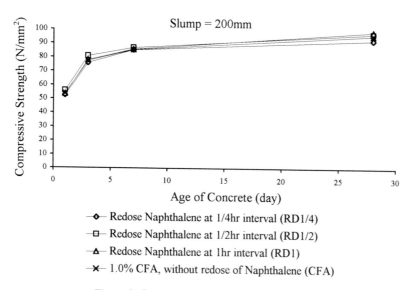

Figure 5: Compressive Strength Development

CONCLUSIONS

1. To maintain the workability of concrete, the required dosage of each redose was similar for a given redosing period. However, the cumulative dosage was found to depend on the time interval between the consecutive redoses. The longer the redosing time interval, the lesser is the cumulative dosage.
2. Carrier fluidifying agent (CFA) can maintain the workability for a period of up to two hours without adverse effect. The workability retention of CFA was afforded both by physical and chemical means. In the physical aspect, the absorbed chemical admixtures were continuously redosed into the concrete as CFA was being dissolved. The size of the CFA pellets has a bearing on the rate of release of the superplasticizer. Chemically, plasticizer reduced the cement hydration, the heat of hydration and the consuming rate of superplasticizer.
3. CFA has minor strengthening effect. The loss in concrete strength from the prolonged setting and the minor air entrainment was compensated by the presence of ZMA.

ACKNOWLEDGEMENT

The authors would like to thank Mr. C.K. Lee. for the useful discussions and Master Builders Technology Ltd for the supply of the superplasticizers used in this study.

REFERENCES

1. Ravina Dan; and Mor Avi, 1986, "Effects of Superplasticizers," Concrete International: Design & Construction, V.8, No.7, July 1986, pp.53-55.
2. Feng N.Q., "The Properties of Zeolitic Mineral Admixture Concrete," Tsinghua University, China, January 1992, pp.36-42.
3. Hattori Kenichi, "Experiences in the Use of Superplasticizers in Japan," Superplasticizers in Concrete, SP-62, American Concrete Institute, Detroit, 1979, pp.37-66.
4. Malhotra V. M., "Superplasticizers: Their Effect on Fresh and Hardened Concrete," Concrete International, May 1981, pp.66-81
5. FIP /CEB, "High Strength Concrete, State of the Art Report," FIP SR 90/1/CEB Bulletin d'Information No. 197, August 1990.
6. Mailvaganam N. P., "Factors Influencing Slump Loss in Flowing Concrete," Superplasticizers in Concrete, SP-62, American Concrete Institute, Detroit, 1979,pp.389-403.
7. Feng N.Q. et al, "Slump Control Admixture and the Method of Production", Chinese Patent No. 87 1 06558.4, The People's Republic of China, December 1990.
8. British Standard Institution, "BS12: 1991, Specification for Portland cement."
9. Feng N.Q., "The Properties of Zeolitic Mineral Admixture Concrete," Tsinghua University, China, January 1992,pp.8-35.
10. British Standard Institution, "BS1881: Part 102: 1983, Testing concrete: Methods for determination of slump."
11. British Standard Institution, "BS1881: Part 106: 1983, Testing concrete: Methods for determination of air content of fresh concrete."
12. British Standard Institution, "BS5057: Part 1: 1982, Concrete admixtures: Specification for accelerating admixtures, retarding admixtures and water reducing admixtures."
13. Feng N.Q., "Properties of Zeolite as an Air-Entraining Agent in Cellular Concrete," Cement, Concrete, and Aggregates, American Society for Testing and Materials, V.14, No.1, Summer 1992,pp.41-49.

RHEOLOGICAL LOSSES PARTIALLY PRESTRESSED STRUCTURAL LIGHTWEIGHT CONCRETE

C Magureanu

Technical University of Cluj-Napoca

Romania

ABSTRACT. In concordance with prognosis surveys, carried out all over the world including Romania, the conclusion is that the concrete, in general, and the lightweight high strength concrete, in particular, will be widely used in the third millennium.

Among the problems less investigated are the rheological losses in the partially prestressed reinforcement of the elements of the lightweight concrete.

The paper contains some quantitative and qualitative considerations relative to the experimental and calculated values of prestress losses from shrinkage and creep in partially prestressed lightweight concrete elements with pretensioned reinforcement.

The theoretical and experimental investigations have been extended over a one year period. The experimental loss of prestress was compared with the calculated values in accordance with the Romanian Code /1/, EUROCODE 2-1992 /2/.

Keywords: Partially prestressed concrete, Losses of prestress Creep, Shrinkage, Lightweight and Normal weight concrete.

Dr.Eng. Cornelia Magureanu is an Associate Professor at Technical University of Cluj-Napoca, Romania. She specialises in the prestressed, reinforced concrete and in ferrocement as well. She is also a specialist in long and short term lightweight, plain, reinforced and prestressed concrete.

Radical Concrete Technology. Edited by R K Dhir and P C Hewlett. Published in 1996 by E & FN Spon, 2–6 Boundary Row, London SE1 8HN, UK. ISBN 0 419 21480 1.

INTRODUCTION

The designing of lightweight partially prestressed concrete elements requires a through quantitative and qualitative knowledge of the tension occurring from shrinkage and creep. On this very purpose and experimental program has been carried out with a focus on centrally prestressed lightweight and normal weight concrete elements having active pretensioned and passive reinforcement. The prestress losses shrinkage and creep were studied for a period of one year.

EXPERIMENTAL DETAILS

In the test program the following parameters were taken into consideration:

The Unitary Stress Within the Concrete

$\sigma_c / f_c = 0.400; 0.342; 0.265; 0.207$ by varying the cross section of central prestressed elements.

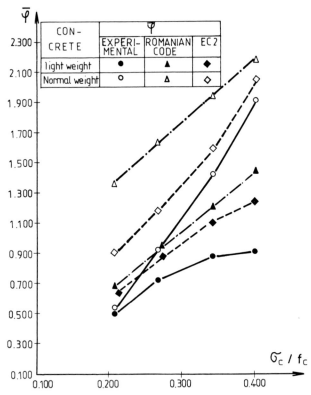

Figure 1 Experimental and calculated creep and shrinkage factor
(after one year) vs. loading degree

Prestressing degrees

Lightweight and normal aggregates concrete elements with and without passive reinforcement and unreinforced elements for determining the shrinkage of concrete.

Type of Concrete

The concrete used the Bc35 grade and was prepared with lightweight aggregates (expanded clay) and ordinary river gravel.

The reinforcement

The prestressed reinforcement was 7 ∅ 4 strands (TBP 12 mm) by means of which a central prestressing was achieved.
The non-prestressed reinforcement (passive reinforcement) was made of deformed bars PC 52- type, with tensile strength 520 N/mm^2 (4 ∅ 6 mm, 4 ∅ 10 mm, 4 ∅ 14 mm, for $\sigma_c / f_c = 0.342; 0.265; 0.207$).

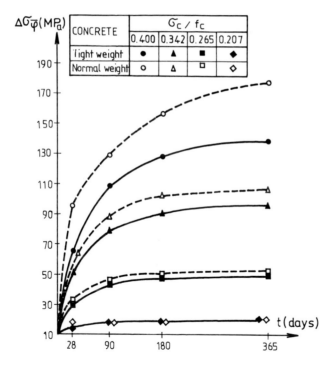

Figure 2 Experimental losses of prestress due to creep and shrinkage

The Manner of Preserving the Experimental Elements

For observing the deformations with time a rigid metallic stand was built to which one end of the elements was connected, the other end being free to deform. The experimental elements were kept in an air-conditioned room within constant conditions of humidity and temperature:

$$U = (65\pm5) \%; t = (20\pm2) \ ^\circ C$$

RESULTS AND DISCUSSIONS

The experimental creep and shrinkage factor $\overline{\varphi}$ were determined over a one year period means of the relations:

$$\overline{\varphi} = \frac{\varepsilon_{cc} + \varepsilon_{cs}}{\varepsilon_{ce}}$$

where: ε_{cc}, ε_{cs} and ε_{ce} are creep, shrinkage and elastic specific strains.

The overall experimental and calculated factory $\overline{\varphi}$ at one year in the case of fully and partially prestressed elements, as a function of the loading degree, σ_c / f_c are represented in Figure 1.

A increase of this long term deformation factor $\overline{\varphi}$ is to be remarked with loading degree.

The $\overline{\varphi}$ factor is greater for normal concrete than the same factor for lightweight concrete, especially at higher degrees of loading.

One could notice that the value of this factor varies proportionally to the value of loading degree, both with normal and lightweight concrete. It is also obvious that the lightweight concrete showed low values of time depending or the deformations factor, these being 1,1 ... 2,1 times (for σ_c / f_c =0.400 ... 0.207) lower than those of normal concrete. This is due to the greater elastic deformation (ε_{ce}) of the lightweight concrete, who is the denominator in relation of the φ factor.

The losses of prestress due to shrinkage and creep ($\Delta\sigma_{\overline{\varphi}}$) in the prestressed reinforcement of the elements under long term tests have been estimated in accordance with Romanian standard and EC2 /1.2/.

The time evolution of prestress losses due shrinkage and creep, calculated on experimental data are shown in Figure 2 with their absolute values.

CONCLUSIONS

1. One could remark that in normal weight partially prestressed concrete elements higher values of prestress losses of 23% ... 5% on the σ_c / f_c =0.400 ... 0.207, were recorded, as compared with lightweight concrete, Figure 2 and Figure 3.

2. The rate of increase for prestress losses is practically the same with the two type of elements (normal and lightweight concrete). After 365 days a more remarkable tendency towards stabilizing this phenomenon was noticed.

3. Comparing the values of losses of prestress due to creep and shrinkage determined on the basis of experimental data with their calculated values in conformity with /1,2/ there is a reasonable concordance.

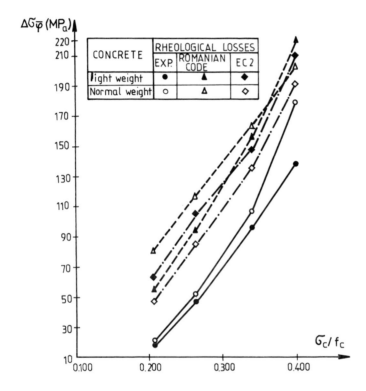

Figure 3 Experimental and calculated loss of prestress after one year, due to creep and shrinkage vs. initial stress - strength ratio

REFERENCES

1. ROMANIAN INSTITUTE OF STANDARDIZATION STAS 10107/0-90; Civil and Industrial Buildings. Design and detailing of concrete, reinforced concrete and prestressed concrete structural members, pp 116.

2. EUROCODE 2 "Design of concrete structures". Part 1: General Rules and Rules for Buildings.

3. MAGUREANU, C. ONET, T. - Rheological losses in prestressed lightweight concrete, Concrete 2000, Ed. By R.K.Dhir and Roderick Jones, 1993, pp 163-170.

4. MAGUREANU, C.- Influence of the long-term and repeated loading on fracture mechanics of partially prestressed beams, Second International Conference on Fracture Mechanics of Concrete Structures, FRAMCOS - 2, Zurich, 1995, July 25-28,

5. TAERWE, L.R.- Simplified Analysis of Stress Redistribution in Partially Prestressed Concrete Section, ACI Journal, 1990,y.30, No.1, January-February, pp 140-171.

6. ESPION, B., HALLEUX, P. - Long Term Behaviour of Prestressed and Partially Prestressed Concrete Beams: Experimental and Numerical Results, ACI-SP 129-2 pp 19-38.

7. TANIGAKI, M., NOMURA, S., OKAMOTO, T., ENDO, K. - Flexural behaviours of Partially Prestressed Concrete Beams reinforced with Braided Aramid Fiber Rods, Transaction of the Japan Concrete Institute, 1989,Vol.11, , pp 215-222

8. DEBERNARDI, P.G. - La precompressione parziale con riferimento alla normativa italiana, L'Industria Italiana del Cemento, 1988,no.4, pp 254-264.

9. CHIORINO, M.A., LACIDOGNA, G.- Design aids for creep analysis of concrete structures. (CEB Model 1990 for creep). Politecnico di Torino, Dipartimento di Ingeneria Strutturale, 1992,40, ottobre,

10. BAZANT, Z.P., KIM, S.S.,-Approximate Relaxation Function for Concrete Creep, Journal of the Engineering Mechanics Division, ASCE, 1979, Vol.106, Dec. pp 2695-2705

11. CEB-FIP MODEL CODE 1990 Final Draft. Bulletin d'Information No.203, 1991,July

MIX DESIGN CONSIDERATIONS FOR GRANULOMETRIC OPTIMISATION OF THE MATRIX OF HIGH PERFORMANCE CONCRETE

W Puntke

Wayss & Freytag AG

Germany

ABSTRACT. A new simple testing and evaluation method is presented, which permits the measurement of all granulometrically relevant properties of the most important components of concrete, including mineral binders and sand up to 5 mm, which can be regarded as the extended matrix.

Subject of the test is the water requirement at saturation point and maximum density of the grain structure. The procedure gives exact, reproducible results after a short familiarisation period and can be used for the evaluation and selection of fine components, such as cement, fly ash, stone dust, sand and silica fume as well as mixtures of these. The method reliably detects the influences of the particle size distribution <u>and</u> the particle shape. The granulometric optimum of the matrix that can be achieved with the materials available is determined using small quantities of trial mixes only, so that the expenditure of energy and time for concrete suitability tests is reduced to the performace of confirmation tests using the optimised matrix.

Diagrams illustrate how the interaction between the particle size, the particle size distribution and the particle shape can be measured by means of the proposed testing procedure.

Keywords: Water requirement; granulometry; particle shape; particle size distribution; grading; optimisation; matrix; mix design; high performance concrete.

Mr Wolfgang Puntke, Civil Engineer, is the head of the Department of Central Building Material Technology of Wayss & Freytag Aktiengesellschaft, Frankfurt/M, Germany. His main duties are to keep track of the state of the art, develop new solutions and further putting them into practise. He serves on technical committees in the German Concrete Association (DBV), the German Committee for Reinforced Concrete (DAfStb), the German Institute for Construction Technology (DIBt) and the European Committee for Standardisation (CEN).

Radical Concrete Technology. Edited by R K Dhir and P C Hewlett. Published in 1996 by E & FN Spon, 2–6 Boundary Row, London SE1 8HN, UK. ISBN 0 419 21480 1.

INTRODUCTION

Ever since concrete has been proportioned systematically, attempts have been made to find a way to proportion the components in such a way that the densest possible grain structure is achieved. In 1907 William B. Fuller and Sanford E. Thompson introduced a method for grading a mixture of cement and aggregate to produce a concrete of maximum density [1]. In 1918 Duff Abrams demonstrated the water/cement ratio law and the relationship between water requirement and fineness modulus of the aggregate [2]. Despite a great number of further proposals for improvement having been made since, these basic rules still represent the state of the art. The density of the grain structure depends on:

- grading and particle shape
- water/cement ratio
- water requirement with respect to workability and compaction procedure

The rules regarding grading set out in [1] have basically proved true. Only adjustments to cater for the desired strength, transport and workability of the concrete are necessary. The influence of the particle shape has been realised and an alternative grading with a higher fine particle content proposed for crushed aggregate. However, the problem of including the particle shape, relevant in each individual case, in the mix design considerations has remained unsolved.

The water/cement ratio law presented in [2] is the decisive design criterion for the density and strength of the hardened cement paste. The relationship between water requirement and fineness modulus of the aggregate, however, must be reconsidered. In the case of fineness modulus the water requirement is derived from the grading only, and the influence of the particle shape is not taken into account.

The present contribution introduces a new testing procedure which permits the measurement of the water requirement and maximum density of fine cohesionless components and includes the reliable determination of the influence of the grading and the particle shape in the most important fine grain range up to 5 mm, which can be understood as the extended matrix .

TESTING PROCEDURE

Contrary to all known test procedures which use exact but arbitrarily chosen compaction energy in order to obtain a dense structure, the new test procedure uses the apparent cohesion as an indicator for the saturation point and the maximum density. Damp but not saturated fine granulates cannot be compacted without load application. Only at the saturation point, when the effect of surface tension of the pore water disappears, can the fine granulate be compacted by tapping lightly. The water requirement at saturation point is determined as follows:

A certain amount of the granular material is weighed into a beaker. Then water is added step by step until the mixture, which is carefully kneaded with a spatula, shows signs of beginning to compact when the beaker is knocked on a hard surface. While the mixture is being stirred repeatedly, additional water is added in drops to achieve the saturation point, which is reached when the surface becomes dense and shiny while the material is being compacted. By weighing the material again the amount of water needed to saturate the voids in the granulates is determined and the saturated pore content of the granulate, n_w, is calculated:

$$n_w = \frac{V_w}{V_G + V_w} = \frac{\dfrac{W}{\rho_w}}{\dfrac{G}{\rho_G} + \dfrac{W}{\rho_w}}$$

n_w =   Water-saturated pore content
V_w =   Water requirement at saturation point and maximum compactness, cm³
V_G =   Volume of Granulate material weighed in, cm³
W =   Water requirement at saturation point and maximum compactness, g
G =   Weight of Granulate material weighed in, g
ρ_G =   Density of Granulate, g/cm³
ρ_w =   Density of Water, g/cm³

As it is generally sufficient to calculate with $\rho_w = 1.0$ g/cm³, the formula can be simplified:

$$n_w = \frac{W}{\dfrac{G}{\rho_G} + W}$$

Due to this evaluation any air pores which may remain in the paste/mortar are irrelevant for the test results. In principle oven-dried materials are used. The absorption water is included in the amount of water required, therefore this procedure is recommended for dense materials with slight water absorption only.

The procedure is simple, gives exact values and can be performed in 15 to 20 minutes. Often too much water is added at the first attempt. This, however, can be recognised after a few blows by the easy compactability and the completely levelled surface. The test must then be repeated using less water and adding it carefully drop by drop. After some practise the saturation point can easily be determined very exactly. The transition from "not yet compactable" to reaching the saturation point = "just compactable" often depends on one drop. A drop from a pipette weighs about 0.05 g. The substances should therefore be weighed to the nearest 0.01 g. In the case of cements the procedure correlates well with the Vicat Test, but it has the advantage that it can also be used for coarser granulates to a maximum size of 5 mm. It has proved successful for the following test objectives:

- Control of quality (uniformity), e.g. of cement, fly ash, sand and silica fume
- Determination of minimum water requirement of fine granulates and their mixtures
- Determination of minimum cementitious material requirement of mortars (sands)
- Optimisation of mixtures of given materials, selection of materials
- Research purposes such as the determination of the influence of:
  - ☐ particle shape,   ☐ particle size distribution,   ☐ improvements in both

PRACTICAL APPLICATIONS

Testing of Basic Materials

The testing of granular mixtures gives information about the water requirement at saturation point, which corresponds with the pore volume or the solid matter content of the densely compacted grain structure. This application of the procedure is suitable for fast quality controls and for checking the uniformity of granulates. In Germany, fly ash traders have used this method for years to control the uniformity of their products. Figure 1 shows test results of silica fume, cement, fly ash, glass spheres and sand. The voids in the coarse aggregate (gravel and chippings) have been tested using the Unit Weight Method.

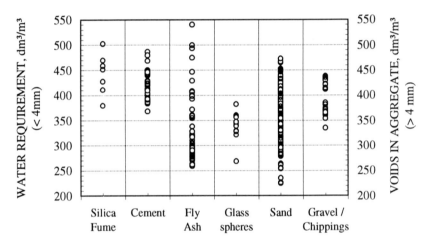

Figure 1   Water requirement at saturation point and, > 4 mm, voids in aggregate

The results give a survey of the resolution achieved by this method. However, it is not possible to assess the results, because all materials differ from each other, both in grading and particle shape.

Water Requirement With Respect to Size and Shape of Particles

For coarse aggregates, it has repeatedly been suggested that the influence of the particle shape on the voids should be determined through the density of dry-compacted uniformly fractionated particle size groups. This method is based on the observation that the voids in a completely compacted grain structure depend exclusively on the grading and the shape of the particles. If the influence of grading is eliminated by producing fractions which are identical regarding grading, the voids are only influenced by the particle shape and the effect can be measured indirectly by means of the voids. With the most favourable particle shape the voids are reduced to a minimum and with the most unfavourable one they reach a maximum. Figure 2 shows that this law is also applicable to fine grained materials.

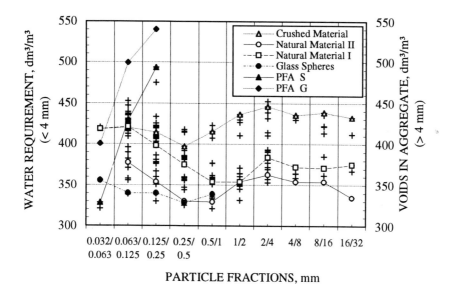

Figure 2   Water requirement at saturation point and, > 4 mm, voids in aggregate

A significant influence of the size of the particle and thus the surface of the particle on the water requirement cannot be observed despite the approximately hundredfold difference in particle size (from 32/63 µm to 2/4 mm). Moreover, the water requirement at saturation point of the fine particles corresponds with voids of the coarse particles determined by the Unit Weight Method. From this can be concluded that the water requirement in the examined size range is determined by the void content and not by the fineness and the corresponding specific surface, as was long believed.

What is striking in the case of the fly ash fractions is the considerable difference between the two represented samples of different origin on the one hand and the much larger water requirement of the "coarser" fractions on the other. The low values of the fraction 32/63 µm can be explained by their different particle shapes, the high values are effected by the water absorption of porous coke particles.

Water Requirement With Respect to Size Distribution and Shape of Particles

Mixtures of materials with similar gradings

Figure 3 shows the water requirement of different types of cement and fly ash as well as of mixtures of these two materials.

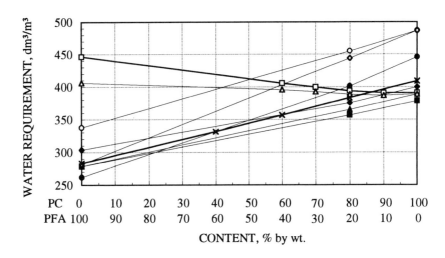

Figure 3   Water requirement of cement, fly ash and mixtures of both

The values which belong together are connected by lines. The water requirements change approximately linearly with the amounts of the materials contained in the mixture. This can be explained by the fact that both materials have a similar grading and the water requirement is initiated by the shapes of the particles according to the amount contained in the mixture.

Fly ashes whose particles are mainly round in shape reduce the water requirement according to the fly ash content of the mixture. However, if the particle shape of the fly ash is more unfavourable than that of the cement, e.g. through the production process, the water requirement increases according to the fly ash content. The water requirement of cements was found to be between 380 dm³/m³ and 480 dm³/m³, (between 38 % and 48 % by volume or appr. 20 % and 30 % by weight). Fly ashes that are approved for reinforced concrete in Germany showed water requirements from 260 dm³/m³ to 360 dm³/m³ (26 % to 36 % by volume or appr. 14 % to 26 % by weight).

Mixtures of materials with different gradings and particle shapes

Figure 4 shows the influence of some types of silica fume on the water requirement of some cements and mixtures of cement and fly ash as well as of one fly ash from smelting combustion < 63 μm. (Special measures are required for testing silica fume, but for the sake of brevity these measures are not dealt with here).
In the mixture of PFA and SILICA both components have the same round particle shape, so that the influence of the grading becomes apparent here. The minimum water requirement shows the effect of the optimum proportions.
In the case of the mixtures of cement and fly ash the content of rounded particles remained constant as the PFA was reduced, when SILICA was increased. So the dif-

ferent water requirements were exclusively a result of the change in particle size distribution. Coarse SILICA types had a lower water reducing effect or none at all. A clear reduction of the water requirement was observed in the case of the SILICA type "SI 1", which resulted from the extension of the grading in the fine particle range.

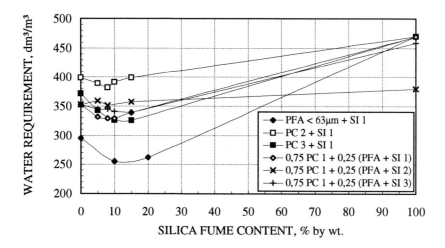

Figure 4   Influence of types of silica fume on water requirement

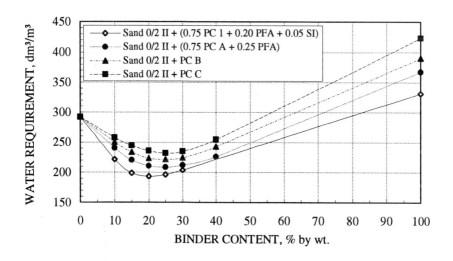

Figure 5   Water requirement of different binders and one type of sand 0/2 mm

Figure 5 shows the influence of various binders on the water requirement when they were mixed with one type of sand 0/2 mm. The binder content necessary for minimum water requirement and maximum density was found to be between 20 % and 25 % by weight. The water requirement for saturation of the mixtures varied between 193 dm³/m³ and 232 dm³/m³.

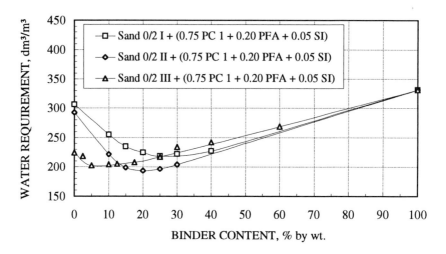

Figure 6   Water requirement of different sands 0/2 mm and one type of binder

In Figure 6 the binder with the minimum water requirement of Figure 5 is shown mixed with different types of concrete sands 0/2. The presentation allows the comparison of the influences of the sands on the water requirement and the maximum density respectively.

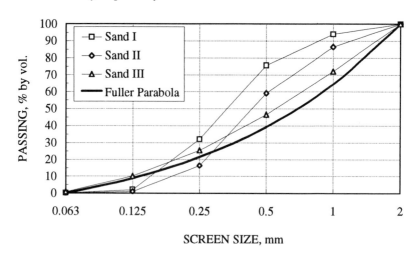

Figure 7   Grading curves of the sands used in figures 5 and 6

The gradings of the sands used for the investigations presented in the figures 5 and 6 are shown in Figure 7. In order to "simplify" the evaluation of the curves, the Fuller parabola is added. All types of sands were well-processed clean concrete sands. The sands I and II were from natural origin, sand III was crushed. Particles larger than 2 mm were removed by screening, the portions of particles smaller than 0.063 mm were below 1 % by weight.

Which type of sand is the best?

This question cannot be answered without the test results shown in figures 5 and 6 and the following evaluation.

Evaluation of the test results

The granulometric optimum of the grain skeleton alone does not automatically result in the densest structure when the mixture has hardened. With regard to the density and the strength of the hardened binder paste the water/binder ratio must be taken into consideration. In order to demonstrate what water/binder ratio can be achieved with what binder content at maximum density, the measured results represented in figures 5 and 6 have been converted accordingly. Figure 8 shows the required binder contents with respect to the water/binder ratio of the mixtures at the saturation point.

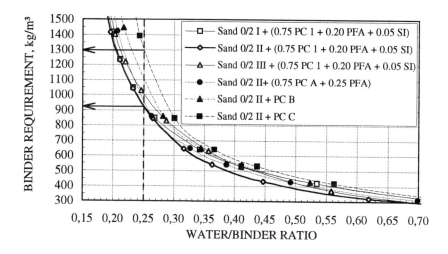

Figure 8   Binder requirement of the extended matrix 0/2 at saturation point with respect to the water/binder ratio

It can be seen that there are considerable differences between the various combinations of materials. With regard to the additional water needed for workability, high performance concretes, for example, require a water/binder ratio of around 0.30 - 0.05 = 0.25 at saturation point. It becomes apparent that some of the mixtures require an extremely high binder content. Moreover it can be recognised that in the lower range of the water/binder ratio the water requirement of the mixtures is mainly dependent on the binder. Here the water reducing influence of the silica fume becomes effective.

With the granulometric optimum (= minimum water requirement at saturation point) in figures 5 and 6 and the densest extended matrix (= mixture of binder + sand with the required water/binder ratio at saturation point), figure 8, the best mixture of the available materials can be selected. The components are determined precisely and the proportions can be read from figure 8 and calculated respectively.

The amount of water required for the saturation point is the minimum amount needed to fill the voids in the dense grain structure. In order to achieve a certain consistency additional water is required. The differences in the mixture's water re-quirement at the saturation point, however, remain as long as the water content is not so high that the granulometric differences lose their importance. Therefore a mixture with the lowest water requirement for the saturation point also requires the lowest water content for workability and compaction, and this is confirmed in figure 9. There the water requirement for workability is shown in relation to the saturation water content of the mixtures "Sand 0/2 II + (0,75 PC A + 0,25 PFA)" and "Sand 0/2 II + PC B" from figure 5 at binder contents of 25 % and 35 % by weight. The spread measures were determined in accordance with DIN EN 459, part 2.

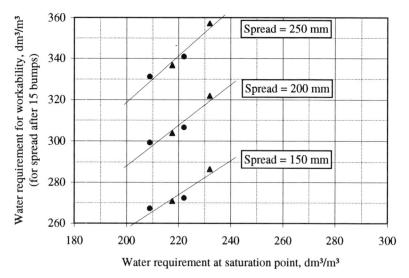

Figure 9   Water requirement for workability in relation to
the water content at saturation point

The amount of the optimised extended matrix required for the concrete mix depends on the shape and the maximum size of the coarse aggregate as well as on the workability required for the compaction procedure and the dimensions of the component parts. In order to find the best solution, concrete trial batches are necessary. With the optimised matrix however, the required decisions with regard to the concrete mix design are reduced to the questions of which and how much coarse material and, if necessary, admixture to add.

Once again, back to the question of which of the sands used for the presented investigations is the best, raised with regard to figure 7. Figure 10 shows the particle size distribution of the matrices 0/2 mm at the granulometric optimum (from figure 6) as well as at the maximum density for the water/binder ratio 0.25 (from figure 8).

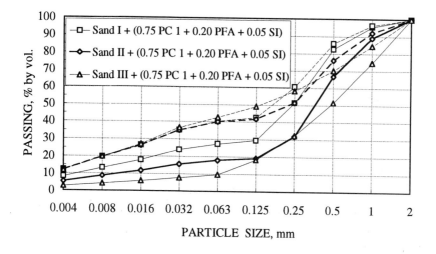

Figure 10   Grading of the extended matrices 0/2 mm including binder and sand at minimum water requirement and at a water/binder ratio of 0.25

The grading curves at the <u>granulometric optimum</u> are drawn with solid lines, those <u>at the water/binder ratio 0.25</u> with broken lines. The curves give no information about the densest structure. From figures 6 and 8 only it is known, that in both cases the maximum density was achieved with sand II, marked with bold lines.

CONCLUSIONS

1.   A new simple testing and evaluation method is presented, which permits the measurement of all granulometrically relevant properties of the most important components of concrete, including mineral binders and sand up to 5 mm, which can be regarded as the extended matrix. Subject of the test is the water requirement at saturation point and maximum density of the grain structure. Diagrams illustrate how the interaction between the particle size, the particle size distribution and the particle shape can be measured by means of the proposed testing procedure.

2. In mixtures of materials with <u>similar gradings</u> it is neither the fineness nor the corresponding surface of the particles that is decisive for the water requirement, but the particle shape exclusively (Figure 2). The water requirement changes approximately linearly with the characteristic particle shape. The mixtures have no optimum, but the effect with respect to water requirement and density can be recognised (Figure 3).

3. In mixtures of materials with <u>different gradings</u> and particle shapes the influence of an extended grading effects a reduction in water requirement down to a clear optimum, which indicates the optimum proportions (Figures 4, 5 and 6).

4. On the basis of the granulometric optimum the best mixture of the best available materials can be selected. The components of the extended matrix are determined precisely and the proportions for minimum water requirement and maximum density can be calculated taking into consideration the water/binder ratio (Figure 8).

5. The amount of the optimised extended matrix (mortar) required for the concrete mix depends on the shape and the maximum size of the coarse aggregate as well as on the workability required for the compaction procedure. In order to find the best solution, concrete trial batches are necessary. With the optimised matrix however, the required decisions with regard to the concrete mix design are reduced to the questions, of which and how much coarse material and, if necessary, admixture to add.

REFERENCES

1. Fuller, William B., and Thompson, Sanford E., "The Laws of Proportioning Concrete," Transactions, ASCE, V. 59, Dec. 1907, pp. 67-143.

2. Abrams, Duff A., "Design of Concrete Mixtures," Bulletin No. 1, Structural Materials Research Laboratory, Lewis Institute, Chicago, 1918, 20 pp.

3. Puntke, Wolfgang, "Grundlagen für die Matrixoptimierung und Ausführung in der Praxis," Matrixoptimierung mikroskopisches Analysieren von Beton, Technische Akademie Esslingen, TAE, 1990, Lehrgang Nr. 12889/81.035.

4. Central Building Material Technology, Wayss & Freytag AG, Frankfurt/M., Germany, internal Reports:
   - Investigations for Selection of Construction Materials for the Størebaelt East Bridge Project, Denmark, December 1992
   - Determination of a Mix Design for Grout to fill the Annular Cavity on the St.-Clair-River-Tunnel Project between Sarnia Ontario, Canada and Port Huron Michigan, USA, April 1993
   - Investigations for Selection of Construction Materials for the Xiaolangdi Multipurpose Dam Project, People's Republic of China, February 1995

HIGH PERFORMANCE CONCRETE STRUCTURES - A SWEDISH RESEARCH PROGRAM

L Elfgren

T Olofsson

M Emborg

J E Jonasson

Lulea University of Technology,

Sweden

ABSTRACT. A joint program for six years 1991/92 - 1996/97 with a total budget of about five million pounds (53 MSEK) is funded by six industry companies and two Swedish Research Foundations. The aim of the program is to develop (a) more efficient structures, (b) better production methods, and (c) more durable materials.
Some results from the structural part of the program will be highlighted here. Main investigations are carried out regarding (I) toughness and ductility, (II) bond, and (III) cracking in young concrete.

Keywords: Concrete, Structures, Ductility, Brittleness, Toughness, Fracture Mechanics, Bond, Cracking, Beams, Columns, High Performance Concrete.

Professor Lennart Elfgren is Head of the Division of Structural Engineering, Luleå University of Technology, Sweden. His research specialities include fracture mechanics, fatigue, fasteners, and combined torsion, bending, and shear. He is coordinator for the projects on structures in the Swedish R&D program on high performance concrete structures.

Dr Thomas Olofsson is an Associate Professor in Structural Engineering at Luleå University of Technology. His research speciality is computer modelling of materials and structures.

Dr Mats Emborg is an Associate Professor in Structural Engineering at Luleå University of Technology. His research specialities include modelling of young concrete, fiber reinforcement and fatigue of concrete structures.

Dr Jan-Erik Jonasson is an Associate Professor in Structural Engineering at Luleå University of Technology. His research speciality is modelling of thermal and moisture conditions in concrete structures.

Radical Concrete Technology. Edited by R K Dhir and P C Hewlett. Published in 1996 by E & FN Spon, 2–6 Boundary Row, London SE1 8HN, UK. ISBN 0 419 21480 1.

INTRODUCTION

A Swedish Research and Development Program on High Performance Concrete was started in 1991. Six major companies are funding it namely Cementa, Elkem, Euroc Beton, NCC, Skanska and Strängbetong together with two Research Councils namely the Swedish Council for Building Research (BFR) and the Swedish National Board for Industrial and Technical Development (NUTEK). The annual funding is about 0.8 MGBP (9 MSEK). Research is carried out at the Cement and Concrete Institue (CBI) in Stockholm and at the universities in Göteborg, Lund, Luleå and Stockholm (CTH, LTH, LuTH and KTH).

In this paper some results will be presented regarding the structural parts of the program, which encompass about one third of it. The other two thirds of the program deals primarily with questions regarding materials and production. General information of the program is given by Jan Byfors 1993 [1] and by Lennart Elfgren, Göran Fagerlund and Åke Skarendahl 1995 [2].

The program on structures was started with a literature study which is summarized in Gabrielsson 1993 [3]. Main projects were then started regarding toughness and ductility, bond, and cracking in young concrete.

TOUGHNESS AND DUCTILITY

General

A basic parameter in fracture mechanics is the *brittleness number B*. It can be defined in the following way. Let us study a tensile test of a concrete prism, see Fig. 1. Up to the maximum load (with *stress f_t,* and *deformation $\delta_E = \varepsilon_t L$*) the prism basicly behaves in an elastic way (*strain $\varepsilon_t = f_t / E$*). After maximum, a narrow fracture zone (FZ) deforms further under falling load. At the same time the material outside the fracture zone is relieved elastically - largely following the first curve back to the origin. The area under the descending curve is defined as the *fracture energy G_F* which is needed in order to separate the prism into two parts. A characteristic *failure zone deformation Δ* can also be defined as $\Delta = G_f / f_t$.

Structures can be defined as *brittle* when the elastic deformation δ_E dominates, whereas the behaviour can be defined as *ductile* when the deformation of the fracture zone D dominates. The *brittleness number B* can be defined as

$$B = \delta_E / \Delta = \varepsilon_t L / \Delta = f_t^2 L / EG_F$$

The reciprocal value *1/B* can be named the *ductility number*. It can be seen that the brittleness/ductility deponds on the *length L*, the *tensile strength f_t,* the *modulus of elasticity E*, and the *fracture energy G_F*. The brittleness number is also proportional to the ratio of elastic to fracture energy:

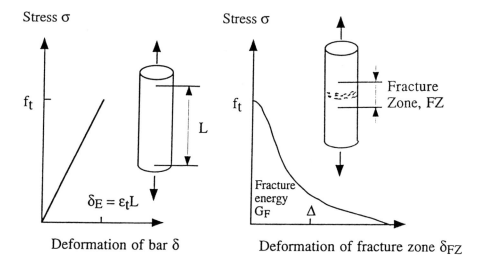

Figure 1. Loading of a concrete prism. Definition of basic parameters for ductility and brittleness. From Bache 1995 [4], modified.

Elastic energy / Fracture energy = $0.5 f_t \delta_E / G_F = 0.5 f_t^2 L / E G_F \sim B$

The factor $E G_F / f_t^2$ is a material parameter which was introduced by Hillerborg 1976, 1983 [5], [6] as the *characteristic length*, l_{ch}. The *brittleness number* was introduced in the form it is given here in the 80-ies by Bache 1995 [4], see Elfgren 1989 [7]. A basic fracture mechanics philosophy is to relate the strength of an object to its brittleness number B or to the components of B i e the length L, the tensile strength f_t, the modulus of elasticity E, and the fracture energy G_F.

This way of describing the tensile fracture is now beginning to be introduced in modern design codes. In e g the CEB-FIP Model Code 1990 [8] values are given for the fracture energy G_F [Nm/m^2] and a bi-linear stress - crack-opening diagram is proposed for concrete in tension. However, in most traditional codes, e g Eurocode EC-2 [9], not much can be seen except some empirical formulae for size effect influences.

Applications to beams, columns and piles

A series of tests on **rectangular reinforced beams** with various concrete strengths has been carried out by Henrik Gabrielsson 1993 [3]. Analysis with a compression field theory gave better results than an analysis with a conventional theory.

Ductility of high performance **prestressed concrete cylindrical elements** tested in pure torsion has been studied by Henrik Gabrielsson, see Figure 2. The elements are used as transmission poles for electric power. It can be seen from Figure 2 that also a high strength concrete can give quite a ductile behaviour if the reinforcement is arranged properly. Similar results regarding **rotational capacity of beams** have been reached in a project carried out at the Royal Institute of Technology (KTH) by Håkan Fransson and Sven Kinnunen.

Fatigue of reinforced and prestressed beams is studied by Robert Danewid and Sven Thelandersson at Lund Institute of Technology. In Lund also **fiber reinforced** structures is studied by Manoucheer Hassanzadeh.

Columns in compression have been tested and analysed by Marianne Grauers 1993 [10] and by Christina Claeson 1995 [11]. A fracture mechanics approach has here been fruitful in the analysis of the results.

Prestressed concrete piles are being studied by Gunnar Holmberg 1995 [12].

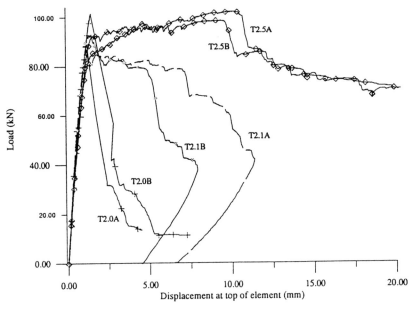

Figure 2. Load-displacement curves for prestressed concrete cylinders loaded in pure torsion. The hollow cylinders had a length of 3 m, an outer diameter of 0.3 m, a wall thickness of 0.07 m and a concrete compression strength $f_{cc} \approx 150$ MPa (100 mm cubes).The characteristic length l_{ch} was 0.2 m for the material in these elements compared to $l_{ch} = 0.4$ m for ordinary commercially used elements with a compression strength $f_{cc} \approx 100$ MPa. Specimen T2.0A and B had no spiral reinforcement while T2.1A and B and T2.5A and B had spiraling 5.5 mm bars with a pitch of 100 and 50 mm respectively. From Gabrielsson and Elfgren 1995 [13].

BOND

An intersting bond splitting model has been developed and tested by Keivan Noghabai 1995 [14]. The model is based on the assumption that the bond stresses around a reinforcement bar give rise to a hydrastic pressure on the concrete. The pressure is a function of the geometry of the bars and the size of the applied force. Some test results are given in Figures 3 and 4 and a comparison with the presented and some other analytical models is given in Figure 5. It can be seen that high strength concrete is more brittle than normal strength concrete but still can be used to good advantage. The results have also be analysed with a Finite Element Method using inner softening bands. Here the localazation of the cracks could be followed. First many small cracks appeared but after a while some of them closed while others grew wider. At failure only a few rather wide cracks remained. As can be seen from Figure 3 and 4 also the influence of various amounts of spiral reinforcement was investigated.

Bond and anchorage of deformed bars is also studied by Jonas Magnusson et al 1995 [15]. Anchor bolts are being investigated by Ulf Ohlsson et al 1995 [16], [17], [24].

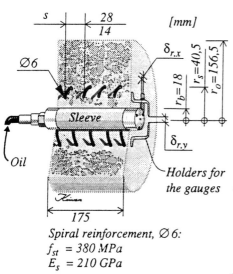

Spiral reinforcement, ⌀ 6:
f_{st} = 380 MPa
E_s = 210 GPa

Component	NSC	HSC	VHSC
Coarse aggregate	818	1149	1060
Fine aggregate	980	770	750
Cement	343	450	540
Silica fume	-	70	54
Superplastisizer	-	6,3	8,7
water/cement ratio	0,56	0,35	0,25
Material parameters			
f_{cc}^{cube} [MPa]	57,0	105,0	157,4
f_{ct}^{cube} [MPa]	3,8	5,0	8,0
G_f^{RILEM} [Nm/m²]	105	145	172
E_c [GPa]	33,8	39,4	41,2

Figure 3. Tests on bond splitting. The tables shows the concrete recipies (dry material kg/m³) used in the splitting tests denominated as Normal Strength Concrete, NSC, High Strength Concrete, HSC, and Very High Strength Concrete, VHSC. From Noghabai 1995 [14].

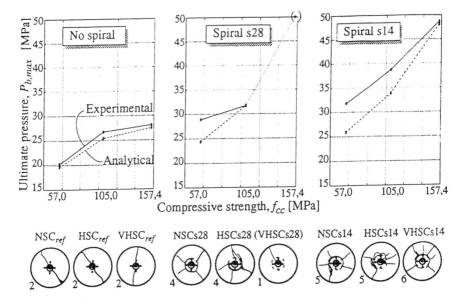

Figure 4. Comparison of test results from splitting tests and an analytical model. The patterns and number of visible cracks on the front surface of the specimens are also illustrated. The results from VHSC s28 are not to be trusted. From Noghabai 1995 [14].

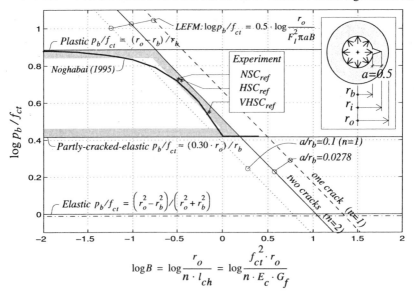

Figure 5. Ultimate relative splitting pressures given by different models for a concrete ring ($r_b = 18$ and $r_o = 156.5$ mm) as a function of the brittleness number B. The coefficient F_I is depending on a, r_b and r_o and is determined with linear elastic fracture mechanics (LEFM). NSC, HSC and VHSC refer to the tests with normal, high, and very high strength concrete described in Figures 3 and 4. From Noghabai 1995 [14], modified.

CRACKING IN YOUNG CONCRETE

Cracking is studied in young concrete structures. Time-dependent gradients of temperature and moisture can give tensile stresses that exceed the tensile strength. An extensive program is being carried out in order to establish materials data and models suitable for computer analysis. In this way tools are beeing made availble with which the hardening technology can be mastered and unwanted cracking can be eliminated by proper procedures. Main results have been presented by Mats Emborg and Stig Bernander 1990 [18], 1995 [19], Jan-Erik Jonasson 1994 [20], and Gustaf Westman 1995 [21]. Some recent results cocerning through cracking at the cooling of a structure are given in Figures 6 and 7. It is of the utmost importance for an accurate thermal stress analysis, that correct modelling of the mechanical behavior is performed. It is essential to study the overall mechanical behaviour and it is not enough to model separate properties as creep, strength development etc. as complicated coupling effects ocur when young concrete is heated.

Figure 6. Relaxation test frame. From Westman 1995 [21].

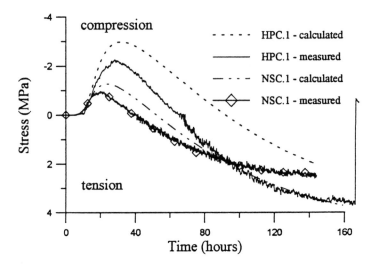

Figure 7. Computed thermal stresses in a 0.7 m thick wall subjected to 100% external restraint if cast using a High Performance Concrete, HPC.1 (water-binder ratio w/B = 0.34, compressive strength f_{cc} = 92 MPa) or a Normal Strength Concrete, NSC.1, (water-binder ratio w/B = 0.40, compressive strength f_{cc} = 51 MPa) compared to corresponding test results with test set up according to Fig.6. From Westman 1995 [21].

Recent tests with an energetically modified cement has also given interesting results regarding e.g. winter concreting. Silica fume is here activated together with cement in a mechanichal-chemical milling process which increases the surface energy of the binder, Vladimir Ronin and Jan-Erik Jonasson 1994 [22]. The activated cement leads to lower binder contents and higher workability.

Steel fibre reinforcement is one way to control cracking and to enhanche toughness. Tests performed by Patrik Groth and Keivan Noghabai 1996 [23] indicate a clear tendency towards a plastic behaviour in bending and splitting tests with a fiber volume content of 1 to 2 %.

CONCLUSIONS

High performance concrete is an efficient material which can be used in many structural applications with good economy. The Swedish R& D program is giving valuable contributions to enable safe and efficient design procedures for structures as beams, columns, transmission poles, piles and hollow core slabs. An essentail prerequisite for modelling of toughness/brittleness of structures is fracture mechanics properties as the fracture energy G_F, modulus of elasticity E_C and tensile strength f_t.

ACKNOWLEDGEMENTS

The contributions from our colleagues in the program are acknowledged as well as the funding from Cementa, Elkem, Euroc Beton, NCC, Skanska, Strängbetong, the Swedish Council for Building Research (BFR) and the Swedish National Board for Industrial and Technical Development (Nutek).

REFERENCES

1   BYFORS, J. An R&D program on high performance concrete - Object, scope and resources (Programmet "Högpresterande betong" - Idé, mål och medel. In Swedish). CBIs Informationsdag 1993, Sammanfattningar, Stockholm 1993, pp 9-25.

2   ELFGREN, L, FAGERLUND, G, AND SKARENDAHL, Å. Swedish R&D Program on High Performance Concrete. To be published by ACI in the proceedings from the Bangkok 1994 Conference on High Performance Concrete, Detroit 1995, 10 pp.

3   GABRIELSSON, H. Shear capacity of beams of reinforced high performance concrete. Licentiate Thesis 1993:21 L, Division of Structural Enginering, Luleå University of Technology, Luleå 1993, 111 pp.

4   BACHE, H. H. Concrete and concrete technology in a broad perspective. Nordic symposium on modern design of concrete structures (ed K. Aakjær), Dept of Build Techn & Struct Eng, Aalborg University, Denmark, 1995, pp 1-45 (ISSN 0902 7513 R9513)

5   HILLERBORG, A. Analysis of one single crack. Chapter 4.1 in Fracture mechanics of concrete (ed F.H. Wittmann), Elsevier, Amsterdam, 1983, pp 223 - 249 (ISBN 0 444 42199 8).

6   HILLERBORG, A, MODÉER, M, PETERSSON, P-E. Analysis of crack formation and crack growth in concrete by means of fracture mechanics and finite elements. Cement and concrete research, Vol 6, 1976, pp 773-782.

7   ELFGREN, L.(Editor). Fracture mechanics of concrete structures. From theory to applications. Chapman & Hall, London, London 1989, 407 pp (ISBN 0 412 30680 8).

8   CEB-FIP MODEL CODE 1990. Design Code, Comité Euro-International du Béton and Féderation International de la Precontrainte, Thomas Telford, London 1993, 437 pp (ISBN 0 7277 1696 4).

9   EC-2 Eurocode 2: Design of concrete structures - Part 1: General rules and rules for buildings. European prestandard ENV 1992-1-1:1991, CEN (Comité Européen de Normalisation), Brussels, 1991, 253 pp.

10 GRAUERS, M. Composite columns of hollow steel sections filled with high strength concrete. Doctoral thesis, Publ 93:2, Div of Concrete Structures, Chalmers Univ of Technology, Göteborg 1993, 140+96 pp.

11 CLAESON, C. Behavior of reinforced high strength concrete columns. Licentiate thesis, Publ 95:1, Div of Concrete Structures, Chalmers Univ of Technology, Göteborg 1995, 54 pp.

12 HOLMBERG, G AND CEDERWALL,K. Fatigue of concrete piles of high strength concrete exposed to impact load. Nordic symposium on modern design of conctrete structures, (ed K Aakjær), Dept of Build Techn & Struct Eng, Aalborg University, Aalborg 1995, pp 313-318.

13 GABRIELSSON, H AND ELFGREN, L. Cylindrical high performance prestressed concrete elements tested in pure torsion. Nordic symposium on modern design of conctrete structures, (ed K Aakjær, Dept of Build Techn & Struct Eng, Aalborg University, Aalborg 1995, pp 319-330.

14   NOGHABAI, K. Splitting of concrete in the anchoring zone of deformed bars. A fracture mechanics approach to bond. Licentiate thesis 1995:26L, Div of Struct Eng, Luleå Univ of Techn, Luleå, 1995, 131 + 56 pp.

15   MAGNUSSON, J, ENGSTRÖM, B AND GYLLTOFT, K. Bond and anchorage of deformed bars in high strength concrete. Nordic symposium on modern design of conctrete structures (ed K Aakjær), Dept of Build Techn & Struct Eng, Aalborg University, Aalborg 1995, pp 75-84.

16   OLOFSSON, T, NOGHABAI, K, OHLSSON, U, and ELFGREN, L. Anchorage and bond properties in concrete. Fracture of brittle disordered materials: concrete, rock and ceramics (eds G. Baker and B.L. Karihaloo), E & FN Spon, London, 1995, pp 525-543 (ISBN 0 419 19050 3)

17   OHLSSON, U AND OLOFSSON, T. Anchor bolts in concrete structures - Finite element calculations based on inner softening bands. Fracture mechanics of concrete structures (ed F H Wittmann), Aedificatio, Freiburg 1995, pp 1545-1554 (ISBN 3 905088 14 2)

18   EMBORG, M. Thermal stresses in concrete structures at early ages. Doctoral Thesis 1989:73D, Div of Struct Eng, Luleå Univ of Techn, 2nd Ed, Luleå, 1990, 285 pp.

19   EMBORG, M AND BERNANDER, S. Assesment of risk of thermal cracking in hardening concrete. Journal of Structural Engineering, Vol 120, No 10, New York, October 1994, pp 2893-2912.

20   JONASSON, J-E. Modelling of temperature, moisture and stresses in young concrete. Doctoral Thesis 1994:153D, Div of Struct Eng, Luleå Univ of Techn, Luleå, 1994, 225 pp

21   WESTMAN, G. Thermal cracking in high performance concrete. Viscoelastic models and laboratory tests. Licentiate thesis 1995:27L, Div of Struct Eng, Luleå Univ of Techn, Luleå 1995, 120 pp. See also EKERFORS, K. Maturity development in young concrete. Temperature sensitivity, strength and heat development (In Swedish with a summary in English). Licentiate thesis 1995:34L, Div of Struct Eng, Luleå Univ of Techn, Luleå 1995, 136 pp.

22   RONIN, V AND JONASSON, J-E. Investigation of the effective winter concreting with the usage of energetically modified cement (EMC) - material science aspects. Report 1994:03, Div of Struct Eng, Luleå Univ of Techn, Luleå 1994, 24 pp.

23   GROTH, P AND NOGHABAI, K. Fracture mechanics properties of steel fibre-reinforced high-performance concrete. Div of Struct Eng, Luleå Univ of Techn, Luleå 1996, 10 pp. (To be presented at the 4th Int Symposium on Utilazation of High-strength/High-performance Concrete, Paris, May 1996).

24   OHLSSON, U. Fracture mechanics analysis of concrete structures. Doctoral Thesis 1995:179D, Div of Struct Eng, Luleå Univ of Techn, Luleå, 1995, 94 pp.

FUZZY COMPREHENSIVE EVALUATION ON HPC

L Zhiguo

Tianjin University

N Changren

Harbin University of Architecture and Engineering

L Zhongxian

Tianjin University

China

ABSTRACT. The term-high performance concrete (HPC) has a determinate intention and non-determinate extension. It is difficult to distinguish and evaluate its properties. In this paper, A new concept "Fuzzy", has been introduced to define the term-HPC and an available theoretical method has been suggested to distinguish and evaluate the complex properties, as well as to optimize the multiple properties of the HPC being used in engineering.

Key words: fuzzy, comprehensive evaluation, high performance concrete

Mr Zhiguo, L. is a master of Eng. and the director of Building Materials Lab. Civil Eng. Dept. Tianjin Univ. P. R. China. He specializes in durability and protection of concrete.

Mr Changren, N. is a professor of concrete science, Building Materials Dept. Harbin Univ. of Arch. and Eng. P. R. China.

Mr Zhongxian, L. is a doctor of Eng. and an assistant Prof. of concrete science,Civil Dept. Tianjin Univ. P. R. China.

Radical Concrete Technology. Edited by R K Dhir and P C Hewlett. Published in 1996 by E & FN Spon, 2–6 Boundary Row, London SE1 8HN, UK. ISBN 0 419 21480 1.

INTRODUCTION

Scientists formulated the concept of High Performance Concrete (HPC)[1] around 1986. A Japanese engineer's report[2] shows that the service life of HPC could be as long as 500 years. The concept of HPC is difficult to define even now that it is being used. There are different opinions about which grades of properties are acceptable.

Scientists and investigators all over world agree in principle that engineers will be more satisfied with HPC's super workability, higher level strength, stronger durability, suitable volume change properties and rather low price. It is understandable that HPC is suitable for civil engineering because it withstands severe environments yet provide more durable service. It is suitable for modern constructional technique such as pre-mixed concrete or concrete pumping with super flowability because it flattens by itself without terrible noise and harmful vibration. The resulting product is high strength and high quality concrete.

In fact, the understanding of the concept of HPC is different. It can be divided into several schools of the thought as (a) P.K.Mehta[3] (b)Y.Malier[4], (c).H.Okamura[5], but the Synthesis of these ideas seems to be more acceptable.

The real question is, what does HPC mean for practical constructional engineering. How should the multiple properties of HPC be distinguished, evaluated and optimized in engineering? We answer this question we have to analyze the characteristics of the concept of HPC.

FUZZY CONCEPT: THE DEFINITION OF HPC AND ITS EVALUATIVE FACTORS.

The definition of HPC and its evaluative factor are fuzzy matter, HPC is non-determinative matter that is quite different form HSC(high strength concrete).That is to say, it is hard to describe the properties of HPC with determinative mathematics. Although each individual property of HPC is relative depending on the structure of the concrete, it is not clear which kinds of properties could best represent the characteristics of HPC. There is no agreement so far, on how we can consider the effect of the combination of the multiple factors on the properties of HPC. Many discussions about this are continuing. All the uncertainty is because the concept and its evaluative factors are fuzzy matter. This means they have determinative intention, but non-determinative extension. In this way we can say that the definition and its evaluation of HPC are the attitude of degrees of satisfaction of the engineer who is responsible for the constructional engineering.

"Fuzzy" is a new branch of mathematics. It deals with the relationships of non-determinative matters. The difference between the objective matters transforming continually in media with the properties of "true or false", or "both this and that" are the characteristics of Fuzzy. There is no relative clear boundary between each concept describing such things. Such a concept has no very clear extension, but has very clear intention. For instance, every one could understand the questions as:

"Concerning the properties, what can be called high performance or low performance?"
"At which level or grade can we define it as high or higher?"
"How can we choose the synthetic properties for the practical construction?"

In order to solve these problems, the concept of fuzzy has to be used since determinative mathematics is helpless. Sometimes we simply can't explain well some concepts such as HPC by traditional methods using certain quantity analysis.

Considering the properties of HPC such as workability, strength, volume change, durability as well as economics, we find the concept of HPC could be qualitatively shown graphically as follows [equation(1)or figure 1]by means of fuzzy:

$$\text{HPC} \in [A(HSC), B(HWC), C(HDC), D(HVC), E(ECF)] \tag{1}$$

where: HSC—high strength concrete HWC—high workability concrete
HDC—high durability concrete HVC— high suitable volume change
                                                 concrete
ECF—economic control factor $A\sim E$—relationship factors

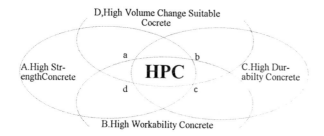

Figure 1   Graphically show the concept of HPC

where the graph abcd may be not in the shape of regular polygon, because probability the weights of each factors are not the same.

FUZZY COMPREHENSIVE EVALUATION ON THE PROPERTIES OF THE HPC

1. The method of evaluation

When we evaluate HPC, we would choose the properties that act as evaluative factors as follows: The multiple property of the HPC effected by W-workability; S-strength; D-durability; V-volume change; E-economics. then, we can get the multiple property of the HPC and its effective factors equation

$$M = f(x_1, x_2, x_3, x_4, x_5) \tag{2}$$

where x1,x2,x3,x4,x5 represent the effect of the factors:workability, strength, volume change, durability, and economics.

For the multiple property of the HPC, we suggest an evaluative set, that is the degree of satisfactions of the engineers dealing with this kind of concrete.

$$M = (c_1, c_2, c_3, c_4) \tag{3}$$

where the c_1, c_2, c_3, c_4 represent the multiple properties of the HPC as "higher", "high", "common", "bad" separately.

In order to evaluate the effects of Xi on M, we give Xi a fuzzy comprehensive evaluation separately, that is Xi monofactor evaluation:

$$X_1 = (a_{11}, a_{12}, a_{13}, a_{14})$$
$$X_2 = (a_{21}, a_{22}, a_{23}, a_{24})$$
$$X_3 = (a_{31}, a_{32}, a_{33}, a_{34})$$
$$X_4 = (a_{41}, a_{42}, a_{43}, a_{44})$$
$$X_5 = (a_{51}, a_{52}, a_{53}, a_{54})$$

where a_j is the degree of the satisfaction of factor xi on M, here in fact a_{ij} is the factors (W.S.V.D.E) evaluative ratio on the multiple properties M.

Therefore, we have the matrix to evaluate the properties on the HPC.

$$X = \begin{bmatrix} a_{11} & a_{12} & a_{13} & a_{14} \\ a_{21} & a_{22} & a_{23} & a_{24} \\ a_{31} & a_{32} & a_{33} & a_{34} \\ a_{41} & a_{42} & a_{43} & a_{44} \end{bmatrix} \tag{4}$$

Practically, for different civil constructional engineering used HPC, the weights of each factor (W,S,V,D,E) is quite different. We can make a decision to support the weight factor by experience or by testing analyses:

effective factors: $x_1\ x_2\ x_3\ x_4\ x_5$
weight   factors: $b_1\ b_2\ b_3\ b_4\ b_5$

then, we have the fuzzy set of the weights:

$$P = \begin{pmatrix} b_1 & b_2 & b_3 & b_4 & b_5 \end{pmatrix} \tag{5}$$

here, we have the fuzzy weight factor evaluation matrix:

$$M = P \cdot X = \begin{pmatrix} b_1 & b_2 & b_3 & b_4 & b_5 \end{pmatrix} \begin{bmatrix} a_{11} & a_{12} & a_{13} & a_{14} \\ a_{21} & a_{22} & a_{23} & a_{24} \\ a_{31} & a_{32} & a_{33} & a_{34} \\ a_{41} & a_{42} & a_{43} & a_{44} \\ a_{51} & a_{52} & a_{54} & a_{55} \end{bmatrix} \tag{6}$$

After the fuzzy algorithmic approach(8), we will have:

$$M = \begin{pmatrix} c_1 & c_2 & c_3 & c_4 \end{pmatrix} \tag{7}$$

Taking the data from eq., (7) by maxim criterion and compare with eq., (3), we can decide the degree of the HPC, whether "higher", "high", "common", "bad", according to the position of the maxim data located.

2. Summary of the fuzzy evaluation

(1) decide the effective factor set: $U = \begin{pmatrix} u_1 & u_2 & u_3 & \cdots\cdots & u_n \end{pmatrix}$ \hfill (8)

(2) support evaluative set: $V = \begin{pmatrix} v_1 & v_2 & v_3 & \cdots\cdots & v_n \end{pmatrix}$ \hfill (9)

(3) take the weighting fuzzy set of the factors effected on the U.

$$R = \begin{pmatrix} r_1 & r_2 & r_3 & \cdots\cdots & r_n \end{pmatrix} \tag{10}$$

(4) give mono-factor evaluate to ,corresponding to V:

$$U_i = \begin{pmatrix} u_{i1} & u_{i2} & u_{i3} & \cdots\cdots & u_{ij} & \cdots\cdots & u_{in} \end{pmatrix} \tag{11}$$

where u_{ij} is the factor (i)attaching grade of the evaluative level (j).then, we have the mono-factor evaluative matrix.

$$U = \begin{bmatrix} u_{11} & u_{12} & \cdots & u_{1j} & \cdots & u_{1n} \\ \vdots & \vdots & & \vdots & & \vdots \\ u_{i1} & u_{i2} & \cdots & u_{ij} & \cdots & u_{in} \\ \vdots & \vdots & & \vdots & & \vdots \\ u_{m1} & u_{m2} & \cdots & u_{mj} & \cdots & u_{mn} \end{bmatrix} \tag{12}$$

(5) The fuzzy comprehensive evaluation is:

$$U = R \cdot U = (r_1 \quad r_2 \quad \cdots \quad r_n) \begin{bmatrix} u_{11} & u_{12} & \cdots & u_{1j} & \cdots & u_{1n} \\ \vdots & \vdots & & \vdots & & \vdots \\ u_{i1} & u_{i2} & \cdots & u_{ij} & \cdots & u_{in} \\ \vdots & \vdots & & \vdots & & \vdots \\ u_{m1} & u_{m2} & \cdots & u_{mj} & \cdots & u_{mn} \end{bmatrix} \qquad (13)$$

After applying the fuzzy algorithmic approach, we have the fuzzy comprehensive evaluation set:

$$U = (c_1 \quad c_2 \quad \cdots\cdots \quad c_n) \qquad (14)$$

Analyze the result of the eq(14) and we can finish the evaluation.

EXAMPLE

There are 3 kinds of concrete; about which the properties are shown in table

Table 1 The properties of HPC

Type of HPC	Slump(slump -flow) mm	R_{28} N/mm²	Permeability N/mm²	Shrinkage $\varepsilon \times 10^{-5}$	Cost RMB/m³
HPC-A	270(560)	52	3.1	68	305
HPC-B	240(520)	64	4.2	60	364
HPC-C	150(400)	50	4.0	84	321

In order to evaluate and optimize the concrete shown in table 1, we suggest that 10 engineers, for instance, be invited to give their evaluations on the properties of the HPC above. The method of evaluation for these engineers is to give marks on each property in the blanks corresponding to c1,c2,c3,c4, which represent engineer's degree of satisfaction about "higher" , "high" ,"common" or "bad" separately. Thus we get the experts' ideas as in table 2 after exchanging these marks into ratios:

Table 2 Experts ideas on the properties of the HPC

Evalu ative	Types of high performance concrete and its evaluations											
	HPC-A				HPC-B				HPC-C			
W	0.7	0.2	0.1	0	0.6	0.3	0.1	0	0.1	0.4	0.5	0
S	0.5	0.5	0	0	0.7	0.3	0	0	0.4	0.6	0	0
D	0	0.1	0.4	0.5	0.5	0.4	0.1	0	0.3	0.4	0.3	0
V	0.3	0.4	0.3	0	0.4	0.3	0.3	0	0	0	0.6	0.4
E	0	0.2	0.4	0.4	0	0.1	0.6	0.3	0	0.5	0.5	0
Evalu ation	c_1	c_2	c_3	c_4	c_1	c_2	c_3	c_4	c_1	c_2	c_3	c_4

Here, we have the function:

$$M = f_{(W,S,D,V,E)} = f_{(x_1 \ x_2 \ x_3 \ x_4 \ x_5)} \tag{15}$$

Then, the mono-factor evaluative set is:

$$X_A = \begin{bmatrix} 0.7 & 0.2 & 0.1 & 0.0 \\ 0.5 & 0.5 & 0.0 & 0.0 \\ 0.0 & 0.1 & 0.4 & 0.5 \\ 0.3 & 0.4 & 0.3 & 0.0 \\ 0.0 & 0.2 & 0.4 & 0.4 \end{bmatrix} \tag{16}$$

$$X_B = \begin{bmatrix} 0.6 & 0.3 & 0.1 & 0.0 \\ 0.7 & 0.3 & 0.0 & 0.0 \\ 0.5 & 0.4 & 0.1 & 0.0 \\ 0.4 & 0.3 & 0.3 & 0.0 \\ 0.0 & 0.1 & 0.6 & 0.3 \end{bmatrix} \tag{17}$$

$$X_C = \begin{bmatrix} 0.1 & 0.4 & 0.5 & 0.0 \\ 0.4 & 0.6 & 0.0 & 0.0 \\ 0.3 & 0.4 & 0.3 & 0.0 \\ 0.0 & 0.0 & 0.6 & 0.4 \\ 0.0 & 0.5 & 0.5 & 0.0 \end{bmatrix} \tag{18}$$

Considering the experiences and the results of discussions of experts, we have the weight factors set:

$$P = (b_1 \ \ b_2 \ \ b_3 \ \ b_4 \ \ b_5) = (0.40 \ \ 0.05 \ \ 0.20 \ \ 0.25 \ \ 0.10) \tag{19}$$

therefore, we have the fuzzy comprehensive evaluative equations:

$$M = P \cdot X_A = (0.4 \ \ 0.05 \ \ 0.25 \ \ 0.2 \ \ 0.1) \begin{bmatrix} 0.7 & 0.2 & 0.1 & 0.0 \\ 0.5 & 0.5 & 0.0 & 0.0 \\ 0.0 & 0.1 & 0.4 & 0.5 \\ 0.3 & 0.4 & 0.3 & 0.0 \\ 0.0 & 0.2 & 0.4 & 0.4 \end{bmatrix} \tag{20}$$

then

$$M_A = (0.445 \quad 0.305 \quad 0.107 \quad 0.060) \tag{21}$$

$$M_B = (0.475 \quad 0.310 \quad 0.225 \quad 0.060) \tag{22}$$

$$M_C = (0.240 \quad 0.320 \quad 0.520 \quad 0.100) \tag{23}$$

As the result, we can believe that, by the principle of maxim criterion and according to the equation (3),(21),(22),(23),the multiples of the high performance concrete HPC-A, HPC-B is of the grade "higher", HPC-C is of the grade "common", while the HPC-B is the best one of these 3 kinds of concrete.

DISCUSSION

With development of computer science, it is the tendency that the optimization always has been done by computer. The main difference between human being and computer is the fuzzy determination, which depends on the mechanisms of the brain of human being and of the operation of computer.

An engineer can effectively optimize the fuzzy HPC properties quickly than computer. Even through, sometimes an experienced engineer couldn't also comprise as more as five or six factors, of which are necessary to be considered for the engineering, to make a correct determination for the properties of HPC, without using the Fuzzy method scientifically and logically. Now there are many problems for an engineer meeting with on site. The reason is because when they made the determination for optimum the properties of HPC, they just only considers one or two properties rather than five or more. Therefore, it is better for us to use the Fuzzy comprehensive method to evaluate HPC. Thus , we can make computer to optimize such complete problems easily and quickly. With out Fuzzy concept and the method, computer will fail to deal with that. So in this way, engineers can use the high performance concrete by computer more effectively on site.

CONCLUSIONS

1. The fuzzy comprehensive evaluation on properties of high performance concrete is an acceptable method for civil engineers. It is possible to help engineers comprise a multi-factor effect on the multiple properties of HPC, and optimize the designations of HPC for constructional engineering.

2. It is very important to determine the factors of the fuzzy objects and suggest the weight actors for the fuzzy comprehensive evaluation, which can be determined by experience or discussions by the experts.

ACKNOWLEDGMENTS

The authors gratefully acknowledge Prof. Wang Guangyuan, Prof.Fan chengmou and Prof. Wang Shifang, Harbin University of Architecture and Engineering, P.R.China for great help in fuzzy applications and concrete science.

REFERENCES

1.   HAJME OKAMURA Development of new concrete materials, Cement and Concrete (Japan) No.475,1986.9

2.   TOSHIO SAITO   Super high strength concrete, Architectural Product-Engineering (Japan)No.291 1990.1, pp130-134

3.   P.K.MEHTA et al Principles underlying production of high performance concrete, cement concrete and aggregates,CCAGPP Vol.12,No.2, Winter 1990.pp70-78

4.   Y.MALIER   High performance concrete-from material to structure   1992.

5.   WANG PEIZHUANG   The result of Fuzzy set and its application   1983

6.   WANG GUANGYUAN   Discussion on the properties and the application of the math-mold in the comprehensive evaluation, Journal of Fuzzy (China) 1984.4

ULTRA STRUCTURES

Chairmen   **Professor C Andradé**
Institute Eduardo Torroja of Construction
Sciences
Spain

Professor G Somerville
British Cement Association
United Kingdom

Leader Paper

Ultra Structures - Indian Experience

Professor D N Trikha
Structural Engineering Research Centre
India

ULTRA STRUCTURES - INDIAN EXPERIENCE

D N Trikha

Structural Engineering Research Centre

India

ABSTRACT. Structures to be regarded as ultra structures should incorporate features which surpass the normally accepted limits of spans, heights, etc. or characterise innovation in structural system or construction. The definition has thus both time frame and locational contexts. The present paper first traces briefly the history of civil construction in India till the advent of modern materials and construction procedures. The paper then cites examples of ultra structures built in the country describing salient features in each case. Examples include description of two important bridges, one nuclear power containment vessel, one multiflue RCC chimney and one outstanding building structure. The paper concludes with prediction of an unprecedented construction activity in the country in the next decade due to current recognition of the importance of first class infrastructure.

Keywords: Ultra Structures, Spans, Heights, Innovations, Concrete, Bridges, Nuclear Power Containment Vessels, Chimneys, Building Structures, Temple.

Prof. Dayanand Trikha is the Director of the Structural Engineering Research Centre at Ghaziabad, India, a national research laboratory of the Govt. of India. Prior to this, he was Professor and Head of the Department of Civil Engineering at the University of Roorkee where he started his teaching career as a lecturer in 1960. He obtained his Ph.D degree from London University and D.I.C. from Imperial College of Science and Technology, London.

Prof. Trikha's main research interests include non-linear post cracking behaviour of concrete structures especially box girder bridges, finite element method, computer aided analysis, dynamic analysis, and materials for low cost durable construction. He has contributed a large number of research papers in national and international journals as well as conferences.

Prof. Trikha is member of Science & Technology Committees of the ministries of Surface Transport and Food of the Govt. of India, member of Himalayan Seismicity Programme Committee of Department of Science & Technology, convenor of Indian Roads Congress Committee on Loads and Stresses and the Founding President of the Indian Society for Construction Materials and Structures.

Radical Concrete Technology. Edited by R K Dhir and P C Hewlett. Published in 1996 by
E & FN Spon, 2–6 Boundary Row, London SE1 8HN, UK. ISBN 0 419 21480 1.

INTRODUCTION

Structures may qualify as ultra structures if they incorporate features which surpass the normally accepted limits as regards material strengths, material behaviour, spans or heights or involve exceptional innovation in construction or structural system. Obviously, the definition is related to a time frame so that technologies which are regarded as radical or ultra at one time may become normal practices in due course. The definition has also a strong regional or locational context as normal practices in one region may appear to be radical in another region due to time lag in the development of infrastructure or industrial practices. The present paper dealing with ultra structures in the Indian context describes a few structures built in concrete in India, each structure being characterised by atleast one distinguishing feature which sets it apart from the normal or the usual. It is, however, necessary to have first a brief historical perspective for a proper appreciation of the choices made.

HISTORICAL PERSPECTIVE

India has been building massive structures for atleast a thousand years. A treatise entitled "Samarangan Sutradhar" written in the eleventh century describes the basic principles of town planning, science of construction, procedures and specifications. The soundness of these principles is amply demonstrated by the still standing old temples, palaces and forts which have successfully withstood the rigours of rain, desert winds, sea winds, earthquakes and other environmental effects for over thousand years. These structures besides being functionally efficient and aesthetically appealing also imbibe principles that result in natural ventilation and cooling even in very hot Indian summer months. The walls are built in stone rubble masonry with or without lime mortar or use of iron clamps. Mortars have sometimes molasses, lentils and a variety of other ingredients to enhance their bonding characteristics. The roofing system progressed from the use of pyramidal domes to stone slabs to arches. Use of bricks (15cm x 10cm x 5cm) is evident much earlier in the Aryan civilization in the excavations of Mohen-ja-Daro and Harappa dating back by about 3000 years.

The history of use of modern construction materials like steel, concrete and prestressed concrete begins with the development of railway net-work in the country in this century. In the first decade, railway bridges were constructed exclusively in steel to cross mighty Indian rivers. Once the art of such crossings was mastered, road bridges were built in concrete in large numbers for over three decades between the two world wars. Prestressed concrete was used for the first time in the construction of Military tank garages at Meerut in 1940. The first P.C. bridge was constructed in 1949 on the Assam rail link followed by the first road bridge, the Polar bridge near Madras, in 1952 with 23 spans of 28.4m each. For next 10 years till 1962 however, there was very little activity in P.C. construction as prestressing equipment, anchorages, H.T. wires etc. had all to be imported. Once these requirements in materials and equipment became available within the country, prestressed concrete became the most popular material for bridging larger spans.

Since then, the country has witnessed huge construction in all sectors simultaneous to the developments in the world. India too, like any other developing country, has severe constraints as regards material and financial resources, although it has a highly developed technical manpower. Its engineers have built exceptionally high concrete dams (Bhakra Dam, 226m high, 1963), very long bridges (5575 m. Ganga Bridge, Patna, 1980), tall chimneys, nuclear containment vessels, barrages, sea shore structures, large building space frames, tunnels, stadia as well as hangers in steel, concrete and prestressed concrete using in-situ or precast construction procedures. They have overcome difficult terrain and environmental conditions, foundation problems in Gangetic alluvial plains and construction difficulties achieving economy as well as durability.

ULTRA STRUCTURES

Concrete has been much preferred over steel as a modern construction material in India primarily for reason of economy. Concrete structures abound in transportation, energy, irrigation and building sectors in the country. Several of these may be termed as ultra structures as they satisfy atleast one of the following conditions:

(i)   The structure is the first time example of a distinct feature

(ii)   The structure is built after over-coming some extra-ordinary construction problems

(iii)   The structure is the tallest/largest etc. ever built in the country

(iv)   The structure introduces a certain technology for the first time in the country.

A few outstanding examples of such structures from different sectors are given below.

TRANSPORTATION STRUCTURES

Two examples have been chosen for description from amongst thousands of road bridges which exist in the country. These are the second Thane Creek Bridge at Bombay and the second Hooghly Cable Stayed Bridge.

Second Thane Creek Bridge at Bombay

This is a 1.837 Km. long bridge consisting of six 4-span continuous units of 53.5, 107, 107 and 53.5 m lengths (Fig.1) and having two independent 3 lane parallel decks with footpaths connecting Bombay to New Bombay across Thane Creek to replace the existing 15 year old prestressed concrete girder bridge which has considerably deteriorated due to corrosion. The super-structure consists of a single rectangular cell prestressed concrete box girder with depth varying from 3.5m to 7.0m.

The continuous units are built adopting in-situ balanced cantilever construction procedure by building cantilevers in segments symmetrically from each of the three intermediate piers and connecting the two arms in a span by an in-situ key segment. The prestressing cables are placed in the deck or the soffit slabs enabling use of only 35 cm thick webs.

The super-structure is carried on open foundations except for the two foundations at one end being of well/caisson type. These foundations are sometimes 22.5 m deep below the high tide level.

The bridge has the following distinguishing features :

i)   It is first bridge of its kind built by cantilever method in India having four continuous spans of this order.

ii)   It is first time that durability has been considered integral with design and construction to protect against highly aggressive environment. Amongst measures adopted for durability enhancement are four stage anti-corrosive treatment to rebars, epoxy-based paint, sacrificial concrete cover, minimum concrete thicknesses, minimum concrete grade and cement content and strict quality control on materials and construction.

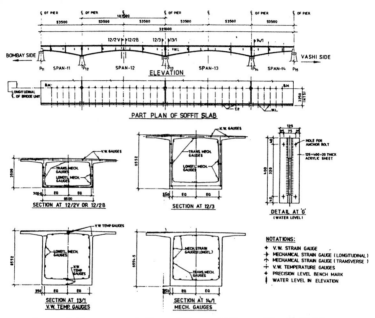

Figure 1 Instrumentation of Second Thane Creek Bridge

iii)   The open foundations have been built in an innovative manner in dry environment in the creek with water depths upto 10m. For this, a cofferdam was used which consisted of a lower lightly reinforced thin concrete shell with a steel cutting edge and height equal to the depth of the bed material overlying the rock strata and an upper reusable double walled steel shell of varying height, see Fig.2. The concrete shell was cast near shore on a pontoon which was towed to the site, the shell lifted off and lowered into position by using a floating gantry and excavating clay from within by grabs. Divers cleared the rock face near the inside periphery and placed sand bags at 60 cms distance alround. Concrete was poured in the space between the shell and the sand bags thus creating a water tight seal between the rock face and the coffer dam. Steel shell extensions (2m at a time) were continuously made while lowering the coffer dam. Once the coffer dam was in position, the inside was dewatered to get a dry environment to cast the open foundation and the pier, as seen in Fig.3. The steel shell was then removed for re-use at the next location.

The above innovation used for the first time in the country has been found so successful that it has been adopted elsewhere for construction in similar situations.

iv)   The bridge's super-structure has been instrumented extensively for monitoring its long term performance. The sensors include V.W. embedded and surface mounted strain gauges, mechanical strain gauges and V.W. temperature gauges. For recording deflections, high precision levels and water levels are being used, as seen in Fig.1.

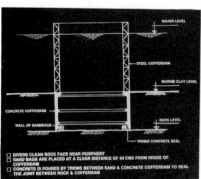

Figure 2 Cofferdam for Second Thane Creek Bridge

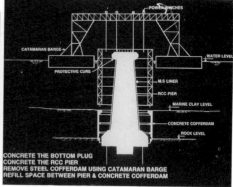

Figure 3 Open Foundation for Second Thane Creek Bridge

v)   Full scale mock-ups as seen in Fig.4, were cast and later demolished to examine actual concrete sizes obtained, locations of reinforcements, concrete cover and its compaction.

Figure 4 Full Scale Mock-up for Second Thane Creek Bridge

Second Hooghly Cable Stayed Bridge at Calcutta

The second Hooghly bridge is a 837m long six lane cable-stayed bridge connecting twin cities of Calcutta and Hooghly across river Hooghly with an average navigational clearance of about 34m, as shown in Fig.5. The total width of the deck is 35m catering to 2x3 lane carriageways divided by a median strip and footpaths on either side.

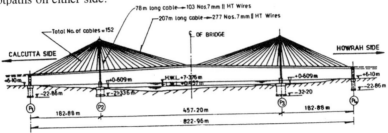

Figure 5 Second Hooghly Cable Stayed Bridge

The bridge has a span configuration of 183-457-183m as determined by navigational requirements. The superstructure is made up of a composite RCC deck slab case monolithic on three main steel girders and 4.1m spaced transverse girders. The 4-legged steel pylons, 122m above the pier top, consist of steel box sections with horizontal beams at the deck and the tower top levels. The deck is supported by parallel wire cables at 12.3m centres. The shortest cables of 78m length consist of 103, 7mm ‖ H.T. wires whereas the longest cables of 207m length have 277 such wires. These wires are housed in HDPE tubes filled with 'Hematite-U' polyurethene to make them corrosion resistant.

The foundations for the anchor piers at the two ends are twin 12.0 and 8.0m dia. circular caissons sunk upto 30m in hard clay and sandy strata. The pylon foundation consists of twin multi-cellular caissons of 23.8 m and 20.6m diameter going 40m below H.T.L. The bottom of the caissons has been sealed with colcrete. To reduce settlements, the caissons have been kept empty of water with arrangement of standby pumps for dewatering. The main / end span ratio is so fixed that upward forces develop only at the end supports. These are held down by 8 cable anchors. Fig.6 shows a typical pier and a caisson.

The bridge has the following distinguishing features :

(i)   It is the first composite cable stayed bridge in the country

(ii)   It could have been the largest span cable stayed bridge in the world, had it been completed in time.

(iii)   National facilities have been mobilised extensively to carry out feasibility studies comprising wind tunnel studies (IISc., Bangalore), earthquake related studies (University of Roorkee) and hydraulic studies (CWPRS, Pune) with RPT, UK acting as the Project Consultants.

(iv)   The large 23.8m diameter multi-cellular caisson foundations going 40m deep have been attempted for the first time in the country constructed under challenging conditions in view of the river being susceptible to large tidal bores in addition to tidal variations. Several innovative techniques were devised to build the foundations.

(v)   The concrete strength of M-50 grade has been achieved at site for the first time.

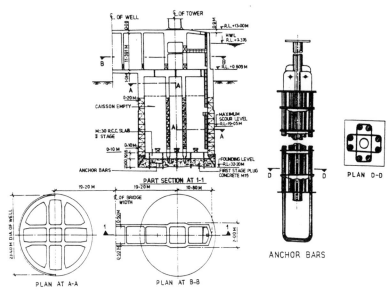

Figure 6 Pier P$_3$ and Caissons F$_3$

CONTAINMENT VESSELS

India is producing about 2000 MW of energy through its ten atomic power generators. India is amongst the first countries in the world to use prestressed concrete containment vessels in place of steel vessels. The construction and design of these vessels have been constantly improved and completely indigenousied over the years resulting simultaneously in development of prestressed concrete industry in the country. The two pressurised heavy water (PHWR) 235 MW containment vessels at Kaiga (Fig.7) described below give the present state-of-the-art of the atomic energy generation technology in the country; the two reactors Kaiga-I and Kaiga-II being still in the construction stage.

The structure is designed on a double containment philosophy. The inner containment (I.C) houses both the reactor and the steam generators and is a cylindrical prestressed concrete vessel of 42.56 m internal diameter with a P.C. dome at top. The 610mm thick wall of the I.C rises 50m above the raft. The outer containment (O.C) is a similar structure in reinforced concrete enclosing an annular space of 2.0 m. There are 4 openings of 4.1m dia. each in the dome for erection and subsequent removal of steam generators. In addition, air lock barrels are provided at ground level in the O.C and I.C. The walls are thickened suitably around all the openings.

The structure has the following design features :

(i)   The internal containment should resist an accidently caused internal pressure of 0.173 N/mm^2 while serving as a primary leak-tight barrier thus limiting leakage to the annular space to acceptable levels. The outer containment designed for an internal pressure of 0.00703 N/mm^2 also serves as a protective shield to the I.C from the environment. The annual space is kept at a slightly negative pressure to prevent any leakage of radioactive air to atmosphere before being scrubbed for letting out.

(ii)   Detailed earthquake studies have been carried out using IAEA's criteria of designing for S-1 level earthquake for Operating Basis Earthquake (OBE) and for S-2 level earthquake for Safe Shutdown Earthquake (SSE). It required studies of seismic - tectonic map of the Kaiga region, identification of faults within 300 KM and carrying out dynamic analysis of the structure for two orthogonal horizontal and a vertical motion.

(iii)   To limit the uplift of the common foundation raft for the I.C and O.C under an S-2 level event, vertical prestressed rock anchors have been provided within the annular space to anchor the foundation raft to the rock.

(iv)   Wind effects for a return period of 1000 years for the O.C and onslaught of floods due to bursting of all the nine upstream dams existing in the region are other important design considerations.

The Kaiga containment has the following distinguishing features.

(i)   The design incorporates all the features gradually evolved over the last thirty years of atomic power generation in the country. The structure is the first of its kind in the country being constructed in an area of high seismicity (ground acceleration of 0.2 g for the S-2 level) and prone to dam burst floods. The double containment philosophy with a common raft has been further refined by introduction of a stressing gallery inside the raft, and complete elimination of the large break-out panel prevalent in earlier structures by increasing the size of the air-lock.

ii)   The I.C wall, prestressed in both horizontal and vertical planes and the dome prestressed in meriodional and circumferential directions use 19K13 system cables. The total prestressing system and cables have been entirely manufactured indigenously and specially tested for acceptance.

iii)   The effect of the plant on the environment has been given as much importance as the safety of the structure under environmental loads.

iv)   The openings in the dome for erection/replacement of steam generators have been introduced for the first time.

v)   Intensive instrumentation comprising embedded V.W. strain gauges, and LVDTS etc. is being introduced in NPC vessels for the first time in the country for long term structural performance monitoring which is proposed to be carried out (see Fig. 7) at different stages such as (a) during concreting (to study effect of hydration of concrete in the initial setting period), (b) during curing and before prestressing (shrinkage effects), (c) during prestressing and (d) after prestressing. This instrumentation is in addition to the one required for initial acceptance tests as desired by AERB of India.

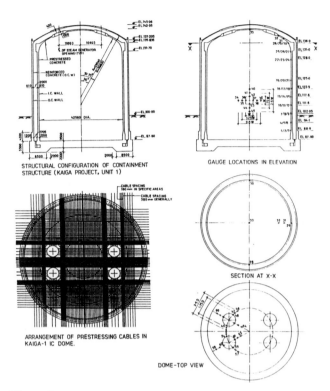

STRUCTURAL CONFIGURATION OF CONTAINMENT
STRUCTURE (KAIGA PROJECT, UNIT 1)

GAUGE LOCATIONS IN ELEVATION

ARRANGEMENT OF PRESTRESSING CABLES IN
KAIGA-1 IC DOME.

SECTION AT X-X

DOME-TOP VIEW

Figure 7 NPC Vessel at KAIGA and its Instrumentation

TAIL STRUCTURES

India has consciously decided not to build more than 25-30 storey high residential or commercial buildings even in metropolitan cities. However, tall structures abound in the form of cooling towers, chimneys and TV towers in which the latest slipform technique using hydraulic jacks and climbing framework has been used. Some examples are given below.

Cooling Towers

Natural Draught Cooling Towers (NDCT) are being increasingly preferred over Induced Draught Cooling Towers for reasons of economy, energy efficiency and environmental benefits. These NDCTs have the usual hyperbolic shape for the tower supported on diagonal columns, often precast. The tallest cooling towers in India are the NDCT at Kakrapar Atomic Power Project, Gujarat (122.3m tall. 101.9m dia. at base) and the NDCT at Gandhi Nagar Thermal Power Station, Gujarat (121m tall, 102m base dia.).

TV Towers

The overall height of the TV tower at Delhi is 235m including 68m high steel mast and having a 1000 person capacity revolving restaurant at a height of 155m. The shaft has base diameter of 15.7m reducing to 6.7m dia. at 147.5m height. It is supported over 279, 50cm dia. piles driven to nearly 20 to 22m depth. The tower has been designed to withstand a maximum wind speed of 206 kmph in addition to EQ loads.

The TV tower at Rameswaram is 323m high comprising a circular RC tower with a square steel mast of 45m height. It has a dia. of 24m at bottom tapering to 6.5m at top. The tower has been designed for a wind velocity of 160kmph (return period of 50 years).

Multiflue RCC Chimneys

Multiflue RCC Chimneys upto 275m height have been constructed in India at Madras Thermal Power Station and at Anpara in UP. The Anpara chimney has a dia. of 30m at base tapering to 20m over a height of 110m after which it remains uniform. The shell has a thickness of 1000mm at base reducing to 400mm thickness at top. The chimney has two flues of 7.5m dia. each and lined with fire clay and acid resistant bricks supported over 26 RCC platforms.

BUILDING STRUCTURES

Concrete is the most preferred material of construction both for residential and commercial buildings in India. At several sites, difficult foundation conditions have been encountered whereas challenges at other places have arisen from architectural designs complicating the structural system or environmental effects. However, Baha'i Temple at New Delhi has been chosen as an example of ultra structure as it has been invariably described as the Taj Mahal of the 20th century. The Temple is aesthetically a marvel, structurally a phenomenon and functionally a perfection as seen in Fig.8.

Figure 8 Baha'i Temple at Delhi

The Temple symbolizes the Baha'i faith as an embodiment of nine major faiths of the world by repeating every component nine times. The Temple is conceived by its architect, F. Sahba, in the form of a half open lotus, a flower much revered in India because of its flowering in all its grandeur and purity even in swamps and ponds. The structure consists of 3 sets of petals made of concrete shells. The outer-most set of nine petals called the entrance leaves open out-wards forming the nine entrances. The second set of nine petals, called the outer leaves, open inwards. The third set of nine leaves are called the inner leaves of which only the tips open giving an impression of a partly opened bud. This portion which raises above the rest and forms the main structure is supported by nine radial beams which meet at a central hub. The inner leaves enclose the interior dome, a canopy made of criss crossing ribs and shells. A glass and steel roof at the level of radial beams provides protection from rain through the half open bud. The entrance and the outer leaves cover the outer hall. Below these two sets of leaves, nine arches rise on a ring. Alround, there are walkways, bridges and stairs leading to nine pools representing the floating leaves of the lotus. The width of the entrance and the outer leaves are respectively 18.2m and 15.4m at the base rising to 7.8m and 22.5m above the podium. The inner leaves of 14m width rise to a height of 34.3m. The shell thicknesses vary along the height.

The structure posed several challenges as described below :

(a)   Geometry

The shell surfaces of both the entrance and the outer leaves are formed out of parts of a number of spheres of different radii with their centres located at different points inside the building. Each corrugation of the inner leaf comprising a cusp and a re-entrant is made up of two torroidal surfaces. Arch soffits lie on parabolic conoids. The structure does not have a single straight line in the entire building. The complexity of the geometry had to be considered both for analysis and for construction. The surfaces and lines had to be defined by mathematical equations and then expressed in a manner understandable to a carpenter and a mason to fabricate the formwork. This exercise itself took almost two years.

(b)   Design

Structural design carried out by Flint & Neill Partnership in London, presented several challenges. A scaled model of 1:125 was wind tunnel tested at Imperial College, and analysis for a variety of structural idealisations was carried out for self weight, super-imposed loads, seismic loads, wind loads and temperature loads using F.E.M. Effects of creep and cracking were examined. Stability was checked by non-linear analysis of ribs not stabilised by shells. The nine outer leaves shells were checked for stability using non-linear large deflection finite element analysis. Support conditions were modified as a result of these analysis.

(c)   Construction

Very stringent construction procedures were laid down in order to ensure that complex doubly curved surfaces and their intersections were accurately produced without any cold joints as no finish of any kind was permitted on the interior. This required fabricating a full scale mock-up of the formwork and replicating the same by first fully fixing the inner formwork as per the true geometry and placing the reinforcement. The outer formwork was gradually built up as concreting progressed. In order to avoid any cold joints, concreting of entire petals was done in a continuous pour, sometimes over a height of 22m lasting for over 48 hours.

Since no surface finish was permitted, it became necessary to prevent long term rusting of reinforcement to save tarnishing of white concrete. This was achieved by galvanizing all reinforcement as well as binding wires, a practice rarely followed in India. Several tests were carried out so that mechanical properties of steel were not affected by galvanising. Quality of course aggregate (dolomite obtained from Alwar mines) and silica sand (obtained from Jaipur) was constantly checked with strict control on water quality. White cement was imported from Korea to ensure proper concrete strength. Concrete mix, 1:1.5:3.5 by weight, with plasticiser and W/C ratio of 0.43 was adopted to avoid crazing and shrinkage cracks. Special precautions were taken to place concrete at a temperature of about 30°C by adding ice and precooling aggregates in air-cooled aggregate storage bins.

As a result, an unblemished surface with only the designed textures and architectural patterns has been achieved on the interior. The outer surface of the shells is cladded with marble panels cut to size and shape in Italy from marble quarried from the Mount Pentilekon of Greece. These panels are held by stainless steel brackets and anchors. The entire construction including the most challenging task of design, erection and dismantling of the form-work was carried out by the Indian contractors E.C.C. Construction Group of Larsen & Toubro Ltd.

CONCLUDING REMARKS

India is poised for a great construction boom in the next decade. The impetus to this heightened construction activity expected to grow at a rate of 12 to 16% has been provided as much by the rising aspirations of the people arising from sound personal economic well-being as from the government's determined emphasis on building/improving infrastructure consequent to the twin policies of globalization and liberalization. Presently, the country's annual cement production of 58.35 million tonnes ranks as the fourth largest in the world after China, Japan and USA. In the transport sector, it is estimated that a sum of $30.0 billion is required for road construction and an equal amount of $30 billion for construction of seven super national highways of total length of about 13000 Km. In the power sector, UNIDO in their 'Global Report 1995' have estimated an investment of $200 billion over the next 15 years to ensure a 6% annual growth rate of GDP. In the housing sector, a minimum investment of about $82 billion would be necessary to provide barest minimum dwellings whose shortfall has increased from 23 million in 1981 to 41 million in 1991.

The investments as envisaged above are beyond the reach of any country and as such, India is pursuing various strategies including B.O.T. bids involving multinational agencies in nation's construction activities. Engineers are also being called upon to meet these challenges through innovations, use of new materials, recycling of materials, optimization of systems and improvements in construction technologies. The next decade is destined to become the decade of **ultra technologies and ultra construction** for India. It is hoped that engineering fraternity world-wide would be partner and comrade in making these developments a great success story.

ACKNOWLEDGEMENTS

The author is grateful to M/s Gammons India Ltd., STUP Consultants Ltd., Mahinder Raj Consultants Pvt. Ltd. and ECC of Larsen & Toubro Ltd. for their help in the preparation of this paper.

MAKING FIELD PLACEABLE VERY-HIGH-STRENGTH CONCRETE IN HONG KONG

H C Chan

A K H Kwan

University of Hong Kong

Hong Kong

ABSTRACT. In order to develop field placeable very-high-strength concrete mixes for Hong Kong, a large number of trial concrete mixes have been cast and tested. Only locally available materials and conventional production and curing methods were used. For the sake of avoiding excessive heat of hydration and drying shrinkage, the binder paste volumes of the mixes were limited to not more than 35%. The mix parameters investigated include: aggregate type, binder paste volume, fine to total aggregates ratio, dosage of condensed silica fume, and combined usage of condensed silica fume and pulverized fuel ash etc. Based on the test results, several selected mixes which may be considered for the production of grade 90~100 concrete are recommended.

Keywords: High-strength concrete, Aggregates, Condensed silica fume (CSF), Pulverized-fuel ash (PFA).

Dr H C Chan is a Reader and the Head of Department of Civil and Structural Engineering, University of Hong Kong. His main research interests include concrete technology, reinforced concrete structures and computer aided design of structures. He has published many journal papers and chapters of books, and served on several technical committees for the local government and engineering institution.

Dr Albert K H Kwan is a Senior Lecturer of Department of Civil and Structural Engineering, University of Hong Kong. He has worked for a number of years in the construction industry before returning to the academic circle. His research interests include concrete technology, tall building structures and earthquake engineering.

Radical Concrete Technology. Edited by R K Dhir and P C Hewlett. Published in 1996 by E & FN Spon, 2–6 Boundary Row, London SE1 8HN, UK. ISBN 0 419 21480 1.

INTRODUCTION

In Hong Kong, the highest strength concrete ever used for in-situ construction is currently only of grade 65. This is relatively low compared to those used in some other places, particularly North America, where concretes of grade well above 100 have already been successfully applied [1]. Hong Kong has a large number of tall buildings, the majority of which are constructed of concrete. Furthermore, the trend is that newly designed buildings are getting taller and taller (a 100-storey concrete building is already being planned and designed). Hence, there is a great demand for higher strength concrete in Hong Kong.

As a first step to develop higher strength concrete, the Authors have recently developed several alternative mixes suitable for making grade 75~80 concrete for in-situ construction in Hong Kong [2]. During the investigation, in which only locally available materials were used, it was found that there is no particular difficulties, apart from more care needed in compaction and curing, in producing grade 75~80 concrete in Hong Kong. It was also inferred from the results obtained that the strength potentials of the local materials have not yet been fully utilized, viz., it should be possible to make even higher strength concrete in Hong Kong, although the difficulties involved would inevitably increase with the strength level.

The present study is a follow-up of the above investigation. Several series of laboratory trials were carried out to explore the possibility and to find out suitable methods of producing field placeable (with slump > 75 mm) very-high-strength concrete (with target mean cube strength > 100 MPa or of grade > 80) in Hong Kong. Only locally available materials were used and conventional compaction and curing methods were employed. Moreover, in order to ensure dimensional stability and avoid excessive heat of hydration, the binder paste volumes of the mixes were limited to not more than 35%, as recommended by Mehta and Aitcin [3].

EXPERIMENTAL PROGRAMME

The experimental programme consisted of two parts. In the first part, the local granitic and volcanic aggregates were compared by testing similar concrete mixes cast separately of the two different aggregates to see which one is more suitable for the production of very-high-strength concrete. Having selected the type of aggregate to be used, in the second part, the effects of various mix parameters including binder paste volume, dosage of CSF and combined usage of CSF and PFA etc. were studied by means of trial mixing. The mix proportions were then optimized to develop suitable mix designs for making very-high-strength concrete in Hong Kong.

Raw materials

Two types of aggregates: granitic and volcanic aggregates, both obtained locally, have been used. Their properties are given in Table 1 and will be discussed later.

Table 1 Properties of the Rock Aggregates

	GRANITIC		VOLCANIC	
	fine	coarse	fine	coarse
Physical properties				
nominal maximum size	5.0	10.0	5.0	10.0
fineness modulus	2.94	-	3.20	-
relative density (oven dried)	2.55	2.58	2.66	2.70
water absorption (%)	1.20	0.98	1.60	0.64
flakiness index (%)	-	17	-	28
elongation index (%)	-	26	-	21
Mechanical properties				
aggregate crushing value (%)	-	26	-	16
aggregate impact value (%)	-	21	-	14
10% fines value (%)	-	180	-	300

The binders used are the same as those used in previous studies [2]. Basically, the cement is a locally ground Portland cement that complies with the requirements of BS12:1991 for strength class 52.5N, the CSF is 94% pure and has a fineness of 18~20 sq.m./g, while the PFA is a local classified ash whose cementing efficiency has been found to be comparable to those in the U.K. [4]. On the other hand, the superplasticizer used is a naphthalene-based admixture which can be used up to more than 2% (dry weight of the admixture as a percentage of the total binder content) without producing excessive air entrainment or retardation.

Concreting, curing and testing methods

All mixing, compaction, curing and concrete testing procedures were carried out in accordance with the Hong Kong Construction Standard CS1:1990 which is similar to the corresponding British Standards except a few minor details. Specifically, the concrete cubes were cured in a mist room maintained at a relative humidity of not less than 95% and a temperature of 27 ± 2 degrees centigrade. 100 mm cubes were used for compression tests. Three cubes were tested at each age for each trial mix. If the range of the three strength results was less than 5% of the mean, the cube strength was taken as the mean of the strength results. However, if the range of the strength results was greater than 5%, the trial mix procedure was repeated until consistent results were obtained.

EFFECTS OF AGGREGATE TYPE ON CONCRETE STRENGTH

The major physical and mechanical properties of the aggregates have been measured using the test methods given in the British Standard BS812:1990 and the results are listed in Table 1. The results fall within the ranges given in Ref.[5] for the local aggregates and thus the aggregates tested may be considered typical. It can be seen that the volcanic aggregate has lower crushing and impact values but higher 10%

fines value than the granitic aggregate, indicating very clearly that the volcanic aggregate is significantly stronger. However, the volcanic aggregate is also more flaky than the granitic aggregate.

The effects of using the two different types of aggregates on the concrete strength were investigated by conducting parallel tests on pairs of concrete mixes consisting of the same binders and mix proportions but with each mix in a pair using a different type of aggregate. A total of 32 mixes which may be divided into two separate series were cast. The first series contained 10% CSF (by weight of total binder content) and a binder paste volume of 30% while the second series contained 20% CSF and a binder paste volume of 35%. In each series, the water/binder ratio varied from 0.50 to 0.18 so as to cover the entire range that may possibly be used. In order to minimize possible variations in materials used and environmental conditions during casting and curing so that direct comparison can be made, the two mixes in a pair were cast in the same day using the same binders and cured side by side under exactly the same conditions.

The 28-day cube strength results are plotted in Figure 1 from which the following points are noted: (1) Within the whole range of mix proportions studied, nearly all concrete mixes cast with the volcanic aggregate have higher strength than the corresponding mixes cast with the granitic aggregate. (2) The difference in concrete strength due to the use of different type of aggregate is more significant at low water/binder ratios when the concrete strength is governed mainly by the strength of the aggregates.

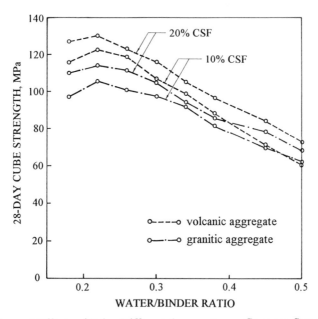

Figure 1 Effects of Using Different Aggregates on Concrete Strength

During the cube crushing tests, it was observed that there were generally more transgranular fracture in concretes made with the granitic aggregate than in those made with the volcanic aggregate, especially at very low water/binder ratios. This reveals that the granitic aggregate is indeed a relatively weak aggregate and its low intrinsic strength could be a limiting factor in the production of high-strength concrete.

With the above mixes included, more than 300 trial concrete mixes using the local granitic and volcanic aggregates have been cast and tested at the University of Hong Kong in the past five years. Virtually all possible combinations of mix parameters have been tried and yet the highest concrete strengths achieved with the granitic and volcanic aggregates were only around 120 MPa and 150 MPa respectively. The Authors believe that there is a certain concrete strength level for each aggregate, called the "ceiling strength", beyond which further improvement in the quality of the binder paste would not result in any further increase in concrete strength, as the concrete strength would then be governed solely by the intrinsic strength of the aggregate particles. On the basis of this argument, it may be said that the local granitic and volcanic aggregates have ceiling strengths of about 120 MPa and 150 MPa.

These ceiling strength values have important implications. Firstly, they are much better indications of the intrinsic strength of the respective aggregates than the crushing values, impact values and 10% fines values. Secondly, concretes having strength close to the ceiling strength of the aggregate used would be very brittle because there will be extensive transgranular fracture during crushing failure. To avoid high brittleness which can be quite dangerous, it is suggested that a safety margin of, say, at least 20% of the ceiling strength should be provided. That is to say, the concrete mix should not be designed to have a target mean strength higher than 80% of the ceiling strength of the aggregate. Based on this principle, it is considered that the local granitic and volcanic aggregates are suitable only for concretes with mean strengths of not more than 96 MPa (80% of 120 MPa) and 120 MPa (80% of 150 MPa) respectively.

The volcanic aggregate was selected and used in all the subsequent mixes.

EFFECTS OF MIX PROPORTIONS ON CONCRETE STRENGTH

Binder paste volume and fine to total aggregates ratio

A series of concrete mixes with varying binder paste volume and fine to total aggregates ratio was cast and tested. The values of binder paste volume studied are 30%, 32.5% and 35%, while the values of fine to total aggregates ratio investigated range from 0.30 to 0.50. The test results are presented in Table 2. It is seen that although within the range of parameters studied, the fine to total aggregates ratio has little effect on the concrete strength, it does have a marked influence on the workability of the fresh concrete mix. Firstly, it is noteworthy that a higher fine to total aggregates ratio, i.e. larger fine aggregate content, will generally require a slightly higher dosage of superplasticizer for a given slump. Secondly, by means of visual assessment of the workability of the fresh concrete mixes, it has been found that while on one hand when the fine to total aggregates ratio is relatively high, the mix would become sticky and difficult to work with, on the other hand if the fine to total aggregates ratio is too low, the mix would become harsh and non-cohesive. Thus, the fine to total aggregates ratio should neither be too low nor too high. Judging from the performance of the mixes cast, it is recommended that the fine to total aggregates ratio should be set within the following ranges:

Binder paste volume	Fine to total aggregates ratio
30%	0.45 ~ 0.50
32.5%	0.40 ~ 0.45
35%	0.35 ~ 0.40

The results tabulated in Table 2 also reveal that the binder paste volume has a significant effect on the concrete strength. When the binder paste volume is increased from 30% to 35%, there is about 8% increase in strength even at the same water/binder ratio. Therefore, when the strength requirement is high, the binder paste volume should not be set too small. One advantage with the use of a slightly larger paste volume is that for a given workability, the water/binder ratio can be further reduced to give an even higher strength.

Table 2 Effects of Binder Paste Volume and Fine to Total Aggregates Ratio

	FINE TO TOTAL AGGREGATES RATIO			
	0.35	0.40	0.45	0.50
Mixes with paste volume = 30%				
water/binder ratio	0.28	0.28	0.28	0.28
superplasticizer (%)	1.8	1.8	2.0	2.0
slump (mm)	shear	shear	65	60
visual assessment of workability	harsh	harsh	fair	good
28-day cube strength (MPa)	76.4	79.6	81.8	80.1
Mixes with paste volume = 32.5%				
water/binder ratio	0.28	0.28	0.28	0.28
superplasticizer (%)	1.2	1.2	1.4	1.4
slump (mm)	shear	shear	50	55
visual assessment of workability	harsh	fair	good	good
28-day cube strength (MPa)	83.4	85.7	84.4	83.0
Mixes with paste volume = 35%				
water/binder ratio	0.28	0.28	0.28	0.28
superplasticizer (%)	0.7	0.7	0.8	0.8
slump (mm)	55	50	55	60
visual assessment of workability	fair	good	good	sticky
28-day cube strength (MPa)	89.2	89.4	88.4	90.5

In each of the subsequent series of trial mixing and testing, three groups of concrete mixes, having binder paste volumes respectively of 30%, 32.5% and 35%, are investigated. In order to achieve the highest concrete strength possible, the water/binder ratios are set as low as possible so long that a reasonable workability can still be achieved. For concrete mixes with binder paste volumes of 30%, 32.5% and 35%, the water/binder ratios are set at 0.24, 0.21 and 0.18 respectively. These water/binder ratios are believed to be the lowest possible for the corresponding binder paste volumes. The corresponding fine to total aggregates ratios are 0.45, 0.42 and 0.38.

CSF content

In order to study the effects of using CSF at various dosages, a series of trial concrete mixes with CSF content (expressed as a percentage by weight of the total binder content) ranging from 5% to 25% has been cast and tested. The results obtained are plotted in Figure 2. It has been found that irrespective of the binder paste volume and water/binder ratio, the concrete strength generally increases with the CSF content until the CSF content reaches an optimum value of approximately 20%, beyond which the concrete strength would remain more or less the same or even decrease as more CSF is added. It has also been observed that a higher CSF content would generally result in a more sticky binder paste and hence a reduction in workability leading eventually to a higher dosage of superplasticizer needed for a given slump. Taking into account all these considerations, a CSF content of more than 20% is not considered beneficial. The recommended dosage of CSF for very-high-strength concrete is 10 to 20%.

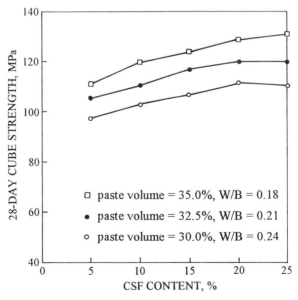

Figure 2 Effect of CSF Content on Concrete Strength

PFA content

In order to study the effects of using PFA, a number of trial concrete mixes incorporating different PFA content (expressed as a percentage by weight of the total binder content) were cast and tested. As CSF will have to be used in the production of very-high-strength concrete, CSF was also added so that the combined effects of CSF and PFA could be investigated. As a whole, two series of concrete mixes, one with 10% CSF and the other with 20% CSF, were tested. The results for the mixes containing 20% CSF are shown in Figure 3 (similar results were obtained for the mixes containing 10% CSF and are thus not shown for brevity). It can be seen that for the mixes investigated, regardless of the amount of CSF present, addition of PFA up to 25% would cause at most a few percent reduction in the 28-day strength, while addition of PFA up to 35% could cause more than 10% reduction in the 28-day strength. As too much PFA may adversely affect the 28-day strength, the PFA content should be kept at not more than 25%. On the other hand, since the use of a slightly higher PFA content can help to reduce the cement content and hence the heat of hydration, it is recommended to use a PFA content of 25%.

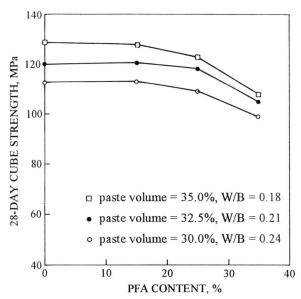

Figure 3 Effect of PFA Content on Concrete Strength

PROPOSED CONCRETE MIXES

Having studied the effects of the various mix parameters, a new series of mixes designed to have 75 mm slump and a mean strength of 110 to 120 MPa was cast and tested. Following the recommendations given previously, 10 to 20% CSF and 25% PFA were used. The mixes tested and their results are given in Tables 3 and 4.

Table 3 Mixes Designed for Possible Application as Grade 90 Concrete

mix number	G90-B-10	G90-B-20	G90-C-10	G90-C-20
binder paste volume (%)	32.5	32.5	35	35
water/binder ratio	0.24	0.24	0.24	0.24
cement content (kg/cu.m.)	347	287	374	310
CSF content (%)	10	20	10	20
PFA content (%)	25	25	25	25
fine/total aggregates ratio	0.42	0.42	0.38	0.38
superplasticizer (%)	2.30	2.50	2.20	2.40
slump (mm)	75	85	90	100
28-day cube strength (MPa)	109.6	111.3	109.5	113.5
90-day cube strength (MPa)	123.4	132.0	124.9	135.3

Table 4 Mixes Designed for Possible Application as Grade 100 Concrete

mix number	G100-C-10	G100-C-20
binder paste volume (%)	35	35
water/binder ratio	0.20	0.20
cement content (kg/cu.m.)	400	331
CSF content (%)	10	20
PFA content (%)	25	25
fine/total aggregates ratio	0.38	0.38
superplasticizer (%)	2.60	2.70
slump (mm)	85	80
28-day cube strength (MPa)	114.0	120.8
90-day cube strength (MPa)	132.7	141.3

All the mixes listed in Table 3 have mean 28-day strengths of about 110 MPa. Among these mixes, the mix G90-B-10, which has a binder paste volume of 32.5%, is probably the most economical. The mix G90-C-10, which has a slightly larger binder paste volume of 35% and therefore a slightly better workability, is also a suitable mix for grade 90 concrete. The other two mixes, which contain much higher CSF contents but only marginally higher strengths, are not recommended.

Both the two mixes listed in Table 4 have a very low water/binder ratio of 0.20. Because of the low water/binder ratio, the binder paste volumes have to be increased to 35%. Among these two mixes, the mix G100-C-10 does not really have sufficient strength to be regarded as a grade 100 concrete, unless the 90-day strength is used to designate the grade of the concrete. The mix G100-C-20 does have sufficient strength to be regarded as a grade 100 concrete, but a CSF content as high as 20% is needed, which means that it will be more costly to produce.

It has been estimated, through consultation with ready mix concrete suppliers, that the cost of producing G90-B-10 or G90-C-10 is approximately two times that of grade 40~50 concrete. In other words, these grade 90 concrete mixes have more or less the same material cost/strength ratio as that of normal strength concrete.

CONCLUSIONS

Overall, an extensive study on developing very-high-strength concrete for in-situ construction in Hong Kong has been carried out. Guidelines for choosing suitable materials and designing concrete mixes for grade 90~100 concrete are presented. Specifically, a strong aggregate should be used, the binder paste volume should be 32.5 - 35%, the CSF and PFA contents should be 10 - 20% and 25% respectively, a water/binder ratio as low as 0.20 - 0.24 is needed and a superplasticizer that can be used up to more than 2% without producing excessive side effects is required. These guidelines may also be applicable to other places if the materials available and the curing conditions are similar. Concern on increased brittleness is expressed, and it is recommended that, pending for further in-depth study, the concrete strength should not be pushed too close to the ceiling strength of the type of aggregate used.

ACKNOWLEDGMENTS

The financial support of the Croucher Foundation for the research work reported herein is gratefully acknowledged.

REFERENCES

1. RANDALL, V.R. AND FOOT, K.B. High-strength concrete for Pacific First Center. Concrete International: Design & Construction, Vol.11, No.4, April, 1989, pp 14-16.

2. KWAN, A.K.H., CAI, Y.B., CHEUNG, Y.K. AND CHAN, H.C. High performance grade 75-80 concrete for in-situ construction in Hong Kong. Transactions, Hong Kong Institution of Engineers, Vol.1, No.2, December, 1994, pp 29-36.

3. MEHTA, P.K. AND AITCIN, P.C. Microstructural basis of selection of materials and mix proportions for high strength concrete. Proceedings of the 2nd International Symposium on Utilization of High Strength Concrete, California, May, 1990, American Concrete Institute, pp 265-286.

4. CHINA LIGHT AND POWER COMPANY LIMITED. PFA Concrete Studies: 1988-1998, Vol.1, 1990.

5. IRFAN, T.Y., CIPULLO, A., BURNETT, A.D. AND NASH, J.M. Aggregate Properties of Some Hong Kong Rocks, Geotechnical Engineering Office, Civil Engineering Department, Hong Kong Government, January, 1992.

CONCRETE COUPLED STRUCTURAL WALLS WITH UNEQUAL WIDTH IN TALL BUILDINGS

J S Kuang

Hong Kong University of Science & Technology

Hong Kong

ABSTRACT. The continuous medium technique has been widely used in the preliminary analysis and design of coupled shear walls in tall buildings by practicing engineers for many years. One of the basic assumptions made when using this approximate method is that the coupling beams deform with points of contraflexure at mid-span. However, when there is a non-negligible difference between the stiffness' of the structural walls connected by the coupling beams, the points of contraflexure of the beams can move off-centre, and hence noticeable error in the analysis may occurred. This paper presents a study on the accuracy of the continuous medium technique for the preliminary analysis of concrete coupled structural walls with unequal width. Comparison of the analytical results between the technique and the wide-column analogy has been made, and recommendations for using continuum method to analyse the coupled shear walls are presented.

Keywords: Coupled shear walls, Coupling beams, Tall buildings, Reinforced concrete.

Dr J S Kuang is Lecturer in Structural Engineering, Hong Kong University of Science and Technology. His research interests include structural analysis and design of tall buildings, reinforced and prestressed concrete, and design for earthquake resistance.

Radical Concrete Technology. Edited by R K Dhir and P C Hewlett. Published in 1996 by E & FN Spon, 2–6 Boundary Row, London SE1 8HN, UK. ISBN 0 419 21480 1.

INTRODUCTION

Reinforced concrete coupled structural walls are widely used in tall buildings to provide lateral resistance against external horizontal loads caused by wind or earthquakes. This type of structure is recognised as one of the most efficient structural systems for such purpose. During the past two decades, a considerable volume of research on the analysis of coupled shear walls has been published, and a summary of different principal methods of analysis has been provided by Stafford Smith and Coull [1].

The continuous medium technique is considered as the most simple and efficient method for analysis of coupled shear walls. One of the basic assumptions made when using this method is that the coupling beams deform with points of contraflexure at mid-span. However, when there is a non-negligible difference between the stiffness' of the structural walls connected by the coupling beams, the points of contraflexure of the beams can move off-centre, and hence significant error in the analysis may occurred. A criterion was presented for assessing when such error becomes noticeable [2]. The purpose of this paper is to assess the accuracy of the continuum method for analysing the coupled shear walls of unequal width by employing wide-column analogy, in which beam points of contraflexure are not limited to mid-section. Recommendations are made for using continuum technique to analyse coupled shear walls.

METHODS OF ANALYSIS

Continuum Method

Consider a coupled structural wall system rigidly fixed at the base as shown in Fig. 1(a). By employing the continuum approach of analysis, the coupling beams may be replaced by a continuous distribution of laminae with equivalent stiffness. It is assumed that the points of contraflexure of the coupling beams coincide with the centre-line of the laminae. A cut is made along the line of contraflexure, and a continuous distribution of shear force will be released, as shown in Fig. 1(b). A deferential equation can be developed by considering the condition of vertical compatibility along the cut line of contraflexure, and the governing equation of the system can then be derived and given by

$$\frac{d^2 T}{dx^2} - \alpha^2 T = -\gamma M_e$$

where T is the axial force at each wall and M_e the applied moment. The parameters in the equation are defied as

$$\gamma = \frac{12 I_b \ell}{h b^3 I} \quad \text{and} \quad \alpha = \sqrt{\gamma(\ell + \frac{AI}{A_1 A_2 \ell})}$$

The solution of the governing equation for the case that the system is subjected to a uniformly distributed load, u, along the structural height is as follows:

$$T = \frac{u\gamma}{\alpha^4}\left[1 + \frac{\sinh(\alpha H) - (\alpha H)}{\cosh(\alpha H)}\sinh(\alpha H) - \cosh(\alpha H) + \frac{\alpha^2 x^2}{2}\right]$$

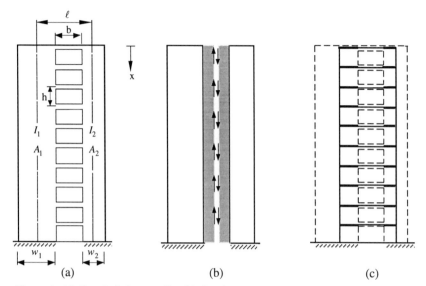

Figure 1 (a) Coupled shear walls; (b) Continuous system; (c) Wide-column frame

In general practice, the parameter αH is a measure of relative stiffness of the coupling beams. High value of αH indicates that coupled shear walls possess strong coupling and behave like a monolithic cantilevered wall; low value of αH indicates that coupled shear walls possess weak coupling and act as two separate cantilevers.

Wide-Column Frame Analogy

Fig. 1(c) shows an analogous wide-column frame for coupled shear walls. In this model, each wall is represented by an equivalent column located at the centroidal axis, to which is assigned the flexural rigidity EI and the axial rigidity EA of the wall. The condition that plane section remain plane may be incorporated by means of stiff arms located at the connecting beam levels. The rigid arms ensure that the correct rotations and vertical displacements are produced at the edges of the walls. In the analogous frame, the lintel beam points of contraflexure are not limited to mid-section. In fact, when the stiffness' of the structural walls are different, the points of contraflexure of the beams can move off-centre. It is clear that there is a difference between the results of analysis when using the two different methods.

NUMERICAL INVESTIGATION

Analyses of a typical concrete coupled shear-wall structure with unequal width of walls ($w_1 \neq w_2$) shown in Fig. 1(a) were carried out. The building has 20 stories with total height of 56 metres, and is subjected to a wind load of 16.5 kN/m^2 along the structural height. The detailed dimensions and material properties are as follows:

Thickness of walls = 0.3 m
Coupling beams: span = 1.5 m, depth = 0.4 m
Storey height = 2.8 m
Elastic moduluds of concrete = 36 kN/mm^2

In the analysis, the example coupled shear-wall structure had been divided into ten groups as the wall width ratio, w_1:w_2, from 1:1 to 1:10. A total of fifty-one example structures had been used for the analysis, and the corresponding αH values are in a range from 3 to 10.

Fig. 2 shows the curves representing the values of αH against the percentage difference of top deflection between continuum method and wide-column analogy are plotted. It is seen that when αH is less than 4, the error from continuum approach dramatically increases. However, this error does not occur at higher αH values, i.e. αH is above 4. Moreover, the percent differences are basically in the similar trend, as the wall width ratio increases from 1:1 to 1:10. It may reveal that the large error occurred from the continuous medium method is happened on the low αH values and its not affected by the unequal width of shear walls.

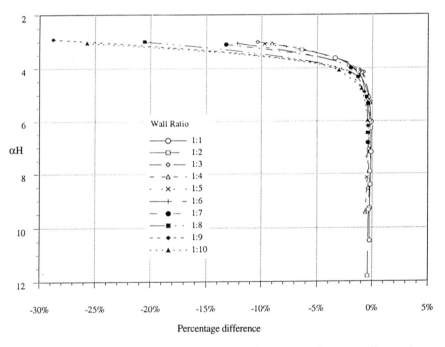

Figure 2 Percentage difference in top deflection between continuous medium and equivalent frame solutions

From Figs 3 and 4, the same phenomenon has been found that the curves plotted of αH against percentage differences of maximum axial force in the walls and shear force in the lintel beams are in rotated L-shape as same as those of top deflection. The percentage difference is within 5% in normal conditions for the value of αH above 4. When αH below 4, the error from continuum approach also dramatically increases. As a results, the change of stiffness of two walls does not affect the axial forces in the walls and shear forces in the beams as αH above 4.

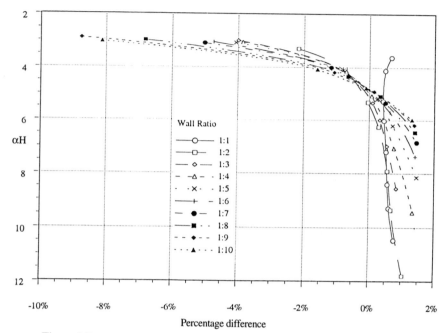

Figure 3 Percentage difference in axial force in walls at base level between continuous medium and equivalent frame solutions

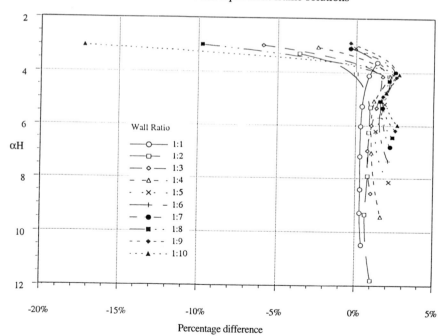

Figure 4 Percentage difference in maximum shear force in coupling beams between continuous medium and equivalent frame solutions

It is revealed that for $\alpha H < 4$ accuracy of the continuum method for analysing coupled shear walls may not be accepted, but is not affected by the unequal width of shear walls.

CONCLUSIONS

The continuous medium approach has been widely used for the preliminary analysis and design of coupled shear walls. The advantage of this approximate method is to provide a simple calculation and important design values in short period of time for predicting the deflections and stresses of the major members. In order to investigate the accuracy of the continuum method, comparison has been made between the results from the method and the wide-column analogy for a coupled shear-wall structure of 20 stories.

The study reveals that the error percent of the continuum method increases as the relative stiffness of coupling beams, αH, decreases. Moreover, the influence of the wall width ratio in the method is not severe except the αH value as low as 4. This error may be mainly due the absent of consideration of the axial deformation and the location of contraflexural points fixed at the mid-span of the lintel beams.

From the results of the analysis with various of wall width, it concludes that the continuum method is acceptable to be used for analysing coupled shear walls with unequal width when the value of αH is not less than 4 as well as the wall width ratio up to 1:10. The value of αH of coupled shear walls generally ranges from 3 to 7. For the case of slab coupling, αH normally falls below 4; thus the wide-column frame analogy should be employed for the analysis.

REFERENCES

1. STAFFORD SMITH, B AND COULL, A. Tall Building Structures: Analysis and Design, John Wiley, 1991, pp 537.

2. MACLEOD, I A. Connected shear walls of unequal width. ACI Journal, May 1970, pp 408-412.

GENERALISED CONTINUUM METHOD FOR STATIC ANALYSIS OF TALL BUILDINGS

P A Lopez-Yanez

Federal University of Paraiba

Brazil

ABSTRACT. A Generalized continuous medium technique (GCMT) for static analysis of multistorey reinforced concrete structures subjected to lateral and/or vertical loads, is presented. The lateral and vertical deformations, as well as joint rotations, are represented by analytical functions.

Joint flexibilities and axial deformations of vertical elements are considered. First order theory is used.

Solving the differential equations is possible to determine deformation functions and member forces.

The method is applicable to the analysis of buildings formed by shear walls, plane frames, coupled shear walls or planar combinations of frames and walls, as well as three-dimensional arranges of planar structures and core walls.

The results are compared with data obtained by Finite Element Method. A good correlation is achieved.

Keywords: Structures, Statics, Wind analysis, Seismic analysis, Multistorey buildings, Reinforced Concrete

Professor Pablo A. Lopez-Yanez is Director of the Structural Engineering Department, Federal University of Paraíba, João Pessoa, Brazil. He specializes in static and dynamic analysis of tall buildings.

Radical Concrete Technology. Edited by R K Dhir and P C Hewlett. Published in 1996 by E & FN Spon, 2–6 Boundary Row, London SE1 8HN, UK. ISBN 0 419 21480 1.

INTRODUCTION

The structural analysis of reinforced concrete tall buildings can be done with advantages using the continuous medium techniques due to the small number of parameters involved.

The continuum techniques have been presented for analysis of coupled shear walls [1,2,3], and frame-wall structures [4]. The axial deformations of vertical elements have been included in some papers [5,6].

Three-dimensional structures formed by frames, shear walls and core walls have been analyzed using these techniques [7,8,9,10,11,12,], and some effects of vertical loads have been included in the mathematical models [13,14].

The purpose of this paper is to present a Generalized Continuous Medium Technique (GCMT), for the analysis of tall building structures subjected to horizontal and /or vertical loads. The analysis is based on the first order theory and considers the joints flexibilities and axial deformations of columns and walls.

PLANAR STRUCTURE ANALYSIS

Introduction

By solving the differential equation, of plane structure, is possible to determine the lateral deformation function and, after that, it could be obtained the joint rotation and axial deformation functions. Then, based on these functions, the axial forces, shear forces and bending moments, can be calculated.

The method can be used for the analysis of planar structures as: single and coupled shear walls, plane frames and plane arrangements of walls and frames.

General Formulation

For developing the general differential equation, a generic frame is presented (Figures 1, 2), and the follow assumptions are considered:
a) all elastic, mechanical and geometrical characteristics are constants along the height;
b) For any story, the joint rotations can be expressed as:

$$\theta j = 0.5 \ (\theta^s_j + \theta^I_j) \tag{1}$$

c) The first derivative of lateral deformation function, $u'_f(z)$, can be written as:

$$u'_f (z) = 0.5 \ (\Delta^s + \Delta^I)/h \tag{2}$$

in which Δ^s and Δ^I are the relative displacements of beams; and

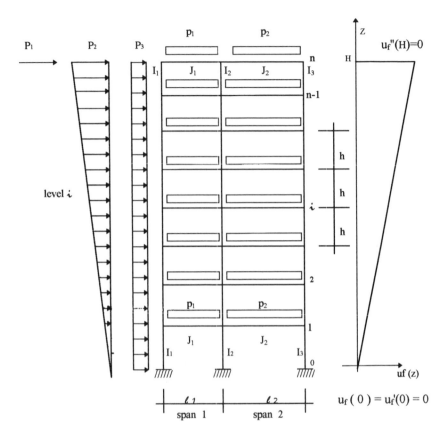

FIGURE. 1 - Generic plane frame and general deformation scheme

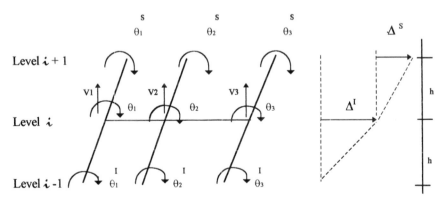

FIGURE. 2.- General deformation scheme of an arbitrary i-level

d) The quotients, $\overline{\gamma}_j$, between vertical displacements of any floor, are constants along the height:

$$\overline{\gamma}_j = v_j / v_1 \tag{3}$$

in which v_1 is the displacement of the first joint.

Based on these assumptions, is possible to write the equilibrium differential equation as:

$$-k_1 u'''_f + \lambda^2_f . k_1 u'_f = Q_f + Q^c_f + Q^q_f \tag{4}$$

where, the stiffness and load constants can be written as:

$$
\begin{aligned}
k_1 &= \{Ie\}^T \{R\} \\
k_2 &= \{\overline{\ell}\}^T \{R_b\} / h \\
k_3 &= (\{\overline{\ell}\}^T [B_l] \{\overline{\gamma}\}) / (h \{\overline{\ell}\}^T [A_e] \{\overline{\gamma}\}) \\
k_4 &= (\{\overline{\ell}\}^T \{D_4\} - \overline{M^F}) / h \\
k_5 &= k_3 \{Ie\}^T \{D_1\} \\
k_{45} &= k_4 + k_5 \\
\lambda^2_f &= k_3 + k_2/k_1
\end{aligned}
\tag{5}
$$

and the shear functions are expressed as:

$$Q_f (z) = - 0.5 P_2 z^2/H - P_3 z + H (P_1/H + P_2/2 + P_3)$$

$$Q^c_f (z) = k_3 \int_0^z \int_z^H Q_f \, dz \, dz \tag{6}$$

$$Q^q_f (z) = - 0.5 \overline{K_3} M^F (2H^* - z) z / h + k_{45}$$

where $H^* = H + 0.5h$ and all the constants are given in Appendix, Q_f is the shear force due to the horizontal loads, Q^c_f is a complementary shear force derived from axial deformations of vertical elements, due to the horizontal lads too, and Q^q_f is the equivalent shear force derived from de vertical loads.

The solution of differential equation (4) can be written, in general form, as

$$u_f(z) = B_1 e^{\lambda_f Z} + B_2 e^{-\lambda_f Z} + B_3 z^5 + B_4 z^4 + B_5 z^3 + B_6 z^2 + B_7 z + B_8 \tag{7}$$

in which $u_f(0) = u'_f (0) = u''_f(H) = 0$ are the boundary conditions and B_i are constant coefficients.

The axial deformation functions of vertical elements, can be determined by solution of following differential equations system

$$-h [A_e] \{v''\} + [B_t] \{v\} = \{R_b\} u'_f - \{D_4\} \tag{8}$$

and the joint rotation functions can be obtained by

$$\{\theta\} = \{R\} u'_f + [B_v] \{v\} - \{D_1\} \tag{9}$$

the constant coefficients are given in Appendix.

THREE - DIMENSIONAL STRUCTURE ANALYSIS

The analysis of three-dimensional structures can be done by the solution of a three differential equations system. The unknowns are two displacements and one rotation for any horizontal plane, represented by three continuous functions.

The three-dimensional formulation of GCMT, that can be used for the analysis of three-dimensional associations of planar structures with or without core walls, is analogous to the formulation presented by LOPEZ-YANEZ, P. A. [9, 11].

EXAMPLE

Considering a 15-storey structure formed by 4 plane frames (Fig. 3), which columns and beams are, respectively, 0.30 x 0.90m, 0.30 x 0.60m and 0.20 x 0.90m. The total height is 45m and the elastic Young's modulus and Poisson's ratio are, respectively, 21000 MPa and 0.20.

A lateral uniform load (10 KN/m) is applied in x-axis and vertical uniform loads (40KN/m) are applied on the beams of frames 1 and 2. The lateral deformations along the x-axis are presented in table 1, in which the results obtained by standard matrix method are presented too.

CONCLUSIONS

The proposed technique (GCMT) can be used for the structural analysis of multistorey concrete buildings.

The GCMT technique have proved to be efficient for solution of various structural types, frequently used in tall buildings, e.g.: plane frames, single and coupled shear walls and three-dimensional arrangements of planar structures and core walls.

The small quantity of parameters is a fundamental advantage of GCMT in comparison with the standard matrix method, because of small computer's memory and small quantity of mathematical operations needed.

The validity and convergence of the method are proven by many examples, not presented in this paper, including seismic and wind loads.

The GCMT allows a fast assessment of a structure adequacy in global resistance for horizontal and/or vertical loads.

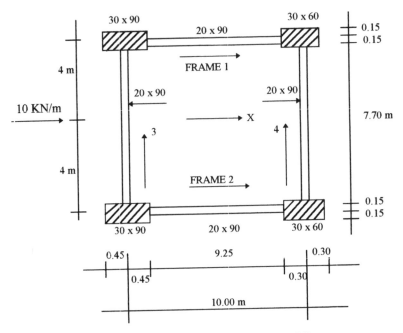

FIGURE. 3 - Example: Plan of building

Table 1. Lateral deformations of frame 1.

STORY	GCMT	MATRIX M.
15	0.0953	0.0975
14	0.0912	0.0933
13	0.0867	0.0886
12	0.0818	0.0836
11	0.0764	0.0781
10	0.0707	0.0722
9	0.0644	0.0658
8	0.0577	0.0590
7	0.0507	0.0518
6	0.0434	0.0442
5	0.0357	0.0363
4	0.0277	0.0281
3	0.0195	0.0198
2	0.0113	0.0115
1	0.0039	0.0041
0	0.00	0.00
	m	m

REFERENCES

1. ALBIGES, M. M.; GOULET, J. Contreventement des batiments. Annales de L'Institut Technique du Bâtiment et des Travaux Publics, n. 149, p. 473-500, May 1960.

2. COULL, A.; SUBEDI, N. K. Coupled Shear Walls with two and three bands of openings. Journal of the Building Science, v. 7, p. 81-86, 1972.

3. COULL, A.; BENSMAIL, L. Stiffened coupled shear walls. Journal of Structural Engineering, ASCE, v. 117, n.8, p. 2205-23, Aug. 1991.

4. CARDAN, B. Concrete Shear Walls combined with Rigid Frames in Multistory Buildings subject to lateral loads. Journal of de American Concrete Institute, v. 58, n.3, p. 299-316, Sep. 1961.

5. CHAN, P. C. K. et al. Approximate analysis of multistory multibay frames. Journal of the structural division, ASCE, v. 101, n. ST5, p. 1021-35, May. 1975.

6. JAEGER, L. G. et al. The structural analysis of tall Buildings having irregularly positioned Shear Walls. Journal of the Building Science, v. 8, p. 11-22, 1973.

7. GLÜCK, J. Lateral load analysis of asymmetric multi-story structures. Journal of the Structural Division, ASCE, v. 96, n.2, p. 317-33, Feb. 1970; discussion by J.K. Biswas and W.K. Tso in Nov. 1970.

8. ANASTASSIADIS, K. Calcul statique des contreventements par la méthode des Troi pivots: Théorie et applications. Annales de L'Institut Technique du Bâtiment et des Travaux Publics, n° 498, p. 49-124, Nov. 1991.

9. LÓPEZ-YÁNEZ, P. A. Técnica do Meio Contínuo para análise de edifícios sob ações laterais. IN: JORNADAS SUDAMERICANAS DE INGENIERIA ESTRUCTURAL, 26. Montevidéu, Uruguai, nov. 1993. Anais v.2, p. 439-50. (In Portuguese)

10. LÓPEZ-YÁNEZ, P. A. Análise sísmica de edifícios altos. IN: JORNADA SUDAMERICANAS DE INGENIERIA ESTRUCTURAL, 26. Montevidéu, Uruguai, nov. 1993, Anais v.2, p. 451-62. (In Portuguese)

11. LÓPEZ-YÁNEZ, P. A. Static analysis of tall buildings subjected to seismic loads. IN: THE INTERNATIONAL CONFERENCE ON EARTHQUAKE ENGINEERING. Amman, Jordan, Oct. 1995, Proceedings.

12. LÓPEZ-YÁNEZ, P. A. Dynamic analysis of three-dimensional multistory structures. IN: THE INTERNATIONAL CONFERENCE ON EARTHQUAKE ENGINEERING. Amman, Jordan, Oct. 1995, Proceedings.

13. RUTENBERG, A. et al Stability of shear wall structures. Journal of Structural Engineering, ASCE, v. 114, n.3, p. 707-16, Mar. 1988.

14. WANG, Q. Stability of shear-wall buildings using method of weighted residuals. Journal of Engineering Mechanics, ASCE, v. 117, n.3, p. 700-706, Mar. 1991.

APPENDIX

COEFFICIENTS OF GENERAL EQUILIBRIUM EQUATION

Considering a generic plane frame (Fig. 1,2), is possible to write two vectors:

$$\{I_e\}^T = \{I_1, I_2, I_3\} . E \tag{A.1}$$

$$\{J_e\}^T = \{J_1, J_2\} . E \tag{A.2}$$

where I_i and J_i are the inertia moments of columns and beams, respectively, and E is the elasticity Young's modulus.

The following matrices have to be calculated (See Eq A. 11):

$$\{I_u\}^T = 2h \{b^{n1}, b^{n2}, b^{n3}\} \tag{A.3}$$

$$[Ij] = \begin{bmatrix} 2(k^{n1} + a^{n1}) + k_i{}^{m1} & a^{m1} & \\ a^{m1} & 2(k^{n2} + a^{n2}) + k^{m1}{}_f + k_i{}^{m2} & a^{m2} \\ & a^{m2} & 2(k^{n3} + a^{n3}) + k_f{}^{m2} \end{bmatrix} \tag{A.4}$$

$$[Jj] = \begin{bmatrix} b_i{}^{m1} & -b_i{}^{m1} & \\ b_f{}^{m1} & -b_f{}^{m1} + b_i{}^{m2} & -b_i{}^{m2} \\ & b_f{}^{m2} & -b_f{}^{m2} \end{bmatrix} \quad ; \quad [J_t] = [J_j]^T \tag{A.5}$$

$$[Jv] = \begin{bmatrix} -t^{m1} & t^{m1} & \\ t^{m1} & -t^{m1} -t^{m2} & t^{m2} \\ & t^{m2} & -t^{m2} \end{bmatrix} \tag{A.6}$$

$$[A_e] = \begin{bmatrix} A_1 & & \\ & A_2 & \\ & & A_3 \end{bmatrix} . E \tag{A.7}$$

$$[C_c] = \begin{bmatrix} 1 & 0 & 0 \\ & & \\ 1 & 1 & 0 \end{bmatrix}_{nv \times nc}$$

(A.8)

(nv is number of spans and nc is number of columns)

$$\{\ell\}^T = \{\ell_1, \ell_2\}$$

(A.9)

$$\{R\} = [I_j]^{-1} \{I_u\} \quad ; \quad \{R_b\} = [J_t] \{R\}$$

$$[B_v] = [I_j]^{-1} [J_j]$$

$$[B_t] = -[J_t] [B_v] - [J_v]$$

$$\{C\}^T = \{I_e\}^T [B_v]$$

$$\overline{\{\ell\}}^T = \{\ell\}^T [C_c] + \{C\}^T [A_e]^{-1}$$

$$\lambda = \sqrt{k_2/k_1}$$
$$\xi_1 = \{H(P_1/H + P_2/2 + P_3) + k_4 - P_2/(\lambda^2 H)\} / (\lambda^2 k_1)$$
$$\xi_2 = \lambda(e^{\lambda H} + e^{-\lambda H})$$
$$\eta_1 = (1 - \lambda H)(\xi_1 e^{-\lambda H} - (P_2 + P_3)/(\lambda^3 k_1))/(\lambda \xi_2)$$
$$\eta_2 = (1 + \lambda H)(\xi_1 e^{\lambda H} + (P_2 + P_3)/(\lambda^3 k_1))/(\lambda \xi_2)$$
$$k^v = \eta_1 e^{\lambda H} + \eta_2 e^{-\lambda H} - (3P_2 + 8P_3) H^3 / (24\lambda^2 k_1) + \xi_1 (0.5 H^2 - \lambda^{-2})$$

$$\{v\} = [A_e]^{-1} \{2k_v \{R_b\} - (H + h) H \{D_4\}\} 0.5/h$$

$$\overline{\{\gamma\}}^T = \{\overline{\gamma}_1, \overline{\gamma}_2, \overline{\gamma}_3\}$$

(A.10)

where $\overline{\gamma}_i = v_i/v_1$, Ai are the cross areas of columns and the constants of matrices A.3 - A.6 are the stiffness of beams and columns. The stiffness matrix of any beam can be written as (see Fig. A.1):

$$[K^{mj}] = \begin{matrix} \theta_i & \theta_f & v_i & v_f \\ \begin{bmatrix} k_i^{mj} & a^{mj} & -b_i^{mj} & b_i^{mj} \\ a^{mj} & k_f^{mj} & -b_f^{mj} & b_f^{mj} \\ -b_i^{mj} & -b_f^{mj} & t^{mj} & -t^{mj} \\ b_i^{mj} & b_f^{mj} & -t^{mj} & t^{mj} \end{bmatrix} \end{matrix}$$

(A.11)

and the stiffness matrix of any column can be expressed analogously replacing m by n, the constants of eq. A.11 can include the shear deformations and the eccentricities due to the width of shear walls.

The constants related to vertical loads are the following:

$$\overline{M^F} = M^F_t + \{ \ell \}^T [C_c] \{D^v\}$$
$$\{D_1\} = [I_j]^{-1} \{D^M\}$$
$$\{D_4\} = [J_t] \{D_1\} + \{D^v\}$$

where, using the fixed end reactions of beams (Figs 1, 2, A.1):

$$M^F_t = (M^{F1}_i + M^{F1}_f) + (M^{F2}_i + M^{F2}_f)$$
$$\{D^M\}^T = \{M^{F1}_i, M^{F1}_f + M^{F2}_i, M^{F2}_f\}$$
$$\{D^v\}^T = \{V^{F1}_i, V^{F1}_f + V^{F2}_i, V^{F2}_f\}$$

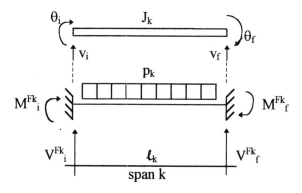

Fig. A.1 - Deformation co-ordinates and fixed end reactions for any span.

The constants M^f_t, $\{D^M\}$, $\{D^v\}$ are determined considering any intermediary level.

REDESIGN OF RC HIGH-RISE BUILDING

M R Reshidat

University of Jordan

Jordan

ABSTRACT. Upon defining the method of construction of a $70 million project of 16 storey hospital building, two structural models have been analysed; the first one is characterised by proposing precast and prestressed girders and topping slabs to form the floor system, and the other one is characterised by proposing cast in-situ reinforced concrete beams and slabs which has lead to monolithic formation with columns. The structural analysis has been carried out for the second system using matrix analysis of structures which is carried out by SAP 90 Computer Program. Calculation of wind loads and equivalent seismic forces were based on British Code, Eurocode 8, and the Jordan Code. Load combinations were based on the ACI-318 recommendations. Furthermore, dynamic analysis was also carried out using spectral as well as modal analysis. Based on the results of the analyses, the structural elements were designed according to the BS8110. The study showed that the second model has increased the seismic resistance of the structure which in turn mitigates and highly reduces expected structural damage which may result from the occurrence of earthquakes in the region.

Keywords: Analysis, Design, Dynamic, Earthquakes, High-Rise Buildings, Reinforced Concrete, Response, Seismic, Spectral, Stability.

Dr Musa Reshidat is an Associate Professor, Department of Civil Engineering, University of Jordan. He graduated from prominent universities; BSc. (Cairo University); MSc (Stanford University), PhD (Purdue University). He has 26 years of intensive experience in academic and professional fields. He is a fellow of ASCE, and a member of ACI, and several engineering organisations.

Radical Concrete Technology. Edited by R K Dhir and P C Hewlett. Published in 1996 by E & FN Spon, 2–6 Boundary Row, London SE1 8HN, UK. ISBN 0 419 21480 1.

INTRODUCTION

At the outset of the construction of a $70 Million project of 16-story hospital building which had been designed ten years ago, the method of construction for the floor systems was based on using precast and prestressed concrete girders and topping slabs. Such girders are considered as simply supported and thus benifits of monolithic formation with columns is not realised. Due to the importance of the hospital and to cope with the recommendations outlined in various international codes of practice for having a seismic resistant building, it was decided to evaluate this design as well as to redesign the building using cast in-situ construction for floor systems. Apart from the construction cost for both schemes, the primary objectives were to achieve stability, safety and seismic resistance of the structure during and after completion of construction. The structural models based on the corresponding method of construction were identical in every respect in terms of geometry, column dimensions and loads, but they differ in modeling the floor beams. The floor areas of this building are about 70,000 m^2 of which the volume of concrete is about 90,000 m^3 and the amount of steel reinforcement is about 5,000 tons. Reinforced concrete works cost 25% of the project value.

Fig.1 shows the plan of the structure which consists of four wings, four end connectors and one central connector forming a crucifix shape. The wings are separated from the connectors by expansion joints. Connectors are formed from cores of shear walls which are stiffened by floor slabs. The wings are formed of frames. Each wing in the longitudinal·direction consists of two exterior frames consisting of beams and columns forming 12 bays of 3.6 m each. In the transversal direction, the wing is one bay of 14.4 m comprised of columns and floors formed of the precast prestressed girders resting on the longitudinal beams; such girders are supporting the floor topping slabs. The girders are spaced 90 cm from each other. The plan shows interior columns for the ground floor only, and they are spaced 7.2 metres from each other.

The comparative study for both models focused heavily on the behavior of the wing in the transverse direction under the action of seismic forces. Of particular concern is the lack of ductility of the structural connection between the slab system and the supporting columns as well as the lack of frame action. The proper connection and detailing of the expansion joints were also considered.

After intensive deliberations among the client, designer, university experts and the contractor, it was decided to replace the floor system of precast prestressed girders and topping slab by cast in-situ reinforced concrete beams and waffle slab. The beams are spaced 3.6 m from each other. This decision required alterations to design of the structure.

DESIGN CRITERIA

The designs are carried out utilising the following general design criteria:

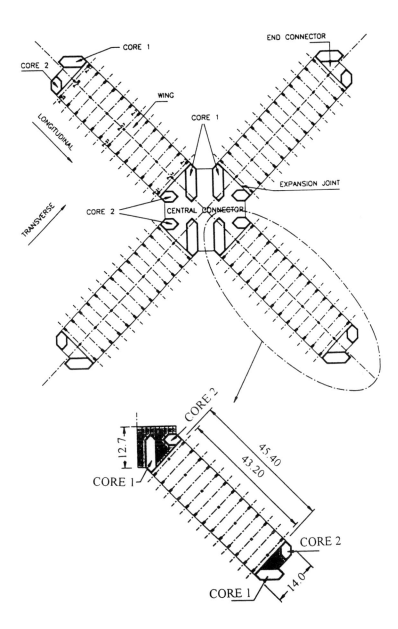

Figure 1. Plan of the Structure; dimensions are in meters.

Dead Loads

In addition to the self-weight, the following loads were considered:

Floor finish = 125 kg/m^2; HVAC loads = 75 kg/m^2; Partitions = 400 kg/m^2;

Radioactive lead thick walls = 570 kg/m^2.

Live Load

Floors = 300 kg/m^2; Public spaces = 500 kg/m^2; Storage and machine rooms = 500 kg/m^2

Horizontal Loads

The horizontal loads are related to wind forces or the equivalent seismic forces. Wind forces are calculated in accordance with the provisions of Part 2 of CP3, for a base wind speed of 120 km/hour [1]. Equivalent seismic forces are estimated in accordance with the provisions of the National Jordan Code for Loads and Earthquake Effects [2].

Material Properties

f$_c$ = 300 kg/cm^2; f$_y$ = 4100 kg/cm^2 for high tensile steel; f$_y$ = 2500 kg/cm^2 for mild steel.

The steel reinforcement is in compliance with BS 4449.

Codes and Standards

All reinforced concrete elements have been designed in accordance with the recommendations of British Standard BS 1110. The ACI Code [3] is employed to define load combinations. The Eurocode 8 is used in spectral analysis considering the existing soil properties of the site. The National Building Code [2] was used to define the applied horizontal acceleration.

ANALYSIS AND DESIGN

Due to the symmetry of the structure and considering the pattern of expansion joints between the wings and the shear wall connectors, only one-quarter of the structure was modeled as shown in Fig. 1. The wing, W, was modeled as a 3-D frame which consists of beam and column elements. The end connector, EC, and the central connector, CC, consist of closed shear walls and horizontal beam-slab system at the level of each floor. They have been modeled by an equivalent 3-D frame that has the same structural characteristics of stiffness.

Response Spectrum

For the seismic analysis, the response spectrum of the Eurocode 8 has been considered. Of particular importance is the provision for the local soil characteristics as well as the energy dissipation capacity of the structure through its more or less ductile behavior. The applied acceleration is directly related to the seismicity zone where the structure is located depending on the intensity of the earthquake. The intensity factor, as defined by the Jordan Code, is equal to 0.5 for the same zone. The intensity of damage according to Merchalli scale is 7 and the corresponding applied horizontal acceleration is 0.05g. The test results of the soil investigation have classified the supporting medium beneath the foundations of the building as type B in accordance with Eurocode 8.

Dynamic Modal Analysis

The dynamic modal analysis was carried out for the first and the second systems using SAP 90 computer program. Only the first five modes of vibration were considered in which they represent over 90% of the entire mass of the structure. Results of modal analysis are shown in Table 1. Furthermore, the first five modes of vibration for the frame (Wing) are shown in Fig. 2. Maximum spectral displacements for the transverse and longitudinal seismic actions are shown in Table 2. Results in terms of seismic shears and bending moments acting on the exterior frame column at different levels are given in Table 3.

Table 1. Characteristics of Modal Analysis: Period and Participating Mass for Transverse Direction.

Type of Design	Mode of Vibration	End Connector		Wing		End Connector-Wing-Central Connector	
		Period. sec.	Mass %	Period. sec.	Mass %	Period. sec.	Mass %
First	1	1.205	48.054	12.188	64.558	1.494	64.604
	2	6.208	21.516	2.608	16.040	0.306	17.716
	3	0.083	13.137	1.066	5.855	0.117	6.396
	4	0.045	6.199	0.576	3.157	0.061	3.373
	5	0.026	2.788	0.364	2.085	0.037	2.242
Second	1	1.205	48.054	1.683	82.207	1.296	65.821
	2	0.208	21.516	0.589	7.152	0.291	16.778
	3	0.083	13.134	0.385	1.089	0.115	6.228
	4	0.045	6.199	0.332	0.221	0.060	3.321
	5	0.026	2.788	0.315	0.030	0.037	2.220

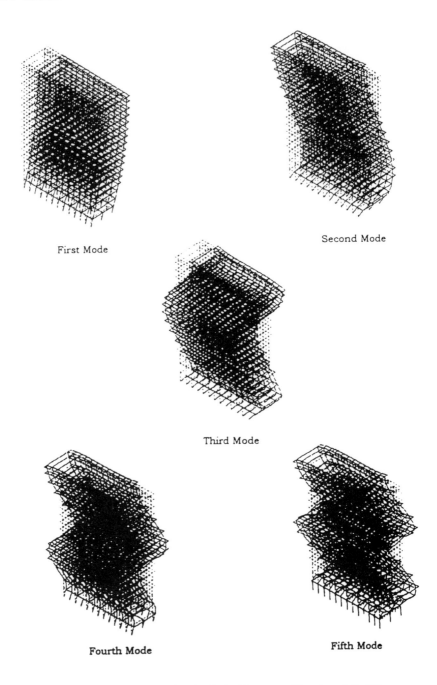

First Mode

Second Mode

Third Mode

Fourth Mode

Fifth Mode

Figure 2. The First Five Modes of Vibration in the Transverse Direction of the Wing.

Table 2. Maximum Displacements

Type of Design	Part of Structure	Spectral Displacements, mm		Static Displacements, mm	
		Longitudinal Direction	Longitudinal Direction	Transverse Direction	Transverse Direction
First	End Connector	41	41	47	67
	Wing	67	67	82	388
	Central Connector	61	61	61	61
Second	End Connector	41	41	47	67
	Wing	61	61	54	74
	Central Connector	45	45	40	40

Table 3. Column Forces due to Seismic Actions.

Type of Design	Level of Exterior Column	Longitudinal Seismic Forces		Transverse Seismic Forces	
		Shear, kN	Moment kN-metre	Shear, kN	Moment, kN-metre
First	Ground floor	259.62	704.11	18.40	154.06
	7th floor	178.55	351.28	0.67	8.31
	Roof	13.32	28.66	12.57	37.02
Second	Ground floor	75.20	249.91	2.65	95.21
	7th floor	50.31	99.77	68.47	135.08
	Roof	3.51	8.12	16.28	71.72

Static Analysis and Design

The structural model incorporates the material properties, geometry, loads, and dimensions of each constituent element of the structure. The model is assumed to be linear and elastic. The computations are carried out by STATIK-2 computer program, version 1.51 which is based on the matrix method of structural analysis.

For the second design, typical results for vertical and shear forces, and bending moments at selected joints of the external columns of the wing in the transverse direction of the wing is given in Table 4. Maximum displacements are given in Table 2.

Having determined the design ultimate strength for each beam, column, or core elements based on the maximum load combinations, the corresponding reinforced concrete sections were designed. The second design is compared with the first one .Of particular concern is the column reinforcement as shown in Table 5.

Table 4. Forces on Selected Joints of the External Column in the Transverse Direction Based on Static Analysis for the Second Type.

Level of Column	Vertical Forces, kN	Shear Force, kN	Bending Moment, kN-metre
Ground Floor	7442.36	90.37	579.84
Seventh Floor	2724.92	59.58	128.22
Roof	127.43	5.76	51.62

Table 5.   Column Reinforcement

Floor Level	Old A_s, cm^2	New A_s, cm^2	$\dfrac{\text{New } A_s}{\text{Old } A_s}$
Ground Floor	120	243	2.02
First Floor	96	200	1.97
Second Floor	96	173	1.80
Fourth Floor	96	110	1.15

DISCUSSION

The results of the dynamic study for both designs showed that both structural models have similar behavior in the longitudinal direction. Due to the existence of expansion joints between the wing and the adjacent connectors, the behavior in the transverse direction of the wing is the dominant and governing in terms of serviceability requirements. In the first design, the wing behaves as a vertical built-in cantilever. As shown in Table 1, the fundamental period of the first mode is slightly over 12 seconds. Practically this behavior is not acceptable and such period falls far beyond the maximum practical value which is 2.5 seconds [5]. This wing behavior is mainly due to the existence of precast prestressed concrete beams which are simply supported or hinged at both ends with supporting beams and/or columns. The wing behavior in the second design is acceptable where the fundamental period is about 1.7 seconds. It is the same reasoning for wing behavior in terms of displacements as shown in Table 2.

The width of expansion joint between the wing and the adjacent shear wall connectors is 20 mm according to the first design. The dynamic analysis in the longitudinal direction showed that the relative movement for the wing and the end connector is 108 mm and 102 mm for the first and second design, respectively (Table 3.) Similarly, the relative movement between the wing and the central connector is 128 mm and 106 mm for the first and the second design, respectively. Based on that, the width of expansion joint as designed [20 mm] is obviously not sufficient. Consequently, strong impact between the wings and connectors will originate causing large stresses which are not only very difficult to evaluate but that would alter the structural behavior to a great extent which is not acceptable.

In the second design, expansion joints are sized to be 120 mm in accordance with the results of dynamic displacements shown in Table 2. It should be noted that for lower levels, such relative values will be smaller than the maximum values at the roof level.

In the transverse direction, the lateral displacement of the wing at the roof level is very large (3082 mm) compared to the corresponding value for the second design (76 mm). Construction of the structure according to the first design may lead to a possible structural collapse during the occurrence of an earthquake.

The reinforcement of columns at lower levels was increased to correspond to the analytical results of the second design. Such column reinforcement at higher levels was decreased accordingly. Comparison between the two designs for only columns of lower levels is clearly shown in Table 5.

Although it was concluded that the second design is superior over the first one, two points have been considered to improve the overall structural behavior. The first point is the selection of proper width of expansion joints which is about 12 cm to avoid rocking or hammering between the wing and the connectors during an earthquake occurrence in the longitudinal direction. The orientation of the expansion joints was recommended not to be a straight line. It was recommended to be in the form of a shear key joint. The second point is characterised by contribution of the connectors to resist lateral forces in the transverse direction of the wing by means of the new orientation of the shear key expansion joints where dowel reinforcement between the connectors and the wings are provided at each floor level. Furthermore, a vertical isolation is introduced between the wing and connectors at the shear key joints to absorb impact during the occurrence of dynamic forces in the transverse direction of the wing. The isolation consists of two steel plates sandwiched by neoprene pads.

CONCLUSIONS

1. The proposed method of construction by casting in-situ concrete for floor system which consists of reinforced concrete beams and waffle slabs forming monolithic formation with columns is superior than the first system which consists of simply support precast prestressed girders and topping slabs.

2. The second design showed a realistic behavior in terms of dynamic or static lateral displacements compared with the first one. Such behavior is due to monolithic formation between floor beams and columns.

3. The second design offered the practical as well as the feasible sizing of expansion joints which conforms to the results of dynamic displacements in the longitudinal direction.

4. It is anticipated that the redesigned structure will possess a substantial increase of seismic resistance which in turn mitigates and highly reduces expected structural damages during the possible occurrence of earthquakes in the region.

ACKNOWLEDGMENTS

The author thanks the team of Iberica de Estudios e Ingenieria, S. A. of Madrid, Spain for their valuable efforts in providing the software packages and design calculations, and also for the technical team in the Department of Engineering Projects headed by Engineer M. Ghanma for their priceless efforts and contribution towards checking and reviewing the design calculations.

REFERENCES

1.   CP3: Chapter V, Part 2: Wind Loads, 1972

2.   PUBLICATION OF MINISTRY OF PUBLIC WORKS AND HOUSING. The National Building Code: Earthquake Effects [In Arabic], Amman Jordan, 1985.

3.   AMERICAN CONCRETE INSTITUTE. Building Code Requirements for Reinforced Concrete ACI 318-89, Chapter 9, Detroit, Michigan, U.S.A., 1989, pp.89-103.

4.   EUROCODE 8. General Rules and Rules for Buildings - Seismic Action and General Requirements for Structures, Second Draft, October 1993, pp.12-20.

5.   WIEGEL, ROBERT L. Earthquake Engineering, Printice-Hall, Inc., Englewood Cliffs, N.J., 1970, pp.404.

RECOGNISING THE STRUCTURAL CHARACTERISTICS OF CONCRETE

N K Subedi

University of Dundee

UK

ABSTRACT. Concrete as a construction material dominated the 20th Century with its application ranging from ultra highrise buildings to structures in the deep oceans. Throughout this period the fundamental structural characteristic on which the design has been based is the compressive strength of concrete. The tensile strength of concrete is largely ignored in design, yet, it is evident that the structural behaviour of most concrete structures is governed by the tensile strength of the material. The failure of most of the concrete structures is led by the weakness of the material in tension. Conventional shallow beams, deep beams or panel walls in which the diagonal tension leads to failure are some of the examples. In highrise concrete shearwall structures the diagonal splitting in tension of the coupling beams governs the overall mode of failure and its ultimate strength under static lateral load.

In this paper the entire basis of the current concept of design of concrete structures is put into perspective. Recognising the characteristics of concrete in the context of structural behaviour is more fundamental to design. Improving the tensile strength characteristics of concrete may well set the scene for research in the next Century.

Keywords: Concrete, Tensile strength, Structural behaviour, Design, Characteristics.

Dr Nutan K Subedi is a senior lecturer in civil engineering at the University of Dundee, Scotland, UK. His main research interests are in the field analysis and design of concrete structures which include deep beams, panel structures, tall buildings, shearwalls, and concrete elements under hydrostatic pressure effects.

Radical Concrete Technology. Edited by R K Dhir and P C Hewlett. Published in 1996 by E & FN Spon, 2–6 Boundary Row, London SE1 8HN, UK. ISBN 0 419 21480 1.

INTRODUCTION

The application of concrete in construction dates back to 500 BC [1] when Ancient Egyptians and Greeks used sand and lime mortar mixed with broken stone in buildings. Concrete as a durable material was developed by Romans when pozzolanic cement using a mix of volcanic ash and lime gave stronger material suitable for arches and domes. The decline of the Roman Empire brought a halt to concrete development. The next significant milestone and revival of interest in concrete was the construction of Eddystone lighthouse in 1756 when John Smeaton used a cement capable of setting underwater. The mix chosen was a cement mortar of lime and pozzolana, similar to what the Romans had used. The first patent on Portland cement was taken out by Joseph Aspdin, a builder from Leeds, in 1824. This was the culmination of the wider use of concrete as a structural material both in Europe and North America. The development of reinforced concrete as a structural material was more gradual through the mid-19th Century and onwards.

Concrete as a basic construction material has dominated the 20th Century with its application ranging from ultra highrise buildings [2] to offshore structures [3] in the deep oceans (Figure 1). Other novel applications of concrete may be found in ships hulls [4] and containment structures [5] for nuclear power plants (Figure 2).

Throughout the development of concrete the fundamental structural characteristic on which the design has been based is the compressive strength of concrete. The tensile strength of concrete is largely ignored in design. However, it is evident that the structural behaviour of most concrete structures is governed by the tensile strength of the material. The failure of most of the concrete structures is led by the weakness of the material in tension.

It is important to recognise the characteristics of concrete in the context of structural behaviour which is fundamental in the design of concrete structures. This paper examines the current concept of design of concrete structures and points out the importance of the tensile strength characteristics of concrete in the context of overall behaviour. Improving the tensile strength characteristics of concrete could not only enhance the load carrying capacity of structures but it could also lead to more economic and efficient form of construction.

STRUCTURAL BEHAVIOUR AND FAILURE MODES

Prestressed Concrete Beams

In prestressed concrete beams one of the mechanisms of failure in shear is the formation of diagonal splitting cracks in the web as shown in Figure 3. The diagonal splitting crack is caused by a tensile stress, σ_t, in biaxial compression - tension state of stress. The resistance of the section in such situation is calculated by limiting the tensile stress in concrete to $0.24 \sqrt{f_{cu}}$ according to BS 8110 [6]. In Table 1, the tensile stresses for different grades of concrete and the percentage of concrete strength they

(a) 311 W Wacker Drive, Chicago : 295 m high concrete building

(b) 473 m high Troll gas production platform during towing

Figure 1 Application of Concrete (a) tall building (b) offshore gravity platform

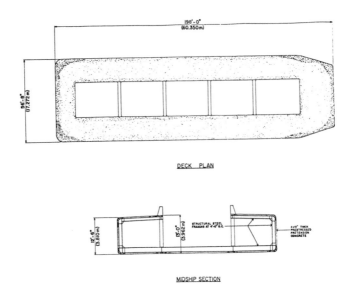

DECK PLAN

MIDSHIP SECTION

(a) 2000 ton Concrete general cargo barge

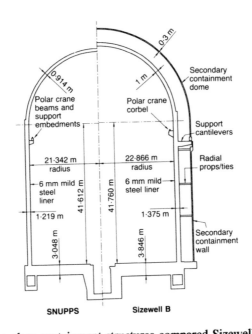

(b) Concrete nuclear containment structures compared Sizewell B and SNUPPS

Figure 2 Application of concrete (a) Concrete barge on Mekong (b) Sizewell B

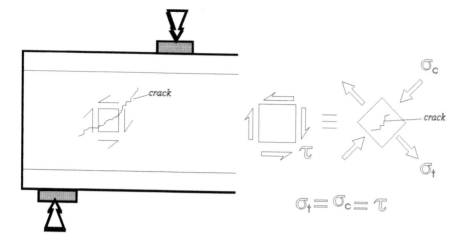

Figure 3 Prestressed concrete beam : diagonal tension crack

represent are shown. The limiting tensile stress $f_{tc} = f_{cu}/21$ is derived from the recommendations in Reference [7] with the 0.67 factor removed for concrete under biaxial tension - compression state of stress.

Table 1 Concrete grades and tensile stresses in N/mm²

f_{cu}	25	30	35	40	45	50	55	60
$0.24 \sqrt{f_{cu}}$	1.20	1.31	1.42	1.52	1.61	1.70	1.78	1.86
% of f_{cu}	4.8	4.4	4.1	3.8	3.6	3.4	3.2	3.1
$f_{cu}/21$	1.19	1.43	1.67	1.90	2.14	2.38	2.62	2.86
% of f_{cu}				4.76				

Table 1 shows that the BS 8110 recommendation of $0.24 \sqrt{f_{cu}}$ gives tensile stress reading as a percentage of concrete strength as the strength of concrete increases. The tensile stress based on Reference [7] suggests a fixed percentage (4.76) of concrete strength.

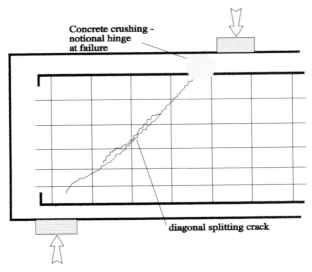

Concrete crushing -
notional hinge
at failure

diagonal splitting crack

(a) Deep beam : diagonal splitting mode of failure

Figure 4 Deep beam example

Deep Beams

In deep beams with moderate amount of main bars, the diagonal splitting is a common mode of failure [8]. The failure is initiated by a diagonal crack which forms in the web part of the shear span as shown in Figure 4. The shear span is primarily subjected to shear stress resulting into biaxial tension - compression state of stress similar to the prestressed concrete beam. The splitting action is initiated by the weakness of concrete in tension which eventually leads to a mechanism resulting into complete failure of the beam.

In the case of deep beams, the ultimate strength of the beam is sensitive to the tensile strength of concrete (in biaxial tension - compression state of stress).The ultimate strength analysis of deep beams carried out using, $f_{tc} - f_{cu}/21$ i.e., 4.76% of f_{cu} (Table 1) has proved to be satisfactory [8, 9]. At the same time, it also highlights that the mode of failure and the ultimate strength of deep beams are governed by the tensile strength characteristics of the concrete.

Coupled Shear Walls of Tall Buildings

The mode of failure and the ultimate strength of coupled shear walls with single and multiple bands of openings and subjected to a lateral load are greatly influenced by the behaviour of their coupling beams [10, 11]. A mode of failure which is often observed is the crushing of the compression wall characterised by the diagonal splitting of the coupling beams as shown in Figure 5. Here again, the diagonal splitting of the beams is caused by the biaxial tension - compression stresses in the beams resulting from the shear force. The main parameter on which the splitting action of the beam depends is the tensile strength of the concrete.

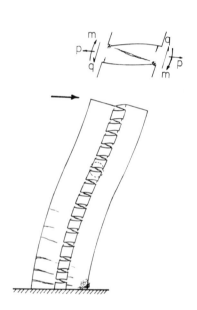

(a) Coupling beams and overall wall deformation near failure

(b) A laboratory model after failure

Figure 5 Coupled Shear wall : failure characterised by the diagonal splitting of coupling beams

The instances of structural behaviour in which the tensile strength of concrete, either in uniaxial or in biaxial state, plays a major part occur in many other structures. In all such situations, even a small increase in the tensile strength characteristics of concrete would have a large beneficial effect to the structure.

CURRENT STATE OF DESIGN

The current concept of design of concrete structures is based on either ignoring or allowing a nominal value for the tensile stress of concrete. Here, the recommendations of BS 8110 and EC2 [12] are examined.

BS 8110 Provisions

(a) In the calculation of curvatures a nominal value for the tensile strength of concrete of 1 N/mm² is assumed.

(b) The main consideration of flexural tensile stress is incorporated in Chapter Four : Design and detailing : prestressed concrete.

For class 1 members, at serviceability limit state, no tensile stress is allowed. At transfer a stress of 1 N/mm² is allowed in the section.

For class 2 members, both at serviceability and at transfer, tensile stresses of 0.45 √f$_{cu}$ and 0.36 √f$_{cu}$ are allowed for pre-tensioned and post-tensioned cases respectively. For grade 60 concrete, the values are 3.5 and 2.8 N/mm² (Table 4.1 BS 8110) representing 5.8 and 4.6% respectively for pre-tensioned and post-tensioned members.

For class 3 hypothetical stresses are suggested.

(c) In shear resistance calculations the allowable stresses arising from diagonal tension is limited to 0.24 √f$_{cu}$ as shown in Table 1.

EC2 Provisions

(a) The tensile strength of concrete is taken into account in the expression for design shear resistance for members which is applicable to both reinforced and prestressed concrete members. The design strength for members without shear reinforcement is given by

$$V_{Rdi} = [\tau_{Rd} \, K(1.2 + 40 \, \rho) + 0.15 \, \sigma_{cp} \,] \, b_w d$$

in which, τ_{Rd} is the basic design shear strength, $= 0.25 \, f_{ctk} \, 0.05/\gamma_c$. Using $\gamma_c = 1.5$, the values are

Concrete :	C20/25	C30/37	C40/50	C50/60
τ_{Rd} :	0.26	0.34	0.41	0.48

(b) Serviceability limit state : Prestressed concrete

Serviceability stresses are checked in section assuming either the uncracked or the fully cracked conditions under characteristic loads. The limiting tensile stresses, f$_{ctm}$, for uncracked conditions are given in Table 2.

Table 2 Concrete strengths and tensile stresses in N/mm^2 : EC2

Strength class	C20/25	C25/30	C35/45	C40/50	C45/55	C50/60
f_{ck}	20	25	35	40	45	50
f_{ctm}*	2.2	2.6	3.2	3.5	3.8	4.1
% of f_{ck}	11.0	10.4	9.1	8.8	8.4	8.2
f_{ctk} 0.05	1.5	1.8	2.2	2.5	2.7	2.9
f_{ctk} 0.95	2.9	3.3	4.2	4.6	4.9	5.3

*$f_{ctm} = 0.30 \, f_{ck}^{2/3}$

Three points are noted with regard to EC2 recommendations

(i) As in BS 8110 the mean value of the tensile strength reduces as a percentage of concrete strength as the strength of the concrete increases.

(ii) EC2 recommendations of mean tensile strengths are higher compared to BS 8110.

(iii) For τ_{rD} the tensile strength is taken from the lower range of the values i.e., f_{ctk} 0.05 (5% -fractile).

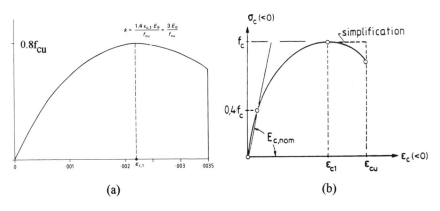

(a)   (b)

Figure 6 Stress-strain relationships for concrete for non-linear analysis
(a) BS 8110 (b) EC2

Provisions for Non-linear Analysis

The tensile strength characteristics of concrete are important considerations in the non-linear analysis, such as Finite Element method, of concrete elements and structures. Most non-linear programs require the input of the concrete uniaxial tensile strength as a proportion of the uniaxial compressive strength.

BS 8110 suggests that, for non-critical sections or overall analysis, the stress-strain relationship be assumed as shown in Figure 6(a). The constitutive relationship is appropriate for the compressive action in the concrete. For the tensile action of the material BS 8110 recommends a maximum value of 1 N/mm². Likewise, for non-critical sections EC2 suggests the stress-strain relationship for the compressive action as shown in Figure 6(b). For the tensile stress properties EC2 suggests the mean values of f_{ctm} (Table 2).

For the analysis of critical sections both BS 8110 and EC2 recommend zero tensile strength.

The current state of design show that there is considerable difference between BS 8110 and EC2 with regard to recommendations for the tensile strength of concrete. In general, BS 8110 is conservative compared to EC2 and there is inconsistency in the approach.

DESIGN PARAMETERS

Research in the past two decades has mainly been focused on improving the quality of concrete with respect to compressive strength and durability parameters. Much time and effort have also been devoted on research into cement replacement materials. As a result, there is a large selection of concrete available today in the market for general and special applications.

It is, however, clear the one of the most important characteristics of concrete, from the structural engineering point of view, is the tensile strength. There appears to have been no research effort directed towards improving the tensile strength. Concrete will be much more an efficient material for structural use if the tensile strength could be improved from the current approximately 10% of its Cube Strength to say 50% in a consistent manner.

Very high strength concretes without simultaneous improvement in its tensile strength characteristics will have restricted applications in structures. Evidence from the current recommendations in BS 8110 and EC2 suggests the as the compressive strength increases the tensile strength is proportionately decreased. As the compressive strength increases, the concrete becomes less and less reliable with regard to its tensile strength. Although the steel reinforcement is supposed to compensate for the lack of tensile strength of concrete, in reality, reinforcement cannot fulfil this role fully unless the structure is reinforced with very closely spaced reinforcement almost like a mat.

FUTURE RESEARCH DIRECTIONS

The current research activities appear to have leaped far ahead with regard to improving the quality of concrete in so far as compressive strength is concerned. If the versatility of the material is to be improved then attention must be paid to improving its tensile strength too. A normal concrete with moderate compressive strength but improved tensile strength is much preferred than a concrete with very high compressive strength but negligible tensile strength.

Research in the next century may focus its attention on improving the tensile strength of concrete. With the basic constituent materials of concrete as they are known today three possible routes might be followed.

(a) Concrete with better binding characteristics
Such a concrete will have the properties of drawing the constituent materials closer than the normal concrete resulting into denser and better adherence property. Although, at present, it is difficult to imagine what chemical agents or additives would provide such a property to a concrete, but, it is envisaged that it should be possible in due course.

(b) Casting and curing of concrete in vacuum
A normal concrete cast and cured in normal atmospheric pressure if tested under higher pressure surrounding will exhibit higher tensile strength. The additional surrounding pressure will act as prestressing and therefore a larger force will be required to pull it apart.

(c) Concrete with metal links
Metal wire structures in the form of regular tetrahedron slightly larger than the aggregate size could act as linkage in the concrete. The metal wire structures will not only provide additional tensile strength due to its presence, but, it will also act as mechanical interlocks for the aggregates forming a continuous medium (Figure 7).

It is considered that concrete which forms more than 50% of all the construction materials today could be recognised as a true versatile material only if its other important characteristic i.e., tensile strength could be improved. This could be a real challenge to the concerned engineers and scientists in the next century.

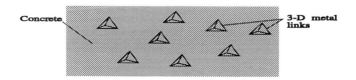

Figure 7 Concrete with metal links

REFERENCES

1. THE CONCRETE SOCIETY. The Story of Reinforced Concrete.

2. BENNETT, D. Introducing High Strength Concrete, Concrete Quarterly, Spring 1990, pp 6-7.

3. NEW CIVIL ENGINEER (NCE). 18 May 1995.

4. SARE, P N AND YEE, A A. Operational Experience With Prestressed Concrete Barges, Symposium on Concrete Ships and Floating Structures, 3&4 March 1977, London, Organised by The Concrete Society at the Royal Institution of Naval Architects.

5. DAVIES, D R AND ROBERTS, A C. Power Station Design, Proc. Instn. Civ. Engrs, Civil Engineering, Sizewell B Power Station, 1995, 108, Feb, pp 15-29.

6. BRITISH STANDARDS INSTITUTION. Structural Use of Concrete, BS 8110 : Parts 1 and 2 : 1985.

7. HOBBS, D W, NEWMAN, J B AND POMEROY, C D. Design of Stresses for Concrete and Structures Subjected to Multiaxial Stresses, The Structural Engineer, 55, No 4, April 1977, pp 151-164.

8. SUBEDI, N K. Reinforced Concrete Deep Beams : A Method of Analysis, Proc. Instn. Civ. Engrs, Part 2, No 85, Mar 1988, pp 1-30.

9. SUBEDI, N K. A Method of Analysis for Reinforced Concrete Deep Beams with Fixed End Supports, The Structural Engineer, Vol 72, No 10, 17 May 1994, pp 155-163.

10. SUBEDI, N K. RC Coupled Shear Wall Structures I : Analysis of Coupling Beams, Jnl Struct Eng, ASCE, Vol 117, No 3, March 1991, pp 667-680.

11. SUBEDI, N K. RC Coupled Shear Wall Structures II : Ultimate Strength Calculations, Jnl Struct Eng, ASCE, Vol 117, No 3, March 1991, pp 681-698.

12. BRITISH STANDARDS INSTITUTION. Eurocode 2 : Design of Concrete Structures, Part 1 : General rules and rules for buildings, DD ENV 1992-1-1 : 1992

CONSTRUCTION OF ARCH BRIDGES USING PRESTRESSED SFRC ELEMENTS AS LOST PANELLING

J Korla

D Magajne

IRMA Institute for Research in Materials and Applications

Slovenia

ABSTRACT. The construction of arch bridges has become uncommon lately, although it is clear that the compressive bearing capacity of cheap building materials is generally very high. The main reason for this is definitely the high price of shuttering system that is in present use. The system presented in this paper demonstrates a new relationship between the price of shuttering and the structure itself. The basic element is a very thin prestressed, post arch-shaped plate, made of polymer modified SFRC. Such elements are used as shuttering systems during the first stage of construction while later they become a part of a general bearing system. This brings about entirely new possibilities for the applications of arch constructions, especially regarding the actual known characteristics of high performance concrete.

Keywords: Arch constructions, Prestressed concrete, Steel fibre reinforced concrete (SFRC), Polymer modified concrete (PMC), High performance concrete (HPC).

Mr Jadran Korla is a Major Research Assistant in Concrete Technology, Institute for Research in Materials and Applications. He specialises in applications for high performance concretes.

Mr Danijel Magajne is a Research Assistant in Concrete Prefabrication Technology, Institute for Research in Materials and Applications. He specialises in applications for inovative concrete elements and structures.

Radical Concrete Technology. Edited by R K Dhir and P C Hewlett. Published in 1996 by E & FN Spon, 2–6 Boundary Row, London SE1 8HN, UK. ISBN 0 419 21480 1.

INTRODUCTION

Arch appear very rarely in todays construction practises. The reason behind this is the high price of the more commonly used shuttering systems. Even for small span bridges over wild rivers or motorways, the technology for the construction of such arches is quite often too expensive.

In recent years, with new data regarding the characteristics of such high performance concretes [1], as SFRC and PMC respectively [2],[3], the possibilities for producing very long and thin prestressed plates have become more and more realistic. It was believed that the shape characteristics of such plates enabled post-shaping procedures to be performed in the early stages of the concrete hardening process without any practical danger of undesired cracking in the tensile zone.

First trials confirmed such possibilities and demonstrated that, depending on the thickness of the prestressed plate, it is possible to apply post-shaping procedures almost up to any normally usable radius.

Few arch bridges of up to 40 m span have been constructed using this technology with much success in Slovenia (Fig.1 and Fig.2) [4]

Figure 1   Soča Bridge

Figure 2   Vrhpolje Bridge

BASIC IDEA

As previously described, the basic element is a 100 to 150 cm wide, extremely thin (2 to 6 cm, depending on final arch radius), flat and very long prestressed plate made of polymer modified steel fibre reinforced concrete (Fig.3). The length of the plate is determined from the transportation system or the bearing capacity of on site placing equipment (Fig.4 and Fig.5).

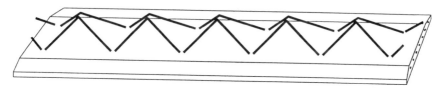

Figure 3   Shows the first phase of the system-producing of flat prestressed polymer modified SFRC plate with needed conventional steel reinforced placed in

The range of prestressing depends on the tensile stresses that appear after post-shaping (final radius). The quality of placement of the SFRC shoud be apropriate through entire element so the stress distribution after post-shaping could reach an equal level.

Figure 4   Transport of elements

Figure 5   Placement of the first element on site

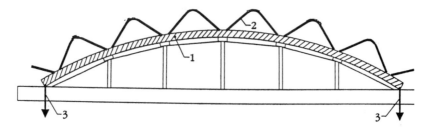

Figure 6 Shows the second phase of the system-post shaping of basic element using
pre-shaped basis and ballast

1 - SFRC plate;   2 - stirrups;   3 - ballast

The post-shaping takes place during the early stages of concrete hardening (2 to 5
days after placing of concrete). The procedure should be rather slow (1 to 3 hours).

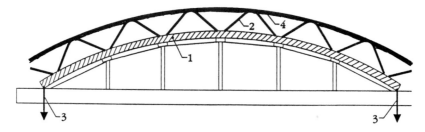

Figure 7 Shows the third phase of the system - the fixing of final shape

1 - SFRC plate; 2 - stirrup; 3 - ballast; 4 - shape-fixing steel bar or profile

Fixing of the final shape is done by welding the steel bar or profile (4) to each of the
stirrups (2). After removal of the ballast (3), the steel bar (4) is under compressive
stress. Additional conventional steel needed is emplaced on site.

Elements prepared in such way (Fig.8) are transported and placed into their final
position where phase 4 is taking place (Fig.9).

After emplacement of the post-shaped elements (1) - (Fig.9) and fixing the joints (2) -
(Fig.9), the further construction of the upper bridge is possible using any known
system.

A very convenient system, used in the case of 40 m Soča bridge in Slovenia, utilizes
transverse placements of extruded hole concrete slabs spanning a Vierendel bearing
system after emplacement of the first concrete between them (Fig.10).

Figure 8   Elements prepared in the factory

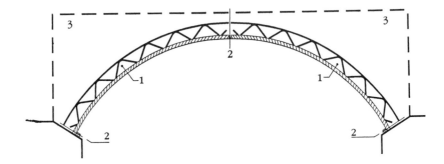

Figure 9   Shows the fourth phase of the system-placing of the basic post-shaped elements and fixing of the joints

1 - basic post-shaped element
2 - joints to be fixed on site
3 - bridge structure to be built later

The raising of the upper structure of the bridge must follow the statical system to avoid the appearance of extreme stresses, especially in the middle of the arch. During the final stage of construction the bottom surface of the arch may be additionally protected against corosion if needed.

Figure 10 Soča Bridge after placements of whole concrete slabs and the first concrete between them

STATICAL SYSTEM

The statical system of the 40 m span Soča arch bridge in Slovenia is shown.

Prestressing and Placing of PM SFRC

The plates used as lost panelling that remains as a component of the main bearing system are made of polymere modified (PM) steel fibre reinforced concrete (SFRC) with compressive strength of 70 to 80 MPa adhesionally prestressed with 7 $\varnothing$ 4,2 mm prestressing steel (Fig. 8).

The prestressing forces (N_1) cause the following compressive stresses (σ_N) in the concrete cross-section (F):

$N_1 = 120$ kN
loses 25%, 7 x 7 $\varnothing$ 4,2 mm prestressing steel cables
$N = 7$ x 0,75 x 120 kN = 630 kN
$F = 562,5$ cm^2
$\sigma_N = 11,2$ MPa

Post-shaping of PM SFRC Plate

Post-shaping (Fig. 6, 8), at the final shape of the arch, causes the following stresses in PM SFRC plate:

$f = 1,44$ m - the arrow of the post-shape element
$L = 21,14$ m - the length of the post-shaped element
$W = 528$ cm^3 - moment of resistancy
$M = 551$ kNcm - moment caused by post-shaping with ballast
σ_u - stress on the upper side of PM SFRC plate
σ_b - stress on the bottom side of PM SFRC plate

$$\sigma_u = \sigma_N - \frac{M}{W} = 0,8 \text{ MPa} \qquad \sigma_b = \sigma_N + \frac{M}{W} = 21,6 \text{ MPa}$$

Placing of Concrete of General Bearing Arch

After the placement of concrete for the general bearing arch in the PM SFRC post-shaped element, the following stresses appear:

$f = 6,00$ m - the final arrow of the bridge
$L = 40,00$ m - the final span of the bridge
$g = 7,9$ kN/m^1 - the weight of fresh concrete
$N = 263$ kN - axial force in arch due to weight of fresh concrete
$F = 687,5$ cm2 - the cross-section of PM SFRC
σ_u^M = maximal stress in PM SFRC - upper edge
σ_b^M = maximal stress in PM SFRC - bottom edge

$$\sigma_u^M = \sigma_u + \frac{N}{F} = 4,6 \text{ MPa}$$

$$\sigma_b^M = \sigma_b + \frac{N}{F} = 25,4 \text{ MPa}$$

The stresses appearing at this point are a critical phase of the system. All further stresses are beared by the concrete of the general bearing arch which, at this point, is under zero stress and still has all of its available bearing capacity.

By all means, the stresses during the critical phase which appear in bottom edge of the PM SFRC plate (25,4 MPa) are not bearable with conventional concrete, according to specifications. It must also to be taken in account that the actual behaviour of PM SFRC could cause even greater stresses due to diminished creep of high strength SFRC as well as the possible appearance of unexpected stresses due to transportation, manipulation, assembly etc...

CHARACTERISTICS OF PM SFRC

The PM SFRC of 40 m span Soča arch bridge is shown.

The tehnology presented in this paper demands some characteristics of concrete which are not attainable by conventional concrete:

- relatively high compressive strenghts (70-80 MPa) combined with relatively low elastic modulus (25-35 GPa),
- high toughness of concrete (PM SFRC),
- the safe acceptance of high compressive stresses during early stages of hardening,
- the capability to accept relevant additional tensile stresses even after the appearance of the first crack (transport, manipulation, montage).

These demands have been reached in the case of the 40 m Soča arch bridge in Slovenia using concrete with following main ingredients:

- Portland cement (45 MPa)
- Aggregate 0/8 mm (crushed calcium carbonate of good quality)
- Silica fume (8% of cement weight)
- Poymere (5% of cement weight)
- Superplasticizer (1,5% of cement weight)
- Air entraining agent (0,1% of cement weight)
- Steel fibres JV 50/30 (0,5 vol%)

CONCLUSIONS

Until now, the presented tecnhology was used primarily for the smaller arch structures spanning up to 40 m span. According to recent experience, it is believed that appropriate modifications and further research will enable us to widen its use.

In addition to its acceptable price, one of the most important characteristics of this technology is that the exploatation surface of the element (inner surface of the arch) is always under strong compressive stress what practically exclude the possibility of crack appearance and therefore should provide relevant durability of the structure.

Further development, as we see it, should be directed towords diminishing the final radius as much as posible and facilitating the production of sections for underground structures for which the speed of construction is of particular importance.

REFERENCES

1. RILEM 3C Coordinating Committee for Concrete Technology. Durability of High Performance Concrete. Proceedings of the International RILEM Workshop, Vienna, 1994, Edited by H. Sommer

2. Zajc A. and Korla J.: Freeze-thaw Resistance With and Without De-icing Salts of Polymer Cement Concrete, Proceedings of ICPIC VIII International Congress on Polymers in Concrete, Oostende, 1995, pp 245-250

3. Šušteršič J.: Toughness of Polymer Cement Fibre Reinforced Concrete. Proceedings of ICPIC VIII International Congress on Polymers in Concrete, Oostende, 1995, pp 649-654

4. Korla J., Magajne D.: Ločne premostitve s prednapetimi tankostenskimi mikroarmiranimi betonskimi elementi kot izgubljenimi opaži. Slovenski kolokvij o betonih: Sanacije betonskih objektov, IRMA, 1995, pp 57-60

SEISMIC CAPACITY ASSESSMENT OF CHUNG-SA BRIDGE AFTER SCOURING BY TYPHOONS OF 1994

Y-F Li

Join Engineering Consultants

I C Chen

Taiwan Area National Freeway Bureau

Taiwan

ABSTRACT. In July and August of 1994, Taiwan was invaded by several aggressive typhoons. Those typhoons brought heavy floods, and the floods scoured the piles of Chung-Sa Bridge, so that the piles at spans No. 15-17 and No. 35-39 were exposed. The Taiwan Area National Freeway Bureau (TANFB) took the strengthening action immediately by putting gabions around the exposed piles to strengthen the stiffness of the exposed piles. This paper focuses on the seismic capacity evaluation of the strengthened Chung-Sa Bridge by means of in-situ ambient vibration measurements and finite-element simulation. Our research shows that the maximum allowable scouring depth of piles at Chung-Sa Bridge is 6.8 m in the site condition at spans No. 35-39, after strengthening the exposed piles by putting gabions around the exposed piles. The condition of the strengthened Chung-Sa Bridge is still on the safe side in accordance with the requirements of the "Seismic Design Specification for Highway Bridges" of Taiwan.

Keywords: Bridge, In-Situ ambient vibration measurement, Natural Frequency, Pile, Scouring, Seismic

Dr Yeou-Fong Li is a Project Section Leader of Bridge Inspection and Evaluation Division of JOIN Engineering Consultants. His majors are focus on structural dynamics, spectrum analysis and in-situ ambient vibration measurement.

Mr I-Chang Chen is Director of Planning and Design Department of Taiwan Area National Freeway Bureau. He is in charge of the maintenance and widening projects of Chung-Shan Freeway at Taiwan.

Radical Concrete Technology. Edited by R K Dhir and P C Hewlett. Published in 1996 by E & FN Spon, 2–6 Boundary Row, London SE1 8HN, UK. ISBN 0 419 21480 1.

INTRODUCTION

Chung-Sa Bridge is located on the border of Yun-Lin and Chang-Hua provinces and is across the Jer-Shui River. It is one of the most important bridges located on Chung-Shan Freeway in the middle area of Taiwan. It was built in 1978 with a total length 2,344.94 m and 67 spans. Each span is 35 m in length. The south and north bound bridges are separated. The superstructure includes 5 simply supported precast prestressed I-section girders at a span with cast-in-place concrete slabs on top; the substructure comprises a dual-column bent and friction piles. There are 24 friction piles under a pile cap, and the elevation of pile caps is 29.2 m. The length of each pile is 24 meters (See Fig. 1).

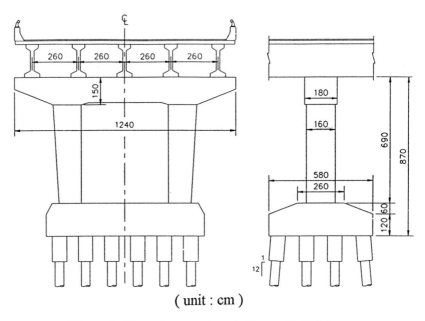

(unit : cm)

Figure 1. Elevation section of Chung-Sa Bridge

The waterway level of the Jer-Shui River was originally higher than the elevation of the pile caps of Chung-Sa Bridge. Due to huge amounts of aggregates and sands of the Jer-Shui River having been removed to supply the needs of construction because of the excellent quality of the aggregates along the banks, the waterway level of the Jer-Shui River is more lower than the elevation of pile caps and some piles are severely exposed. Therefore, TANFB decided to take strengthening action to ensure the safety of Chung-Sa Bridge in 1994. Steel sheet piles are used to stop and change the direction of water flow, so the exposed piles can be protected from scouring continually and can be strengthened smoothly. Unfortunately, 6 aggressive typhoons invaded Taiwan in the summer of 1994. Those typhoons brought a lot of flood water to the Jer-Shui River on which Chung-Sa Bridge is built; the floods destroyed part of the strengthening facilities of Chung-Sa Bridge and scoured the piles of Chung-Sa bridge heavily, so that the piles at spans No. 15-17 and No. 35-39 were severely exposed. The maximum scouring depth was about 8 m from the original design river bed level; and the maximum exposed length of pile was much higher than the allowable scouring depth of the original design, which is about 4.5 m. In consequence, TANFB took strengthening action immediately, including filling the river bed around the exposed piles to the elevation of 25.4m with gabions, placing concrete anchor blocks downstream of the bridge, putting four layers of 60 cm high gabions around the exposed piles from the elevation of 25.4 m to 27.8 m (see Fig. 2) to strengthen the stiffness of the piles, and grouting concrete into the gabions on the upstream side to prevent the gabions from being damaged by flood. Once, the vehicle speed was limited up to 70 km/hr to protect the piles against the impact loading due to vehicles travelling at high speed on the bridge.

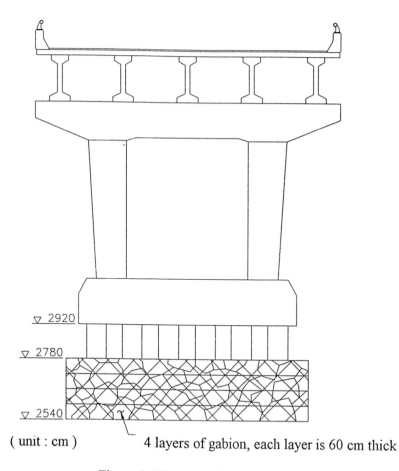

▽ 2920

▽ 2780

▽ 2540

(unit : cm)   4 layers of gabion, each layer is 60 cm thick

Figure 2. Pile strengthening sketch

Because Taiwan is located in the earthquake zone of the circumferential Pacific Ocean, TANFB would like to know whether or not Chung-Sa Bridge can resist an earthquake attack after the strengthening work. Therefore, JOIN Engineering Consultants organized a team to assess the seismic capacity of Chung-Sa Bridge. We used on-site ambient vibration measurements to find the natural frequencies of the Chung-Sa Bridge and then used the results to verify the exactness of the finite-element model. In the meantime, we used SAP90 to proceed the modal analysis and calculate the dynamic response of the structural members of Chung-Sa Bridge under the predefined earthquake force which is given by the design code of Taiwan. We have checked the pile and pier members, and we have proved those members are safe when the predefined earthquake force is applied on the bridge strengthened by gabions. Then, we used the model to find the maximum scouring depth of Chung-Sa Bridge. Knowing the maximum scouring depth, we recommend to install a monitoring and warning system on the bridge, so TANFB can take proper action within a very short time to ensure the safety of the drivers on Chung-Shan Freeway.

IN-SITU AMBIENT TEST AND ANALYSIS RESULTS

The in-situ ambient test uses very sensitive instruments to record the structure dynamic response induced by wind, micro-vibrations of ground and other interruption. Then according to the recorded response, the natural frequencies (or fundamental periods) and the fundamental mode shape of structures can be found. In the following, the in-site ambient test process and analysis result of Chung-Sa brideg are introduced.

Analysis results

The fundamental period and mode shape of the bridge can be found by using fast Fourier transformation to transfer the time domain data, which is recorded by sensors, to frequency domain. Table 1 shows the fundamental period of piers at spans No. 35-41 obtained from the in-site ambient test.

In-situ ambient test

The in-situ ambient test was performed in the north bound bridge by JOIN Engineering Consultants with the assistance of the Center of Bridge Engineering Research of National Central University on October 5 and 6, 1994. On the first day, the responses of the piers at spans No. 35-38 were measured. Then, on the second day, the tests included measuring the pier responses at spans No. 39-41, the response of ground motion, the responses of the piers at spans No. 36 and at span No. 37 at the same time and the mode shape of bridge decks which are continuous at the interval of 3 spans. Tests were done twice for each layout of the sensors, and the duration of record data is about 2 minutes. For some specific piers, in the first test, six sensors were installed on the locations of the pile cap, cap beam and bridge deck slab separately in the longitudinal and transverse directions, and the responses in both directions were recorded at the same time. The procedures in the second test are the same with those in the first test except that the sensors put on the bridge deck were moved 17.5 m toward the north.

Table 1. Fundamental periods of the piers

Pier No.	Longitudinal Fundamental Period (sec)	Transverse Fundamental Period (sec)	Vertical Fundamental Period (sec)
35	0.524	0.373	0.248
36	0.476	0.412	0.250
37	0.485	0.398	0.243
38	0.495	0.425	0.235
39	0.500	0.407	0.240
40	0.488	0.389	0.237
41	0.495	0.366	0.238

ANALYTIC MODEL AND ANALYSIS RESULTS OF CHUNG-SA BRIDGE

In order to reflect the real behaviors of the bridge, some properties of the bridge must be taken into considerations in building up the finite-element analysis model, such as geometrical arrangement of all structural elements, the distribution of mass, cross section properties of elements, material properties, and the fixed-end position of the piles.

In this paper, we use in-situ ambient test to find the natural frequencies of Chung-Sa Bridge and then use the results to correct the fixed-end positions of piles and verify the exactness of the finite-element model. Moreover, we use the verified bridge structure model, which is built up with SAP90, to proceed the static analysis and spectrum analysis of the bridge, and to check the strength and seismic capacities of Chung-Sa Bridge. The stress distribution of all structural elements can be obtained from the SAP90, and the seismic capacities of the piles were evaluated by the LPILE program which is designed to give the user the capability to solve problems related to the selection of different types of piles for a variety of purposes.

Analytic model

The superstructure of Chung-Sa Bridge includes 5 simply supported precast prestressed I-section girders with cast-in-place concrete slabs on top of a span. Diaphragms are constructed between girders to resist the torsions induced by unsymmetric loading and to maintain the stability of the superstructure. The deck expansion joints, which are finger plate type, are provided at the interval of every three spans. Restrainers are installed between expansion joints to prevent the superstructure from falling down when earthquakes happen. Moreover, bearings and earthquake stops are set up to transfer shear force and to resist the horizontal force induced by earthquake. The substructure of Chung-Sa Bridge is composed of a dual-column bent and frictional piles. There are 24 frictional piles, 4 rows in the transverse direction and 6 columns in the longitudinal direction under a pier foundation. In building up the finite-element model of Chung-Sa Bridge, these inventory information of Chung-Sa Bridge can be divided into four parts and be summarized as follows:

(a) Simulation of geometric configuration

The region between two expansion joints can be considered as a vibration unit; thus, the bridge can be separated into many vibration units and each vibration unit has 3 spans. Fig. 3 shows the finite-element simulation of a vibration unit. Three dimensional beam-column elements are used in the finite-element mesh. The superstructure at each span is modelled by 4 elements and each pier is modelled by 3 elements.

(b) Distribution of mass

For simplicity, the distribution of mass is regarded as lump mass. It must be noticed that whether some important mode shapes can be found in this kind of simulation. In general, 3 lump masses for each span is enough. Besides, each node has only 3 translational degrees of freedom, and the rotational degrees of freedom can be ignored.

(c) Cross-section properties

The real area of the cross section is used. The value of moment of inertia, I, is calculated from the real cross section, but J is a estimated value in some complex cross sections.

(d) Simulation of pile

The piles are modelled by beam-column elements. And the interaction between soil and piles is modelled as a spring system.

Modal analysis

Modal analysis is performed in the analytic finite-element model to find the natural frequencies and mode shapes of Chung-Sa Bridge; then, compare the calculated natural frequencies and mode shapes with those obtained from in-site ambient test to

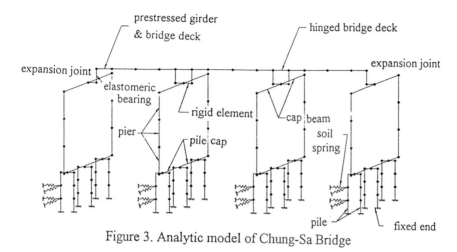

Figure 3. Analytic model of Chung-Sa Bridge

verify the exactness of the analytic finite model. After verifying the exactness of the analytic model, the seismic capacities of Chung-Sa Bridge can be evaluated by means of static analysis and dynamic spectrum analysis in the verified model.

Modal analyses of the analytic model are separated into two cases. In the first case, the piles are in the condition of being scoured by flood but strengthened by gabions; in the second case, the piles are assumed in the condition of maximum allowable scouring depth. The calculated natural frequencies of case 1 are very close to those obtained from in-site ambient test; see table 2. The first mode shape is in the longitudinal direction (see Fig. 4), and the second mode shape is in the transverse direction (see Fig. 5). The relationships of the fundamental period of Chung-Sa Bridge with the scouring depth and fixed end depth of piles are shown in Fig. 6.

Table 2. Comparison of the fundamental periods obtained from in-site ambient tests and those obtained from the analytic model for Chung-Sa Bridge

	Direction	Fundamental Period (sec)	Measured Position
		Span No. 35-37	(Analytic Mode)
Ambient Test	Longitudinal	0.500	Cap Beam
	Transverse	0.400	Cap Beam
Analytic Model	Longitudinal	0.506	First Mode
	Transverse	0.401	Second Mode

Design seismic force and seismic capacity assessment of piers

(a) Design seismic force

At present, the design seismic forces are decided according to the "Seismic Design Specification for Highway Bridges" of Taiwan. Both in static analysis and in dynamic spectrum analysis, the combination effect of dead load, transverse seismic force and longitudinal seismic force on the bridge must be considered and they are shown as follows:

case 1 :

$$DL + 0.3 \ EQ(T) + 1.0 \ EQ(L) \tag{1}$$

case 2 :

$$DL + 0.3 \ EQ(L) + 1.0 \ EQ(T) \tag{2}$$

in which : DL is the dead load. EQ(T) is the transverse seismic force. And EQ(L) is the longitudinal seismic force.

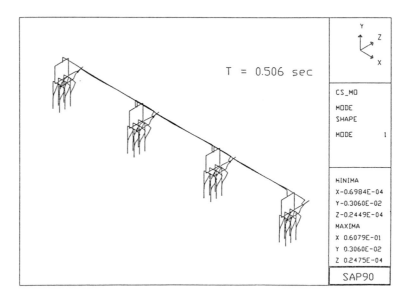

Figure 4. The first mode, in the longitudinal direction, of Chung-Sa Bridge corresponding to the fundamental period, T = 0.506 sec

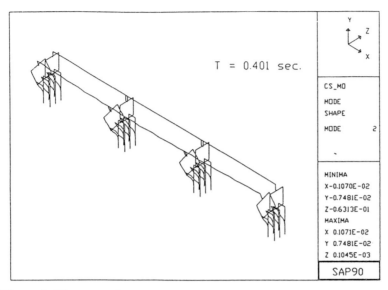

Figure 5. The second mode, in the transverse direction, of Chung-Sa
Bridge corresponding to the fundamental period, T = 0.401 sec

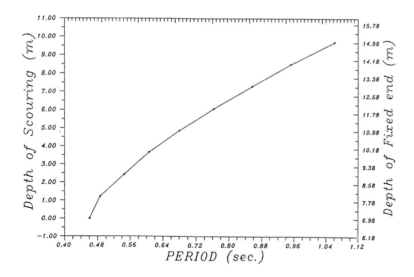

Figure 6. The relationship of the fundamental period and the fixed-end
depth of the pile; and the relationship of the fundamental period
and the scouring depth of the pile

But Chung-Sa Bridge was built in 1974 and it was designed in accordance with the "Design Specification for Highway Bridge" of Taiwan in 1987. In the original design of Chung-Sa Bridge, the horizontal design seismic force is considered as

$$EQ = K_h \cdot W \tag{3}$$

K_h is the coefficient of horizontal design seismic force. W is the dead load of the structure. But the design seismic force in "Seismic Design Specification for Highway Bridges" of Taiwan, on January 1995, is :

$$EQ = Z_d \cdot C \cdot W \tag{4}$$

Z_d is the coefficient of design ground acceleration, and C is the normalized site-specific acceleration response spectrum.

In the static analysis, Eq.(3) is used, and the value of K_h is 0.15. On the other hand, Eq.(4) is used in the dynamic spectrum analysis. The original design philosophy of Chung-Sa Bridge is elastic design, so we cannot evaluate the ductility of the bridge. Hence, the value of Z_d is decided as 0.06 by comparing Eq.(4) with Eq.(3). The value of C is in accordance with the second type ground. Modal superposition is done by means of CQC method(Complete Quadratic Combination method).

(b) Seismic capacity assessment of piers

The internal forces of piers, which were scoured by flood but strengthened by gabions, induced by the combinations of dead loads, transverse seismic forces and longitudinal seismic forces both in the static analysis and in the dynamic spectrum analysis, are shown in Table 3. The coordinate of x is in the longitudinal direction of the bridge, and the coordinate of y is in the transverse direction of the bridge. P is the notation of axial force; H_x and H_y are the horizontal forces in the x and y directions; M_x and M_y are the moments in the x and y directions. In the meantime, we can plot the interaction diagrams for the axial force(P) and moments(M_x, M_y) of the piers in accordance with the strengths of the piers. The real cross sections of the piers are shown in Fig. 7. The analyses of the piers are by means of the ultimate strength design method. Then, comparing the calculated internal forces of piers with the interaction diagram of P-M curve (Fig. 8 and Fig. 9), we can decide whether the piers are safe or not. The safety factor F_{sm} is defined as the ratio of M_{nr}, the nominal moment obtain from the P-M curve, to M_r, the calculated internal moment of piers. The values of F_{sm} are 4.5 and 2.6 separately in both longitudinal and transverse direction as shown in Fig. 8 and Fig. 9.

But we must consider if resonance will occur during earthquakes due to the natural frequency change resulting from scouring.

Table 3.   Combination effect of the loads applied on the pier

| | Loading | | | | | Loading Combination | | | |
| | Dead Load | Static Analysis | | Spectrum Analysis | | Static Analysis | Spectrum Analysis | Static Analysis | Spectrum Analysis |
	DL	EQ(L)	EQ(T)	EQ(L)	EQ(T)	1.3*[DL+0.3EQ(T)+1.0EQ(L)]	1.3*[DL+0.3EQ(T)+1.0EQ(L)]	1.3*[DL+0.3EQ(L)+1.0EQ(T)]	1.3*[DL+0.3EQ(L)+1.0EQ(T)]
P	368.304	0.067	88.906	0.676	99.512	513.56	518.48	594.40	608.42
Hx	1.829	58.525	0.161	58.015	0.552	78.52	78.01	25.41	25.72
Hy	59.71	0.013	56.719	0.133	62.358	99.76	102.12	151.36	158.74
Mx	23.06	0.002	136.346	0.028	153.629	83.16	89.93	207.23	229.71
My	3.452	295.821	0.628	304.891	2.1	389.30	401.66	120.67	126.13

UNIT : t,m

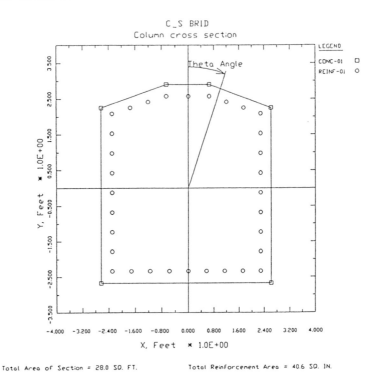

Figure 7. The cross section of the pier and the locations of reinforced bars

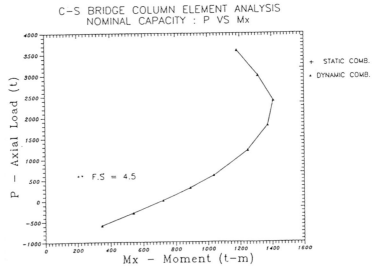

Figure 8. Interaction diagram of the axial force and x-direction moment of

the pier (in the longitudinal direction)

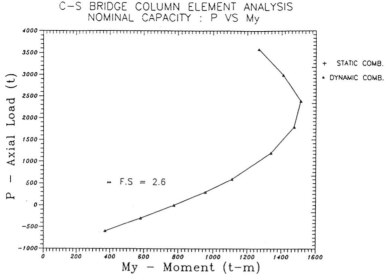

Figure 9. Interaction diagram of the axial force and y-direction moment of
the pier (in the transverse direction)

BEARING CAPACITY AND SEISMIC CAPACITY OF PILES

Bearing capacity of piles in the vertical direction

The boring data of the hole No. 199-1-20 on April, 1974 are shown in Fig. 10. And
Table 4 shows the information of soil properties resulting from the boring data in Fig.10.

Table 4. Soil properties and soil spring constants

Sublayer No.	Distance From The Top of The Pile (m)	Soil Type	SPT N-Value	Soil Spring Constant K (kN/m³)
1	1.4 ~ 3.8	Gabion	5	1.0605 E04
2	3.8 ~ 5	Sand	13	2.757 E04
3	5 ~ 10	Sand	20	4.242 E04
4	10 ~ 15	Silty Sand	27	5.726 E04
5	15 ~ 20	Sand	32	6.787 E04
6	20 ~ 24	Sand	36	7.635 E04

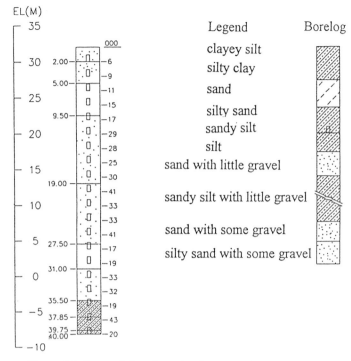

Figure 10. Record data from boring tests, in April of 1974

The calculated end-support bearing capacity of a single pile is 86.4 tons, which is much higher than the dead load, 31 tons, so the bearing capacity of the pile is enough. But the end-support bearing capacity, frictional bearing capacity of piles and the Young's modules of the soil are dependent on the standard penetration test value N. Please note that the conclusion mentioned in the above is only suitable for the local area in the vicinity of the boring hole No. 199-1-20.

Bearing capacity of piles in the horizontal direction

The failure modes of vertical piles suffered by horizontal forces are dependent on the properties of the soils around the piles, the restrain condition at the pile heads, and the sizes and lengths of the piles. The failure modes of short piles are often controlled by shear failures of the soils around the piles, due to the lower passive earth pressures provided by soil than the stiffness of the short piles. On the contrary, the failure modes of long piles are often controlled by the allowable bending moments of the piles, due to the flexibility of long piles.

The influences on the bearing capacities of the vertical piles suffered by horizontal forces can be summarized as follows:

1. The passive earth pressures provided by the soils around the vertical.

2. The allowable horizontal displacement of substructures.

3. The allowable bending moments of the vertical piles.

The standard penetration test value N in the region of the piles strengthened by gabions should be smaller than that of sand, because the voids in the gabions were not filled by sands or other aggregates. By the analysis of LPILE program, it is shown that the value of N in the region of the piles strengthened by gabions has little influence on the horizontal bearing capacity of piles. For the sake of convenience, it was assumed to be 5.0 . The analysis results of LPILE program show that the horizontal displacement on the top of the pile is 1.5 cm, when the pile is the application of a force, 21 tons, in the horizontal direction (see Fig. 11).

(a) Pile strength assessment

The assessment of the pile strength includes the checks of the allowable stresses of concrete and pre-stressed tendons. When piles are in the application of axial forces and bending moments at the same time, compressive and tensile stresses will be induced in the cross sections of the piles. (The compressive stress is regarded as positive, and the tensile stress is regarded as negative.) In the assessment of pile strength, we compare the induced stresses with the allowable stresses of concrete and pre-stressed tendons, in order to ensure whether the induced stresses are within the allowable region.

The comparison of the induced internal stresses in the concrete cross sections of the piles with the allowable stresses of concrete is shown in Fig. 12, in the case that the piles are at the present state, scoured by flood but strengthened by gabions immediately. The results of static analysis and spectrum analysis show that the piles at the present state are safe. The check of concrete allowable stresses of the piles are shown in Fig. 13 and Fig. 14, separately for the static analysis and for the spectrum analysis, in the assumption of various scouring depths of the piles. The maximum allowable scouring depth of the piles is about 6.8 m in the static analysis, and is about 9.3 m in the spectrum analysis. The spectrum analysis can reflect the variation of structure periods, so the induced internal stresses of the piles are lower.

(b) Seismic capacity assessment of the piles

We can find the internal forces in the cross section of each pile head by using the SAP90 finite element model; then, use the LPILE program to calculate the horizontal displacement on the top of the pile and maximum bending moment of the pile, by inputting the horizontal forces applied on the top of the pile and the soil properties around the pile.

In accordance with the analytic results for the bridge at present state under predefined earthquake force, which is given by the design code in Taiwan, the analytic maximum horizontal displacement of the piles is 0.3 cm, and the analytic internal force applied on the top of the pile is 4.8 tons. However, when the horizontal displacement reaches 1.5 cm, which is the allowable horizontal displacement specified in the design code, it is needed that a force, 21 tons,

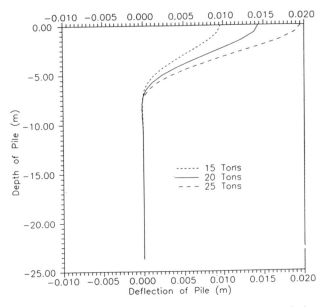

Figure 11. Relationship of the horizontal deflection of the pile and the magnitude of horizontal force applied on the top of the pile

STRENGTH OF PILE SECTION

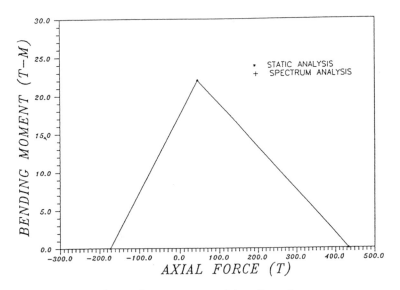

Figure 12. Strength assessment of the pile at the present state

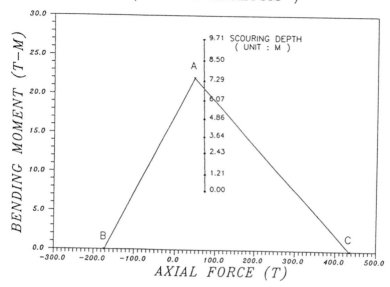

Figure 13. Strength assessment of the pile in the condition of various
scouring depths (static analysis)

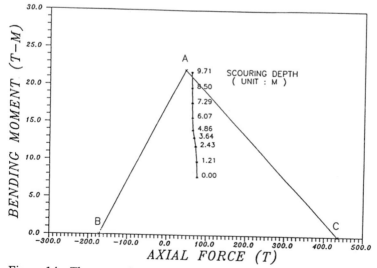

Figure 14. The strength assessment of the pile in the condition of various
scouring depths (dynamic spectrum analysis)

applies on the position of pile head. Hence, the safe factor of the horizontal displacement of the pile is 5.0, and the safe factor of the horizontal force is 4.4 .The displacements of pile heads in the condition of various scouring depth are shown in Fig. 15. The maximum scouring depth is 7.8 m obtained in the static analysis, and is 8.6 m in the spectrum analysis.

Liquefaction of soil

When the saturated loose sands are vibrated by earthquake or other kinds of motion, the effective stresses of the saturated loose sands will disappear, due to the significant increase of pore water pressure. This phenomena is called liquefaction of soil. Liquefaction often induces settlements and inclinations of the structures, resulting in severe damage to the structures.

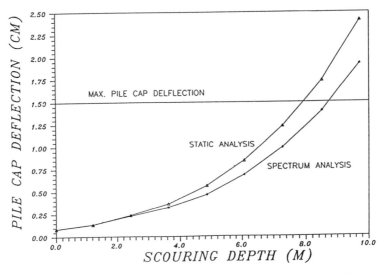

Figure 15. Relationship of the horizontal displacement of the pile cap and

the scouring depth

The liquefaction potential of the soil in the area around Chung-Sa Bridge had been researched by JOIN Engineering Consultants in the case of the "Seismic and Liquefaction Potential Assessment in the Section of Yun-Lin and Chia-Yih of High Speed Railway". This report shows that there is much probability of occurring soil liquefaction in the area 2 km downstream from Chung-Sa Bridge, which the high speed railway will pass through. This report also shows that the variation of soil properties around Chung-Sa Bridge is insignificant, so the possible damage to Chung-Sa Bridge resulting from soil liquefaction should be noticed.

CONCLUSIONS

From the experimental results of in-situ ambient tests, finite-element simulations (including static analysis and dynamic spectrum analysis) and the bearing capacity assessment of the piles of Chung-Sa Bridge, the following conclusions can be reached :

(1) Assuming there is no interaction between horizontal and vertical forces, the piles at present state, which are scoured by flood but strengthened by gabions immediately, have enough vertical bearing capacities to resist dead loads. But the end support bearing capacities, and frictional bearing capacities of the piles are greatly dependent on the standard penetration test value N.

(2) According to the seismic capacity assessment of the piers, the piers of Chung-Sa Bridge are safe at their present state. The safety factor is 4.5 in the longitudinal direction, and 2.6 in the transverse direction.

(3) In accordance with the seismic capacity evaluation of the piles, it is shown that when the displacement on the top of the pile reaches 1.5 cm , which is the allowable horizontal displacement on the top of the pile specified by the design code in Taiwan, it needs a force 21 tons horizontally applied on the top of the pile. But the displacement on the top of the pile is 0.3 cm and the horizontal force applied on the top of the pile is only 4.8 tons, when the bridge is applied by predefined earthquake force, which is given by the design code in Taiwan. Hence, the safety factors are 5.0 and 4.4 separately for the horizontal displacement and for the horizontal applied force. From the displacement control assessment, the maximum allowable scouring depth is 7.8 m in the static analysis, and is 8.6 m in the spectrum analysis.

(4) In the pile strength assessment, it is shown that the piles are safe too, and the maximum allowable scouring depths are 6.8 m and 9.3 m separately in the static analysis and in the spectrum analysis.

(5) Comparing the conclusions of (3) and (4) mentioned above, the maximum allowable scouring depth is conservatively considered as 6.8 m from the pile cap.

(6) Due to the natural frequency change of Chung-Sa Bridge resulting from scouring, there is the probability of resonance when Chung-Sa Bridge is attacked by earthquakes.

(7) According to the report, "Seismic and Liquefaction Potential Assessment in the Section of Yun-Lin and Chia-Yih of High Speed Railway", made by JOIN Engineering Consultants. the probability of liquefaction the occurring at the area around Chung-Sa Bridge is great . Therefore, the liquefaction problem must also be taken into consideration in the seismic capacity assessment of Chung-Sa Bridge.

ACKNOWLEDGMENTS

We greatly appreciate that TANFB provided us with the inventory information of Chung-Sa Bridge and its support in our in-situ ambient tests. Furthermore, we are deeply grateful for the professional techniques and prudence of the Center of Bridge Engineering Research of National Central University in doing in-situ ambient tests. We also appreciate the helpful comments and suggestions of Professor Yih-Chao Tsai at National Taiwan University .

CONCEPT AND PRACTICAL DESIGN CRITERIA OF A CONCRETE FLOATING AIRPORT

Y Amano

Nagoya University

T Kawamura

P.S. Corporation

T Tanabe

Nagoya University

Japan

ABSTRACT.   In this report, the concept of concrete floating airport and its practical design criteria will be investigated. Main emphasis will be also in the analytical model of floating structure of which structural oscillation is affected by the wave characteristics as elastic panel and is considered to give affects on the structural dimensioning quite extensively. While usual analytical method treats the floating body as a rigid one, the influences of rigidity to displacement and bending moment are very large. Treating it as elastic panel, this method can tell more details of the structure and lead us to more reasonable design.

Keywords: Concrete Floating Airport, Wave Characteristics, Structural Oscillation, Influences of Rigidity, Elastic Panel

Mr Yoshikatsu Amano   is a Graduate Student of Civil Engineering, Nagoya University, Japan. His main resarch is the analysis of very large floating structures using Finite Element Method.

Mr Tetsuo Kawamura   is a General Manager of Technical Department, P.S. Corporation, Osaka Branch. He has taken part in construction of a lot of pre-stressed concrete structures.

Dr Tada-aki Tanabe is a Professor of Civil Engineering, Nagoya University, Japan. His main research is directed to the dynamic failure mechanism of RC structures besides the thermal stress. He is the chairman of JCI Committee on the Thermal Stress of Massive Concrete Structures besides a member of various committees of JSCE, JCI and ACI.

Radical Concrete Technology. Edited by R K Dhir and P C Hewlett. Published in 1996 by E & FN Spon, 2–6 Boundary Row, London SE1 8HN, UK. ISBN 0 419 21480 1.

INTRODUCTION

Floating type airports which would be built by unifying large plates components either build by steal or concrete are drawing attention in Japan due to the difficulty of finding good in land location since high density of the population and noise, and pollution problem always comes up as hindrance. Therefore there are several on going construction project of airports in the offshore of main island of Honshu by reclaiming a new land of the size of several hundred to thousands of hectares, with a depth of about 20 to 30m under the sea level.

Due to the huge construction cost of reclaiming as well as environmental hazards, an idea of constructing those airports by floating structures is becoming more a matter of consideration. However, we have to solve several engineering problem as well as enviromental ones, if we proceed to make it real.

In this paper, the general conceptual problem as well as some specific engineering problem will be surveyed, and new finding which is important to design will be presented.

THE ADVANTAGE OF CONCRETE FLOATING AIRPORTS

Why Do We Need Offshore Construction?

If it is possible to get a large land for an airport near the city, of course it is better to use that land. But now in Japan as well as in other countries, there may not be an enough location to build an airport near a large city. So many countries have developed sea shore, and there already exist several reclaimed airports. The advantages of constructing airports on the sea are the followings.

1) It is easy to get a space for an airport and to expand a runway.
2) As it is surrounded by the sea not by the residential area, the noise pollution doesn't occur, and the security can be kept easily.
3) It doesn't become a factor of traffic congestion.

Why Floating Structure is Useful?

As mentioned above, some airports were constructed by reclaiming, but on the deeper sea shore (more than 20m depth of water) where we are planning to develop in future, this type of construction may not be suitable due to huge construction cost of reclaiming. At present, though a floating airport dose not exist, there are some large floating bridges which are on the lakes in America and on the west coast of Norway. So in the future, the concept of constructiong floating airport is not unrealistic. The advantages of floating airports are the followings.
1) The floating structures influence environmental hazards less than reclaimed ones.

2) It can be constructed on weak soil and is not concerned with the depth of water. Therefore when the depth of water is high, it is more economical to construct a floating airport.

3) Floating offshore structure will not have settlment problem which was one of the primary problems of the construction of Kansai reclaimed offshore airport.

4) We can adopt pre-cast methods, so the fabrication of floating body on the sea becomes easy and construction period becomes much shorter.

5) The floating breakwaters can reduce the wave-height. Hence the very strong waves don't attack directly to the air port structures.

6) It is not much effected by earthquake and liquefaction.

The floating airports have above advantages, but the influences of waves, seaquake, mooring, and winds etc. are very complex. This effects have to be clarified by analysis as well as by experiment.

Advantage of Concrete over Steel

There are two types of floating body, one is pontoon type, and the other is semi-submersible type as shown in Figure 1. For the latter type, the superstructure is better to be light weight, so the steel is considered to be suitable for this type [1]. However in order to build an airport, a large amount of material is necessary, so we propose the former type of airport made of concrete which is a very available material. The advantages of the airport made of concrete compared with that of steel are the followings.

1) Construction cost maybe much lower for concrete
2) For the steel airports, maintenance cost of corrosion protection and coating is high while these are low for concrete.
3) Concrete is a good dampening material for vibration and noise
4) When an airport is damaged, concrete can also be repaired easily.
5) The influences of temperature is small.

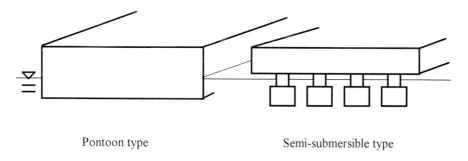

Pontoon type   Semi-submersible type

Figure 1   Types of Floating Body

FUNDAMENTAL ISSUES IN DESIGNING CONCRETE
FLOATING STRUCTURES

In the following, some specific issues to be clarified in constructing floating airports will be explained.

Numerical Modeling

We treat a floating concrete body as an elastic beam, so that the bending oscillation can be considered, and it's motion is assumed to be a simple harmonic with the incident wave frequency ω. External forces considered in this analysis are the followings.

1) Loads due to incident and scattered wave.
2) Hydrostatic pressure due to a change of vertical displacement.
3) Mooring forces by elastic springs.

We apply the hybrid element method introduced by Mei et al [2]. Firstly, we divide water region into two parts with a fictitions vertical boundary S_R which surrounds the floating body nearby (Figure 2). The velocity potential which is exterior to S_R is then represented analytically by the eigenfunction expansion and is matched with the inner region represented by the finite element analysis at S_R.

In order to get good results, we must divide the water region into fine mesh. So the key problem faced in solving this type of problem is the need of high degrees of freedom. Noticing that the freedoms of beam (displacement) are not related to those of water (velocity potencial), we can adopt incompatible nodes between the beam elements and water elements (Figure 2). Thus we can reduce the degrees of freedom without dividing water into coarse mesh.

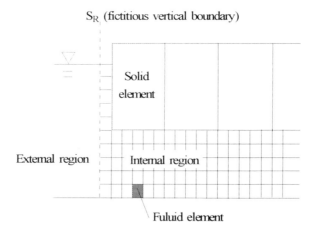

Figure 2   Meshes Used in This Analysis

Influences of Rigidity to Vertical Displasement and Bending Moment

It is obvious that the properties of oscillation of the floating bodies are related to the rigidity. So, to investigate the influence of rigidity is very important for the design. The section proposed in the conception of pre-stressed concrete floating airports[3] is shown in Figure 3. Its width is 60m and moment of inertia of area I is about 2500m^4. Young's modulus E of concrete is $2.1*10^6$ kN/m^2, so the unit width flexural rigidity EI becomes $8.1*10^9$ kNm^2. To investigate the influences of rigidity, we simulate 3 types of free body which rigidities EI are $8.1*10^7$ (Type 1), $8.1*10^9$ (Type 2) and $8.1*10^{11}$ (Type 3) kNm^2. The results are shown in Figure 4 (maximum vertical displacement at each point) and Figure 5 (maximum bending moment at each point). Dimensions of floating body and properties of incident wave are shown in Table 1.

Table 1. Analysis Model

Dimensions of the Floating Body			Properties of the Incident Wave			
Length	Height	Draft	Height	Period	Length	Depth
2100m	12.0m	7.0m	4.6m	9.6s	115m	20.0m

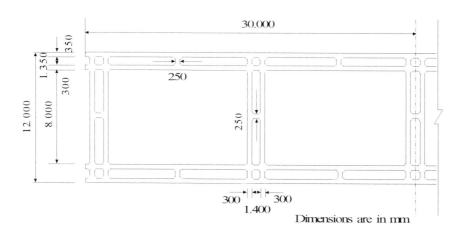

Figure 3   Section of the Floating Body

According to Figure 4 and Figure 5, the rigidity influences the vertical displacement and bending moment very much. The higher the rigidity is, the smaller the vertical displacement becomes. But the bending moment becomes quite large. As shown by the analytical results above, the rigidity is a very important influence factor, it's unreasonable for the usual analytical methods to treat the floating body as rigid, by contrast this method can tell more details of the structure and lead us to more reasonable design.

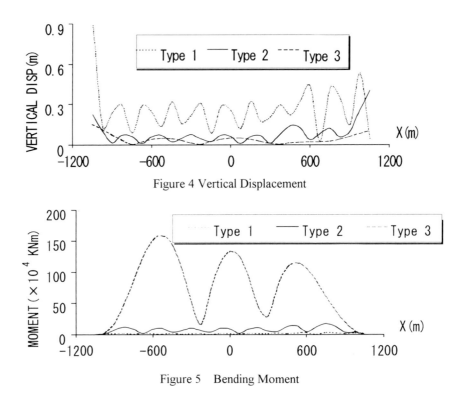

Figure 4 Vertical Displacement

Figure 5   Bending Moment

Influence of Hinges

For constructing large floating structures, we may introduce hinges between each blocks. So it's necessary to investigate the influence of hinges. We prepare 3 simple models of those structures as shown in Figure 6. The hashed parts as shown in the Figure 6 which has low rigidity can be consideres as hinges. The results are shown in Figure 7 and Figure 8.

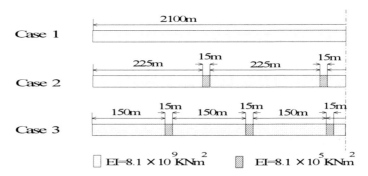

Figure 6   Simple Models of Structures which is Introduced Higes

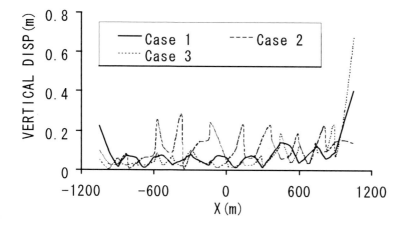

Figure 7   Vertical Displacement

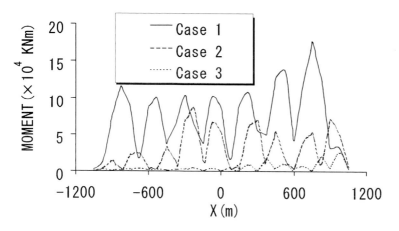

Figure 8   Bending Moment

We can see that the location of hinges influences vertical displacement and bending moment very much. If we use hinges, bending moment becomes small, but vertical displacement at the wave front side becomes quite large.

The relation between wave height and stress is shown in Figure 9. In Case 1, the stress extend 20 N/mm^2 when the wave height becomes about 3m. So we need to introduce hinges or take some countermeasure to decrease the stress.

Above results are one of the typical out come of the developed analytical method. With the accumulation of those detailed data, we think the construction of a floating airport by precast concrete panels will become possible.

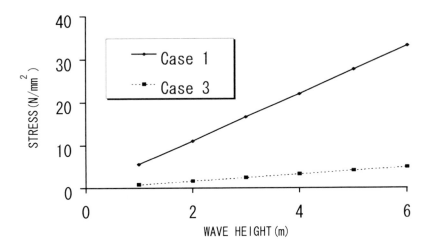

Figure 9   Wave Height and Stress

CONCLUSIONS

1. The oscillation analysis method which treats floating structure as a flexible body was proposed.

2. The rigidity influences the vertical displacement and bending moment very   much. The higher the rigidity, the smaller the vertical displacement. But the bending moment becomes quite large.

3. The location of hinges influences the vertical displacement and bending moment considerably. For design, it's important to chose the location of hinges carefully.

REFERENCES

1. TADANOSUKE MUROTA. One Example of Technology assessment of Very Large Offshore Structure: Technological Investigation of Construction of Floating airports, Bulletin of the Society of Naval Architects of Japan, Vol. 638 (1981.8) - Vol. 652 (1983.10)

2. CHIANG C. MEI, The Applied Dynamics of Ocean Surface Waves, John Wiley & Sons.

3. A COMMITTEE OF OFFSHORE STRUCTURE, A Conception of Pre-stressed Concrete Floating Airports Concrete Floating Airports

FOUNDATION OF BUILDINGS ON SEISMIC GROUNDS

M Kłapoć

G Bryś

Technical University Zielona Gora

Poland

ABSTRACT. The solution is based on the indirect foundation by using reinforced concrete or steel piles which are specially constructed. All the existing theoretical solutions show that the new proposed way of indirect foundation allows:
- to abate the frequency of free vibrations of a building from 2 to 10 times,
- to abate the transmission of amplitudes of soil vibrations to the construction of a building considerably and even reduce to negligible quantities,
- to eliminate the influence of slipping and creeping,
- to execute quick and effective vertical rectification of a building.

The advantages of indirect foundation will have an effect on safety and comfort of the buildings placed on seismic and paraseismic grounts. A practical application of the proposed solutions of indirect foundation of buildings requires a testing model which would verify theoretical assumption.

Keywords: active seismic grounds. Building protection, piles which are not completely plunged into the ground.

Dr eng. Mikołaj Kłapoć assistant professor at the Civil Engineering Department of the Engineering College in Zielona Góra, Poland. He specializes in rheology, paraseismic problems and industrial building problems. He has published a lot of articles and some time ago he was an UN expert in the field of building engineering. At present he is vice-chairman of the Section Board of the Provincial Building Engineer Association. He has also acted as vice rector of the College.

Dr eng. Gerard Bryś Lecturer at the Building Structure Department of the Engineering College in Zielona Góra, Poland. He specializes in numercial analysis of load - carrging ability of structures.

Radical Concrete Technology. Edited by R K Dhir and P C Hewlett. Published in 1996 by
E & FN Spon, 2–6 Boundary Row, London SE1 8HN, UK. ISBN 0 419 21480 1.

INTRODUCTION

The buildings, which are raised on seismic grouns, are designed with a certain definite reserve of structure rigidity or certain structure elements, which make it possible to accept the additional loads, which are resulted from seismic influences. This is one of the main methods of building protection against the paraseismic influences [6]. In scientific research and in engineering there appear solutions to protect buildings against the seismic influences by the application of corresponding damping insulations or special shape selection and structural solutions of the objects. The theoretical considerations concerning the applications of antiseismic insulations tend in the direction of total object insulation against the accesion of seismic or shocking wave by insulatine either the foundation from the building [2] or individual object elements [1]. The other proposed solutions are e.g. buildings on reinforced concrete shaft with special spacing of particular storeys, which are supported with a console on a shaft [3], the buildings with steel structure with additionally applied braces on eccentricities [7], or buildings with carrying walls with special structure [4]. The mentioned structural solutions are characterised by great material consumption or unsatisfactory protection against simultaneous reactions of shock and ground deformations. The method of indirect foundation is presented in this paper. It is reserved by a patent [5], which will enable practically complete elimination of deformation influences and ground vibrations.

DESCRIPTION OF A FOUNDATION METHOD

The solution relies on application of an indirect foundation of buildings on bearing piles, which are not fully plunged in the ground (Figure 1). The bearing pile is plunged in the ground, so that its higher section stands out over the ground level. The foundation part (2) of the building is made on pile complex (1). Each pile ends with the head (3), on which the foundation part of the building is realized with the help of the steel tube (7), which is embedded in the concrete of the pile head by using the steel mandrel (8), which is placed inside the tube (7) and is anchoraged in the foundation part. Such articulated connection enables unbounded vertical displacement and rotation of the foundation part of the building in relation to the heads of the foundation piles.

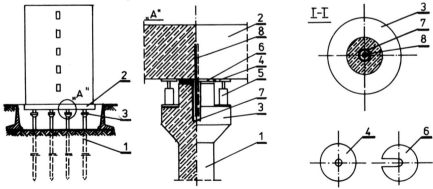

Figure 1 Scheme (diagram) of proposed solution

For the vertical rectification of the building there are the hydraulic lifts (5), which are supported on the head consoles and compensationing plates (6). The space between the ground and the foundation part of the building can be used for garages, household quarters. The presented method of indirect foundation makes possible full elimination of all disadvantages, connected with utylizing the buildings, which are raised on mining and seismic grounds. Rapid and efficient vertical rectification is possible. This eliminates the influence of terrain inclination, excreeping and increeping the ground. Moreover, the presented method enables a very good reduction of paraseismic and seismic concussions on building structure by the confinement of vibration propagation on the building.

ESTIMATION OF INFLUENCE FOR PROPOSED FOUNDATION ON DYNAMIC LOADS OF BUILDING STRUCTURE

For the definition of free vibration frequency of a building, which is founded on piles, the statical scheme showed on Figure 2 was adopted.

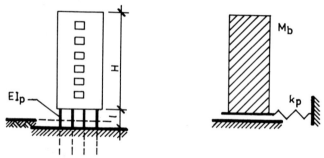

Figure 2 Schematic diagram of investigated structure

On the basis of a principle for parallel and series connections, the following formula for free vibration frequency is recived

$$f_w = f_p \cdot \frac{\sqrt{n_p}}{\sqrt{1 + n_p \cdot \left(\frac{f_p}{f_b}\right)^2}}$$

with

n_p - complete number of piles,

$$f_p = \frac{1}{2\pi} \cdot \sqrt{\frac{3EJ_p}{M_b l^3}}$$

f_b - free vibration frequency of direct founded building e.g. per PN-65/B-02170: $f_b = 50/H$ kHz,

M_b - complete building mass,

H   - building height,

EJ_p - flexural rigidity of a single pile,

l   - „computational" length of a single pile.

For estimation of quantity of paraseismic and seismic influences on loads of building structure the amplification coefficient was applied

$$\xi = \frac{1}{1-\mu^2}$$

with

$$\mu = \frac{f_w}{f_{wl}}$$

f_{wl} - force frequency.

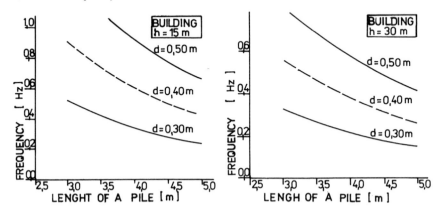

Figure 3 Dependence of free vibration frequency on pile parameters

The calculation results for two buildings were presented in Figure 3 and 4: the one with the height of 15 m and the mass 4821 tons and the second with the height 30 m and the mass 12000 tons. For both cases the foundation on 60 piles was accepted. The dependence of vibration frequency fron pile length and diameter was presented in Figure 3. In Figure 4 the dependence of amplification coefficient fron pile length and diameter, for building with the height 15 m and 30 m was presented respectively.

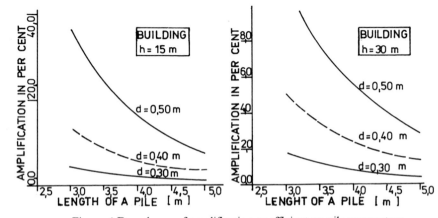

Figure 4 Depedence of amplification coefficient on pile parameters

The presented minimize results of calculations prove, that it is possible to minimalize the paraseismic and seismic influences by the appropriate selection of pile rigidities and lengths. In some cases the influence of reactions on building structure cannot even be taken into account. Extreme quantities of amplification coefficient oscillate for a building with the height 15 m from 38% to 2%, and for a building with the height of 30 m from 12% to 0,2%.

CONCLUSIONS

The realized calculations, concerning the influence of the proposed foundation on dynamic loads of the building structure, indicate that this influence is significant and profitable because:
1. free vibration frequency of a building is considerably reduced. In the considered cases it is reduced from 2 to 10 times,
2. the transmission of vibration amplitudes on building structure is considerably reduced (38% - 0,2%), which influences very advantageously on the safety of the object.

The proposed method of the foundation eliminates the influence of excreeping and increeping of the ground and enables vertical rectification of the building.

REFERENCES

1. CIESIELSKI R. Actual scientific-technical problems in antiseismic engineering. Seismic influences on buildings. Cracow Engineering College , 1978

2. DELFOSE G. C. Protection contre les seismes: le system GAPEC. Note intern du Laboratoire de Mechanique et d' Acoustique. Marseille, avril, 1976.

3. GORDALADZE W.M., KLIGERMAN I. S. Reinforced concrete structure of 18 -storey building. ZNIIEP, Gosgrazdaya, Kiev, 1980.

4. KŁAPOĆ M. The problems of building prevention against the negative influence of terrain deformation and seismic phenomena. Selected problems of mining grounds protection. (November) , 1984.

5. KŁAPOĆ M. KORENTZ J. PARUS R. Foundation pile for building foundation, particularly on mine grounds. Patent office PPR, (November), 1989.

6. LEDWOŃ J. A. Building on mine grounds. Arkady, Warsaw, 1983.

7. POPOV E. P. BARTERO V. Systematic analysis of some steel building frames. Proceeding of the ASCE, vol. 106, 1980.

EVALUATION OF SEISMIC RESISTANCE OF SPATIAL SHELL STRUCTURES

A S Zhiv

E V Kosygin

Vladimir State Technical University

Russia

ABSTRACT. The paper considers one of the approaches to the evaluation of seismic resistance of spatial shell structures for seismic areas. Evaluation of seismic resistance of some types of spatial shell structures has been done experimentally to define by spectral analysis the amount of energy being transmitted to structures under the action of dynamic motion. Up-to-date physical procedures and proper handling of oscillograms have been used when evaluating the dynamic forces acting on thin shell structures during occuring earthquakes. The approach allowed to explain the "phenomenon of selectivity" when such systems as shells are more responsive to such constituents of vibrations the frequency of which is closer to their own frequency.

Keywords: Shells, Spatial structure, Frequences of natural oscillations of structures, Frequences of soil vibrations, Convolution of spectra, Pseudovelocity.

A.S.Zhiv is a Doctor of Technical Sciences, Professor at Vladimir State Technical University, Russia. He specialises in the field of experimental investigation and application of spatial structures under static and dynamic load.

E.V.Kosygin is Candidate of Technical Sciences, Associate Professor at Vladimir State Technical University. He specialises in the field of optimization of construction on compressible soils.

Radical Concrete Technology. Edited by R K Dhir and P C Hewlett. Published in 1996 by E & FN Spon, 2–6 Boundary Row, London SE1 8HN, UK. ISBN 0 419 21480 1.

The paper presents the authors' observations of the behaviour of reinforced concrete spatial structures erected in
seismic zones of Kazakhstan. Prismatic, thin-slab shell
structures, cylindrical, spherical and hyperbolic shells
of heavy and lightweight concrete have been calculated for
seismic resistance and the calculations have been supported
by experimental results. Some spatial structures were designed of reinforced and cast-in-place concrete.

In the shell design great significance has been attached to
the numerical evaluation of forces and displacements originating in reinforced concrete shell structures of different
spans under seismic loading characteristic of earthquakes
of magnitude of 9 (in accordance with 12-point scale adopted in Russia). Such seismic conditions were typical for
most of the construction sites in the design area.

Experimental investigations conducted on models have shown
that at the seismic magnitude of 9 one might not expect
great increase both of section dimensions of shell with
spans up to 30m and the amount of reinforcement. It probably accounts for the fact that under seismic conditions the
test shell structures were not subjected to oscillations
of critical velocities.

It may be suggested that with the increase of the size of
shell spans amplitudes of their forced oscillations resulting from vertical displacements of the base under earthquake conditions would fit the displacements from the action of static useful loads. It is reasonable to assume
such state to be the ultimate one for the given structure.
We took as our example square on plan sloping shells of positive Guassian curvature and calculated deflections and
amplitudes of different spans under permanently acting static useful loads 4 kN/m^2 and simulated seismic loading of
9. The test results analysis showed that at the accepted
boundary conditions the maximum shell spans should be restricted to 36–40m when erected under the considered seismic conditions (Fig.1).

The data have been obtained by considering only low-frequency oscillations. However, experimental investigations on
models and full-size structures point to the appearance of
some additional spectra of oscillations depending on the
rigidity of the supporting contour, thus suggesting the installation of dampers in shells.

The supporting contours of shells (bowstring arches or
girders) damp oscillations in themselves as they respond
only to low-frequency oscillations as dampers. The use of
the damping properties of the supporting contour provides
a possibility to reduce the amount of energy being transmitted to a structure under the dynamic motion. The strain
energy method has been applied to the evaluation of seis-

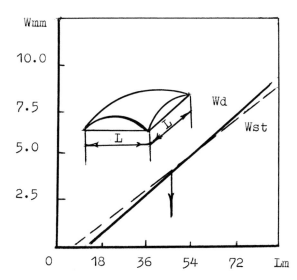

Figure 1   Dependance of amplitudes Wd and deflections Wst on the increase of shell spans

mic resistance of structures subjected to oscillations. The
method implies the evaluation of the amount of energy transmitted to structures during earthquakes and the amount of
potential energy structures can absorb without severe damage.

The amount of energy E transmitted to a structure at the
earthquake has been defined by the formula:

$$E = 1/2 \, m_n \, Y_n^2 \, V_n(\omega_n)$$

where, m_n = reduced masses
         Y_n = distribution coefficients in spectral
                analysis (abstract numbers)
     $V(\omega_n)$ = pseudovelocity

The potential energy of the elastic system with several
degrees of freedom performing oscillations without severe
damage has been defined by the formula:

$$E = 1/2 \sum_n \omega_n^2 m_n \, q_n^2 (t)$$

where, ω_n = frequencies of natural oscillations
        $q(t)$ = generalized coordinate corresponding to
                the main n-mode of oscillations

Using the data obtained by instrumental recording when conducting micro-seismic zoning of different construction
sites, the pseudovelocity spectra of three spatial structures and their energeitics have been calculated. The
considered structures were the sloping shell of positive
curvature, the prismatic thin-slab shell and the hyperbolic paraboloid shell measuring 18x18m, 12x24m and 18x18,
accordingly. The structures were constructed in different
seismic areas of Kazakhstan.

The results of the calculations are presented in Fig. 2
and the Table.

The data analysis shows that higher amplitudes of oscillations can not greatly influence the pseudovelocity value and consequently, the amount of energy transmitted to
a structure at the dynamic motion. Taking into consideration the actual rigidity of the supporting contour (the
shaded section) the values of pseudovelocity come close
to the critical values, thus pointing to the necessity of
damping oscillations in large-span shells erected in seismic zones.

In the simplified calculations of structures for seismic
loading the motion of foundation soils is viewed as conventional and following the Harmonic (sinusoidal) Law.

In fact the process of movement of foundation soils can
be thought of as the result of the imposition of finate
or non-finate numbers of harmonic (sinusoidal) components
(by expansion into a Fourier series).

The experimental data obtained with the help of seismometers installed on the shells erected in different seismic
zones of Kazakhstan show that with the wide range of soil
vibrations vertical oscillations of shells initiated by
microseisms are consistent with harmonic oscillations or
close to them, and thereby are different from soil vibrations. Investigations of microseisms allowed to arrive at
the conclusion that the registered surface waves to which
the natural vibrations of the soil layers refer, do not
depend on the processes in the source of vibrations but
are determined by the physical properties of soils and
their bedding conditions in the upper layers (Fig.3).

The experimental data analysis shows that the lower frequences of shell oscillations were close to one of the
frequences of the microseism spectra observed. The similar
phenomenon is called the convolution of spectra and is
widely known in the field of physics /2/.

As the seismic motion transmitted to structures is acci-
dental let us consider an effect it has on the structures
under investigation.

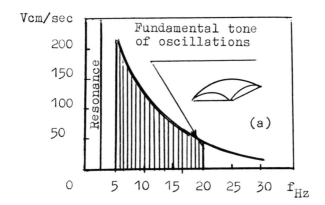

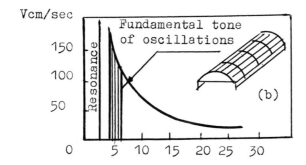

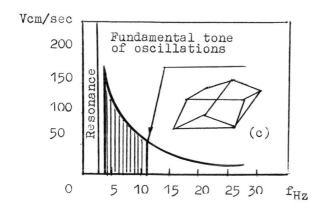

Figure 2 Pseudovelocity spectra of structures:

(a) 18x18m positive Gaussian curvature, (b) 12x24m prismatic thin-slab shell and (c) 18x18m hyperbolic shell

Table 1   Shell energetics

Structure	Upper earthquake energy limit m/2 x 10^6	Ultimate energy of structure m/2 x 10^6
18x18m sloping shell	19.406	37.9
12x12m prismatic thin-slab shell	31.624	105
18x18m hyperbolic paraboloid shell	13.085	68.02

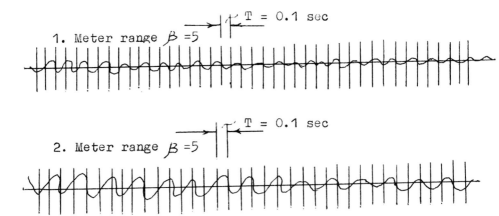

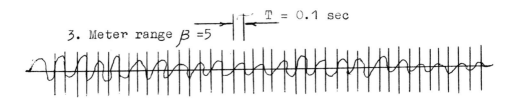

Figure 3   Vertical displacements according to micro-
           seism records
1 - 18x18 m sloping shell of positive curvature
2 - 18x18 m hyperbolic shell
3 - 12x24 m prismatic thin-slab shell

Let the pulse to be seen in the presented oscillograms (Fig.3) be called the output pulse of the system, the input pulse being seen in the oscillogram or accelerogram of the soil registered by seismometers. Any structure has its own spectrum of natural oscillations which can be called the impulse response of the system. Using this line of reasoning it follows that any input pulse changes its shape at the output. Therefore, if we know the impulse response of the system we can calculate the output pulse as well using the convolution of spectra. In actual practice the reverse is possible: if given the output pulse and the input pulse is determined and thus the level of the dynamic impact on the system is defined.

With the microseisms recordings presented in Fig.3 the convolution of spectra has been calculated for a number of structures. The calculations made it clear that microseisms with the frequency of oscillations of 5 Hz proved to have the most steady-state amplitudes and were responded by the shells the lower frequency oscillations of which were close to this value.

The earthquakes registered in these seismically active regions lent support to the validity of the proposed approach. The authors believe that the approach provides an opportunity to think seismically when doing a seismic design and to design safe shell structures capable of withstanding earthquakes of magnitudes characteristic of the design region.

REFERENCES

1. BOLT B.A. and UHRHAMMER R.A. Resolution techniques for density and heterogeneity in the Earth. Geophys. I.Rou Astr.Soc. 1975 pp. 19-35.

2. MAX J. Methodes et techniques de traitement du signal et applications aux mesures phisiques. Masson Paris New York Barcelone Milan Mexico Rio de Janezo 1981 pp 52-57.

3. ZHUNUSOV T.ZH., ZHIV A.S. Investigation and designing of reinforced concrete shell structures for seismic zones in Kazakhstan. International Conference IASS., TASS, Section 2. Mir Publishers, 1977 pp.402-408

MAXIMISING STRENGTH/DURABILITY

Chairmen   **Mr L H McCurrich**
Fosroc International Limited
United Kingdom

Professor S Nagataki
Tokyo Institute of Technology
Japan

Professor A E Sarja
Technical Research Centre of Finland
Finland

A COMPARATIVE STUDY OF THREE TYPES OF CONTROLLED PERMEABILITY FORMWORK LINERS

A E Long

P A M Basheer

P Brady

A McCauley

Queen's University of Belfast

UK

ABSTRACT. The use of controlled permeability formwork (CPF) systems is one of the options to improve the quality of the near surface concrete. The structural materials research group at Queen's University of Belfast has investigated one such system, the Zemdrain formwork liner, in the past and results have been reported elsewhere. The main objective of the work reported in this paper was to compare the performance of Zemdrain formwork liner with two other types available in the market and study the effect of the reuse of these liners. Air permeability, water absorption, pull-off tensile strength and abrasion resistance were used to compare their performance. In general, the different systems resulted in a significant improvement in the near surface properties of concrete, however their relative benefits varied. Although the reuse of the fabric liners resulted in a reduction in their performance, there was still a significant enhancement in properties relative to a normal cast concrete surface.

Keywords: Controlled permeability formwork, concrete durability, permeability, sorptivity, surface strength, abrasion resistance.

Professor Adrian E. Long is the Director of the School of the Built Environment at the Queen's University of Belfast. He has 33 years of research experience, mostly relating to structural concrete and concrete technology. He has more than 150 publications to his credit on concrete and concrete structures.

Dr P.A. Muhammed Basheer is a lecturer at the Queen's University of Belfast. He has over 8 years research experience in the general area of concrete durability, particularly on inspection, assessment and protection of concrete and concrete structures. This has resulted in nearly 60 technical publications.

Mr Pat Brady was a postgraduate student at the Queen's University of Belfast. The work reported in this paper was carried out by him for his MSc thesis.

Mr Anthony McCauley was a postgraduate student at the Queen's University of Belfast. He has recently completed his Msc degree in Civil Engineering.

Radical Concrete Technology. Edited by R K Dhir and P C Hewlett. Published in 1996 by E & FN Spon, 2–6 Boundary Row, London SE1 8HN, UK. ISBN 0 419 21480 1.

INTRODUCTION

The near surface concrete provides the first line of defence to the ingress of deleterious substances and, hence, to the durability of structural concrete. As a consequence, it is the permeation characteristics of this layer of concrete which control the rate of deterioration. It is commonly recognised that by reducing the water-cement ratio of the near surface layer, the durability properties can be improved. This can be achieved by using a Controlled Permeability Formwork (CPF) system.

The first use of some form of controlled permeability formwork was reported in 1950 where softboard liners were used on the inside of formwork panels. The permeable liner soaked up excess air and water from the concrete surface, however this practice was abandoned due to problems arising when stripping the formwork. The Japanese have since developed a permeable textile formwork liner [1] (RC. Cloth is one such liner), which can be attached to the vertical or inclined surface of conventional formwork. The liner allows surplus air and water to percolate through it, while preventing the escape of cementitious particles, thus lowering the water-cement ratio and increasing the strength. Parallel research carried out by Du Pont has resulted in the development of a similar, but less expensive, polypropylene liner known as "Zemdrain". The principle of operation is similar to that of the Japanese liner, both systems have been shown, through site and laboratory investigation, to produce a highly durable near surface concrete [1, 2].

This paper presents the results obtained from two investigations into the performance and reusability of CPF systems. The surface durability was assessed in terms of air permeability, sorptivity, water permeability, surface tensile strength and abrasion resistance of the near surface concrete.

EXPERIMENTAL PROGRAMME

Test variables

Four types of concrete surfaces were made with the following types of formworks :
(i)   Conventional formwork (Normal plywood formwork)
(ii)   Zemdrain lined formwork
(iii)   RC Cloth lined formwork
(iv)   Softboard lined formwork
The work was carried out in two series. The first investigated the effectiveness of the three CPF systems by comparing them with the conventional formwork system. A total of six specimens of 950x300x150mm size were cast with a w/c of 0.65 and this gave a total of twelve test surfaces. This allowed each of the four test surfaces to be assessed on the basis of the different specimens.

The second series of the work was to study the effect of the reuse of the formwork liners. Hence, four specimens of 700x300x150mm were cast with a w/c of 0.55. This resulted in eight test surfaces; three with the Zemdrain liner, three with the RC cloth liner, one with the Softboard and the last with the conventional formwork. The Zemdrain and RC Cloth were reused twice to achieve the three test surfaces for each of these two liners.

A mix ratio of 1:1.65:3 between ordinary portland cement, zone 3 sand and 10mm basaltic aggregate were used in both test series.

Preparation of Test Specimens

The formwork liners were attached to the formwork in accordance with standard practices. In the case of the Zemdrain liner, it was stretched and nailed onto the timber formwork in compliance with the manufacturer's specifications as described in reference 3. The RC cloth consists of two layers, a permeable layer and a drainage layer. The permeable layer is made up of a white polyester woven fabric and this allows water and air to escape from the near surface concrete. The drainage layer is a blue polythene knitted fabric and allows the air and water, that has passed the first layer, to drain away. Both the layers are attached together to conventional formwork with drainage holes (6mm diameter). In order to eliminate problems during the release of the softboard lined formwork, a permeable fabric sheet was tacked to the conventional formwork using panel pins. The other surfaces of the mould were lightly oiled prior to casting.

All specimens were removed from the mould 24 hours after casting and placed in a controlled environment at 20 ^{0}C an 55% RH for 14 days. They were then transferred to an environmental cabinet at 40 ^{0}C and 20% RH and there they remained until they were 28 days of age. The 40 ^{0}C drying was expected to have produced a reasonably low moisture content in the near surface concrete. The specimens were then placed in the controlled environment at 20 ^{0}C and 55% RH for the duration of the testing programme.

TEST METHODS

The following tests, all of which are related to concrete durability, were applied to the test specimens:

Permeation Tests using the 'Autoclam'

The air permeability, water permeability and sorptivity tests were carried out at three locations along the depth of the specimens using the 'Autoclam' [4]. A detailed description of the Autoclam permeability system can be obtained in reference 5. For test series 1 the air permeability test was carried out on the 28th and 47th day after casting. The water permeability and sorptivity tests were carried out on the 48th day after casting. For test series 2 the air permeability test was carried out on the 28th day and the sorptivity test was carried out on the 29th day after casting.

Surface Tensile Strength

The surface tensile strength was determined by carrying out the pull-off test [6] on the 48th day (Test series 1) and 29th day (Test series 2) after casting. Steel disks of diameter 50mm were bonded to the concrete surface by means of an epoxy resin adhesive. Then by applying an increasing tensile load via a hydraulic jack, the disk was pulled off the test surface and the force applied to cause the failure was recorded and the corresponding tensile strength calculated.

Abrasion Resistance

For test series 2, when the test specimens were 100 days old the abrasion resistance was measured by employing an apparatus developed at Queen's [7]. The depth of abrasion at 8 different locations on the test surface was measured and an average value determined in order to assess the abrasion resistance.

PRESENTATION AND DISCUSSION OF RESULTS

Comparison of CPF liners

<u>Surface Finishes:</u> The cast surfaces were inspected immediately upon striking the formwork 24 hours after casting. The surface cast against the normal formwork face was light grey in appearance and had an extensive coverage of blowholes. The Zemdrain formwork face showed an unblemished surface free from any blowholes or other defects and was slightly darker in appearance. The surface produced using the RC. cloth was very similar to that obtained from Zemdrain, but more textured resembling the textile fabric of the liner. The surface obtained by casting against the softboard system was also blowhole free, but a layer of fine particles was evident, which when removed revealed a rough indented finish. This may be attributed to the excessive absorption of water by the softboard, thus leaving insufficient water for full hydration of the cement particles near to the surface. The absence of blowholes in concretes produced with the CPF systems is due to the removal of the excess air and water in concrete by the formwork liners during the vibration of fresh concrete.

<u>Bleed Water:</u> The volume of bleed water collected in test series 1 increased steadily for the first 20 minutes and tapered off after 100 minutes (Figure.1). There was no significant difference between Zemdrain and RC cloth in removing bleed water. No bleed water drained off from the softboard system as the thick softboard layer absorbed all the excess water.

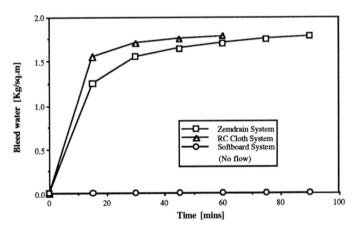

Figure 1. Average volume of water removed by formwork liners

<u>Permeation tests with the Autoclam:</u> The air permeability indices obtained from the three CPF formwork faces are substantially lower than that obtained from concrete cast against conventional formwork (Figure 2). The softboard system showed a reduction in the range of 3 to 5 times and the RC cloth and Zemdrain liners gave reductions ranging between 6 and 9 times that of the conventional formwork surfaces. This trend in performance was observed for both the 28th day and 47th day tests, however the 47th day results showed a further 10% reduction in air permeability relative to the 28th day results.

The concrete surfaces made using the CPF systems produced sorptivity indices on an average 85% better than that of the conventional formwork (Figure 3). The Zemdrain liner performed on an average 20% better than the other two systems. The same trend of results was found for the water permeability tests.

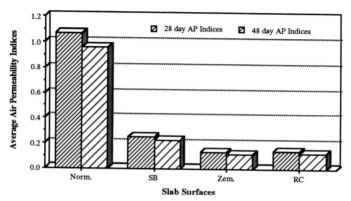

Figure 2. Influence of various CPF liners on Air Permeability

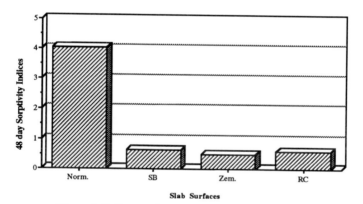

Figure 3. Influence of various CPF liners on Sorptivity

Surface Strength: From the results presented in Figure 4 the CPF systems resulted in significantly higher surface strength than the conventional formwork (average for all CPF systems 40%). The minor variations for the CPF faces can probably be put down to variations in mixes from specimen to specimen.

Effect of the reuse of CPF liners

Figures 5 to 8 report the effect of the reuse of Zemdrain and RC cloth in comparison to the conventional formwork for concrete with a w/c of 0.55. With respect to permeation properties there is no significant difference between Zemdrain and RC cloth and the effect of the reuse is seen to be to increase the permeation properties. However, this increase is modest compared to the reduction achieved by the use of these liners. The softboard did not perform as well as the other two at this w/c, however the actual air permeability value is much less than that obtained for the 0.65 w/c concrete made with softboard lined formwork.

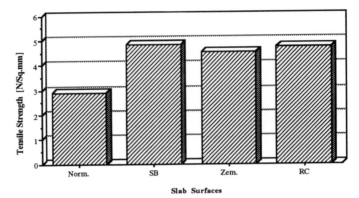

Figure 4. Influence of various CPF liners on Surface Strength

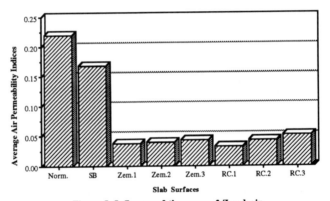

Figure 5. Influence of the reuse of Zemdrain
& RC Cloth on Air Permeability

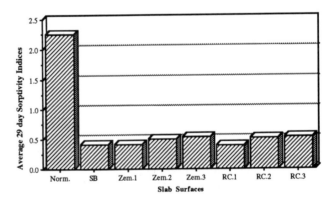

Figure 6. Influence of the reuse of Zemdrain
& RC Cloth on Sorptivity

The pull-off tensile strengths in Figure 7 illustrate that the reuse of Zemdrain and RC cloth reduced the surface strength of 8%. However, there was still an increase in surface strength by 18%. On average the single use of CPF systems resulted in an increase in surface strength by 40%, which is similar to the results in the first test series.

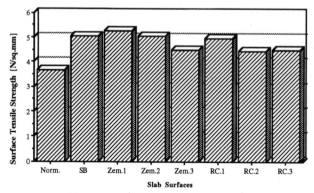

Figure 7. Influence of the reuse of Zemdrain
& RC Cloth on Surface Strength

Compared to conventional formwork, the softboard system gave a 50% reduction and the other two systems gave an average of 90% reduction in abrasion depth (Figure 8). The depth of abrasion increased significantly when Zemdrain and RC cloth were reused. A second reuse resulted in a decrease in abrasion resistance of some 50% of the original value, however there was still an improvement of 65% compared to the conventional formwork.

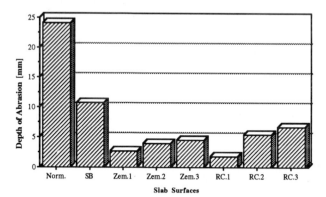

Figure 8. Influence of the reuse of Zemdrain &
RC Cloth on Abrasion Resistance

CONCLUSIONS

On the basis of results presented in this paper, the following conclusions have been drawn :
1.   The use of a controlled permeability formwork system improves the durability characteristics of concrete significantly. At a high water-cement ratio, there is no significant difference between the use of Zemdrain, RC cloth and the

softboard with a fabric liner. However, the performance of softboard is not as good as the other two at a medium water-cement ratio concrete. Zemdrain was found to be slightly better than RC cloth in reducing the permeation properties and improving both the surface strength and the abrasion resistance.

2. The reuse of both Zemdrain and RC cloth resulted in reduction in the improvement in the properties normally obtained by using these liners. However, these reductions in performance are relatively modest. It may be noted that these formwork liners were used carefully in the laboratory and such care in handling may not be possible on site. Therefore, the effectiveness of reusing of these formwork liners may depend on site practices and workmanship. There is also a risk that the liner could be reused to such an extent that no significant benefits are achieved and this would be difficult to control on site.

ACKNOWLEDGEMENT

Zemdrain was supplied by CIS Construction products. Support for the research included in this paper was provided by Department of Civil Engineering, Queen's University of Belfast. This and the funding provided by both the DuPont Nemours (Lexumbourg) and the Engineering and Physical Sciences Research Council, are gratefully acknowledged.

REFERENCES

1. Harrison, T., Introducing controlled permeability formwork, Concrete Quartely, British Cement Association, 1990, pp 6-7.
2. Price, W.F. and Widdows, S.J., The effects of permeable formwork on surface properties of concrete, Magazine of Concrete Research, Vol. 43, No. 155, 1991, pp 93-104.
3. Sha'at, A. A-H., Long, A.E., Montgomery, F.R. and Basheer, P.A.M., The influence of controlled permeability formwork liner on the quality of cover concrete, ACI, SP 139-7. 1993, pp 91-105.
4. Basheer, P.A.M., A brief review of methods for measuring the permeation properties of concrete in-situ, Proc. ICE, Buildings and Structures, London, February 1993, pp 74-83.
5. Basheer, P.A.M., Montgomery, F.R. and Long, A.E., 'Clam' tests for measuring in-situ permeation properties of concrete, Journal of Non-destructive Testing and Evaluation, Vol. 12, 1995, pp 53-73.
6. Long, A.E., and Murray, A.McC. The pull-off partially destructive test for concrete, Proc. Intl. Conf. on In-Situ/Non-Destructive Testing of Concrete, Ottawa, Canada, October 1994, Editor: V.M. Malhotra, ACI SP-82, pp. 327-350.
7. Montgomery, F.R., Long, A.E. and Basheer, P.A.M., Assessing surface properties of concrete by in-situ measurements, Proceedings, IABSE symposium, Durability of structures, Lisbon, Portugal, 6-8 September 1989, IABSE, Zurich, Switzerland, 1989, pp 871-876.

CONTROLLED PERMEABILITY FORMWORK

D Wilson

CIS Construction Products

UK

F L Serafini

Du Pont de Nemours (Luxembourg) S.A

Luxembourg

ABSTRACT. Durability of concrete is perhaps one of the great problems to be confronted by engineers. Not only in terms of the performance of the material, but in terms of weathering. Durability of the material is of limited value without durability of appearance.

Durability of concrete is therefore predominantly a surface issue. Environmental mechanisms act on the surface of the concrete affecting both appearance and performance. The permeability of the concrete is indirectly determined by the water cement ratio and the appearance by surface porosity and the presence of blowholes.

Controlled permeability formwork (CPF) improves every aspect of concrete durability by removing excess air and water from the concrete / formwork interface. CPF also prevents the formation of a surface laitence. The technical benefits of CPF are now widely accepted and this paper will review the main benefits, whilst detailing some benefits which are not always appreciated.

Keywords: Controlled Permeability Formwork (CPF), Durability, Water/cement ratio, Curing, Appearance, Surface contamination, Economics

David J.Wilson is Technical Sales Manager with CIS Construction Products, Burton-on-Trent, England

Franco L.Serafini is a Development Engineer with Du Pont de Nemours (Luxembourg) S.A., Luxembourg

Radical Concrete Technology. Edited by R K Dhir and P C Hewlett. Published in 1996 by E & FN Spon, 2–6 Boundary Row, London SE1 8HN, UK. ISBN 0 419 21480 1.

INTRODUCTION

The lack of durability of appearance of concrete structures is readily apparent. The surface quality of concrete surfaces after several years exposure to the environment shows rapid deterioration.

A growing appreciation of the deleterious effects of carbonation, chloride ingress and exfoliation caused by freezing and thawing have caused the concrete industry to take the question of durability seriously. The quality of cover zone concrete is critical, as this provides the first line of defence to the reinforcing steel and can easily be more important to long term durability than the bulk structural properties of the concrete.

As we become more concerned about the quality of our potable water supplies, any potential contamination of water from the concrete must be of concern. Leaching of hazardous materials from the concrete has been researched, but presence of residues of release and curing agents in the concrete surface has all but been ignored. New research is showing that we do so at our peril.

CONTROLLED PERMEABILITY FORMWORK

The Need

A technique is needed which addresses the four main areas of concern

- durability of appearance
- durability of material
- effective curing
- surface contamination

Until the advent of CPF, solutions were sought to each of these areas of concern separately. Coatings have been the main method of improving appearance, but their indeterminate life and variable performance have made their use problematical. The most work has gone into durability of material with the use of blended cements and admixtures. These products generally improve the bulk properties of the concrete but not those of the outer portion of the cover zone, generally recognised as the most important area. Correct curing practice is still not applied on most sites.

The use of release agents on the surface of formwork to enable its easy removal has several drawbacks. The agent retards the surface reaction giving a poorer quality concrete, but more importantly residues of release agents have been shown to penetrate up to 5mm into the concrete surface [1]. These bio-degradable products do not degrade in the concrete and provide a nutrient source for the micro-organism growth .

CPF has shown that in each of these four areas, it offers real benefits.

The Concept

The use of CPF can ensure that the concrete in the cover zone, particularly the outer 20mm, has optimum density and consistency.

A controlled permeability formliner (CPF) is engineered to resist concrete pressure but with a multitude of micro pores so that it acts as a filter which retains the fines of the concrete mix, but allows the trapped air and the excess water to pass through it and escape.

The form liner is placed over conventional formwork so that during the concrete compaction (vibration) the trapped air and excess water from the concrete mix, which moves towards the concrete surface, is evacuated through the CPF, as shown in Figure 1. Indeed, the water leaching out will continue afterwards and, depending on the case, can continue for up to 2 - 3 hours.

As a result the concrete surface is free of blowholes and has a very closed structure as the water/cement ratio near the concrete surface is significantly lowered (instead of becoming higher), Figure 1. This gives a quantum improvement in most concrete properties measured near the surface.

Release agents are not needed with the liner which helps avoid possible contamination of the concrete surface. This makes CPF particularly attractive for environmental problems or special applications such as reservoirs for drinking water and food silos.

The Benefits

The effects of CPF on concrete properties have been tested in numerous research laboratories, universities and specialised institutes throughout Europe and also in North America and the Middle East.

The tests involved a range of concrete grades from C25 to C65, with water/cement ratios ranging from 0.40 to 0.65, with different cements and blends: PC (Portland Cement), GGBS (Ground Granulated Blastfurnace Slag), PFA (Pulverised Fuel Ash). Different slumps (50 to 210mm), the addition of admixtures (plasticizers, super plasticizers, retarders water proofers and air entrainment) and the addition of silica fume have also been assessed. Work has been done on vertical and inclined slabs using different curing techniques and the effects of hot-dry and hot-wet conditions have been examined.

In all cases significant improvements were experienced on all the properties affected by the quality of the concrete surface. The improvement observed varied from concrete to concrete and also depended on the test considered, the design w/c, the type of cement used and on the casting conditions.

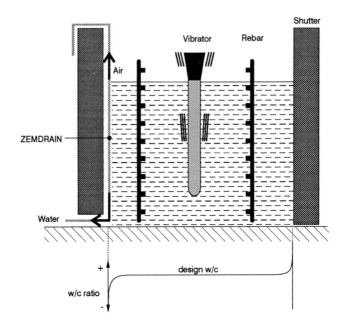

Figure 1. Basic concept of CPF

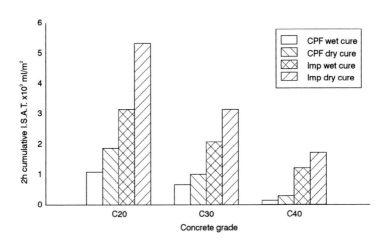

Figure 2. The effect of formwork type and curing
on 2 hour cumulative I.S.A.T [7]

As a general rule, it was found that the more critical the conditions (cement type, inclination, slump, curing) and the more critical the test, the more improvement could be observed. The sensitivity of surface properties to poor curing conditions, which is typical for concretes with slow hydration, was very much reduced.

The largest improvement was observed in the increase in tensile strength near the surface, which is not surprising as both water/cement ratio and the release oil used on regular shutters (especially if used in excess), have a negative effect on this parameter Surprisingly, also a quantum improvement was observed on a high grade concrete (w/c = 0.4, C65 with air entrainment and 5% silica fume). For a very critical test (salt freeze-thaw cycling) the amount of scaled material after 56 salt freeze-thaw cycles, was reduced by over 95% [2].

Describing all the results of the above tests goes beyond the intent of the present paper, so we will just summarise the overall results obtained in the given tests in Table 1, and later we will expand on some other aspects and on results obtained using different cement blends.

End Use Benefits

The improved concrete properties obtained with CPF make it particularly suitable for use in those cases where critical environmental conditions make it difficult to achieve the required life expectancy using more traditional techniques. Consideration will now be given to a number of application areas and how CPF is of benefit will be examined.

Bridges And Transport Projects

Durability is of major concern to highway authorities throughout the world and the UK is no exception. A major study [3] commissioned by the Department of Transport in the UK, studied 200 bridges to assess their durability characteristics. The results were quite disturbing and subsequently advice notes have been issued regarding durability, which have highlighted two major damage causing mechanisms.

- salty water from de-icing salts being splashed onto bridge abutments, piers, parapets and deck soffits. This spray also contaminates the soil adjacent to carriageways causing chloride attack to concrete sub-structures.

- poor curing.

The reduced permeability and absorption characteristics of CPF cast concrete help to significantly increase the time to activation of steel corrosion, by factors of 2 to 10 [4]. Figure 2 shows the benefits in terms of the reduced absorption characteristics for three strength grades and the effects of wet curing. Table 2 shows the improvements in the effective chloride diffusion coefficients. It is evident that CPF used in combination with blended cements can give the optimum protection against the ingress of chlorides.

Table 1. Summary of test results

Test Considered	Average	(Range)
Surface hardness (Hammer - Schmidt)	+21.7	(10-43)%
Surface breaking strength	+165	(27-522)%
Surface abrasion resistance	+32.6	(31-34)%
Water absorption (2hrs cumulated ISAT*)	-81.4	(47-93)%
Water penetration depth	-55.00	(31-89)%
Water sorptivity	-52.30	(20-84)%
Water permeability coefficient	-85.00	(61-99)%
Autoclam pressure decay test(Q.U.B.)		
air permeability	-69.9	(41-99)%
water permeability	-65.6	(58-81)%
Oxygen diffusion coefficient	-25.5%	
Chloride diffusion	-57.60	(25-90)%
Salt freeze-thaw erosion	-79.70	(36-99)%
Carbonation, accelerated(CO_2) chamber	-71.60	(39-99)%
Carbonation,normal exposure	-82.60	(80-87)%
Surface porosity (0-4 mm depth)	-46%	

Table 2. Efffective chloride diffusion test results [4]

Mix	Formwork	Cure	Surface Chloride (% conc)	Effective Diffusion Coefficient ($m^2s \times 10^{-12}$) 40°C	20°C*
OPC	IMP	Wet	0.16	10.00	2.50
	IMP	None	0.18	8.88	2.22
	CPF	Wet	0.28	3.02	0.76
	CPF	None	0.27	3.04	0.76
OPC/PFA	IMP	Wet	0.16	9.99	2.50
	IMP	None	0.24	5.62	1.40
	CPF	Wet	0.18	2.05	0.51
	CPF	None	0.15	4.00	1.00
OPC/50%GGBS	IMP	Wet	0.19	10.56	2.64
	IMP	None	0.20	8.05	2.01
	CPF	Wet	0.11	1.75	0.44
	CPF	None	0.12	2.48	0.62
OPC/70%GGBS	IMP	Wet	0.18	5.03	1.26
	IMP	None	0.24	5.62	1.40
	CPF	Wet	0.11	1.85	0.46
	CPF	None	0.11	1.80	0.45
OPC/AEA	IMP	Wet	0.20	3.47	0.87
	IMP	None	0.15	4.49	1.12
	CPF	Wet	0.19	1.37	0.34
	CPF	None	0.22	1.29	0.32

*Assuming Arrl.enius relationship applies and dividing measured value at 40°C by 4

Marine Structures

Coastal structures have to withstand the erosive forces of the sea, both from abrasion, continuous cycles of wetting and drying and the chloride rich environment.   The splash zone is regarded as being the most severe with regard to the accumulation of surface chlorides.   This is due to the wetting and drying cycles which result in a progressive build up of chloride by a process of wetting with seawater, evaporation and salt crystallisation. In some areas, freeze/thaw action can aggravate the problem. This phenomenon occurs world-wide, but the rate of failure of concrete in the harsher environments like the Middle East is more rapid than in milder climates.   The benefits of CPF in hot-wet and hot-dry climates have been previously published [5]. On going tests on in-situ structures in the U.A.E.,see Figure 3 are already showing that after 3 years exposure in the splash zone, that the benefits of CPF are very evident.   The same benefits can be anticipated in the UK, although over a longer time period. The benefits of early age curing by the CPF liner are also significant.

Wastewater Structures

The safe treatment of all kinds of wastewater is of major public concern. As we treat more and more aggressive elements the quality of the surface concrete is very important.   As well as having no blowholes, a dense impermeable surface which is resistant to wear and abrasion and to aggressive gases and effluents is required. Figure 2 shows how the absorption of the surface is reduced by using CPF, whilst Figure 4 shows the improvements in abrasion resistance for a 0.45 and 0.65 W/C mix, showing that the use of CPF with a weaker mix gives better properties than the stronger mix cast conventionally.   In exposed areas the freeze/thaw characteristics of concrete are very important, the use of CPF avoids any problems, Figure 5.

Potable Water

Water is our most precious asset and the quality of our drinking water supplies is of major public concern.   The integrity, durability and impermeability of structures containing water are therefore essential.

Conventionally cast concrete tends to produce a low denisity porous surface with numerous blowholes, and contaminated with residues of release agents. CPF liners do not require release agents so no surface contamination occurs, helping to avoid the possible occurrence of micro-organisms growth.   The reduced porosity of CPF surfaces, Figure 6 and improved abrasion resistance and surface strength [6] make the CPF surface ideal for this application. In those circumstances where it is necessary to coat concrete the use of CPF can be of major benefit. The surface produced does not require grit blasting or a faring coat and provides the best possible base for most coating types.

Non-structural cracks as well as being unsightly can be a significant problem. Although not the whole answer, the use of CPF overcomes many of the problems which lead to crack formation.

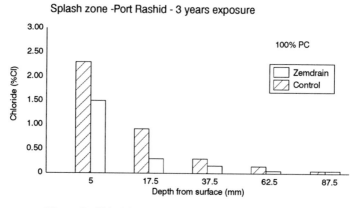

Figure 3. Chloride at given depth (%Cl/cement)

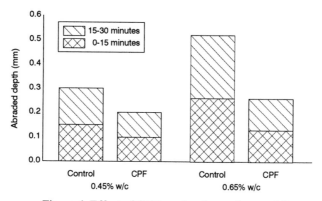

Figure 4. Effect of CPF on abrasion resistance [6]

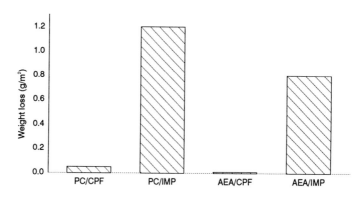

Figure 5. Effect of CPF on frost resistance after 25 F/T cycles [4]

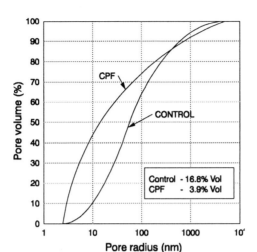

Figure 6.Pore Distribution (surface)

ECONOMIC & ENVIRONMENTAL CONSIDERATIONS

CPF systems have been used on over 100 UK projects and considerably more worldwide. The concrete produced has consistently benefited from improved appearance, durability and curing, irrespective of grade, mixes and admixtures. On all of these contracts the perceived extra cost of using CPF has been considered insignificant by clients whose main interest is the whole life cost of their structures.

Under most circumstances using CPF adds to the initial construction costs, but contractors can make savings by using lower grade plywood, avoiding the use of release and curing agents and with reduced cleaning and maintenance of formwork. The most significant savings are in producing an as struck surface to which cosmetic repairs are not required. On a motorway pier replacement contract these savings were so significant that CPF was used at no extra cost to the client. Surface impregnation with silanes is expensive and hazardous and the use of CPF is a better, more economical solution. For coating contracts savings (£15/m^2 on one reservoir contract) were made when the grit blasting and faring coat operations were deemed unnecessary.

The main savings are over the longer term with reduced maintenance and increased structural life. As always these benefits are difficult to quantify, but,as any remedial work will involve disruption of service and extremely costly repairs, the benfits to all structures will be considerable. From an environmental point of view, the use of CPF is very beneficial. By making the maximum use of the initial construction and ensuring longer life structures, repairs may not be necessary and scarce resources will be preserved.

CONCLUSIONS

The arguments presented in this paper demonstrate that the use of CPF gives a quantum improvement in the surface properties of concrete. This is the only technique which can give durability of appearance and performance to a structure in one operation. CPF can give concrete back its reputation for quality.

This can only be achieved if the industry wishes to do so, as it involves a change in attitude and practices. This is going to happen despite the conservatism of the construction industry, if for no other reason, than that in the past nobody could be blamed if they used commonly accepted practices resulting in poor quality concrete: in future it will be very difficult to justify ignoring techniques like CPF and to continue producing vulnerable concrete.

Despite the attempts to improve the environment it is obvious that with an increasing population that the quality of air rain and water is not improving and the environment will continue to be aggressive to concrete. So the case for the use of CPF will increase with time.

This site proven technology for producing longer life concrete is here today. If engineers do not specify and use "Controlled Permeability Formwork", then their client's first maintenance bill will not be long in coming.

REFERENCES

1. FRANKE, P. LGA Institute of Biotechnology Report. Micro-organisms on Concrete Surfaces (Hygienic Problems in Drinking Water Reservoirs).

2. SINTEF REPORT No. STF65 A91049, Oct 1991

3. WALLBANK, E.J.:The Performance of Concrete in Bridges, A Survey of 200 Highway Bridges, HMSO, London 1989 96pp.

4. PRICE, W.F.: The Improvement of Concrete durability using Controlled Permeability Formwork. Proc. 5th. Intl.Conf. "Structural Faults and Repairs", 1993, Vol 2., Pg 233-238.

5. PRICE, W.F. & WIDDOWS, SJ, : Concrete in Hot Climates. Proc.of 3rd Intl Rilem Conf. Pg 207-220.

6. LONG, A E, BASHEER, M, & CALLANAN, A : Controlled Permeability Formwork. Construction and Repair, Nov/Dec 1992, pp 36-40

7. PRICE, W.F. & WIDDOWS, S.J., The effects of permeable formwork on the surface properties of concrete. Magazine of Concrete Research, 1991, 43, No.155, June, pp93-104.

A NEW ADMIXTURE FOR
HIGH PERFORMANCE CONCRETE

Y-O Tanaka

S Matsuo

A Ohta

M Ueda

NMB Ltd

Japan

ABSTRACT: We are currently studying high performance concrete (HPC). We believe that the most important aspects of HPC to develop are its flowability and its compaction ability. For this purpose, an advanced superplasticizer (ADSP) to attain excellent flowability and segregation resistance is required. We have developed a polycarboxylate based ADSP containing cross-linked polymer, SP-A.

Flowability and segregation can be explained by DLVO therory and steric effect theory as well as depletion effect, Tom's effect and tribology effect.

Keywords: High performance concrete (HPC), Advanced superplasticizer (ADSP), Polycarboxylate, Cross-linked polymer, Depletion effect, Tom's effect, Tribology effect

Mr Yoshi-o Tanaka, PhD is a Research staff member at R&D Laboratories, NMB Ltd. His research interests are chemical synthesis and materials science.

Mr Shigemi Matsuo is a Concrete engineer at R&D Laboratories, NMB Ltd. He is now working on the development of high performance concrete.

Mr Akira Ohta is a Research chemist at R&D Laboratories, NMB Ltd. His research interest is the synthesis of various polymer dispersing agents and evaluation of their performance.

Mr Minoru Ueda is a Research chemist at R&D Laboratories, NMB Ltd. His research interest is the chemical analysis of admixtures, cement and concrete.

Radical Concrete Technology. Edited by R K Dhir and P C Hewlett. Published in 1996 by
E & FN Spon, 2–6 Boundary Row, London SE1 8HN, UK. ISBN 0 419 21480 1.

INTRODUCTION

With recent concrete technology, new design and placement methods have been developed to cope with the increase in the variety of concrete structures and their scale, as well as to improve strength, durability, and high flowability of concrete. In addition, attracting laborers to job sites is becoming difficult due to increases in the number of projects carried out under special environments and the changes in Japanese social structure. Conditions surrounding the concrete business are becoming severe. Therefore, greater diversification and improvement in concrete performance and technology leading towads high performance concrete (HPC) are needed. Given these conditions, chemical admixtures for concrete have taken a greater role in the progress of concrete technology. In particular, the influence of advanced superplasticizers (air entraining and high-range water reducing agents and plant added superplasticizers, ADSP) has been significant.

HPC is defined by three criteria: strength, durability and flowability . We believe that the most important aspects of HPC to develop are its flowability and its compaction ability. For the purpose, an ADSP to attain excellent flowability and segregation resistance is required. We have developed a poycarboxylate-based ADSP containing crosslinked polymer, SP-A. The first ADSPs were introduced into Japan in 1987 and were called the 4th generation of admixtures, after AE agents, AE water reducing agents, and superplasticizers. ADSPs exhibit high water reduction and good slump retention.

The General Project by the Ministry of Construction called The Technology Development of Ultra-Light and Ultra-High Rise Reinforced Concrete Buildings, (the so-called New RC) has taken a great role in the development of ADSP. By using ADSP, the production of concrete with a unit water weight of 160 kg/m^3 or 165 kg/m^3 and with a lesser slump change over time became possible despite a 22% water/cement ratio.

Thus, ADSP have had a significant role in providing concrete with durability following their introduction into the Japanese market in 1987.

THE PURPOSE OF ADSP

There are three typical purposes of ADSPs: as a measure to reduce the unit water weight of ordinary concrete; to produce high strength concrete; and to produce high flowable concrete.

TYPES OF ADSP

The ADSP on the market at present are roughly classified, for convenience, into four types based on their principal water reducing component, in other words, the main component which disperses cement particles. This classification is further subdivided based on slump retentive components.

Chemical structure of ADSP currently on the market

At present, the member companies of the Chemical Admixture Association have marketed various types of ADSP. These admixtures can be roughly divided into four types based on their main component: 1) naphthalene based, 2) melamine based, 3) polycarboxylate based, and 4) amino sulfonate based admixtures. The presumed chemical structures of the main components are given in Figure 1.

Naphthalene based

Beta-naphthalene sulfonate formaldehyde condensate (BNS) is commonly called naphthalene sulfonate, polyalkylaryl sulfonate, or alkylnaphthalene sulfonate.
Action derivative polymer, reactive polymer, or modified lignin (secondary components exhibiting slump retention) are added to naphthalene-based ADSP.

Melamine based

Melamine sulfonate formaldehyde condensate (M S) is commonly called melamine, modified methylol melamine, or sulfonated melamine. In addition, slump retentive components as well as MS are added to melamine-based ADSP.

Polycarboxylate based

Various polycarboxylate-based water reducing agents have been developed, from which ADSP have been produced and marketed. Polycarboxylate-based water reducing agents are classified into two types: olefin-maleate based and acrylate-acrylic ester based. Polycarboxylate-based ADSPs with new functions can be developed because unlike BNS and MS, the chemical structure of polycarboxylate-based water reducing agents can accept side chains exhibiting new fuctions by grafting and copolymerization. New functions can be added by controlling the chemical structures and their molecular weights [1].

Aminosulfonate based

Aromatic amino sulfonate-based polymer compounds are said to be three dimensional condensates.

Standards for ADSP in Japan

Quality standards for ADSP are stipulated in the " Guidelines and Explanatory Notes on the Mix Proportion, Production, and Placement for Air Entraining High-range Water Reducing Agents (draft)" Attachment 1 JASS 5T-403 (Quality Standards for Air Entraining High-range Water Reducing Agents for Concrete) published by the Japan Architect Society in June 1992. JASS 5T-403 is a collection of quality specifications and test methods [2]. Approximately one year later, the Japan Society of Civil Engineers

(A) Beta-naphthalene sulfonate (BNS)

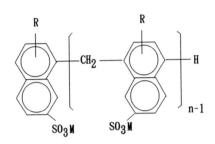

(B) Melamine sulfonate (MS)

(C) Action derivative polymer

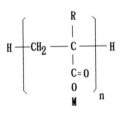

(D) Polycarboxylate (Acrylate 1)

(E) Polycarboxylate ether (Acrylate 2)

(F) Cross-linked polymer (Acrylate 3)

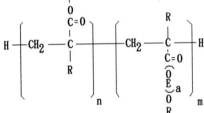

(X) ◄ Crosslinked point

(G) Amino sulfonate polymer

Figure 1 Chemical structure of ADSP

produced " The   Guidelines for the   Placement of Concrete Using Air Entraining and
High-Range Water Reducing Agents (draft)", and published " The Guidelines for Concrete Placement Using Air Entraining and High-Range Water Reducing Agents" in
Concrete Library   No. 74 in July, 1993. In these guidelines, an ADSP was defined as
an admixture that is added into a mixer with other materials when concrete is mixed
and that exhibits air entrainment, higher water reduction than an AE water reducing
agent, and good slump retention [3]. These quality specifications for this ADSP have
been certified by the Japan Industrial Standards (JIS A 6204) in March, 1995.

Besides these standards authorized by academic societies, there are also the following:
" Quality Standards for Air Entraining High-Range Water Reducing Agents (draft)" by
the Honshu-Shikoku Bridge Authority; " A Criterion for Air Entraining High-Range
Water Reducing Agents for High Strength Concrete"   by the Housing and Urban De-
velopment Corp.; and "The Criterion for Air Entraining High-Range Water Reducing
Agents for High Strength Concrete" by the New RC of the Construction Ministry.

FLUIDIZING MECHANISMS OF ADSP

Dispersion theory

In order to improve the water reducing effect in concrete, the dispersibility of cement
particles should be increased and the dispersibility should remain stable. It is said that
the dispersion stability of inorganic particles is due to electrical repulsion and the
steric effect of adsorbed surfactants[4]. The basis of the dispersion stability of cement
particles is the same as that of inorganic particles, although the surface conditions of
the particles is changed by hydration.

DLVO theory

The explanation of dispersion stability in terms of electronic repulsion according to the
DLVO theory by Derjaguin, Landau, Verwey, and Overbeek is well known. The dispersion stability of particles is determined based on the curve shape of the total potential
energy (V_T) made up of the electrostatic repulsion force (V_R) obtained when two particles approach and the London-van der Waals' attraction (V_A). When the distance between two particles corresponds to the point on the curve where V_T is at a maximum
(V_{max}), the two particles are dispersed.   As V_{max} increases, the dispersibility is increased and exhibits a close co-relationship with the zeta potential [4].

Steric effect theory

Dispersion stability due to the steric effect can be explained by the entropy effect theory proposed by Mackor. The total potential energy V_T between two particles is given
by $V_T = V_A + V_R^S$. V_A is the van der Waal's attractive potential energy and V_R^S is
the steric repulsive potential energy calculated as an entropy term from the structure

and conformation of adsorbed surfactant on the particles. The dispersion stability is maintained by this steric repulsive force. Fisher and Ottwill conclude this theory as the osmotic pressure effects [4].

The water reducing effect on cement components such as concrete and mortar is obtained by increasing the dispersion of cement particles. Water reducing agents (dispersing agents) can be roughly divided into 2 types: 1. water reducing agents which increase the zeta potential of cement particle surfaces and increase the electrical repulsive force; and 2. those which increase the repulsive force by sterically expanding the adsorption layer on the surface of the cement particles.

BNS- and M S- based water reducing agents are adsorbed on the surface of cement particles in rod-shaped chains and in several layers. The cement particles are dispersed due to the strong repulsion caused by the negative ions of the sulfonic group in the chemical structure of the agents. The size of this electronic repulsive force can be estimated by measuring the zeta potential of the cement particle surface. The dispersibility and water reduction can also be estimated [5]. With polycarboxylate-based water reducing agents, the water reducing effect is obtained by dispersing the cement particles through two factors: 1. the electrical repulsive force due to negatively charged ions of the carboxylic groups in the chemical structure of the agents; and 2. the steric effects of the main and / or side chains (graft chains). Therefore, polycarboxylate-based water reducing agents can obtain equal water reduction at quite small dosages compared with BNS- or MS-based agents that obtain dispersion only through electrical repulsion [1].

Amino sulfonate-based water reducing agents can reduce water content by steric and electrical repulsion due to the sulfonate group in their chemical structure, the wetting effect due to hydroxyl group (-OH), and the steric effect due to their three-dimensional structure. Aminosulfonate-based water reducing agents can obtain equal water reduction at a mean dosage between that of BNS and polycarboxylates.

Flowability enhancement and segregation resistance

To enhance the flowability and segregation resistance of concrete, the following four effects are important: depletion effect, depletion cogulating effect, Tom's effect, and tribology effect in addition to DLVO theory and steric effect theory.

Depletion effect

DLVO thory or steric effect theory states that since a polymer dispersing agent adsorbs to cement particles and disperses them, there must be a depletion effect that occurs when a non-adsorbing polymer with a molecular weight in the tens of thousands ingresses between particles and disperses them due to volume repulsion [6]. (Figure 2)

Depletion coagulating effect

With the depletion effect, a polymer with a molecular weight from hundreds of thousands to millions cannot ingress between the cement particles and so the particles become coagulated, which provides segregation resistance. (Figure 3)

Tom' effect

Since linear polymers line up along the direction of concrete flow, friction resistance decreases [7]. (Figure 4)

Tribology effect

Low molecular weight compounds having lubrication properties reduce the friction resistance between particles [8]. (Figure 5)

Dispersibility retention

Dispersibility retention of BNS-based agents

The zeta potential of polycarboxylate just after mixing is approximately half of BNS. In the event that BNS and polycarboxylate are added until they exhibit similar water reduction, the zeta potential of polycarboxylate-based agents is roughly half that of BNS. It is assumed that steric repulsive forces are greatly affected by the graft side chains. BNS shows high zeta potential just after measurement, but this greatly declines over time. The change in zeta potential over time affects the dispersibility retention in concrete. As BNS is adsorbed to cement particles in layers, these adsorption layers are immediately covered with cement hydrates. Therefore, electrical repulsive force is lost and dispersing retention disappears. Because BNS-based water reducing agents have a small effect on maintaining the dispersion stability of cement particles, a BNS-based ADSP is added with an action derivative polymer, reactive polymer, or modified lignin, etc. as a slump retentive component which retains dispersibility.

Dispersibility retention of polycarboxylate-based agent

With polycarboxylate ether-based ADSP, cement particles are dispersed by the negative ions of their carboxyl groups. This dispersing effect is maintained by the side chains which exhibit steric extension. The even distribution of these side chains greatly contributes to the dispersion of cement particles.

Slump retention of polycarboxylate-based cross-linked polymers

Based on these basic study, partially cross-linked acrylate polymer (acrylate 3) was synthesized from a copolymer of acrylic acid and polyethlene glycol monoalkyl ether acrylate (acrylate 2). The change in zeta potential and molecular weight distribution measured by GPC analysis proves chemically that cross-linked polymer in an alkali

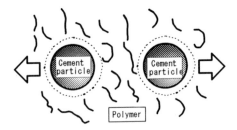

Figure 2 Depletion effect

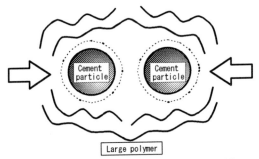

Figure 3 Depletion coagulation effect

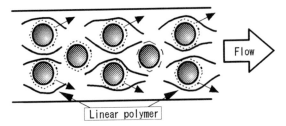

Figure 4 Tom's effect

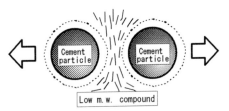

Figure 5 Tribology effect

environment is hydrolyzed and converted into a water reducing acrylate polymer which retains dispersibility [1, 9, 10].

Each step of the slump retention of polycarboxylate containing cross-linked polymer is as follows.

a) Cement particles coagulate.

b) When polycarboxylate containing cross-linked polymer is added at this stage, polycarboxylate possessing steric side chains adsorbs onto the cement particles. Due to the electrical repulsive force of the negative ions in the carboxyl groups and the steric repulsive force of the side chains, the cement particles are dispersed.

c) After a definite time, as hydrates is deposited, part of the polycarboxylate adsorbed onto the surface of cement particles is covered with hydrates.

However, since the water reducing agent possesses steric side chains, most of the remainder of the side chains is still uncovered. The protruding side chains maintain dispersibility.

In addition, from this stage onwards, the polycarboxylate cross-linked polymer is gradually converted into polycarboxylate exhibiting gradually dispersibility, and the cement particles are further dispersed. Thereby, the slump of the concrete remains for a long time.

PROPERTIES AND APPLICATIONS OF HPC USING ADSP SP-A

During the last few years, our ADSP SP-A has been developed, and applied at various job sites such as the Trans-Tokyo Bay project, Akashi Suspension Bridge project and so on.

CONCLUSIONS

As a result of intense research aimed at the development of an advanced superplasticizer with high dispersibility, long-termdispersion stability, excellent flowability and segregation resistance, we have developed a polycarboxylate-based ADSP containing cross-linked polymer, SP-A.

Based on our findings, we hypothesized the differences in fluidizing mechanism for the different types of chemical structures and functional groups. The new ADSP is being used at several big construction projects.

REFERENCES

1.   a) Tanaka, Y, and Ohta, A, Chemical Structures and Mechanism of Air Entraining High-Range Water-Reducing Agents, NMB Research Transaction, 1992, No. 9, pp. 5-11.
b) Moriya, Y , and Ohta, A., The trend of Air Entraining and High Range Water Reducing Agents, J. Soc. Mat. Sci., 1994, Vol. 43, No. 491, pp. 919-929.

2.   Japan Architect Society's Mix proportion, Production, and Application Guide (draft) and Explanatory Notes, 1992.

3.   The Japan Society of Civil Engineering's Application Guide (Draft) for Concrete using Superplasticizers, Concrete Library No. 74, 1993.

4.   Sato, T, and R. Ruch, Surfactant Series 9, Stabilization of Colloidal Dispersions by Polymer Adsorption, Marcel Dekker, N.Y., 1980 and references cited therein.

5.   Kodama, K, Mechanism of High Range Water Reducing Admixture, Cement Concrete, 1982, No.427, pp. 8-16.

6.   Napper, D H, Polymeric Stabilization of Colloidal Dispersion, Academic Press, New York, 1983.

7.   Sumitomo Seika Co., Technical Report, Polyethylene oxides, 1991.

8.   Matsubara, K, Tribology, Sangyo Tosho, Tokyo, 1981.

9.   Ohta, A, Tanaka, Y, and Uryu, T, Polycarboxylate-Based Cement Dispersing Agents, Polymer Preprints Japan, 1989, Vol. 38, No. 7, pp. 2310-2313.

10.   Ueda, M, Kishitani, K, Tanaka, Y and Nakajima, M, Characteristics of High Range AE Water Reducing Agent Using Cross-Linked Polymer, Proceedings of Archit. Soc. Jpn. Annual Conference, 1992, pp. 565-566.

INFLUENCE OF NEW ENVIRONMENTALLY BENEFICIAL LIGNIN-BASED SUPERPLASTICIZER ON CONCRETE PERFORMANCE

J G Zhor

T W Bremner

University of New Brunswick

Canada

ABSTRACT. Lignosulphonates originating from the pulp and paper industry are well known raw materials for the production of water-reducing admixtures for concrete. Traditional sulphite and kraft pulping processes which are noted for their contribution to air and water pollution can now be replaced by more environmentally friendly technologies. One of them, the ALCELL® organosolv pulping process, has minimal impact on the environment and produces a pure lignin as one of the co-products to the pulp. A modified sulphomethylolated ALCELL® lignin was found to be an efficient water-reducing admixture for concrete. It was evaluated according to CSA Standard CAN3-A266.6-M85 in both non-air-entrained and air-entrained concrete. The workability, unit weight, air content and setting time of the fresh concrete as well as compressive strength, length change, spacing factor and durability of hardened concrete were determined. After 1500 cycles in the automated freeze-thaw apparatus the durability specimens were placed at a natural marine exposure station for further monitoring. The results of this study indicated that the new ALCELL® lignin-based admixture meets the requirements of CSA Standard CAN3-A266.6-M85 and can be classified as a Type SPR- superplasticizing admixture with set-retarding characteristics. Even with a spacing factor of 0.79 mm it was found that superplasticized concrete was resistant to freezing and thawing.

Keywords: Air-entrainment, Air-void spacing factor, Compressive Strength, Durability, High range water-reducing admixture, Length change, Setting time, Slump loss, Superplasticizer.

Mr Jiri Zhor is a doctoral candidate in the Department of Civil Engineering at the University of New Brunswick in Fredericton, N.B., Canada. After 10 years of experience in areas of concrete structures, materials, repairs and testing, he has been engaged in research on concrete admixtures since 1986. His current research interests are on the chemistry of superplasticizers for concrete and properties of fresh concrete.

Dr Theodore W. Bremner is a professor of civil engineering at the University of New Brunswick in Fredericton, N.B., Canada. He is chairman of ACI Committee 213 Lightweight Aggregate Concrete and a member of several other ACI technical committees. His research interests include structural lightweight concrete, durability of concrete structures and concrete in the marine environment.

Radical Concrete Technology. Edited by R K Dhir and P C Hewlett. Published in 1996 by E & FN Spon, 2–6 Boundary Row, London SE1 8HN, UK. ISBN 0 419 21480 1.

INTRODUCTION

One of the traditional resources for the manufacture of water-reducing admixtures for concrete are the waste products from the pulp and paper industry, namely lignin and its derivatives. Until recently, sulphite pulping was the major source of lignosulphonates which after extended modification were used as normal range water-reducing and retarding admixtures for concrete. Researchers reported [1,2,3] several attempts to enhance the lignosulphonates so that they would meet the requirements of a superplasticizer (high range water-reducing admixture). To date no purely lignosulphonate-based concrete superplasticizer has been placed on the market.

An important component in the development trends of pulping technologies is the environmental aspect. Due to increasing environmental demands during the last three decades the traditional sulphite pulping is gradually being replaced by a more environmentally friendly sulphate (kraft) pulping process. The current dominant position of kraft pulping is likely to change in the near future. A potential replacement for sulphate pulping is a sulphur-free organosolv pulping process, which has been long known but only recently taken to commercial scale production [4]. In Canada a proprietary organosolv pulping method known as the ALCELL® process was developed and the first commercial pulp mill based on this technology should be completed by 1997 [5]. The currently operated pre-commercial plant has proved that this process has a minimal impact on the environment [6]. An industrial implementation of this pulping technology would clearly constitute a substantial environmental benefit. Among several interesting and marketable co-products, the ALCELL® process generates a significant amount of pure lignin, quite distinct from either lignosulphonates or kraft lignin. Successful marketing of ALCELL® lignin and other co-products enhances the economic feasibility of the process.

The modified methylsulphonated ALCELL® lignin was found very suitable for the preparation of water-reducing admixtures [7]. The relationship between chemical characteristics and performance of ALCELL® lignin-based admixtures was also studied and reported by the authors elsewhere [8]. Consequently the new ALCELL® lignin-based superplasticizer was developed and tested in concrete. In this research a series of laboratory tests on both non-air-entrained (NA) and air-entrained (AE) concrete was conducted according to the CSA standard CAN3-A266.6-M85 "Superplasticizing Admixtures for Concrete". The tests were run simultaneously on a reference concrete (REF), concrete with the newly developed ALCELL® lignin-based superplasticizer (AL) and concrete with commercial naphthalene-based superplasticizer (CS). The results of a substantial part of the testing program are presented and discussed in this paper.

EXPERIMENTAL PROGRAM

The entire experimental program was designed in accordance with CSA Standard CAN3-A266.6-M85. For the most part, the program is also in accordance with ASTM Standard C 494 - 92 "Chemical Admixtures for Concrete". The individual parts of the program were performed in accordance with specific CSA and ASTM Standards.

Materials

Ordinary portland cement (OPC) CSA-A5 Type 10 (ASTM C 150, Type I) and potable water were used. Blagdon coarse aggregate (size 5 to 25 mm) and Zealand fine aggregate (size 0 to 5 mm) were graded according to CAN3-A266.6-M85. Two high range water reducing (superplasticizing) admixtures were incorporated in some of the mixes: Experimental ALCELL® lignin-based superplasticizer (AL) and commercial naphthalene-based superplasticizer (CS). A commercial vinsol resin-based air-entraining agent (AEA) was used.

Concrete Mix Proportions

Two basic mix proportions were used, one for the non-air-entrained (NA) concrete and one for the air-entrained (AE) concrete. Both proportions were designed using ACI 211.1-91 Standard Practice and are presented in Table 1.

Table 1 Concrete mix proportions

MATERIAL	MIX PROPORTIONS, $kg.m^{-3}$	
	Non-air-entrained concrete (NA)	Air-entrained concrete (AE)
Cement	307	307
Water (mix without superplasticizer)	175	160
Coarse aggregate (SSD)	1150	1128
Fine Aggregate (SSD)	734	694
Superplasticizer: R mixes (REF): None	None	None
L mixes: AL (1300 ml/100 kg of cement)	$4000\ ml.m^{-3}$	$4000\ ml.m^{-3}$
C mixes: CS (1000 ml/100 kg of cement)	$3100\ ml.m^{-3}$	$3100\ ml.m^{-3}$
Air-entraining agent (AEA): R mixes: (47 ml/100 kg of cement)	None	$145\ ml.m^{-3}$
L mixes: (118 ml/100 kg of cement)	None	$362\ ml.m^{-3}$
C mixes: (47 ml/100 kg of cement)	None	$145\ ml.m^{-3}$

Note: The actual amounts of water and air-entraining agent were adjusted for each mix in order to maintain the slump and air content of the fresh concrete within the limits recommended by CSA Standard CAN3-A266.6-M85.

Mixing of Concrete

All concrete mixes were produced using an Eirich EA21 laboratory concrete pan mixer with a capacity of 100 litres. A series of 65 litre batches of NA and 75 litre batches of AE concrete mixes were prepared. Each series included three mixes: one reference (REF) and one with each superplasticizer (AL and CS), all mixed on the same day. The mixing procedure was in accordance with CSA Standard CAN3-A266.6-M85. The 10 minute procedure included a major mixing period of 4 minutes, a major rest period of 3 minutes, a remixing period of 1 minute and final rest period of 2 minutes.

Testing of Fresh Concrete

Workability was determined by measuring the slump in accordance with Test Methods CAN/CSA-A23.2-5C and ASTM C 143 - 90a. Initial and final slump were determined 5 and

25 minutes after the termination of the mixing procedure, respectively. Between the measurements the concrete was covered with plastic sheeting and remained undisturbed. From the two slump values the slump retention (slump loss) was calculated.

Unit weight (density) of compacted fresh concrete was determined by measuring the mass of known volume of concrete in accordance with CAN/CSA-A23.2-6C and ASTM C 138 - 92. The volume of a cylindrical metal container was 14 litres.

Air content was determined by three different methods: pressure method (CAN/CSA-A23.2-4C and ASTM C 231 - 91b), volumetric method (CAN/CSA-A23.2-7C and ASTM C 173 - 78) and gravimetric method (ASTM C 138 - 92).

Time of setting was determined by measuring the penetration resistance on mortar extracted from the concrete mixture in accordance with ASTM C 403 -92. Times of initial and final set were determined from the plot of penetration resistance versus elapsed time as the times when penetration resistance equals 3.5 MPa (500 psi) and 27.6 MPa (4000 psi), respectively.

Testing of Hardened Concrete

Compressive strength was tested using cylindrical specimens (150 x 300 mm) prepared in accordance with CAN/CSA-A23.2-3C and ASTM C 192 - 90a. The testing was performed on a Riehle 300,000 lbs hydraulic testing machine in accordance with CAN/CSA-A23.2-9C and ASTM C 39 - 86. Prior to testing all specimens were capped with sulphur mortar in accordance with ASTM C 617 - 87. The specimens were tested at ages of 1, 3, 7, 28, 180 and 1000 days.

Length change was tested using concrete prisms (76x76x406 mm) with measuring studs prepared in accordance with CAN/CSA-A23.2-3C and ASTM C 192 - 90a. The length changes were measured on a comparator equipped with a Mitutoyo IDF-150E Digimatic Indicator. The testing was done in accordance with CAN/CSA-A23.2-14A and ASTM C 157 - 91. The moist curing period was 14 days (including 2 days in the moulds) followed by a 14 day drying period. The initial length was measured at the time of demoulding (2 days). The length change relative to the initial length was measured at ages of 4, 7, 14 and 28 days and was expressed as a per cent based on the specimen gauge length.

Air-void spacing factor and other air-void system parameters were specified by microscopical determination using a Point-Count Device meeting the requirements of ASTM C 457 - 90 and a Stereoscopic Microscope model A0580 by American Optical Corp. The modified point-count method in accordance with ASTM C 457 - 90 was used. The concrete samples were prepared from a longitudinal slice taken from the centre of a 150x300 mm test cylinder.

Durability factor was calculated from relative dynamic modulus of elasticity changes in concrete prisms (76x102x406 mm) exposed to repeated cycles of freezing and thawing. The specimens were first placed in automatic freezing and thawing equipment Logan H-3185 and exposed to 1500 cycles in accordance with ASTM C 666 - 92. Then the specimens were removed from the equipment, transferred to a natural marine exposure station and at mid-tide level exposed to another 200 cycles of freezing and thawing over about 2 years. During the procedure, relative dynamic modulus of elasticity was periodically determined via measurement of fundamental transverse frequency by impact resonance tests in accordance with ASTM C 215 - 91. The Hand-Held FFT Signal Analyzer SA-77 by Rion Co., Ltd. was used for the frequency measurements.

RESULTS AND DISCUSSION

Workability and Slump Retention

The results of the slump measurements and calculated slump retention are shown in Table 2. The graphical representation of the average values of slump retention is given in Figure 1. In all cases the slump retention is well above the CSA Standard minimum of 50%. In both non-air-entrained and air-entrained concrete the experimental admixture contributes to a higher slump retention (lower slump loss) than the commercial one. This pattern is more pronounced in the non-air-entrained concrete.

Table 2 Properties of fresh concrete

MIX CODE	WATER kg.m⁻³	W/C	SLUMP, mm 5 min	SLUMP, mm 25 min	SLUMP RET. %	DENSITY kg.m⁻³	AIR CONTENT, % Press.	AIR CONTENT, % Vol.	AIR CONTENT, % Grav.	SET. TIME, h:min Initial	SET. TIME, h:min Final
NAR1	180	0.59	90	75	83	2382	2.5	2.0	1.5	3:50	5:05
NAL1	156	0.51	95	70	74	2404	3.0	2.5	2.3	6:20	7:30
NAC1	156	0.50	90	55	61	2408	2.5	2.5	2.1	3:40	5:00
NAR2	179	0.59	95	80	84	2379	2.5	2.0	1.9	4:10	5:15
NAL2	156	0.51	100	80	80	2409	3.0	2.5	2.3	6:40	7:55
NAC2	157	0.51	100	75	75	2406	2.5	2.5	2.0	4:00	5:10
NAR3	180	0.59	85	60	71	2381	3.0	2.0	1.7	3:50	5:10
NAL3	157	0.51	90	65	72	2411	3.0	3.0	2.0	6:50	8:20
NAC3	157	0.51	95	60	63	2404	3.0	2.5	2.3	4:15	5:45
AER1	171	0.56	90	80	89	2288	5.5	5.5	5.9	3:55	6:10
AEL1	146	0.48	90	60	67	2310	5.5	5.0	5.7	6:40	8:10
AEC1	148	0.48	100	65	65	2274	6.5	5.5	5.8	3:45	4:50
AER2	167	0.56	80	60	75	2253	7.0	6.0	7.2	3:50	4:55
AEL2	147	0.47	85	50	59	2324	5.0	6.0	6.2	6:45	8:05
AEC2	147	0.47	95	60	63	2317	5.5	6.5	6.1	3:40	4:45
AER3	170	0.56	90	60	67	2275	6.5	6.0	6.2	4:00	5:15
AEL3	147	0.47	90	55	61	2336	5.5	5.0	5.4	6:35	7:50
AEC3	147	0.47	95	55	58	2339	5.5	5.0	5.3	3:40	5:00

Air Content Measurements

The values of the air-content measurements are summarized in Table 2. ASTM and CSA Standards require using only the pressure method, however three different methods of measuring the air-content were used to enable comparisons to be made. As the results show, all three methods give similar air-content values. In the case of the non-air-entrained concrete all air content values are within CSA Standard requirements, i.e. under 3.5% and within ±0.5% difference from the air content of the reference concrete. For the air-entrained concrete the standard recommended ±0.5% difference from the reference concrete value is exceeded (mainly because of the synergic effect of superplasticizer and air-entraining agent), but the air-content range of 5 to 7 % was maintained.

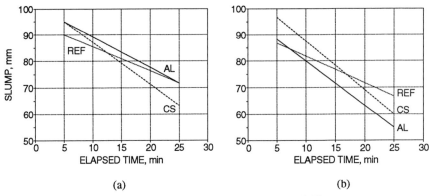

(a)   (b)

Fig. 1 Slump retention of non-air-entrained (a) and air-entrained (b) concrete

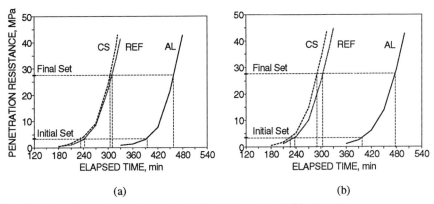

(a)   (b)

Fig. 2 Setting time of non-air-entrained (a) and air-entrained (b) concrete

Setting Time Evaluation

The results of the setting time determination also are given in Table 2. The average values of penetration resistance versus elapsed time obtained for non-air-entrained and air-entrained concrete are shown in Figure 2. From the presented results it is obvious that the commercial superplasticizer accelerates both initial and final setting by 5 to 15 minutes, while the experimental admixture retards both setting times by 150 to 180 minutes.

Strength and Dimensional Stability

The values of compressive strength and length change are presented in Table 3 and 4. The average values of the compressive strength and length change versus time are plotted in

Figure 3 and 4, respectively. The compressive strength development was measured over a period of 1000 days. During this time period all superplasticized concrete specimen had a compressive strength of more than 120% of the reference concrete values on any given day. The percent of the length change after 28 days determined the dimensional stability in terms of drying shrinkage after a drying period of 14 days. The differences between the drying shrinkage of the reference concrete and superplasticized concrete are in all cases below the 0.010% maximum.

Table 3   Properties of hardened non-air-entrained concrete

MIX CODE	COMPRESSIVE STRENGTH, MPa						LENGTH CHANGE						
	1 day	3 days	7 days	28 days	180 days	1000 days	Initial L,mm	7 days		14 days		28 days	
								L,mm	ΔL,%	L,mm	ΔL,%	L,mm	ΔL,%
NAR1	21.0	26.4	29.7	33.6	39.1	43.1	7.912	7.877	0.008	7.895	0.005	7.823	0.026
NAL1	26.0	31.0	33.4	39.0	48.1	50.2	8.302	8.316	0.003	8.291	0.003	8.191	0.032
NAC1	26.8	32,4	34.7	42.2	48.3	50.6	8.154	8.119	0.008	8.141	0.004	8.049	0.030
NAR2	19.0	23.7	26.7	32.3	34.1	35.2	8.208	8.176	0.008	8.192	0.005	8.110	0.028
NAL2	24.8	30.4	34.2	40.2	45.0	45.8	7.958	7.922	0.009	7.945	0.004	7.842	0.033
NAC2	26.1	32.0	34.1	42.6	47.7	49.7	7.873	7.796	0.010	7.814	0.007	7.729	0.031
NAR3	20.7	25.7	27.5	34.5	37.7	39.2	7.071	7.077	0.001	7.077	0.002	6.997	0.027
NAL3	28.5	33.0	36.0	40.0	44.6	50.2	8.023	8.032	0.002	8.029	0.002	7.918	0.030
NAC3	28.3	33.6	35.4	41.5	45.9	48.3	8.010	8.024	0.003	8.017	0.002	7.900	0.032

Table 4   Properties of hardened air-entrained concrete

MIX CODE	COMPRESSIVE STRENGTH, MPa						LENGTH CHANGE			DURABILITY FACTOR, %				RDF, % 300 Cycles
	1 day	3 days	7 days	28 days	180 days	1000 days	Initial L,mm	28 days		Cycles				
								L,mm	ΔL, %	300	600	1500	1700	
AER1	16.2	20.4	21.2	27.7	32.0	34.0	8.283	8.179	0.030	99	97	94	93	--
AEL1	22.2	27.7	30.6	35.3	40.2	40.5	8.417	8.305	0.032	98	98	93	91	99
AEC1	22.2	26.9	29.7	35.4	39.8	41.3	8.329	8.238	0.026	100	99	96	96	101
AER2	18.2	21.9	27.7	29.8	34.2	35.1	8.189	8.085	0.030	100	100	97	96	--
AEL2	27.6	33.4	35.8	39.1	49.2	51.1	7.666	7.554	0.032	99	99	98	97	99
AEC2	25.8	31.5	36.0	38.9	45.8	47.8	8.278	8.160	0.034	99	99	97	97	99
AER3	18.9	22.0	26.4	30.8	33.2	34.7	7.541	7.426	0.033	99	98	95	94	--
AEL3	26.4	33.7	35.5	40.5	48.0	48.7	8.176	8.054	0.035	97	86	76	75	98
AEC3	25.9	32.2	35.8	39.7	46.1	48.3	8.237	8.117	0.034	99	99	96	96	100

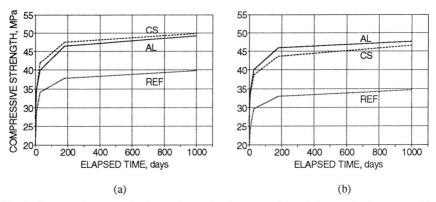

(a)   (b)

Fig. 3 Compressive strength of non-air-entrained concrete (a) and air-entrained concrete (b)

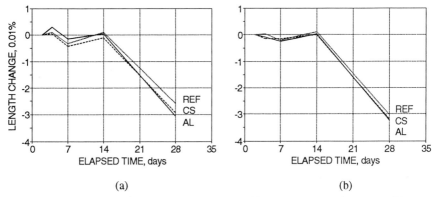

(a)   (b)

Fig. 4 Length change of non-air-entrained concrete (a) and air-entrained concrete (b)

Air-Void System and Durability

The air-void system parameters of several air-entrained concrete samples were determined and are presented in Table 5 together with related values of selected durability factors. The complete listing of durability factors is given in Table 4. The graphical representation of the average values of relative dynamic modulus of elasticity during the period of 1700 cycles is shown in Figure 5. The spacing factor values are in the range of 0.20 to 0.80 mm, i.e. in all cases above the generally recommended maximum value of 0.20 mm (e.g. ACI 201.2R-92 Guide to Durable Concrete). Nevertheless, the durability factor of all tested samples after 300 cycles is more than 97%, and after 1700 cycles only 1 sample has a durability factor below 80 %. Therefore all tested concretes are considered to be durable in terms of CSA and ASTM Standard requirements. The results imply either that the recommended spacing factor limit of 0.20 mm is unnecessarily low, or that 300 cycles of freezing and thawing is not enough to guarantee a durable concrete.

Table 5   Air-void system parameters and durability factors of air-entrained concrete

MIX CODE	AIR CONTENT, %		NUMBER OF VOIDS PER mm	SPECIFIC SURFACE, mm^2/mm^3	SPACING FACTOR, mm	DURABILITY FACTOR, %		
	Fresh concrete	Hardened concrete				300 cycles	1500 [1] cycles	1700 cycles
AER1	5.6	4.4	0.281	25.47	0.2058	99	94	93
AEL1	5.4	3.3	0.081	9.86	0.6572	98	93	91
AEL2	5.7	3.8	0.068	7.25	0.7931	99	98	97
AEC2	6.0	3.0	0.100	13.24	0.4685	99	97	97

[1] The number of cycles after which the samples were removed from freeze-thaw apparatus and placed at the natural marine exposure station.

Table 6   Overview of the test results with the ALCELL® lignin based admixture and physical requirements of CSA CAN3-A266.6-M85 and ASTM C 494 - 92 Standards

CONCRETE PROPERTY	RESULTS WITH ALCELL® LIGNIN BASED ADMIXTURE		PHYSICAL REQUIREMENTS FOR CONCRETE ADMIXTURE	
	Non-Air-Entrained Concrete	Air-Entrained Concrete	CSA / CAN3-A266.6-M85 Type SPR	ASTM C 494 - 92 Type G
Water content, % of reference	87	87	max. 88	max. 88
Slump retention, %	76	63	min. 50	not required
Time of initial set retardation, h:min	2:40	2:45	1:00 to 3:00	1:00 to 3:30
Time of final set retardation, h:min	2:45	2:35	not required	max. 3:30
Compressive strength, % of ref.x1.05(CSA) / % of ref.(ASTM)				
1 day	137 / 131	150 / 143	min. 130	min. 125
3 days	131 / 125	155 / 148	min. 130	min. 125
7 days	143 / 136	142 / 135	min. 125	min. 115
28 days	124 / 119	137 / 130	min. 120	min. 110
180 days (6 months)	130 / 124	145 / 138	min. 100	min. 100
365 days (1 year)	N/A	N/A	min. 100	min. 100
1000 days	130 / 124	142 / 135	not required	not required
Flexural strength, % of reference				min.110, 100
3, 7 and 28 days	N/A	N/A	not required	and 100
Length change (shrinkage) [1], % of ref.	119	106	max. 135	max. 135
or increase over reference	0.005	0.002	max. 0.010	max. 0.010
Relative durability factor [2],	not required	109 / 99		
% of ref. x1.1(CSA) / % of ref.(ASTM)			min. 100	min. 80
Air-void spacing factor [2], mm	not required	0.73	max. 0.23	not required

[1] When length change of reference concrete is 0.030% or greater % of reference limit applies; increase over reference limit applies when length change of reference is less than 0.030%.

[2] Applicable only when tested in air-entrained concrete. According to CSA Standard either the air-void spacing factor or the relative durability may be used to determine acceptability.

N/A - Not Available

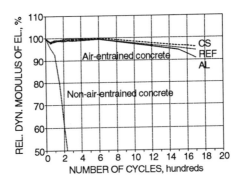

Fig. 5 Resistance of concrete to freezing and thawing cycles

EVALUATION ACCORDING TO THE STANDARD REQUIREMENTS

Three sets of both non-air-entrained (NA) and air entrained (AE) concrete that include both reference and superplasticized concretes were evaluated for the Type SPR- Superplasticizing Admixture with Set-Retarding Characteristics according to the CSA Standard CAN3-A266.6-M85. Concurrently, the evaluation for the Type G - Water Reducing, High Range and Retarding admixture according to ASTM C 494 - 92 was conducted. Average values obtained for the concrete with the ALCELL® lignin admixture are presented in Table 6. The results show that except for the higher value of air-void spacing factor (optional test according to CSA) and flexural strength results (test required only by ASTM but not done) all properties are within the required limits of both CAN3-A266.6-M85 and ASTM C 494 -92. Testing also confirmed that the commercial superplasticizer meets the CSA Standard requirements of a superplasticizer with normal setting characteristics.

CONCLUDING REMARKS

The testing results confirmed that the new ALCELL® lignin concrete admixture can be classified as a superplasticizing admixture with set-retarding characteristics - type SPR according to CSA Standard CAN3-A266.6-M85 or as a high range water-reducing and retarding admixture - type G according to ASTM Standard C 494 - 92 (the flexural strength test was not carried out).

The only property out of recommended limits was the CSA optional parameter of air-void spacing factor. Its value exceeded the standard recommended maximum by a significant amount, which should indicate low durability of concrete. Nevertheless, the freezing and thawing test confirmed excellent durability of concrete as the durability factor of all tested samples after 300 cycles was more than 97%. This discrepancy between the required air-void spacing factor values and durability factor values was also found by other investigators [9]. It indicates that either the value of the air-void spacing factor as an indicator of concrete durability should be reevaluated, or the freezing and thawing testing procedure required for durable superplasticized concrete should be reconsidered.

The authors believe that both areas have to be addressed. The recommended spacing factor value indicating a durable concrete should be specified individually according to the water-cement ratio and concrete admixtures applied. For superplasticized air-entrained concrete with water-cement ratio less than 0.5 the recommended spacing factor value can be increased from 0.20 mm to 0.50 mm. The authors also recommend that the requirements on freeze-thaw resistance be increased from 300 to 600 cycles.

ACKNOWLEDGEMENTS

The authors wish to thank Mr C. Keats of Gemtec Ltd. for carrying out the microscopical analysis. The financial support from the Natural Sciences and Engineering Research Council of Canada and from Repap Enterprises, Inc. is also acknowledged.

REFERENCES

1.   RIXOM, M.R. AND WADDICOR, J. Role of lignosulphonates as superplasticizers. *In* Developments in the Use of Superplasticizers, Editor V.M.Malhotra, ACI SP-68, Ottawa, 1981, pp.485-498.

2.   MIYAKE, N., ANDO, T. AND SAKAI, E. Superplasticized concrete using refined lignosulphonate and its action mechanism. Cement and Concrete Research, 1985, Vol.15, No.2, pp.295-302.

3.   BIALSKI, A.M. AND NAYAK, K.V. Superplasticizers from lignosulphonates. Proceedings of the 4th International Symposium on Wood and Pulping Chemistry, Paris, 1987, Vol.1, pp.83-90.

4.   STOCKBURGER, P. An overview of near-commercial and commercial solvent-based pulping processes. Tappi Journal, 1993, Vol.76, No.3, pp.71-74.

5.   REPAP ENTERPRISES, INC. Competing in the global marketplace. Annual report 1994, Repap Enterprises Inc., Montreal, 1995, 76 pp.

6.   PYE, E.K. AND LORA, J.H. The ALCELL® process - A proven alternative to kraft pulping. Tappi Journal, 1991, Vol.74, No.3, pp.113-118.

7.   BREMNER, T.W. AND ZHOR, J. High Purity ALCELL® Lignin as a Concrete Admixture. Final report for NSERC, University of New Brunswick, Fredericton, October 1993, 10pp.

8.   ZHOR, J., BREMNER, T.W. AND LORA, J.H. Effect of chemical characteristics of ALCELL® lignin-based methyl-sulphonates on their performance as water-reducing admixtures. Proceedings of the 4th International Conference on Superplasticizers and Other Chemical Admixtures in Concrete, Editor V.M.Malhotra, ACI SP-148, Montreal, 1994, pp.333-351.

9.   ATTIOGBE, E.K., NMAI, C.K. AND GAY, F.T. Air-void system parameters and freeze-thaw durability of concrete containing superplasticizers. Concrete International, 1992, Vol.14, No.7, pp.57-61.

A NEW HIGH PERFORMANCE FIBRE FOR CONCRETE REINFORCEMENT

L Della Sala

R Cerone

University of Basilicata

Italy

ABSTRACT. The following paper presents the result of a research aimed to the study of a new metallic fibre for concrete reinforcement. The principal characteristic of this fibre is that it is provided by a special anchorage device that improves very much the interaction with cementitious matrix. The tensile load substained by the fibre is transferred to matrix by a wedge effect induced by the particular shape of the anchorage reducing the consequent pullout displacements. So fibre can be loaded till it stretches without great displacements take place and this enhances concrete strength and reduces crack width and propagation. Different tests have been performed; at the end of the paper is presented a diagram that has been obtained as an average of different tests executed to compare the new fibres with the conventional deformed fibres.

Keywords: Deformed fibre, Fibre-matrix interaction, Anchorage device, Stress distribution, Flexural behaviour, Crack width.

Professor Lucio Della Sala is professor of Structural Engineering at University of Basilicata, Italy. His principal interest is the study of masonry ancient buildings and their restoration. He also is interested in general problems concerning composite materials.

Dr Rosario Cerone is a civil engineer whose main research subject is the study of the behaviour of steel fibre reinforced concrete. In particular, he is concerned with numerical modelling of fibre reinforced cementitious composites and with the study of new technologies for the enhancement of steel fibre reinforced concrete performances.

Radical Concrete Technology. Edited by R K Dhir and P C Hewlett. Published in 1996 by E & FN Spon, 2–6 Boundary Row, London SE1 8HN, UK. ISBN 0 419 21480 1.

INTRODUCTION

As well known, deformed steel fibres result effective to prevent brittle failure of concrete and to improve its toughness but not always they allow to control fracture width if not used in great amount. In fact, for fibre amount less then 1 percent by volume, when the crack limit of concrete is reached, are induced great displacements that in some cases could not be accepted.

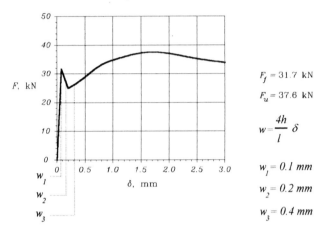

Figure 1 Flexural behaviour of steel fibre reinforced concrete

In figure 1 is presented a load versus displacement diagram obtained by a flexural test of a 15x15x60 cm concrete beam reinforced with conventional hooked ends fibres in amount of 0.5 percent by volume. Can be observed that when the fissuration limit of concrete is reached, the crack width w overcomes the limits imposed by design codes (see Italian Code) making not possible to overcome this limit in exercise conditions.

In general, this is due to the fact that fibres induce in the matrix some stress concentrations which plasticize it locally around the fibre inducing a reduction of the stiffness of fibre-matrix system.

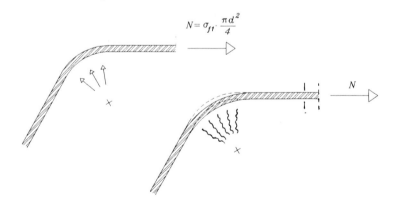

Figure 2 Stress concentrations in the matrix near fibre bends

This aspect determines some restrictions to the stress level of the fibres because in the exercise conditions it is necessary to limit the deformations of the composite in the range imposed by design codes. Besides, the reduced stiffness of fibre-matrix system does not allow a full integration between the fibres and the ordinary steel bar reinforcement. For this reason, generally, solicitations within the composite are calculated mantaining an adequate safety margin from the fissuration limit state reducing the overall efficiency of the fibres.

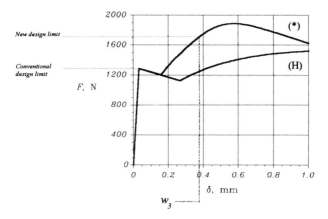

Figure 3 Comparison between conventional fibres (H) and new fibres (*)

It follows that crack width control makes surely possible a better usage of fibres because the upper limit of solicitation becomes the one corresponding to the limit state of crack opening.

DESCRIPTION OF THE NEW FIBRE

Keeping this priciple in mind, had been studied various modalities of stress transfer from fibre to matrix and, at the end, had been achieved a solution which seems to give very good and interesting results.
Beforehand, must be stated that all the valutations concern only the technical aspect of the problem, not the economical ones, but had been estimated that the new fibre could be manufactured at costs comparable with the current ones.

The metal fibre for concrete reinforcement that is now presented is obtained by a piece of metal wire, preferably of steel, to the whose ends are present two metal elements formed and dimensioned to improve fibre anchorage to the cementitious matrix. In a preferred embodiment, the anchorage device has approximately a fusiform shape; it is constituted by an expansion near fibre ends that realizes a wedge effect which inhibits fibre pullout. The particular shape of the anchorage device is, besides, especially convenient because it determines a field of compression stresses around the wire that improves the shear resistance of the interface between the anchorage device and the wire itself.

The anchorage device can be realized in different ways both during the wire drawing process and by hot application of a metal covering around the wire; in the second case could be used different kinds of metal provided that they could be effectively fixed to the wire forming the body of the fibre. The shape must be designed in an opportune way in function of the stress level of the fibre. In particular, its dimension in the fibre axis direction must ensure a good bonding with the body while its transversal dimension must give a frontal section sufficient to avoid pullout. For this purposes the lenght of the anchorage device must be included between 1 and 6 diameters of the wire while its cross section must have an extention between 2 and 20 times the cross section of the wire.

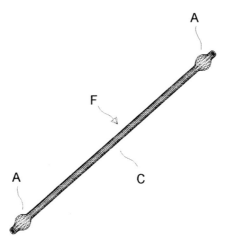

Figure 4 View of the new fibre

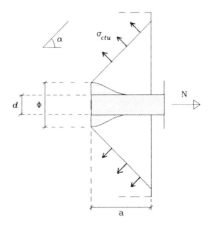

Figure 5 Stress transfer between fibre and matrix

Fibre may have a straight or slightly curve longitudinal profile; in the second case the bending radius must exceed 100 times the wire diameter. Figure 6 shows a longitudinal section of the fibre; the overall length L, measured along its axis, can assume values icluded in the range of 10-80 millimeters; wire diameter changes between 0.3 and 1.2 millimeters.

The length L and the diameter d are chosen in function of the particular technical requirement in such a way that the aspect ratio L/d of the fibre is included between 25 and 100.

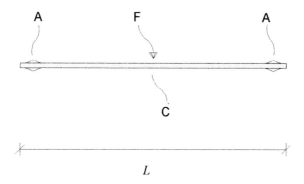

Figure 6 Longitudinal profile of the fibre

Figure 7 shows the particular of the anchorage device; d is the diameter of the wire. The maximum thickness ϕ must be calculated in such a way to confer a maximum frontal section **3** equal to 2 - 20 times the cross section of the wire (1.5 - 4.5 times d). The lenght λ of the anchorage device varyies between 1 and 6 times d, the lenght e of the final part of the fibre measures 0 - 5 times d and the distance t between the end of the fibre and the maximum section **3** measures 0 - 8 times d.

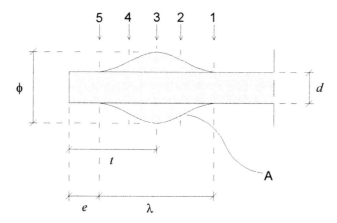

Figure 7 Detail of the anchorage device

Figure 8 shows different transversal sections of the anchorage device.

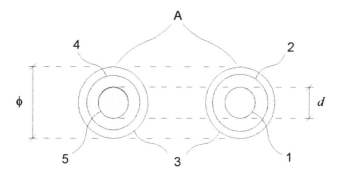

Figure 8 Transversal sections of the anchorage device

MECHANICAL CHARACTERISTICS

The fibre according to the invention shows a more high pullout resistance with respect of the deformed ones obtained by the same wire; this is due to the fact that the bonding with the matrix is ensured by the joint of the anchorage device with the matrix itself.

The improvement of mechanical performances may be relevant because fibre-matrix interaction is positively modified under different aspects. Principally, fibres according to the invention are characterized by the following factors:

- a reduced anchorage length;

- a more homogeneous stress field in the matrix;

- compression stresses are diffused in a greater volume of matrix and this reduces the effects of stress concentrations and plasticizations;

- plastic deformations do not reduce fibre efficiency in terms of ultimate pullout resistance.

Three substantial advantages, confirmed by the experimental evidences, may be found. First of all, the reduction of the anchorage length allows a greater degree of usage of the fibre. In fact this length varies between 1 and 6 diameters while the one of the conventional deformed fibres lies between 10 and 15 diameters. This determines an increase in the effective lenght and, thus, in the effective aspect ratio.
The second advantage is due to the fact that fibres could be solicitated up to the tensile strength of the wire and so the material is fully utilized.
The third advantage, as just said, consists in a reduction of tensional concentrations in the matrix; this allows a considerable reduction of fibre displacements necessary to its full activation making possible a very good control of fracture width. This fact has considerable implications especially in the case of structural elements subjected to flexural loads. In fact, when the fissuration limit state is reached, if fibres are well proportioned, also in small amounts, they are able to substain the overall solicitation

without great displacements take place. This makes possible to use in the exercise conditions the fibre reinforced concrete over the above mentioned limit. This means that the calculation of load bearing capacity of bended elements can be performed mantaining an adequate safety margin from an assigned crack opening limit state.

It must be added that the fibre according to the invention gives another advantage during mixing and placing because its simple shape avoids entanglement and balling. This confers a very good workability of the fresh mix and allows a good and uniform distribution of the reinforcing elements within the matrix.

PREFERRED EMBODIMENT

The new fibre may be realized in the shapes and dimensions previously defined using different kinds of materials. In a preferred embodiment can be used drawn steel wire 900 MPa tensile strength for the fibre body and zinc for the anchorage device. This combination of materials seems to be particularly useful in that, as well known, zinc bonding with steel is very good. In fact the melting temperature of the zinc does not determine substancial variations to steel structure and properties; besides, zinc atoms diffuse in the steel structure forming with iron a heavy hardness alloy.

For this combinations of materials the production process of the bimetallic fibre must foresee an adequate passivation treatment in alkaline environment to reduce the effect of chemical reactions of the metals, especially of zinc, with the concrete and thus with the purpose to improve fibre efficiency. The amount of zinc used to form the ancorage devices of a single fibre is comparable with the one used for conventional hot zinc plating.

Obviously, the unit cost of the presented fibre may be higher than the one of a conventional deformed drawn wire fibre, but it must be pointed out that the increasing of the effective lenght and of the exercise strength may determine an improvement of the overall efficiency. In particular may be allowed a reduction of the dimensions of the structural elements and/or fibre amount. This makes them proposable; besides exist always the advantages that result from the containing of crack width.

FIRST TEST RESULTS

Figures 9 and 10 refer to diferent laboratory tests performed with the aim to compare the flexural behaviour of different kinds of deformed fibres. Must be specifyied that all comparison made are not intended to confront simply the performaces of the presented fibre with the existing fibres. In fact the improvement of concrete properties is quite different consisting for deformed fibres in a great increase in toughness while for the presented bimetallic fibre in a better control of crack width.

The fibres according to the invention used in these tests have been prepared using drawn steel wire 0.5 mm in diameter 1100 MPa tensile strength. The anchorage has been realized in zinc in the shape described in the previous figures; the dimensions ϕ, λ, e had been set to 1.15 mm, 1.9 mm and 0.3 mm respectively while fibre length had been fixed equal to 27 mm.

Flexural tests have been performed on small beams 40x40x160 mm. These dimensions had been assumed to reduce the number of fibres used in each test. However, the results obtained can be compared with the corresponding of normal test beams if is calculated the maximum crack width in function of the mid span deflection. Assuming that after the crack limit is reached the beam is divided in two rigid parts, the crack amplitude w corresponding to an assigned displacement δ can be estimated by the following expression

$$w = \frac{4h}{l} \cdot \delta$$

in which h and l represent respectively the thickness of the beam and the span length.

To avoid the influence of random distribution of the fibres, the reinforcing elements had been placed in an assigned position on P.V.C. frames fixed in the mid span of the beam; some fibres had been placed with an inclination from the beam axis. The details of fibre position are shown in figure 9.

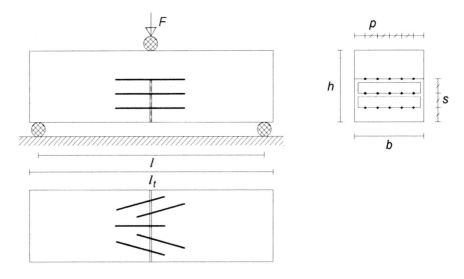

Figure 9 Test configuration

The number of fibres had been calculated so that a fibre amount of 0.5% by volume had been reproduced in the mid span section. Dimensions s, p, l had been set respectively equal to 8 mm, 7 mm and 150 mm.

Figure 10 represents the average load versus deflection diagrams obtained by 2 series composed each of 3 specimens. The initial elastic stage had been normalized to the case of absence of the plastic frame. The curve (*) is related to the fibres according to the invention while the curve (H) is related to conventional hooked ends fibres. It can be seen that (*) fibres make possible a better crack width control. In fact when the fissuration limit is reached, mantaining the crack load constant, in the case (*) are induced displacements less than 0.2 mm while in the case (H) are registered displacements greater than 0.4 mm. These displacements induce a crack width w, according the previous formula, of about the same order of magnitude and so in the second case the crack load becomes an upper limit of exercise solicitation because over this force crack propagates in width with an high gradient.

Instead in the case (*) it is possible to overload the specimen over the fissuration limit having still a considerable safety margin from global failure either in terms of displacements or in terms of strength.

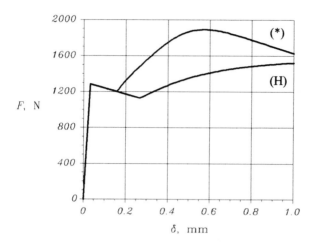

Figure 10 Comparison between hooked ends fibres and new fibres

CONCLUSIONS

As said before, the new fibre in here presented determines some important advantages with respect of the conventional deformed fibres. The most significative enhancements can be quantified from the previous figure. In fact, can be derived that the new fibre determine, on an average for the dimensions before specified, an improvement in ultimate flexural load of about 25 % and a reduction of displacements and crack width higher than 60 %. This could have important implications in practice because it could allow a reduction of the dimensions of the structural elements and a better integration with ordinary reinforcing bars.

Obviously this considerations could be carried out if the costs of productions are comparable with the ones of the conventional fibres in such a way to make them proposable on the market.

HOW TO IMPROVE THE TENSILE CAPABILITY OF HIGH STRENGTH CONCRETE

M A Imam

University of Mansoura

Egypt

ABSTRACT. The widespread production and use of high strength concrete (HSC) with its superior properties is in fact a revolutionary stage of the concrete technology. Furthermore, when steel fibres are incorporated in HSC, much more improvements and numerous benefits can be achieved. To-date, the exploitation of these benefits has been somewhat restricted in some areas of applications in which concrete is subjected to tension, shear, or flexure. This paper presents the results of 28 mixtures prepared with HSC with and without steel fibres [f_c = 82 - 117 MPa]. The attention is paid to the significant effect of steel fibres in increasing the splitting tensile strength of HSC. The gain of splitting strength due to the incorporation of steel fibres has been investigated. The validity of the existing relations for measuring the splitting strength of steel fibre high strength concrete is checked. In addition, an empirical expression has been proposed to predict the splitting strength of high strength concrete containing steel fibres. The predicted splitting strength values using the proposed equation show a close agreement with both the own test data as well as the data compiled from literature.

Keywords : High Strength Concrete, Steel Fibres, Splitting Strength, Tensile Strength, Composites.

Dr Mahmoud Imam is an assistant Professor in the Civil Engineering Department, Mansoura University, Egypt. In 1995, he got his Ph.D. from the Civil Engineering Department, Catholic University of Leuven, Belgium. His main research interests include high strength concrete, steel fibres, shear and diagonal tension in reinforced concrete structures.

Radical Concrete Technology. Edited by R K Dhir and P C Hewlett. Published in 1996 by E & FN Spon, 2–6 Boundary Row, London SE1 8HN, UK. ISBN 0 419 21480 1.

INTRODUCTION

High Strength Concrete (HSC) with its exceptional performance characteristics is one of the most visible results of a continuous research for excellence in concrete technology. The benefits of HSC are now fully apparent and more than compensate for the increased costs of raw materials and quality control. Furthermore, when steel fibres are incorporated in HSC, much more improvements and numerous benefits can be achieved. Now, the question is, why would we want to add fibres to HSC? The simple answer is that HSC is a brittle material, with low tensile strength and strain capacities[1]. The role of fibres is to alter the behaviour of the composite once the matrix has cracked, by bridging across these cracks and so providing some post-cracking ductility. In addition, fibres may increase the strength of the composite by transforming loads and stresses across the cracked matrix.

When fibres are wisely used, they can help us to produce concrete with increased tensile strength and strain capacities, fatigue and impact resistance, energy absorption, crack resistance, and durability. However, fibres give us the opportunity to utilize the concrete for a variety of applications such as pavements (highways, roads, parking areas, runways, and bridge decks), industrial floors, shear failure zones in structures, shotcrete, repair of concrete structures, and lining of tunnels.

The question then arises : do fibres truly increase the tensile strength of HSC? The answer of this question is emphasized in the work presented herein. In this paper, experimental data about splitting tensile strength (f_{sp}) of 18 different HSC-mixes as well as 10 mixes of Steel Fibre High Strength Concrete (SF-HSC) are reported. Based on these data, an empirical expression has been proposed for predicting f_{sp} of SF-HSC. The main objectives of this investigation are :
1 To determine whether or not the existing relationships between compressive and tensile strengths are applicable and valid for SF-HSC.
2 To propose a new expression for predicting the splitting tensile strength of SF-HSC.
3 To clearly show the role of steel fibres in increasing the tensile strength of HSC.

EXPERIMENTAL DETAILS

Mix Proportioning

Twenty eight different HSC mixes with and without steel fibres were investigated. The cylinder compressive strength of concrete mixes ranged from 82.0 to 117.4 MPa. When proportioning HSC mixes, the basic considerations are the selection of a combination of materials that will produce a high quality concrete with the desired workability, strength and durability. These objectives are more difficult to attain for HSC since optimum performance from each material is required. The mix proportioning method adopted in this study as well as the concrete composition for the different mixes are reported elsewhere [2].

Materials

The cement used was portland cement type CEM I, with 28-days compressive strength ≥ 52.5 MPa according to European Standard NBN B12-001. The fine aggregate was a 0/5 natural river sand with a fineness modulus of 2.6. The coarse aggregates were

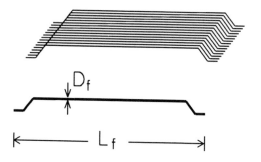

Figure 1 Used steel fibres.

natural river gravel and stone (porphyry). Two sizes (4/14 and 4/16) of gravel, and three sizes (2/7, 7/10, and 7/14) of stone were used.

To achieve workable mixes with the desired quality and strength, a superplasticizer in the form of an aqueous solution was used with a dosage of 2.5 to 3.5 percent by weight of the cement and silicafume. Five different types of superplasticizer as a, b, c, d and e from two different manufacturers were used in this experimental program. The silicafume used in this study was ELKEM microsilica, with a specific surface 18.0 m^2/g, and 260 g/l density. Hooked steel fibres (Figure 1) were used. Fibres were joined together by water-soluble glue to insure good dispersion in the concrete. The volume fraction V_f used was 1.0 %. The fibre length, diameter, and aspect ratio were 50 mm, 0.5 mm, and 100 respectively.

Experimental Procedures

Concrete mixes were designed, treated, and controlled under the same conditions. The constituents were mixed in dry state for about one minute to ensure the uniformity of the mix. Mixing water and superplasticizer were added gradually and simultaneously during mixing. For mixes without fibres, all contents were mechanically mixed for two minutes, while for fibrous mixes, the contents -except the fibres- were mixed first for one minute, adding the fibres during mixing process, and then mixed for another one minute. The consistency of fresh concrete for mixes without fibres was measured by the conventional slump test, while both slump and V.B. (consistometer; ISO 4110-1979) tests were used for mixes with fibres. Vibrating table was used during placing of concrete to ensure full compaction. To determine the engineering properties of concrete, the following tests and specimens were used:
- Compression test : 150 x 300-mm steel cylinders (tested at 28 days)
- Splitting test   : 150 x 300-mm steel cylinders (tested at 28 days)
All test specimens were demolded after 24 h and then exposed to continuous moist curing until testing (Fog room: 20±2 °C and 95±2 percent relative humidity). The determination of different strengths was based on the average of 3 specimens.

Workability

HSC mixes exhibited very sticky characteristics, resulting in a loss of workability. Adding steel fibres to HSC, even in a small amount, leads to a substantial reduction in workability. Therefore, the use of a chemical admixture such as superplasticizer with appropriate dosage is of great necessity. The choice of superplasticizer type and dosage should be based - in addition to price - on the water reducing effect, the rate of workability loss, and the retardation of set. Table 1 shows that the mixes made without fibres had slump of 1.0 to 24.0 cm. The slump values of fibre concrete mixes ranged from 0.0 to 7.0 cm, while the results of the V.B. test for those mixes were 21.0 to 60.0 sec. It was observed that slump and V.B. measurements can not be related for the same fibre concrete mix. Despite sometimes a low slump, most of fibre concrete mixes were quite placeable using vibration.

RESULTS AND DISCUSSION

Towards the answer of the question which has been previously posed " Do fibres increase the concrete tensile strength ? ", it is now generally accepted that the fibres currently used do not significantly enhance the first-cracking stress of fibrous concrete [1]. This does not necessarily mean that the tensile strength of the matrix is not influenced by the presence of fibres. In fact, steel fibres provide a significant increase of the ultimate splitting tensile strength (peak-strength). To clear this aspect, three different stages of the cracking process as shown in Figure 2 should be distinguished :

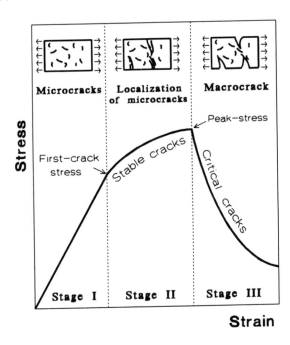

Figure 2 Cracking process in fibrous concrete

Table 1 Test results

Mix	Slump mm	V.B. sec.	f_c MPa	f_{sp} MPa	Mix	Slump mm	V.B. sec.	f_c MPa	f_{sp} MPa
I- Mixes Without Fibres					M15	17	-	105.0	6.74
					M16	20	-	103.8	7.71
M1	140	-	82.2	6.70	M17	165	-	108.9	7.74
M2	190	-	82.0	5.94	M18	115	-	117.4	7.05
M3	185	-	100.4	6.34	**II- Mixes With Fibres**				
M4	80	-	92.2	6.07					
M5	240	-	111.7	6.71	F1	70	26	110.1	11.42
M6	230	-	98.2	7.25	F2	0	56	96.6	10.23
M7	120	-	105.7	7.18	F3	0	41	89.9	10.31
M8	140	-	104.4	7.25	F4	10	21	99.0	10.50
M9	23	-	102.6	7.28	F5	25	23	111.4	12.00
M10	55	-	105.1	7.18	F6	0	46	95.2	10.77
M11	22	-	107.4	7.42	F7	0	47	111.4	11.33
M12	10	-	99.8	6.75	F8	0	55	104.2	10.65
M13	76	-	110.5	6.64	F9	0	60	112.6	11.23
M14	110	-	102.9	6.63	F10	0	45	110.0	11.30

Stage I: In this stage, randomly oriented microcracks exist in a concrete specimen even before loading. During this stage, the stress-strain curve is almost linear and the material may be considered as a continuous medium. The behaviour is inherent to the material and thus independent of geometry or boundary conditions. This stage continues until the initiation of the first crack which is expected to depend more on plain concrete parameters than on fibre parameters.

Stage II: The microcracks tend to concentrate following the directions of the principal strains (Localization of microcracks). Thus, the first crack initiation is generally followed by stable crack propagation. Steel fibres bridge these stable cracks and limit their growth. As a result, the formation of a critical crack (macrocrack) is delayed and a greater peak stress (ultimate strength) is usually attained. The stress-strain curve in this stage is non-linear and reflects the inclusion of steel fibres in concrete.

Stage III : This stage starts with the onset of one or more critical cracks (macrocracks). Steel fibres are no longer capable to carry extra load and hence the strain greatly increases with corresponding decrease in the post-peak stress until complete failure occurs. A highly non-linear branch of stress-strain curve can be observed in this stage. Unlike stage I, stages II and III are strongly dependent on the boundary conditions and thus represent the structural behaviour. The behaviour of concrete in the latest two stages as well as the stage's length mainly depend on fibre parameters such as type, size, and amount.

This aspect of study was recently proved by researchers [1,3,4]. According to the test results reported in Table 1, the splitting strengths of 10 different SF-HSC mixes were ranging from 10.2 to 12.0 MPa. If the results of mixes without fibres are compared with those of fibrous concrete mixes (Table 1), it can be seen that the steel fibre used (V_f =1%) provides an increase of splitting strength of 50 to 70 percent. Predictive relations for splitting strength of SF-HSC are seldomly reported in literature. Narayanan et al [5], proposed a relationship connecting the split cylinder strength (f_{sp}) of steel fibre reinforced concrete with its compressive strength (f_c) as:

$$f_{sp} = 0.7 + \sqrt{F} + \frac{f_c}{20 - \sqrt{F}} \tag{1}$$

where:
f_{sp} = splitting strength in MPa
F = fibre factor = $(L_f/D_f)\ V_f\ d_f$; $L_f/\ D_f$= fibre aspect ratio, V_f = fibre volume fraction
d_f = bond factor = 0.5 for round fibres, 0.75 for crimped fibres, and 1.0 for indented fibres

As mentioned in the same reference [5], Eq. (1) has been found to give a quick and safe estimation of split cylinder strength of fibre concrete. Applying this equation to the available test results (Table 2) to ensure its validity, it results in an underestimation of the splitting strength of SF-HSC.

Recently, Wafa and Ashour [6] reported test results (Table 2) and an empirical equation for predicting the splitting tensile strength of HSC containing steel fibres. According to their data, when 1.25 percent of hooked steel fibre (aspect ratio 75) is added to plain HSC, an increase of its splitting strength with about 59.8 percent is observed. Note that, the validity of the equation proposed by Wafa and Ashour is limited to the hooked steel fibre type, and it is expressed as :

$$f_{sp} = 0.58 \sqrt{f_c} + 302\ V_f \qquad MPa \tag{2}$$

where V_f is the fibre volume fraction.

On the basis of the strength of a composite material [ref. 7], the efficiency of fibres in concrete [ref. 8], the results of interface bond of fibres [ref. 4], the own test results [Table 1], and available test results in literature [6], the following relationship for predicting splitting strength of SF-HSC is proposed:

$$f_{sp} = 0.8 \times f_c^{0.44} + \alpha\ I_F \qquad MPa \tag{3}$$

where:
α is a non dimensional constant depending on the fibre type as follows:
$\alpha = 1.0$ for hooked fibres,   0.9 for deformed fibres, and   0.5 for smooth fibres.
I_F is a dimensional factor depending on fibre characteristics (L_f/D_f, V_f), and concrete strength (f_c). The value of I_F can be adopted as:

$$I_f = 0.6 \left(\frac{L_f}{D_f} V_f \right)^{0.3} f_c^{0.44} \qquad MPa \tag{4}$$

Table 2 Experimental and predicted splitting strength of fibrous HSC.

Test Results		Fibres		Existing Equations				Proposed Equation	
f_c	f_{sp}	V_f	L_f/D_f	Narayanan Eq.(1)		Wafa Eq. (2)		Eq. (3)	
MPa	MPa	%	Ratio	f_{sp} MPa	Error %	f_{sp} MPa	Error %	f_{sp} MPa	Error %
I Own test results									
110.0	11.30	1.0	100	7.49	-33.7	9.10	-19.4	11.07	- 2.0
111.4	12.00	1.0	100	7.56	-37.0	9.14	-23.8	11.14	- 7.2
112.6	11.23	1.0	100	7.63	-32.1	9.17	-18.3	11.19	- 0.4
110.1	11.42	1.0	100	7.49	-34.4	9.11	-20.3	11.08	- 3.0
104.2	10.65	1.0	100	7.18	-32.5	8.94	-16.1	10.81	+1.5
111.4	11.33	1.0	100	7.56	-33.2	9.14	-19.3	11.14	- 1.7
99.0	10.50	1.0	100	6.91	-34.2	8.79	-16.3	10.57	+0.7
96.6	10.23	1.0	100	6.78	-33.7	8.72	-14.8	10.46	+2.2
89.9	10.31	1.0	100	6.43	-37.6	8.52	-17.4	10.13	- 1.7
95.2	10.77	1.0	100	6.71	-37.7	8.68	-19.4	10.39	- 3.5
Average					**- 34.6**		**-18.5**		**-1.5**
Stand. Dev.					2.1		2.6		2.8
II Wafa and Ashour									
93.5	6.45	0.00	--	5.37	-16.7	5.61	-13.1	6.26	- 2.9
95.1	8.10	0.25	75	5.99	-26.0	6.41	-20.9	8.63	+ 6.5
97.3	8.38	0.50	75	6.33	-24.4	7.23	-13.7	9.35	+11.5
96.5	8.81	0.75	75	6.46	-26.7	7.96	- 9.6	9.74	+10.6
97.1	10.31	1.00	75	6.64	-35.6	8.74	-15.3	10.11	- 1.9
93.9	10.47	1.25	75	6.60	-37.0	9.39	-10.3	10.24	- 2.2
97.8	10.04	1.50	75	6.93	-31.0	10.27	+ 2.3	10.68	+ 6.4
Average					**-28.2**		**-11.5**		**+4.0**
Stand. Dev.					7.0		7.1		6.2
Total Average					**-32.0**		**-15.6**		**+0.8**
Stand. Dev.					5.6		5.9		5.2

* The fibres used are hooked-end type.

Table 2 shows the experimental results as well as a comparison between the values of splitting strength predicted by the existing equations (Eq. (1) and Eq. (2)) and the proposed equation [Eq.(3)]. It can be seen that the predicted values using the proposed equation give a close agreement with both the own test data as well as the data compiled from literature. The average errors of the predicted splitting strength are 32 and 15.6 percent for both the existing equations [Eq.(1) and Eq.(2)] respectively, whereas the proposed equation yields an average error less than 1.0 percent (0.8 %). It is worth noting that the first term in Eq. (3) represents the tensile strength of the matrix, while the second term (αI_F) is the contribution of steel fibres in the splitting strength of the composite. In other words, Eq. (3) is valid and applicable for predicting the splitting strength of both steel fibre high strength concrete as well as high strength concrete without fibres ($\alpha I_F=0$). More details concerning the prediction of the splitting strength of HSC without fibres can be seen in Reference 9.

CONCLUSIONS

1 The first-crack strength depends primarily on plain concrete characteristics rather than fibre parameters. After the occurrence of the first-crack, the fibres effectively work, carry the entire applied load, and tend to suppress the localization of microcracks into macrocracks and consequently the tensile strength increases.

2 Steel fibres provided a significant increase of splitting tensile strength. A new equation [Eq.(3)) is proposed to evaluate the splitting strength of Steel Fibre High Strength Concrete. The predicted values using the latter equation show a close agreement with both the own test data as well as the data compiled from literature.

REFERENCES

1 SHAH, S., "Do Fibers Increase the Tensile Strength of Cement-Based Matrixes ?" ACI Material Journal, November-December 1991, pp. 595-602.

2 IMAM, M., "Shear-Moment Interaction of Steel Fibre High Strength Concrete" Doctoral thesis, Civil engineering Department, Catholic University of Leuven, Belgium, April 1995, 199 pp.

3 ROSSI, P., ACKER, P., and MALIER, Y., "Effect of Steel Fibres at Two Different Stages: The Material and The Structure" Materials and Structures, RILEM, 1987, No. 20, pp.436-439.

4 NAAMAN, A., and NAJM, H., " Bond-Slip Mechanisms of Steel Fibres in Concrete" ACI Materials Journal, March-April 1991, pp 135-145.

5 NARAYANAN, R., and DARWISH, Y., " Use of Steel Fibres as Shear Reinforcement " ACI Structural Journal, May-june 1987, pp 216-227.

6 WAFA, F., and ASHOUR, S., " Mechanical Properties of High-Strength Fiber Reinforced Concrete" ACI Material Journal, V. 89, No. 5, Sep.-Oct. 1992, pp. 449-455.

7 LIM, T., PARAMASIVAM, P., and LEE, S., "Bending Behavior of Steel-Fiber Concrete Beams" ACI Structural Journal, November-December 1987, pp. 524-536.

8 VANDEWALLE, L., and MORTELMANS, F., " The Advantage of Long, Thin Steel Fibres" An International Conference on Modern Techniques in Construction 27-28 March 1990, Singapore, pp 630-643.

9 IMAM, M., VANDEWALLE, L., and MORTELMANS, F., "Indirect Tensile Strength of Very High Strength Concrete" Proceedings of the International Symposium on Utilization of High Strength Concrete, Lillehammer, Norway. 20-23 June 1993, pp. 1114-1121.

DESIGN OF A STABLE AIR VOID SYSTEM IN CONCRETE BY OPTIMIZATION OF THE COMPOSITION OF THE AGGREGATE

M Glavind

Danish Technological Institute

B Jensen

Unicon Concrete

Denmark

ABSTRACT. The present article describes a project with the aim of developing a method to design a stable air void system in concrete by optimizing the composition of the aggregate among other things by carrying out a packing analysis. The composition of the aggregates is chosen so that the size of the cavities between the aggregates ensures an optimum air void system. In addition, the amount of cement paste must be less than what is required to fill in the cavities between the aggregates. This decifit of cement paste makes room for air voids.

Keywords: Concrete mix design, Air voids, Packing of aggregates, Decifit of cement paste.

Dr Mette Glavind is a consultant at the Concrete Centre, Danish Technological Institute. Her main working areas cover mix design of concrete, fibre reinforced concrete, environmental aspects and life cycle analysis of concrete.

Mr Ib B. Jensen is Head of technical department at Unicon Concrete covering 8 ready mixed concrete plants. His main working areas cover mix design of concrete and quality management.

Radical Concrete Technology. Edited by R K Dhir and P C Hewlett. Published in 1996 by E & FN Spon, 2–6 Boundary Row, London SE1 8HN, UK. ISBN 0 419 21480 1.

BACKGROUND

Current practice in design of air void systems

In standards and codes of practice it is required that a concrete has a fine air void system in order to be frost resistant. The air void system is normally established by addition of air entrainment to the concrete. In practice however, it is very difficult to establish and maintain the air void system during mixing and casting of the concrete.

The effect of a specific air entrainment depends on a number of different parameters and the amount of the air entrainment always has to be established by testing. The parameters which have to be considered are among other things the type and composition of the aggregates, the mixing time and -method, the temperature, the water-cement ratio, the amount and type of fly ash and silica fume and the casting method.

The relation between the different parameters is not known. Therefore, it is impossible to ensure a desired air void system without testing a number of mix designs and thereby choose the best one.

Adopted hypothesis

The adopted hypothesis in the present project is that it is possible to design a stable air void system in a concrete by optimization of the composition of the aggregates. The composition of the aggregates must be chosen so that the size of the cavities between the aggregates ensures an optimum air void system. In addition, the amount of cement pasta must be less than what is required to fill in the cavities between the aggregate. This decifit of cement paste makes room for air voids.

It is expected that physically produced air voids between the aggregate skeleton are more difficult to break down under mechanical influence than chemically produced air voids. Therefore, a concrete with physically produced air voids should be more stable during pumping and vibration.

THEORETICAL CONSIDERATION

The background for the proposed hypothesis is the principle of using geometrical packing of aggregate for concrete mix design. The principle will be described briefly below.

When choosing a concrete mix design it is always desirable to compose the aggregates as densely as possible in order to minimize the amount of cement paste necessary to fill in the cavities between the aggregates. Apart from an obvious economic advantage, a minimum amount of binder in a concrete results in less shrinkage and creep and a more dense and therefore probably a more durable and strong concrete. Further, a high packing of the aggregates ensures a good workability of the concrete.

The Concrete Centre, Danish Technological Institute disposes of a computer program to calculate the packing of aggregates. The computer program is based on a model that was developed in the light of the principle of packing of binary mixtures, Stovall et al. [1]. The basic model has been developed further and modified by incorporating an experimentally determined packing, Glavind et al [2]. For a multicomponent system, the packing can be calculated as

$$\text{Packing} = \min_{i=1}^{n} \left(\alpha_i + (1-\alpha_i) \sum_{j=1}^{i-1} g(i,j)\,\phi_j + \sum_{j=i+1}^{n} f(i,j)\,\phi_j \right)$$

where
  α_i is the mono-disperse packing
  $g(i,j)$ and $f(i,j)$ are interaction functions
  ϕ_j is the volume fraction

An example of a result of a packing analysis is shown at figure 1. The packing is shown as contour lines in %. The marked point at the figure has a packing of 84% and the corresponding material composition is 30% 0-8 mm, 25% 2-8 mm and 45% 8-16 mm. Packing calculations and concrete mix design are treated in Glavind et al. [3].

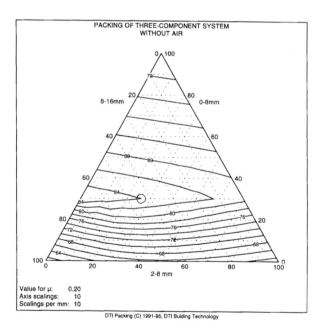

Figure 1 Packing calculation of three aggregate materials

When carrying out a test series at a Danish concrete factory to investigate the relation between packing of the aggregates and workability of the concrete it was discovered that for a constant volume of cement paste, different packing combinations gave different air void volumes, Glavind et al. [4]. These observations were the background for formulation of the proposed hypothesis.

INTRODUCTORY INVESTIGATIONS

Already in 1907 Fuller and Thompson [5] investigated the importance of the size distribution of the aggregates and the properties of the concrete.

The concept of packing in connection with concrete mix design was applied as early as 1911 (and maybe before) by Suenson [6], and comparatively recently the packing concept has been applied by for instance Bache [7], Larrard et al. [8] and Just Andersen [9]. Powers, [10], also has to be mentioned for his extensive work with regard to concrete mix design on the basis of packing.

The authors are without any knowledge of packing calculations in connection with design of an air void structure. The hypothesis that the air void structure is influenced by the aggregate composition has been confirmed by a literature screening and an analysis of data from concretes made during the past years.

Literature screening

A list of selected observations from the literature screening is shown below. For more details the reader is referred to the report from the project, Glavind, [11].

- the larger the amount of particles with a grain size between 0.125 mm and 0.6 mm, the larger the total air content
- the larger the amount of particles with a grain size between 0.125 mm and 0.5 mm, the larger the specific surface
- the influence of the grading of the sand on the air void structure diminishes when the cement paste content increases.

Analysis of data originating from concretes made during the past years

207 air void test reports made as the daily concrete production control have been analyzed. The main conclusion from this analysis is as given below. For more details the reader is referred to Frandsen, [12].

- the smaller the amount of particles with a size less than 0.5 mm, the more air entrainment must be added in order to obtain a constant air void content
- the larger the amount of sand with a grain size smaller than 0.5 mm, the larger the number of air voids smaller than 0.35 mm.

LABORATORY TESTING

A testing program is initially being carried out to test the hypothesis of the project. The testing program is performed on model concrete made of glass spheres and cement paste. The reason for using glass spheres as aggregates is that they are spherical-shaped and can be supplied almost mono-disperse.

Aim

The aim of the testing program is to test if an air void system can be composed from the packing of aggregates and a decifit of cement paste. Further the testing program will investigate the relation between the sizes of the cavities between the aggregates and the size distribution of the air voids. Finally, the testing program will investigate the stability of the air voids.

Materials

Portland cement (white), fly ash, silica fume, glass spheres size 0.2 mm, 0.5 mm, 1.0 mm, 2.0 mm, 7.0 mm and 10.0 mm.

Packing

The packing of the six glass sphere sizes is shown at figure 2. Note that six sizes of glass spheres have been combined into three sizes. The marked point comes close to a grading curve for a concrete which functions well in practice.

Mix design

Important parameters for the mix design are as follows:

- a packing of 78 %
- an equivalent water-cement ratio of 0.5 (activity factor of 0.5 for fly ash and 2.0 for silica fume)
- 15 % fly ash of cement
- 5 % silica fume of cement

Mixes nos 1, 2, 3 and 4 have been designed to obtain an air content of 4% by adding 18% cement paste. The composition of the aggregate in mix no 2 has been repeated with 20% and 16% cement paste to obtain an air content of 0% and 6%, mix no 2x and 2xx, respectively.

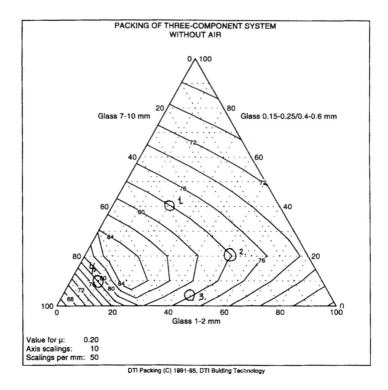

Figure 2 Packing calculation of six glass sphere sizes

Investigations

The following properties of the fresh mixes have been tested:

- slump
- density
- air content.

Further, cylinders have been cast and the air void system analyzed on impregnated thin-sections of the hardened model concrete:

- total air content
- air void sizes
- air void geometry.

Results

Figure 3 shows the accumulated size distribution of air voids and cavities for mix no 1, 2, 2xx, 3 and 4.

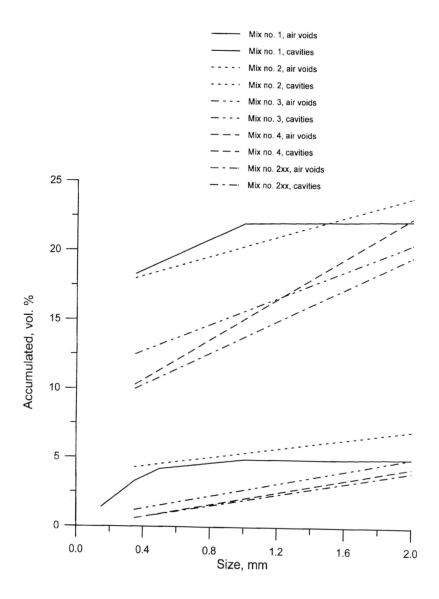

Figure 3 Accumulated distribution of air voids and cavities

Table 1 shows a survey of the results of total air content measured with pressure-meter on the fresh concrete, calculated from measured density and measured on the hardened concrete together with the designed air content. Further, table 1 shows the designed and the measured packing.

Table 1 Survey of results of air content and packing

Mix. no.	1	2	2x	2xx	3	4
Air content						
Designed, %	3.7	3.7	0	5.7	3.7	3.7
Fresh concrete, %	4.5	5.5	4.3	6.5	5.0	2.0 [1]
Calculated, %	4.5	3.7	2.7	5.1	3.0	5.3
Hardened concrete, %	4.8	7.1	- [2]	6.3	6.0	6.7
Packing						
Designed, %	78	78	78	78	78	78
Measured, %	77.8	76.0	- [2]	- [2]	78.5	75.2

[1] The mix separates which has the effect that the measured air content is too low.
[2] No measurements yet.

Discussion

From table 1 it appears that it is possible to obtain an air void content of 4% or more without adding air entrainment. Generally, the measured air content is higher than the designed air content. This can be explained by the fact that any deviation from the theoretical packing will result in a lower packing and thereby in a higher air content. It can also be seen from table 1 that the obtained packing for all mixes except mix no 3 is lower than the designed packing.

There are indications that the air content in mix no 2x as designed is lower than for the rest of the mixes.

However, it has to be emphasized that the uncertainties of the measurements might be too high to be able to detect significant differences in an air content of 2%.

With regard to the size distribution of cavities and air voids figure 4 indicates that there is a relation. Mixes nos 1 and 2 have many small-sized air voids and cavities, while mixes nos 3 and 4 have many large-sized air voids and cavities. Mix no 2xx should in theory have the same size distribution of cavities as mix no 2, but is very similar to mix no 4.

Figure 4 shows a photo of an impregnated thin section where examples of the shape of the air voids can be seen. It can be seen that the air voids on the contrary to chemically produced sphere-shaped air voids are rhomb shaped which is natural as they are entrapped by sphere-shaped aggregate. Further, it is indicated that the air voids are stable because they cannot escape from the cavities between the aggregate on the contrary to chemically produced air voids which are situated as islands in the cement paste and therefore more easily can be squeezed out of the concrete.

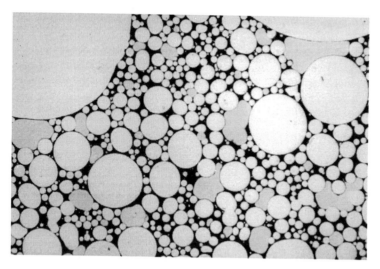

Figure 4 Thin section of model concrete.
Air voids are the darkest not spherical shaped areas.

More tests have currently been initiated to be able to extract significant conclusions from the laboratory testing, i.e. more detailed measurements of size distributions of cavities and air voids and repetition of all mixes in order to detect uncertainties. Further, the mixes will be repeated with "normal" concrete aggregate, and investigations related to the stability of the air voids will be initiated. The investigations will be completed for presentation at the International Congress "Concrete in the Service of Mankind", June 1996".

FULL SCALE TESTING

When the results of the laboratory testing and the modelling are available full scale testing will be carried out at a concrete plant to test the practical applicability of the air void design method. Results of the full scale testing will be presented at the International Congress "Concrete in the Service of Mankind, June 1996".

EXPLOITATION POTENTIAL

In order to obtain the building owner's approval of a concrete mix design the manufacturer has to carry out full scale trial casting. The biggest problem relates to obtain an optimum air void system as described in Background. If the required air void system is not obtained at the first trial casting, it has to be repeated one or more times. The cost of carrying out full scale trial casting can be up to 140,000 ECU for large construction works and about 5-15,000 ECU for normal sized construction works.

The results of the project will probably form the basis of the development of a computer program which can design a concrete with the desired air void system. The computer program will aim at being practical applicable for concrete producers. With this computer program the numbers of full scale trial casting can be reduced and it is estimated that this will lead to cost savings of 10-50%.

However, it has to be underlined that it is possible that the investigated design method for air void systems might imply that the changes that occur in the grading curve for the aggregate will be smaller than is the case today and maybe it also implies more fractionated aggregates. The two mentioned possible preconditions cannot be fulfilled without major investments which will delay the exploitation of the project result.

CONCLUSIONS

The preliminary conclusions from the present project are as follows:

- the hypothesis that an air void system can be designed from the composition of the aggregates has been confirmed
- it is indicated that there is a relation between the composition of the aggregates and thereby the cavities between aggregate and the air void size distribution
- the shape and the location of the physically produced air voids indicate that they are more stable than chemically produced air voids.

REFERENCES

1. STOVALL, T, LARRARD, DE F and BUIL, M. Linear Packing Density model of grain mixture. Powder Technology 48, 1986, pp 1-12.

2. GLAVIND, M. and STANG, H. Geometrical packing model as a basis for composing cement paste containing clay for high strength concrete. Proceedings from the Third International Symposium on Brittle Matrix Composites, BMC3, 1992, pp. 508- 518.

3. GLAVIND, M. OLSEN, G.S. and MUNCH-PETERSEN, C. Packing calculations and concrete mix design. Nordic Concrete Research, Publication no. 13, 2/1993, pp. 21-34.

4. GLAVIND, M. and OLSEN, G.S. Relation between packing of the aggregate and the properties of fresh concrete (in Danish), Concrete Centre, Danish Technological Institute, 1994, pp. 11.

5. FULLER, W.B. and THOMPSON, S.E. The laws of proportioning concrete. Trans. ASCE 59, 1907, pp. 67-143.

6. SUENSON, E. Building Materials III: Stone, potters, mortar, concrete, artificial stone, glass (in Danish), 1911, pp. 522.

7. BACHE, H.H. New concrete - New technology (in Danish), Beton-Teknik, 8/04/1992.

8. LARRARD, de F. and SEDRAN, T. Optimization of ultra-high-performance concrete by the use of a packing model. Cement and Concrete Research, Vol. 24, No. 6, 1994, pp. 997-1009.

9. JUST ANDERSEN. A Study of particle packing and rheology. Submitted in fulfilment of the requirements for the degree of industrial researcher. The Danish Academy of Technical Sciences, 1990, pp. 236.

10. POWERS, T.C. The properties of fresh concrete. John Wiley & Sons. Inc. New York, 1968.

11. GLAVIND, M. Optimization of concrete work. Packing analysis as a tool for dimensioning air void structure, Phase 1.1 Literature screening (in Danish). Concrete Centre, Danish Technological Institute, 1995, pp. 14.

12. FRANDSEN, J. Optimization of concrete work. Packing analysis as a tool for dimensioning air void structure. Phase 1.2 Analyses of old air void test reports (in Danish). Unicon Concrete, 1995, pp. 6.

FROST DURABILITY AND DEICER SALT SCALING RESISTANCE OF HIGH-PERFORMANCE CEMENT PASTES REINFORCED WITH STEEL AND CARBON MICRO-FIBRES

R Pleau

M Azzabi

M Pigeon

Laval University

N Banthia

University of British Columbia

Canada

ABSTRACT. The paper presents the results of an experimental study aiming to assess the frost resistance of 45 cement pastes. The influence of four parameters was studied: the water to binder ratio (0.25, 0.35, and 0.45), the dosage of silica fume (0, 10, and 20% in weight), the type of micro-fibers (steel or carbon), and the fiber dosage (0, 2, and 5% in volume). All mixtures were subjected to 300 rapid freezing and thawing cycles in water and to 50 daily freezing and thawing cycles in presence of deicer salts. They were also subjected to a microscopic determination of the characteristics of the air-void system, and to flexural strength measurements. The results obtained clearly indicate that the use of micro-fibers improve very significantly both the frost durability and the deicer salt scaling resistance of cement pastes.

Keywords: micro-fibers, durability, freezing and thawing, deicer salt scaling, air entrainment, steel, carbon, flexural strength

Dr Richard Pleau is an Assistant Professor in the School of Architecture at Laval University (Québec, Canada). His research interests include micro-reinforced cementitious composites, fiber-reinforced concrete, high-performance concrete, frost durability, and air entrainment.

Maha Azzabi is a graduate student in the Department of Civil Engineering at Laval University (Québec, Canada). Her Ph.D. thesis have for object the durability of micro-reinforced cementitious composites.

Dr Michel Pigeon is the Director of the Centre Interuniversitaire de Recherche sur le Béton Sherbrooke/Laval and a Principal Investigator of the Concrete Canada Network of Centres of Excellence on High-Performance Concrete. His research interests include durability, additives, roller-compacted concrete, shotcreting, and repairs.

Dr Nemkumar Banthia is an Associate Professor in the Department of Civil Engineering at the University of British Columbia (Vancouver, Canada). His research interests include all aspects related to fiber-reinforced concrete.

Radical Concrete Technology. Edited by R K Dhir and P C Hewlett. Published in 1996 by E & FN Spon, 2–6 Boundary Row, London SE1 8HN, UK. ISBN 0 419 21480 1.

INTRODUCTION

Cement paste is a brittle material. However, the use of micro-fibers at a relatively high dosage (2 to 5% by volume) was found to significantly improve the tensile (or flexural) strength and the toughness of cementitious materials [1-4]. These fibers can be made with mineral (asbestos, wollastonite), vegetal (cellulose, sisal), or synthetic (steel, carbon, glass, etc.) materials. Due to their very small size (typically 20 μm in diameter by a few mm in length), the micro-fibers are very numerous (up to 500 000 fibers per cm^3), and the average spacing between them is usually less than 200 μm. Consequently, the micro-fibers can restrain the formation, the coalescence and the propagation of microcracks which probably account for the better mechanical properties.

Although many papers have been published on the mechanical properties of micro-reinforced cementitious composites, very little data is available as regards with their frost durability. However, considering that micro-fibers improve the tensile strength of the cement paste and restrain the crack propagation, it is reasonable to believe that these fibers can also exert a positive influence of the frost durability. In conventional concrete, the frost resistance is achieved by intentionally adding microscopic air voids into the concrete mixture. In micro-reinforced cement composites however, air entrainment must be avoid as much as possible because the presence of air can significantly reduces the tensile strength of the composite. The frost resistance of micro-reinforced composites is thus a matter of concern which need to be addressed.

Recent studies have clearly shown that carbon fiber micro-reinforced mortars can satisfactory sustain 300 freezing and thawing cycles without air entrainment [5-7]. It seems that the frost damages decrease with the increase of the fiber and/or sand volume fraction. The frost durability and deicer salt scaling resistance of high-performance non air-entrained micro-reinforced mortars was also recently investigated by the authors [8]. In this study, mortars having a water to binder ratio of 0.35 were made with two types of fibers (steel and carbon), three fiber dosages (0, 1, and 2% by volume) and two sand to binder ratios (1 and 2 by weight). Both steel and carbon fibers were found to improve the resistance of mortars to rapid freezing and thawing cycles in water (ASTM C 666 Standard) and, particularly, to surface scaling in presence of deicer salts (ASTM C 672 Standard). But this improvement was due, at least in part, to the fact that the air content increases very significantly with the amount of fibers (up to 43% in volume). In the mixtures containing fibers, the spacing factor was never higher than 380 μm which is probably lower than the critical spacing factor value for that kind of mortar. The air entrainment associated with the presence of fibers into mortars was also observed in other recent studies [4,7,9], but experimental data indicate that this effect is not very important into cement pastes [4-9].

TEST PROGRAM

This experimental study had for objective to assess the influence of micro-fibers on the frost resistance of cement-based materials. All the tests were carried out on micro-reinforced cement pastes to avoid the air-entraining effect observed on micro-reinforced mortars. In order to cover a wide range of matrix properties, the mixtures tested were made using three water to binder ratios (0.45, 0.35, and 0.25) and three dosages of silica fume (0, 10, and 20% by weight) used as partial replacement for Portland cement. Silica fume facilitates the fiber dispersion and improves bonding at the interface between the fibers and the cement paste [1,4]. Steel and carbon micro-fibers were used at three different dosages (0, 2, and 5% by volume). A total of 45 mixtures were moist cured for 14 days and subjected to both rapid freezing and thawing cycles in water and surface scaling tests in presence of deicer salts. The characteristics of the air-void system and the 14-day flexural strength were also measured on each mixture.

MATERIALS, MIXTURE COMPOSITION, AND EXPERIMENTAL PROCEDURES

Table 1 gives the chemical composition of the ordinary Portland cement (ASTM type I) and of the silica fume used for all the mixtures. The Portland cement have a Blaine fineness of 3692 cm^2/g and the average diameter of the silica fume particles was approximately equal to 0.1 μm. The steel micro-fibers have a cross section of approximately 5x25 mm and their length was roughly equal to 3 mm. The 10 mm long carbon micro-fibers were round in shape and have a diameter of approximately 20 μm. Both fibers have a tensile strength of about 600 MPa, but the modulus of elasticity of steel (200 000 MPa) was much higher than that of carbon (30 000 MPa). The specific gravity of carbon (1.65) is also much smaller than that of steel (7.85). A naphtalene-based superplasticizer admixture (having 42% by weight of solid materials) was used at a dosage ranging from 0 to 40 L/m^3 to produce mixtures having a good workability. All the mixtures were batched in a 10 L mortar mixer according to the following sequence: 1°) the cement and the silica fume were first mixed together, 2°) the water (containing the superplasticizer) was progressively added to obtain an homogenous cement paste, and 3°) the fibers were slowly added to the mixture to facilitate fiber dispersion. The composition of mixtures is given in Table 2. After mixing, all the specimens were cast into molds on a vibrating table and covered by a plastic sheet and a wet towel. Those specimens were demolded 24 hours after mixing and keep in lime-saturated water at 23°C until testing.

Table 1   Chemical Composition of Cement and Silica Fume (% in mass)

Constituent	Portland cement	Silica fume
CaO	64.0	0.8
SiO$_2$	19.6	92.6
Al$_2$O$_3$	4.1	0.5
Fe$_2$O$_3$	2.8	1.8
MgO	2.8	0.6
SO$_3$	2.3	0.3
Na$_2$O	0.5	0.0
K$_2$O	0.5	1.0

For each mixture, the characteristics of the air-void system were determined on two 75x75 mm polished section according to the ASTM C 457 modified point count method. The flexural strength was measured by a third-point loading test (ASTM C 293) on two 23x36x195 mm prisms after 28 days of moist curing. Two similar 23x36x195 mm prisms were subjected to 300 rapid freezing and thawing cycles (at a freezing rate of 8.6°C/h) after 14 days of moist curing according to the requirements of the ASTM C 666 procedure A test method. The deterioration caused by frost action was assessed by measuring the residual length change of the test specimens. The ASTM C 672 deicer salt scaling tests were carried out on two 110x150x25 mm slabs covered with a 3% NaCl solution and subjected to 50 daily freezing and thawing cycles from 23°C to -18°C. The slabs were moist cured for 14 days and air dried at

Table 2   Composition of the Tested Mixtures

W/B	Silica fume (%)	Fibers	Fiber volume (%)	Water (kg/m³)	Cement (kg/m³)	Silica fume (kg/m³)	Super-plastic. (L/m³)	Unit weight (kg/m³)
		none	0	585	1324	0	0.0	1909
		steel	2.1	563	1250	0	0.0	1976
	none	steel	5.1	530	1185	0	5.1	2122
		carbon	2.0	530	1191	0	10.0	1776
		carbon	4.9	504	1134	0	15.6	1736
		none	0	587	1174	122	0.0	1892
		steel	2.1	550	1100	121	0.0	1933
0.45	10	steel	5.1	523	1053	117	5.1	2100
		carbon	1.9	517	1044	116	9.7	1720
		carbon	4.7	479	974	109	15.1	1656
		none	0	566	1006	252	3.0	1831
		steel	2.1	574	942	236	2.3	1918
	20	steel	5.2	530	947	237	5.2	2132
		carbon	1.9	508	915	229	12.0	1698
		carbon	4.6	455	825	206	16.8	1582
		none	0	512	1479	0	5.2	1997
		steel	2.1	496	1430	0	7.2	2096
	none	steel	5.2	480	1390	0	12.5	2294
		carbon	1.9	467	1352	0	11.8	1864
		carbon	4.9	442	1294	0	19.6	1838
		none	0	513	1330	148	7.4	1999
		steel	2.1	490	1276	142	10.4	2084
0.35	10	steel	5.2	465	1218	136	15.5	2242
		carbon	1.9	437	1148	127	14.1	1751
		carbon	4.7	412	1097	122	21.4	1732
		none	0	501	1156	289	9.0	1956
		steel	2.1	488	1125	281	12.6	2074
	20	steel	5.1	451	1054	264	17.5	2192
		carbon	1.9	437	1021	256	16.4	1764
		carbon	5.0	402	951	238	24.1	1694
		none	0	421	1709	0	10.3	2142
		steel	2.1	416	1688	0	12.5	2308
	none	steel	5.4	410	1686	0	18.0	2540
		carbon	2.0	389	1599	0	16.8	2039
		carbon	4.9	358	1506	0	29.4	1978
		none	0	420	1546	171	14.8	2154
		steel	2.1	399	1475	164	17.5	2220
0.25	10	steel	5.1	370	1387	154	25.3	2336
		carbon	1.9	357	1334	148	23.5	1896
		carbon	4.8	339	1295	144	35.5	1896
		none	0	401	1324	331	26.0	2084
		steel	2.1	393	1309	327	26.3	2224
	20	steel	5.0	354	1197	299	35.1	2282
		carbon	1.9	355	1193	298	31.0	1912
		carbon	4.7	317	1091	273	41.9	1804

23°C for an additional 14-day period prior testing. The damage caused by scaling was assessed by measuring the mass of the scaling residues. Since it was observed that the size of the scaled-off particles decreases significantly with the increase of the fiber content, the size-distribution of the scaling residues was later determined by a sieve analysis (with sieve openings ranging from 6 mm to 80 μm).

TEST RESULTS AND DISCUSSION

Characteristics of the Air-Void System

The characteristics of the air-void system are summarized in Table 3. Although their air contents are very low (1 to 5%), the spacing factors of the mixtures made without fibers are surprisingly low (360 to 704 μm) because their specific surfaces (16.1 to 40.9 mm^{-1}) are much higher than the specific surface usually obtained in non air-entrained concretes ($\approx$ 10 mm^{-1}). These relatively low spacing factors could probably be explained by the high energetic efficiency of the vertical-axis mortar mixer which was used at a high rotating speed in order to facilitate the fiber dispersion. For mixtures containing steel micro-fibers, the air content (1.2 to 8.8%) generally increases with the fiber volume, and the spacing factor (244 to 527 μm) is generally lower than the spacing factor of the corresponding mixture without fibers although the difference is often not very important. However, Table 3 clearly shows that carbon micro-fiber reinforced mixtures have a much higher air content (9.8 to 22.8%) and a much lower spacing factor (151 to 275 μm) than the other mixtures although very little differences are observed between the mixtures containing 2% and 5% of carbon micro-fibers. For all mixtures, silica fume does not appear to significantly influence the characteristics of the air-void system.

Flexural Strength

The modulus of rupture obtained from the third-point loading tests are given in Table 4 and shown in Figure 1. This Figure shows that the adding of micro-fibers can roughly twice the flexural strength of the cement pastes and that, most of the time, the carbon fibers provide higher flexural strengths than the steel fibers. Carbon fibers are much longer than steel fibers and, consequently, their bond to the cement matrix is probably stronger which could explain, at least in part, their better behavior. The examination of many micro-reinforced paste broken surfaces under a scanning electron microscope had clearly shown that both the steel and carbon micro-fibers were mainly pulled-out rather than broken during the test which strongly suggests to fiber debonding is the main fracture mechanism. The higher flexural strengths which are often obtained by increasing the silica fume content (see Figure 1) could also be explained by a better bond between the fibers and the matrix. Due to their very small size (0,1 μm in diameter), the silica fume particles could significantly improve the bonding resistance by filling the spaces left between the much coarser cement grains at the fiber-matrix interface. Figure 1c indicates however that, at a very low water to binder ratio, the increase of the silica fume content have a detrimental effect on the flexural strength of both reinforced and unreinforced cement pastes. This is probably simply due to the fact that the paste thus become too sticky which tend to significantly reduce the workability of the mixture. A reduced workability can also be held responsible for the strength losses sometimes observed at high dosages (5%) of carbon fibers (see Figure 1).

Table 3   Characteristics of the Air-Void System

W/B	Silica Fume (%)	Fibers	Fiber volume (%)	Air content (%)	Specific surface (mm^{-1})	Spacing factor (μm)
		none	0	2.3	16.7	704
		steel	2.1	3.8	17.9	527
	none	steel	5.1	8.5	16.5	392
		carbon	2.0	11.6	22.4	246
		carbon	4.9	12.4	18.7	275
		none	0	1.2	32.0	484
		steel	2.1	3.2	20.8	481
0.45	10	steel	5.1	6.3	19.6	372
		carbon	1.9	12.6	19.4	273
		carbon	4.7	22.8	14.9	225
		none	0	1.5	27.0	512
		steel	2.1	2.5	34.9	326
	20	steel	5.2	3.5	28.4	342
		carbon	1.9	11.8	22.8	241
		carbon	4.6	10.6	21.2	264
		none	0	3.8	26.2	360
		steel	2.1	3.9	35.6	264
	none	steel	5.2	1.8	45.0	293
		carbon	1.9	14.2	22.7	220
		carbon	4.9	13.6	22.1	223
		none	0	2.3	25.5	459
		steel	2.1	4.8	20.3	413
0.35	10	steel	5.2	6.7	17.7	401
		carbon	1.9	16.7	20.1	230
		carbon	4.7	20.5	17.7	217
		none	0	1.1	40.9	395
		steel	2.1	2.8	31.6	343
	20	steel	5.1	5.7	23.4	335
		carbon	1.9	12.0	23.9	228
		carbon	5.0	12.3	24.8	211
		none	0	5.0	22.5	372
		steel	2.1	4.4	27.0	329
	none	steel	5.4	1.2	60.4	253
		carbon	2.0	9.8	40.2	151
		carbon	4.9	11.0	29.7	191
		none	0	3.8	16.1	557
		steel	2.1	4.2	33.5	271
0.25	10	steel	5.1	8.8	23.2	277
		carbon	1.9	11.5	28.6	194
		carbon	4.8	11.5	27.3	199
		none	0	2.2	30.0	393
		steel	2.1	2.6	27.0	374
	20	steel	5.0	5.3	33.4	244
		carbon	1.9	10.6	27.0	219
		carbon	4.7	9.8	28.8	206

Table 4 Summary of Test Results

W/B	Silica Fume (%)	Fibers	Fiber Volume (%)	Modulus of rupture (MPa)	Spacing factor (μm)	Length change (μm/m)	Mass of residues (kg/m^2)
		none	0	2.5	704	destroyed	7.84
		steel	2.1	5.1	527	685	0.62
	none	steel	5.1	6.6	392	139	0.05
		carbon	2.0	5.4	246	144	0.68
		carbon	4.9	6.2	275	390	0.71
		none	0	1.9	484	destroyed	2.47
		steel	2.1	4.9	481	560	0.99
0.45	10	steel	5.1	6.9	372	125	0.41
		carbon	1.9	6.2	273	278	0.27
		carbon	4.7	7.7	225	320	0.35
		none	0	2.8	512	destroyed	13.1
		steel	2.1	4.6	326	1765	2.06
	20	steel	5.2	6.9	342	213	1.03
		carbon	1.9	8.4	241	427	0.33
		carbon	4.6	9.1	264	380	0.12
		none	0	7.1	360	destroyed	0.03
		steel	2.1	9.7	264	126	0.01
	none	steel	5.2	12.2	293	73	0.51
		carbon	1.9	7.7	220	229	0.16
		carbon	4.9	6.3	223	773	0.43
		none	0	3.8	459	destroyed	4.84
		steel	2.1	8.1	413	775	0.26
0.35	10	steel	5.2	9.7	401	73	0.18
		carbon	1.9	9.4	230	80	0.01
		carbon	4.7	5.6	217	171	0.14
		none	0	5.5	395	destroyed	8.84
		steel	2.1	6.7	343	937	0.68
	20	steel	5.1	11.3	335	366	0.36
		carbon	1.9	9.2	228	267	0.11
		carbon	5.0	12.3	211	146	0.08
		none	0	7.1	372	274	0.11
		steel	2.1	8.9	329	133	0.03
	none	steel	5.4	14.0	253	126	0.22
		carbon	2.0	9.9	151	146	0.07
		carbon	4.9	8.9	191	181	0.04
		none	0	5.8	557	destroyed	0.06
		steel	2.1	10.4	271	120	0.07
0.25	10	steel	5.1	13.2	277	77	0.02
		carbon	1.9	12.1	194	117	0.01
		carbon	4.8	12.1	199	79	0.02
		none	0	7.3	393	destroyed	0.26
		steel	2.1	9.6	374	165	0.02
	20	steel	5.0	12.0	244	153	0.01
		carbon	1.9	14.3	219	206	0.02
		carbon	4.7	11.8	206	247	0.02

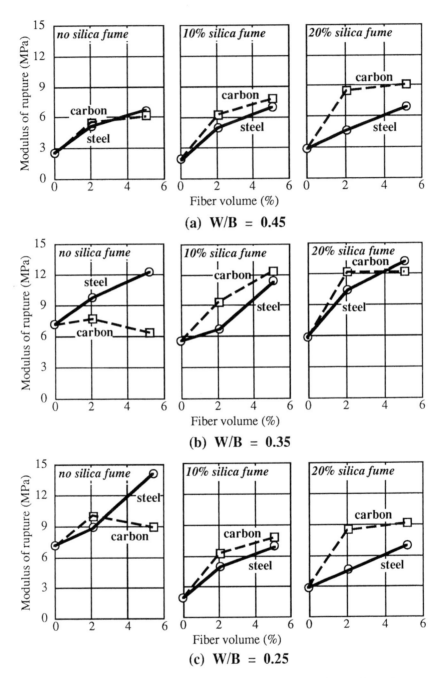

Figure 1 Relationship between the modulus of rupture and the fiber volume for cement pastes having different water to binder ratios, containing various amounts of silica fume, and reinforced with steel and carbon micro-fibers

Resistance to Rapid Freezing and Thawing Cycles

The residual length changes measured after 300 rapid freezing and thawing cycles are given in Table 4. As it can be seen in this Table, all but one mixtures not containing micro-fibers were completely destroyed by frost action long before the completion of the 300 freezing and thawing cycles. On the other hand, all mixtures containing micro-fibers have satisfactory sustained the 300 freezing and thawing cycles since the residual length changes are always lower than 1000 μm/m (except for one mixture having a 1 765 μm/m length change). Table 4 also indicates that the residual length change is generally lower for the mixtures made with carbon fibers as compared with those made with steel fibers. For the mixtures made with steel fibers, the residual length change decreases significantly with the increase of the fiber volume. However, little differences are observed between the mixtures containing 2% and 5% of carbon fibers.

It is clear, of course, that the presence of fibers have significantly improve the frost resistance of the cement pastes. This improvement could be due, at least in part, to the fact that the micro-reinforced pastes have relatively low spacing factors. But an attentive examination of Table 4 reveals that many micro-reinforced pastes are frost resistant even though their spacing factor is quite similar to that of the corresponding unreinforced paste which was completely destroyed by freezing and thawing cycles. Thus, it seems reasonable to conclude that the good frost resistance provided by micro-fibers is not only due to their positive influence on the characteristics of the air-void system but, also, on their ability to limit the propagation of cracks through the cement paste.

Deicer Salt Scaling Resistance

The mass of scaling residues after 50 daily freezing and thawing cycles are given in Table 4, and Figure 2 shows the relationship between the mass of scaling residues and the fiber volume. According to the test results, all the mixtures having a 0.25 W/B ratio suffered very little scaling no matter the volume of fibers simply because, at such low W/B values, the amount of freezable water is almost negligible. But Figure 3 indicates that, for higher W/B ratios, the adding of micro-fibers improves very significantly the scaling resistance of cement pastes since all the mixtures not containing micro-fibers show a poor scaling resistance while all the mixtures containing micro-fibers have performed satisfactory with a mass of scaling residues smaller than 1 kg/m^2. Little differences are observed between steel and carbon micro-fibers and it seems that a fiber volume of 5% provides only little improvement (if any) as compared with a 2% fiber volume.

As mentioned in the previous section, the better scaling resistance provided by micro-fibers can not be solely due to the fact that they generally yields a lower spacing factor value. The positive influence of micro-fibers is probably mainly related on their ability to limit the propagation of cracks. This statement is supported by the fact that the presence of fibers influences very significantly the size of the scaled-off particles. Since it was observed that the size of these particles vary widely from one mixture to another, all the residues were subjected to a sieve analysis and the median diameter of particles was used as an index of the size of the scaling residues. The relationship between the median diameter of scaled-off particles and the fiber volume is shown in Figure 3. This Figure clearly indicates that the median diameter of the scaled-off particles is much smaller for mixtures containing micro-fibers as compared with similar mixtures without fibers. It seems that fibers act as crack arresters and restrain the propagation of cracks over long distances. The differences between steel and

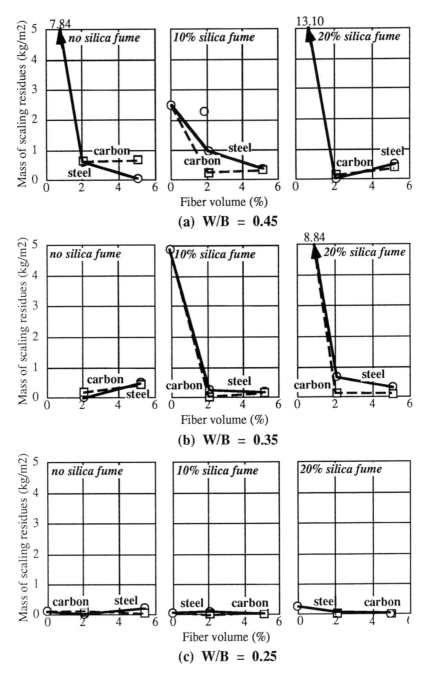

Figure 2   Relationship between the mass of scaling residues and the fiber volume for cement pastes having different water to binder ratios, containing various amounts of silica fume, and reinforced with steel and carbon micro-fibers

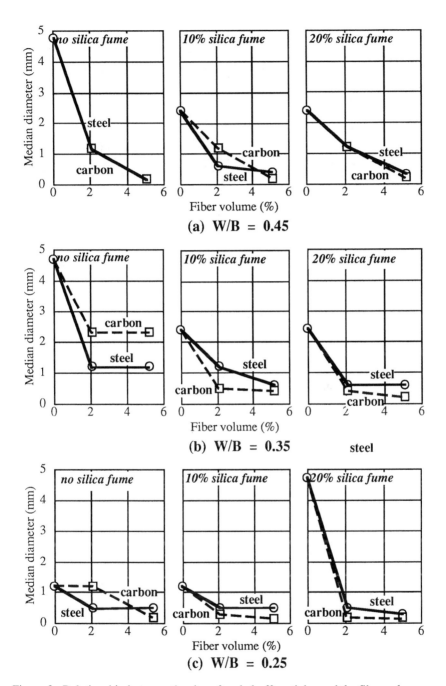

Figure 3   Relationship between the size of scaled-off particles and the fiber volume
for cement pastes having different water to binder ratios, containing various
amounts of silica fume, and reinforced with steel and carbon micro-fibers

carbon micro-fibers are usually small, and a fiber volume of 5% still provides only little improvement (if any) as compared with a 2% fiber volume. Figure 4 also indicates that the median diameter of scaled-off particles generally decreases with the increase of the silica fume content probably because silica fume increases the brittleness of the paste and reduces the critical spacing factor value [10].

CONCLUSION

Recent studies have shown that carbon fiber micro-reinforced mortars can satisfactory sustain the 300 rapid freezing and thawing cycles of the ASTM C 666 test method, even without air entrainment [5-8]. It was suggested, however, that the good frost resistance of micro-reinforced mortars could be simply due to the fact that, for still not well understood reasons, micro-fibers facilitate air-entrainment (even in the absence of an air-entraining admixture) and yield spacing factors values smaller than the critical spacing factor associated to a given mixture [7,8]. The data reported in this paper indicate that both steel and carbon micro-fibers significantly improve, not only the frost resistance against rapid freezing and thawing cycles, but also the deicer salt scaling resistance of micro-reinforced cement pastes. It was found that non air-entrained micro-reinforced cement pastes have relatively low spacing factor values (often below 400 μm) even though their total air content is very low (usually less than 5%). This is due to the fact that those mixtures contain many very small air voids but very few large voids. However, the improvement provided by the micro-fibers can not be solely due to their ability to facilitate air-entrainment because many mixtures containing micro-fibers were frost resistance while similar plain mixtures (without micro-fibers) having similar spacing factors were severely deteriorated by frost action. The positive influence of micro-fibers is most probably due to their ability to limit the propagation of cracks through the cement paste.

ACKNOWLEDGEMENTS

The authors are grateful to the Natural Sciences and Engineering Research Council of Canada for its financial support.

REFERENCES

1. BANTHIA, N., and SHENG, J., "Strength and Toughness of Cement Mortars Reinforced With Micro-Fibers of Carbon, Steel, and Polypropylene", Proceeding of the Second Canadian Symposium on Cement and Concrete, Vancouver, Canada, S. Mindess Editor, 1991.

2. OHAMA, Y., AMANO, M., and ENDO, M., "Properties of Carbon Fiber Reinforced Cement With Silica Fume", Concrete International, Vol. 7, March 1985, pp. 58-62.

3. SAROUSHIAN, P., and MARIKUNTE, S., "Reinforcement of Cement-Based Materials with Cellulose Fibers", Thin Section Fiber Reinforced Concrete and Ferrocement, Publication SP-24, American Concrete Institute, 1990, pp. 99-124.

4. BOISVERT, J., PLEAU, R., and PIGEON, M., "Propriétés Mécaniques des Mortiers Armés de Micro-Fibres d'Acier et de Carbone", Colloque Béton Renforcé de Fibres Métalliques, Béthune, France, 1994, pp.227-236.

5. AKIHAMA, S., SUENAGA, T., and NAKAGAMA, H., "Carbon Fiber Reinforced Concrete", Concrete International, Vol. 10, No. 1, 1988, pp. 40-47

6. SOROUSHIAN, P., NAGI, M., and OKWUEGBU, A., "Freeze-Thaw Durability of Lightweigth Carbon Fiber Reinforced Cement Composites", ACI Materials Journal, Vol. 89, September-October 1992, pp. 491-494.

7. CHEN, P.W., and CHUNG, D.D.L., "Concrete Reinforced with up to 0.2 vol% of Short Carbon Fibers", Composites, Vol. 24, No. 1, 1993, pp. 33-52.

8. PIGEON, M., AZZABI, M., PLEAU, R., and BANTHIA, N., "Durability of Steel and Carbon Fibers Micro-Reinforced Cementitious Composites", accepted for publication in Cement and Concrete Research, 1995.

9. PIERRE, P., PLEAU, R., and PIGEON, M., 1995, "The Flexural Strength of Micro-Reinforced Cementitious Composites: Pastes vs Mortars", Supplementary Papers of the 2nd CANMET/ACI International Symposium on Advances in Concrete Technology, Las Vegas, USA, pp. 451-468.

10. PIGEON, M., and PLEAU, R., "Durability of Concrete in Cold Climates", Chapman & Hall, London, England, 1995, 244 p.

DESIGN OF CONCRETE AND REINFORCED CONCRETE CONSTRUCTION IN AGGRESSIVE ENVIRONMENTS

N V Savitsky

Pridneprovskaya State Academy of Building and Architecture

Ukraine

ABSTRACT. The methodology of reliability calculation of load-bearing concrete and reinforced concrete constructions which are used under influence of corrosion-active environments has been considered. Physical and mathematical models, that are used for analysis in conditions of "environment-load" have been described. The examples of the use of the proposed method are given.

Keywords: Reliability, Durability, Constructions, Aggressive environment, Anti-corrosive protection, Design.

Dr Nickolay V Savitsky is Professor of the Reinforced Concrete and Stone Constructions Departament, Director of the Test Laboratory of Building Materials and Products, Pridneprovskaya State Akademy of Building and Architecture, Dnepropetrovsk, Ukraine. Scientific interests unite fundamental and applied research in the fields of physics-chemical mechanics of materials, technology of the concrete, solid state physics, theory of the reinforced concrete, theory of reliability. He specialises in the fields of diagnosis and estimation of technical state of concrete and reinforced concrete constructions that are in use, of the optimal design of constructive-technological parameters and anticorrossive protection, of repairs and restoring of the damaged reinforced concrete constructions.

Radical Concrete Technology. Edited by R K Dhir and P C Hewlett. Published in 1996 by E & FN Spon, 2–6 Boundary Row, London SE1 8HN, UK. ISBN 0 419 21480 1.

INTRODUCTION

Building code of any country of the world does not contain any quantitative principles for designing of concrete and reinforced concrete constructions in aggressive environments [1, 2]. Optimal design anticorrosive protection on the whole is complex of technological problems is closely connected with tasks of constructive character. Only proceeding from functional requirements to certain construction it is possible to solve as tasks of the durability provision of the construction on the whole as separate elements - concrete and reinforcement.

On the base of the carried out experimental and theoretical researches the method of the anticorrosive protection design of load-bearning concrete and reinforced concrete constructions by parameter of reliability have been worked out [3].

THE FUNDAMENTALS AND PRINCIPLES OF ESTIMATION AND ENSURING OF RELIABILITY

The proposed method is based on the following premises:

1. In accordance with reality it is permitted and accounted the development of corrosive processes in concrete. Calculation and design of constructions in aggressive environments are made taking to the account the concrete corrosive process kinetics.

2. The time is introduced into calculation obviously: construction durability is determinated under given parameters or its parameters are determinated under the given service life.

3. Influence of aggressive environment is estimated by criterion of the functional characteristics change (strength, deformation, crack resistanse and etc.).

4. In accordance with changeable character of concrete, constructive and environmental parameters probabistic method of designing was adopted .

5. The principle of equal reliability of constructions for aggressive and non-aggressive conditions is laid: the constructions that are designed for aggressive environments to the end of service life must have the same reliability as the constructions that are designed for non-aggressive conditions.

The general system approach to the solution of the problems connected with estimation of reliability of concrete and reinforced concrete constructions in aggressive environments includes the following elements:

1. The mathematical models: of the mass-transfering process with consideration of the heterogeneous chemical reactions; of the process of the

damage accumulation in concrete; of the estimation constructions stress-strain state ; probabilistic models of the constructions functioning.

2. Analytical, numerical, numerical-analytical and engineering methods for solutions of the above-mentioned models.

3. Experimental definition: permeability characteristics and chemical activity of concrete in mass-transfering models; physics-mechanical characteristics of concrete as functions of the corrosive damage accumulations; statistical characteristics of changeability of the concrete, constructions, environment parameters.

4. Standardization: of construction's service life; of the aggressive environment parameters (corrosion load); of the force load; of the permitted limits of the constructions functional properties.

MODEL OF CONCRETE AND AGGRESSIVE MEDIUM INTERACTION

The process of interaction of medium and the concrete with consideration of stressed state under definite assumptions has been considered as superimposition of elementary acts of damage accumulation by stress and medium. The calculation of the fields of stress is preceeded by calculation of the moisture, heat, concentration, and corrosion fields. The solution of the system of differential equations of mass-transfering jointly with limit conditions allows to get the most complete data about distribution of these fields.

The basic well-known results of analitical solutions of some problems of mass-transfering, which are of the most practical interest for forecast calculations are the following: the depth of the concrete carbonation, duration of protective action of the concrete under influence of ions-activators, depth of the corrosion damage of the concrete under the influence of the liquid acid aggressive mediums [4].

We offer the physical and mathematical model of concrete corrosion under diffusion, complicated by chemical reactions, which allows to take to the account the main features of the corrosion process: dependence of the speed of the process on the concentration of the components of the aggressive medium and active components of the cement stone; multicomponency of the aggressive medium; parallelism of chemical reactions; dependence of the concrete penetrability on the initial charachteristics of the structure; stressed state, temperature, extent of the development of the corrosion process; dependence of the chemical reaction speed constant on mical-mineralogical and substantial compound of the cement, the type and compound of the attendant ions in the multicomponental system of the aggressive solution and temperature; multimeasureness of the process; possible change of components concentration and of the temperature of the aggressive medium in time according to any law. Parameters which are used in the model (effective coefficient of diffusion, constant of the speed of chemical reaction) are the average values, which determine by integrals thermodynamical and kinetic parameters of the corrosion process.

An equation from the system of equations, which describes diffusion of the aggressive medium active component is given as an example:

$$\frac{\partial C_{a,i}}{\partial t} = \sum_{j=1}^{N} \frac{\partial}{\partial x_j} \left[D_{a,i}\left(S,T,\sigma,C_{ab,k},\ldots\right) \frac{\partial C_{a,i}}{\partial x_j} \right] -$$

$$-k_{i,l}\left(H,T,\sigma,\ldots\right) \prod_{i=1}^{M} C_{a,i}^{\alpha_i} \prod_{l=1}^{L} C_{b,l}^{\beta_l},$$

where, $C_{a,i}$, $C_{a,b}$, $C_{ab,k}$ = correspondingly, concentration i - th active component of agressive medium, l-th active component of concrete, k - th product of interaction

$D_{a,i}(S,T,\sigma, C_{ab,k},\ldots)$ = effective diffusion coefficient i-th active component of agressive medium

$k_{i,l}$ $(H,T,\sigma,\ldots)$, α_i, β_l = correspondingly, constant and exponents of the chemical reaction i-th active component of aggresive medium and l-th active component of concrete

S, H = correspondingly, concrete strustural parameters and characteristics of chemical activity

T = temperature

σ = stress

x_j = spatial coordinates

t = time

The numerical solution of the system of differential equation of mass-transfering jointly with limit conditions allows to get the fields' distribution of aggressive medium, active components of concrete, products of the interaction in concrete with which the change of concrete physical and mechanical characteristics are connected.

CALCULATION MODELS OF REINFORCED CONCRETE ELEMENTS FOR DETERMINATION OF THE STRESS-STRAIN STATE

The perspective direction in development of the methods of calculation of concrete and reinforced concrete constructions under conditions by complex actions of load and aggressive mediums is using model notions of constructive elements and equations of mechanical state of the materials, what in formalistic type reflect a result of environment action.

Stress-strain state of the reinforced concrete elements on all stages during development of the process of concrete degradation and on any level and regime of loading is defined using the conditions of equilibrium. It is posible, using model   notions   to   evaluate   strength,   crack-resistance   and deformability of the elements from unified position.

As an example, the bending reinforced concrete elements, working in sulphate solutions, were examined in order to study the efficiency as technological so constructive parameters of reinforced concrete protection that provide durability of normal cross-section strength. The influence of the medium was from   compressed   side   of   cross-section.   As   variable technological parameters the chemical and mineralogical compound of the cement (sulphate resistant cement and portland cement containing C_3A and C_3S, respectively, 5 and 50 %, 7 and 65%), penetratability of concrete, characterised by water resistance grade of concrete (W4, W6, W8) or water-cement ratio (W/C-0.60, 0.55, 0.45) were considered. As variable construction parameters the height of rectangular cross-section (h = 10, 30, 50 cm), grade of reinforcing (coefficient of reinforcing μ = 0.45, 1.43%) were considered. Concentration of sulphate-ions was changing in limits of 1 to 20 grams / liter (evaluated in SO_4^{2-})

The forecast of strength change of normal cross-section   was carried out using   developed   models and   experimental   data. Relative   change   of strength was determined for the comparison of the results. Some results of calculations are displayed on fig.1. The analisis of the results witnesses that equal extent of concrete corrosive damage (under equal concentration of sulphate-ions and technological parameters) is more essential for strength change of elements with the height of cross-section equal to 10 cm and   for elements with the heightened content of reinforcement.

CALCULATION OF RELIABILITY AND DURABILITY OF REINFORCED   CONCRETE CONSTRUCTIONS,   INTERACTING WITH THE AGGRESSIVE   MEDIUM

It is known that indices which characterise the corrosive strength of concrete, the   force parameters (value of prestressing) and geometrical parameters of construction, the external influences such as medium and load are   of random nature. Therefore probability   forecast   is   more theoretically substantiated in comparison with determinated forecast for one or another property change of construction which determines durability.

The problem of construction reliability   determination according to some property comes to creation of m-dimensional function which shows density of   random   process   distribution . The   reliability   is   determined   by m-dimensional   integral   from   the   mentioned   function   of   density

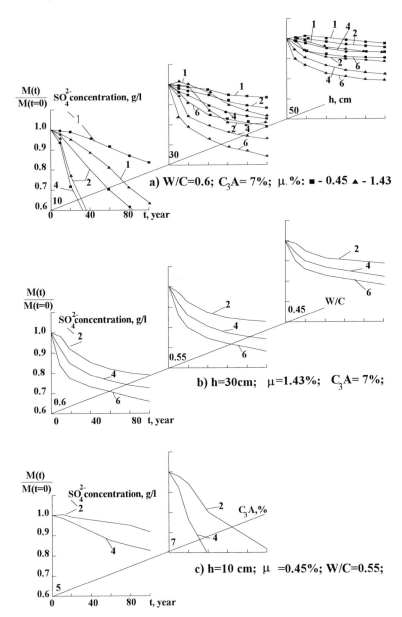

a) W/C=0.6; C$_3$A= 7%; μ,%: ■ - 0.45 ▲ - 1.43

b) h=30cm; μ=1.43%; C$_3$A= 7%;

c) h=10 cm; μ =0.45%; W/C=0.55;

Figure 1 Influence of constructive and technological parameters on change of bending reinforced concrete element strength under action of sulphate solutions

distribution. We offered a numerical-analitical method for creating of complex functions distribution .   Service life is determined by duration of construction operation until their properties leave permitted limits.

As an example the results of determining the concrete protective coat reliability under conditions of influence by acid HCl concentration 0.01 g/l are given . The parameters of the concrete are the following : cement expense - 300 kg/m^3, with the variability coefficient equal to 0.20, relative contents of calcium oxyde in cement - 0.641 under variability coefficient equal to 0.02. The acid concentration variability coefficient is equal to 0.15. The thickness of the protective coat is equal to 30 mm with the variability coefficient equal to 0.185. Reliability of protective coat concrete was determined according to the criterion of its destruction danger. The results of calculation are shown on fig. 2.

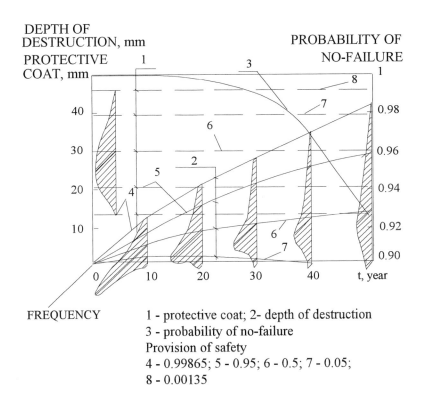

1 - protective coat; 2- depth of destruction
3 - probability of no-failure
Provision of safety
4 - 0.99865; 5 - 0.95; 6 - 0.5; 7 - 0.05;
8 - 0.00135

Figure 2 Reliability of concrete protective coat under
action of acid

PRACTICAL APPLICATION OF RESULTS

During design of construction in aggressive environments usually such tasks are solving:

1. To appoint technological and constructive parameters that provide determinated service life.

2. To define the costruction service life under determinated parameters and external conditions.

3. To evaluate corrosive danger of aggressive medium for construction with determinated parameters.

On the base of the offered method there is an opportunity to solve this tasks. The propositions for building code on problem solving connected with providing of concrete and reinforced concrete reliability in conditions under action of aggressive medium at three levels of design: research, engineering, and prescriptional have been proposed [5].

As an example the task of corrosive danger evaluation of sulphate solutions for bending reinforced concrete elements have been solved at prescriptional level of design. There were adopted the following premises: corrosive danger of sulphate was evaluated by changing (reduction) of normal cross-section strenght for slight degree of severity - not more than 5%, for moderate degree - 5-15%, high degree - 15-25% for the service life equal 100 years. Water resistance grade of concrete equal W4 (water-cement ratio W/C 0.6). Table 1 gives limiting values of deleterious sulphate (SO_4^{2-} , mg/litre) in water of predominantly natural composition for the assessment of the severity of chemical attack accordingly to CEB-FIB Model Code [1], Building Code [2] and by offered method at prescriptional level of design.

Thus there is an opportunity to evaluate corrosive danger of a medium not only for concrete or reinforcing (the way how it is still being done) but immediately for concrete and reinforced concrete constructions according to the functioning requirements to these constructions. Besides the technologycal parameters (chemical, mineralogical and substantial compound of the cement, penetratability of concrete and etc.) as means of concrete anti-corrosive protection the construction parameters (shape and dimensions of cross-sections, extent of reinforcing, value of prestressing, mechanical characteristics of concrete and reinforcement, schemes of element bearing, schemes of load and aggressive mediun application and etc.) are used.

Alternative design of anticorrosive protection of the some concrete and reinforced concrete constructions with account of concrete corrossive process kinetics were made. So we managed to abolish the secondary protection such as pasted insulation of foundation surfaces and retaining walls under action of extremely aggressive sulphate solution and also of enclosing constructions of vegetable storehouse with the regulated gas medium. By this means a great economical benefit is accomplished.

Table 1   Limiting values of deleterious sulphate (SO_4^{2-} , mg/litre) in water of predominantly natural composition for the assessment of the severity of chemical attack

Cement	Concrete		Reinforced concrete			Degree of Severity
	MC*	BC**	height of cross-section, cm	coefficient of reinforcing, %		
				0.45	1.43	
	200-600					slight
	600-3000					moderate
	>3000					high
Portland cement $C_3A=7\%$ $C_3S=65\%$	250-500	10	250-600	250-350	slight
	500-1000		600-1000	350-550		moderate
	>1000		1000-1500	550-700		high
		30	250-800	250-400		slight
			800-3500	400-1200		moderate
			3500-8000	1200-3000		high
		50	800-2000	400-800		slight
			2000-9000	800-3000		moderate
			9000-20000	3000-8000		high
Sulphate resistant cement $C_3A=5\%$ $C_3S=50\%$	1500-3000	10	800-1300	400-600	slight
	3000-4000		1300-2500	600-1300		moderate
	>4000		2500-4200	1300-1600		high
		30	1300-3000	600-1600		slight
			3000-12000	1600-5000		moderate
			12000-25000	5000-12000		high
		50	3000-6000	1600-2000		slight
			6000-25000	2000-10000		moderate
			25000-50000	10000-25000		high

* - [1], ** - [2]

CONCLUSIONS

1. The general system approach for solving the problems connected with the estimation of reliability of load-bearing concrete and reinforced concrete constructions in aggressive mediums has been developed.

2. Basic statements and principles of calculation and design of constructions aimed for work in conditions of aggressive mediums have been offered.

3. Physical and mathematical models of concrete corrosion under diffusion, complicated by chemical reactions, which afford to take in to account the most important features of corrosion process, have been offered.

4. The methods of calculation of stress-strain state of shaft reinforced concrete elmenents on the basis of model assumptions and equations of mechanical state of materials have been developed.

5. The results of quantity modeling approved the hypothesis about the essential influence of constructive parameters as means of the primary protection of reinforced concrete elements in conditions of corrosion process development in concrete.

REFERENCES

1.   CEB-FIP MODEL CODE 1990. First Draft. CEB Bulletin D'information No. 195. - 1990.

2.   BUILDING CODE 2.03.11 - 85. Anticorrosive protection of building costructions: Gosstroy of the USSR, Moscow,1986, 48p.(in Russian).

3.   SAVITSKY, N V. Fundamentals of reinforced concrete reliability calculation in aggressive mediums . Thesis for a doctor's degree. Dnepropetrovsk Civil Eng. Inst. - NIIZB, 1994, 400 p. (in Russian).

4.   ALEKSEEV, S N, IVANOV, F M, MODRY S , SHISSL P. Durability of reinforced concrete in aggressive environments, Moscow, Stroyizdat, 1990, 320p. (in Russian).

5.   GUZEEV, E A, ALEKSEEV, S N, SAVYTSKY, N V. Consideration of aggressive influence in Building Code. Concrete and Reinforced Concrete, 1992, No.10, Pp 8-10 (in Russian).

INFLUENCES OF STEAM CURING ON STRENGTH, SHRINKAGE AND CREEP OF OPC AND SLAG CONCRETES

M Boukendakdji

University of Blida

Algeria

J J Brooks

P J Wainwright

University of Leeds

UK

ABSTRACT. Compressive strength, shrinkage and creep of water-cured and steam-cured concrete with and without slags have been compared. Two slag replacement levels of 50 and 70% using slags from two countries (U.K. and South Africa). Creep was measured under a stress of 0.2 of the creep cylinder strength at the age of 14 days for water-cured concretes and 1 day for steam-cured concretes. The influence of steam-curing is to reduce long-term strength, ultimate drying shrinkage, basic and total creep of all types of concrete and to increase short-term strength. The ultimate drying shrinkage of steam-cured slag concretes were similar to or lower than that of PC concrete. For steam cured concrete, the effect of replacement of Portland cement by slag is to decrease the ultimate basic and total creep.

Keywords: Compressive strength, Shrinkage, Basic creep, Total creep, Slags, Steam-curing, Water-curing.

Dr BOUKENDAKDJI M. is a lecturer at the Department of Civil Engineering, University of Blida (Algeria). He obtained an MSc in Construction Engineering and a PhD from the University of Leeds.

Dr BROOKS J J, is a Senior Lecturer at the Department of Civil Engineering, University of Leeds, UK. His main research interests are long-term deformations of concrete and masonry. He is also a co-author of the text books: 'Creep of Plain and Structural Concrete' and 'Concrete Technology'. Dr Brooks is currently Chairman of a sub-committee dealing with Creep of concrete containing special ingredients for ACI Committee 209.

Dr WAINWRIGHT P J. He is a Senior Lecturer at the Department of Civil Engineering, University of Leeds. His main research interests included the use of waste materials and by-product materials in concrete, thermal cracking and durability aspects of concrete, He is a member of the RILEM Committee on Sustainable Application of Mineral Raw Materials in Construction and of the CEN Committee on Recycled Aggregates.

Radical Concrete Technology. Edited by R K Dhir and P C Hewlett. Published in 1996 by E & FN Spon, 2–6 Boundary Row, London SE1 8HN, UK. ISBN 0 419 21480 1.

INTRODUCTION

Whereas there is much published data available on the long-term deformations of concretes cured under normal conditions, there is a limited amount of available data concerning concrete cured under steam or accelerated curing conditions, and there is no known data for slag concrete. For the purpose of prestressed concrete design, creep and shrinkage account for the larger portion of prestress loss. Since most fabrication plants utilize accelerated strength concrete there is a need to have more information about drying shrinkage and creep of this type of concrete.

Two types of creep are considered. Basic creep applies to sealed or water-stored concrete, and total creep refers to drying concrete after allowing for drying shrinkage. Creep at a constant initial stress-strength ratio is used since it allows for any change in strength due to replacement of cement by slag, or for a change in curing type.

The effect of steam-curing was found to reduce creep up to 50% compared with moist-cured concrete [1, 2]. In general, the effect of steam-curing is to decrease shrinkage [2].

This paper deals mainly with strength and long-term deformations of slag concrete cured under steam, a more results of water-cured slag concrete having been reported previously [3]

EXPERIMENTAL DETAILS

The materials used to make five concretes were ordinary Portland cement (PC), ground granulated blastfurnace slag and quartzitic aggregate, conforming to the 10 mm. coarse and F category fine grades of B.S.882: Part 2: 1983. Slags were obtained from the U.K. and South Africa.

The first concrete was a PC control mix having mass proportions of 1:1.62:2.49 with a water/cement ratio of 0.43 and the cement content was 437 kg/m^3. The four slag concretes had 50 and 70% of the mass of PC replaced by equal masses of slags, with the same aggregated proportions and water/cementitious ratio as the control concrete.

For each type of concrete, 100 mm cubes were cast for compressive strength together with 255x76 mm dia. cylinders for strength, shrinkage and creep in compression. For water-cured concretes, all moulded specimens were cured for 24 hours under wet hessian, then demoulded and cured in water for 13 days. The steam curing cycle is shown in Figure 1.

Strength and time dependent deformations were measured in two storages environments : in water at 20 ± 2 °C, and in air at 65 ± 5% R.H. and 20 ± 2 °C . Creep and shrinkage were measured as previously [4]. Creep was measured under a stress of 0.2 of the creep cylinder strength at the age of 14 days for water cured concretes and 1 day for steam cured concretes.

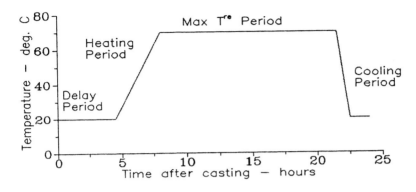

Figure 1 Steam curing cycle

TEST RESULTS AND DISCUSSION

Compressive Strength

Table 1 lists the 100-mm cube strength for water-cured and steam-cured concretes stored in the two environments. Compressive strength at 1 day is inversely proportional to slag content, as the higher the slag content the lower the compressive strength. This observation is similar to that of compressive strength of water-cured concrete at the age of 7 days.

Approximately 67% to 78% of the 28-day compressive strength of water-cured concrete at 20 °C was achieved after 1 day of steam curing. At the age of 28 days, the steam-cured PC and slag concretes were weaker than the continuously water-cured concrete; This difference is greater the higher the slag content and is greater in the case of S.A.slag concretes compared to U.K.slag concretes.

As the object of steam curing is to obtain a sufficiently high early strength, compared with that cured at room temperature, without an appreciable loss of gain in strength at later ages, it seems that from these results concrete containing 50% slag content can be cured successfully under steam. On the other hand, it seems that the cycle of curing adopted in this test programme may not be very suitable to concrete containing a higher slag content. For high slag content mixes, a delay period of longer than 4.5 hours is suggested since, according to Price [5], the temperature applied before setting may cause a decrease in strength at later ages compared with that stored at lower temperature and, generally, the initial and final setting times increase with the addition of slag [6].

As for 14-day water-cured concrete, specimens stored dry after 1-day of steam curing were generally stronger than those stored wet (Table 1).

Table 1: Development of cube compressive strength (MPa)

Curing	Environment	Age	PC	UK Slag		SA Slag	
				50%	70%	50%	70%
		7	45.5	38.5	38.5	37.5	23.6
		14	56.0	53.0	48.5	48.0	40.0
	Stored	28	62.0	67.5	62.5	.63.0	54.0
	in	56	63.0	72.0	72.5	76.0	62.0
Water	water	84	68.3	75.5	70.8	77.0	68.8
cured		112	64.0	76.5	81.5	79.5	72.6
		28	67.5	68.5	62.5	68.5	57.5
	Stored	56	69.0	80.0	80.0	77.5	65.5
	in	84	75.5	76.5	72.5	78.5	70.0
	air	112	72.0	82.5	85.8	83.0	76.0
	Stored	1	49.0	42.0	39.6	44.0	35.1
Steam	in	28	58.1	54.2	48.3	49.4	38.5
cured	water	71	63.7	58.6	57.3	56.3	42.0
	Stored in	28	68.3	58.7	53.5	55.5	42.0
	air	71	73.6	63.8	59.3	61.8	46.6

The ultimate compressive strength of steam-cured concrete is not as great as that of concrete continuously moist cured at lower temperature; however, in actual practice concrete is often given very little moist curing so that the advantage of steam curing may be considerably greater than would be apparent from comparison with 28 day moist curing. The improvement of early strength of steam cured slag concrete is an importent attribute in the manufacture of precast and prestressed components.

Shrinkage

The shrinkage of water-cured and steam-cured PC and slag cement concretes are shown in Figure 2. It can be observed that the initial rate of shrinkage is higher for slag-cement concrete, but the later-age rate is less.

Extrapolation of the regression curves on the hyperbolic expression indicates that the ultimate drying shrinkage values of steam-cured slag concrete are similar to or lower than that of PC concrete (Table 2), the maximum decrease being about 38%, for the 70% S.A.slag concrete

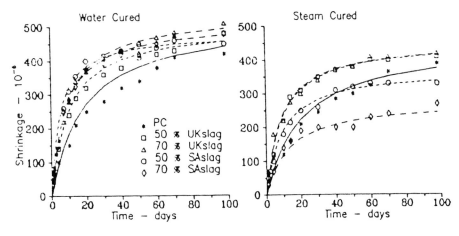

Figure 2 Shrinkage of PC and slag-cement concretes

In fact, the effect of steam-curing was to decrease the ultimate drying shrinkage of PC and slag concretes by about 14% and 23%, respectively. Steam curing reduces long-term shrinkage probably because of a lower hardened cement paste content due to the acceleration of hydration at a higher temperature.

Table 2: Ultimate shrinkage, basic creep and total creep (10^{-6})

Curing		PC	UK Slag		SA Slag	
			50%	70%	50%	70%
Water	Shrinkage	526	505	537	480	515
cured	Basic Creep	267	151	220	236	215
	Total Creep	530	480	396	366	352
Steam	Shrinkage	454	456	454	367	278
cured	Basic Creep	213	163	147	141	139
	Total Creep	397	271	319	257	283

Basic Creep

The creep-time characteristics of water-cured and steam-cured PC and slag concretes are shown in Figure 3. The 70-day basic creep of steam-cured slag concretes are similar to or lower than that of PC concrete. The maximum decrease is approximately 51% for the 70% S.A.slag concrete.

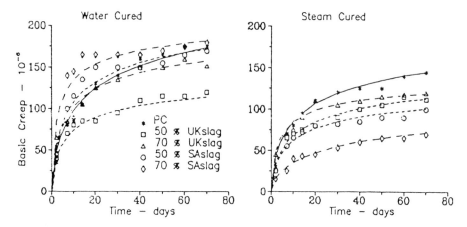

Figure 3 Basic creep of PC and slag-cement concretes

The effect of steam curing is to decrease the ultimate basic creep of all concretes, obtained by regression on a hyperbolic-power expression (Table 2). For PC concrete, this can be explained by the increase in strength under load, which was smaller in the case of water-cured PC concrete (Table 1).

The lower basic creep of steam-cured slag concrete cannot be explained in terms of the increase in strength, since it was similar or higher in the case of water-cured concrete. As for drying shrinkage, this decrease in basic creep may be attributed to the lower cement paste content due to the accelerated hydration in the case of steam-cured concrete. This suggestion of decrease in the volume of cement paste content is indicated by the results of strengths (Table 1), since the values of steam-cured concrete were lower.

Total Creep

As for water-cured concrete (Figure 4), the 70-day total creep of steam-cured slag concrete was similar to or lower than that of PC concrete.

As for steam-cured concrete stored in water, the effect of steam curing was to decrease the ultimate total creep by up to 45% (Table 2). This reduction is probably partly because of the decrease in shrinkage of the steam-cured concrete and, therefore, reduction in drying creep and total creep, and partly because of the lower volume of the cement paste content in the case of steam-cured slag concrete.

This reduction of creep of steam-cured slag concrete could be of practical importance in situations where creep is to be minimised, for example, in prestressed concrete members.

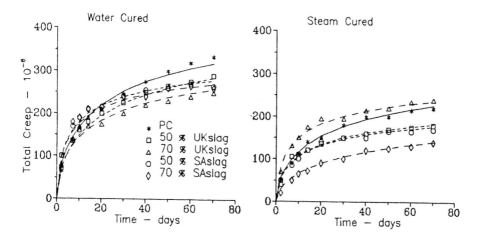

Figure 4 Total creep of PC and slag-cement concretes

CONCLUSIONS

1. The effect of steam curing is to reduce long-term strength, ultimate drying shrinkage, basic and total creep of all types of concrete.

2. It seems that concrete containing 50% slag content can be cured successfully under steam. However, for a high slag content, a delay period of longer than 4.5 hours is suggested.

3. For steam-cured concrete, the ultimate drying shrinkage of slag concretes were similar to or lower than that of PC concrete.

4. The effect of replacement of Portland cement by slag is to decrease the ultimate basic and total creep.

REFERENCES

1. AMERICAN CONCRETE INSTITUTE COMMITTEE. Pressure steam curing, ACI Journal, Vol 60, No 8, August 1963, pp 953-986.

2. BROOKS, J J, WAINWRIGHT, P J AND NEVILLE, A M. Time dependent behaviour of high-early strength concrete containing a superplasticizer. ACI Special Publication, No 68, 1981, pp 81-100.

3. BROOKS, J J, WAINWRIGHT, P J AND BOUKENDAKDJI, M. Influences of slag type and replacement level on strength, elasticity, shrinkage and creep of concrete. Fourth Inter. Conf. on the use of Fly Ash, Silica Fume, Slag and Natural Pozzolans in Concrete, Vol 2, Istambul, Turky, 1992, ACI Spec. Publ., pp 1325-1341.

4. NEVILLE, A M and LISZKA, W Z. Accelerated determination of creep of lightweight aggregate concrete. Civil Engineering and Public Works Review, London, 68, No 803, June 1973, pp 515-519.

5. PRICE, W H. Factors influencing concrete strength. ACI Journal, Vol 47, Feb. 1951, pp 417-432.

6. WAINWRIGHT, P J. Properties of fresh and hardened concrete incorporating slag cements, in Cement Replacement Materials. Editor SWAMY, R N. Concrete Technology Design, Vol 3, Surry University Press, 1986, pp 100-133.

HARDENING AND SPLITTING OF CONCRETE UNDER THE ACTION OF MICROWAVE ENERGY

V S Mkrttchian

Armenian State University of Engineering

Armenia

S V Mkrtchian

University of Hildesheim

Germany

ABSTRACT. This article deals with the total combination of theoretical and technical tasks in creating plant and the automatic systems of control for using microwave energy in hardening and splitting of concrete. The application of microwave energy gives an opportunity of generating heat inside the concrete in stead of usual heat feeding into the wave by means of heat conductivity. It is fixed that if the heating realized by microwave energy is localized the heated area of concrete dilates in comparison with the remaining part of it where rises high stretching tension causing cracks.

Keywords: Hardening, Splitting, Microwave Energy, Sliding Mode, Discontinuous Control and Corrective Devices (DC & CD).

Professor Vardan S. Mkrttchian is the President of Armenian Automatic Control Council, State University Engineering of Yerevan, Armenia. He specializes in application of microwave energy at radical concrete technology and automatic systems with Discontinuous Control and Corrective Devices for thermo-damp process of concrete production.

Mr Souren V. Mkrtchian is a Student in University of Hildesheim, Germany. He is specialized in the modeling of concrete in thermo-damp processes.

Radical Concrete Technology. Edited by R K Dhir and P C Hewlett. Published in 1996 by E & FN Spon, 2–6 Boundary Row, London SE1 8HN, UK. ISBN 0 419 21480 1.

INTRODUCTION

The usual process of concrete hardening in which the ready ware acquires strength is effected in normal conditions of the environment. In this case the process goes on very slowly 28 days. For the purpose of accelerating the process of hardening the ware is processed with steam under atmospheric pressure - the time of hardening is reduced to 4 hours, however, in this case the final strength of concrete.

At the Armenian State University of Engineering and University of Hildesheim (Germany) the authors (of the article) carry out research on microwave energy application for the acceleration of the concrete hardening process. The application of microwave energy gives an opportunity of generating heat inside the material instead of usual heat feeding into the ware by means of heat conductivity.

In the meantime the authors have carried out research on splitting of concrete ware and concrete coating by means of microwave energy. It is fixed that if the heating realized by microwave energy is localized the heated area of concrete dilates in comparison with the remaining part of it where rises high stretching tension causing cracks.

This paper summarizes a modeling and practical experiment in hardening and splitting of concrete under action of microwave energy and presents the underlying principles.

MATHEMATICAL MODELING OF HARDENING
Source equation.
The heat energy, influencing over temperature mode generation in concrete at hardening got by the influence of microwave energy, is added from simultaneous affect to it by heat energy passed from microwave radiation and from energy emanated from because of exothermic. The source equation of energy balance will be [1]:

$$q_{absorb} + q_e \pm q_{ak} - q_{evap} - q_{ef} = 0 \qquad (1)$$

Theoretical calculation.
In the conditions of simultaneous emanation of microwave energy to hardening concrete the quantity of energy received by concrete is calculated from the quantity of absorbed radiation energy (q_{absorb}) and from the heat lost to environment (q_{env}):

$$q_{warm} = q_{absorb} - q_{env} \qquad (2)$$
$$q_{absorb} = q_{whole} * \alpha_B \qquad (3)$$
when microwave energy is supplied in free space (at every side).
$$q_{absorb} = q_{whole} * \alpha_B \tau_1 / (1 - (1-\alpha_B)\varphi_d) \qquad (4)$$
at single-directional radiation
$$q_{absorb} = q_{whole} * \alpha_B \tau_1 \tau_2 (1-\varphi_1\varphi_2)/(1-(1-\alpha_B)\varphi_d) \qquad (5)$$

at bi-directional radiation.

The mechanism of heat loses mostly is due to processes of evaporation of moisture and also is due to convective and radiant heat exchange

$$q_{env} = q_{evap} + q_{conv} + q_{rad} \qquad (6)$$

besides each component of (6) is defined as follows:

$$q_{evap} = r\beta_c(P_p - P_s) \qquad (7)$$
$$q_{conv} = \alpha_B(t_{sur} - t_s) \qquad (8)$$
$$q_{rad} = \alpha_{conv}(t_{sur} - t_s) \qquad (9)$$

The values of the coefficients α and β are defined in the following formulae:

$$\beta_c = 8,5 * 10^{-6} M_m t_s^{1,95} V^{0,9}(t_s - t_{max}) * 0,135 / RL^{0,1} v^{0,57} t_{sur}^2 \qquad (10)$$
$$\alpha_1 = 5\lambda/L(L^3 gPr/v^2)^{0,104} * (1 - t_{sur}(P_s - 0,378 * P_0)/t_0(P_s - 0,378 * P_p))^{0,104} \qquad (11)$$
$$\alpha_B = \theta_B \sigma(t_{sur}^2 + t_s^2)(t_{sur} + t_s) \qquad (12)$$
$$\alpha_{conv} = 3,95 (V/L)^{0,5} \qquad (13)$$
$$\alpha_{env} = 0,027 \lambda Pr^{0,38}/L^{0,1} * (V/v)^{0,9} * t_s^{1,87}(t_s - t_{max})^{0,135}/t_{sur}^2 \qquad (14)$$

Based on (2) - (14) equations there have been made theoretical calculation of heat exchange components for three considered cases of the hardening of concrete. Here are the results showed in the table:

TABLE 1. Heat exchange of hardening concrete

Compone nt of heat exchange	Time, hour					
	9 hours			15 hours		
	from every side	single directio nal	bi directio nal	from every side	single directio nal	bi directional
q_{absorb}	528	459	402	528	459	402
q_{env}	-42	-40	-20	312	173	79
q_{warm}	570	499	382	216	286	329

From these calculations follows that heat loses from the surface of concrete are the lowest at dual-directional radiation that is considered in designing of the whole drying device.

THEORY OF SPLITTING

After drying up of the material this component decreases very much, but never equals to zero so that this material always has nonzero lost and an absorption of microwave energy is taken place with emanation of heat. If the warming is localized then the warmed part of concrete is enlarged relatively to the neighbor parts and big pushing apart stress is being raised in them bringing to cracking of concrete. The necessary power for cracking of concrete is calculated according to the known formula.

$$P_a/E_0^2 = 0,0555*10^{-9}f\,\varepsilon''tg\sigma_1K^2 \tag{15}$$

EXPERIMENTAL DETAILS

Equipment. The process of generation, transmission and distribution of microwave energy is realized at the laboratory of Armenian State University of Engineering. The laboratory has conditions for work power system. A system comprising installations of generators, transformers, switchgear, lines, accessories and structures, is used for the generation, conversion, transformation, transmission and distribution of microwave energy. The concrete was exposed to warming in two ways. In the first case the concrete was radiated by movable horn. In the second case the concrete plate was placed into stove operating as cavity resonator. In each case the environment pressure in which the material is placed must be equal to atmosphere or even more than atmosphere. In the place of resonator it is used a manufactured in former USSR stove "Electronica" SP23 ZIL which has its maximum power in its output equal to 10 kVt.

The material concrete was carefully mixed and vibroconcentrated after flood into form. Usually concrete contained cement locally manufactured and marked to 400, volcanic white sand and filler (local road metal) in the proportion of 1:3:5 and the relation of water to cement was equal to 0,66.

Results. The results of the experiments proved that the durability of the samples having passed the usual hardening process within 24 hours (see table 2).

TABLE 2 The concrete hardening process at microwave energy warming in comparison with usual hardening process.

The number of experiment	The time of hardening (hour)	The strength of compression kg/sm^2	
		Hardening under the action of microwave energy	Common process of hardening, 28 days
1	2	430	440
2	2	432	441
3	2,5	439	443
4	2,5	440	443,5
5	3	450	452
6	3	452	454
7	4	460	462
8	4	462	463

The experiments were realized at the frequency of 2.9 Hz. Also there have been made checking of temperature in the drying concrete during the whole process of hardening in the different places of concrete. Samples after warming during 2-4 hours were not been processed in the future but were been dried naturally.

THE PROCESSING MODE, CONTROL SYSTEM AND TECHNICAL SOLUTIONS

The author's experiments resulted in creating an apparatus for realizing microwave irradiation of concrete, getting the conditions of reprocessing, making us the concrete hardening process optimization algorithm based on methods of sliding mode. The author's research and research of his pupils showed in the class of control system that is the usage of automatic systems with discontinuous control and corrective devices (DC & CD) [2, 3, 4]. The path of the state vector in sliding modes belongs to majority of dimensions; less than the space of conditions and the differential. Equations describing movement in sliding mode correspondingly are for an exponent lower than differential equations of automation system of hardening and splitting of concrete in thermo-damp processes. Besides the movement in sliding mode doesn't depend on control and is defined by the properties of the object and equations of surfaces of the break. In some conditions when the sliding modes are deliberately put into they are invariant to influences and change dynamic properties of control object - equipment in which the process of hardening and splitting of concrete is taken place. A special device for breaking concrete into pieces by horning irradiation in created by the author. A special system for operator's protection from microwave energy is created as well. The authors have also created a unique device for parting the broken pieces of concrete attached together by reinforcement.

CONCLUSIONS

1. Under the influence of microwave energy concrete becomes more firm and the hardening time is shortened.

2. Splitting of concrete plates under the influence of microwave energy makes this difficult process more safe and saves environment from pollution.

3. The processes of hardening and splitting can be made more optimal by controlling in sliding mode using the technique of discontinuous control and corrective devices (DC & CD).

4. Received practical results and technical solutions were successfully used by authors and their companions in arms during the elimination of Spitak earthquake legacy in Leninakan.

REFERENCES

1. HALMOS E. E. York harnesses sun to cure concrete block, Concrete Products, Vol. 82, No. 1, 1979.
2. MKRTTCHIAN V. S. Concrete flagstone in high frequent electronic field. Mag. Information of Higher Schools, 1988, Vol. 3, No. 184, March, pp. 85-90.
3. MKRTTCHIAN V. S. Hardening and Splitting of Concrete under the action of Microwave Energy. Conference Documentation, The Concrete Future, Kuala Lumpur, Malaysia, ISBN uf 981-00-4915-3, p. 127.
4. MKRTTCHAN V. S. Microprocessor control of heat and process in sliding modes. Agropromizdat, Moscow, 1987, p. 182 (in Russian).

CONDITIONAL MEANINGS

q_{absorb} - warm stream, absorbed by body, W/m^2;

q_e - warm stream, yielded because of hydratation of cement, W/m^2;

q_{ak} - warm stream, accumulated by body, W/m^2;

q_{evap} - warm stream, used for evaporation of moisture, W/m^2;

q_{ef} - effective warm stream, radiated from warmed body, W/m^2;

q_{warm} - warm stream going for warming the heat recipient, W/m^2;

q_{env} - warm stream going to environment through the super transparent fence or from concrete surface, W/m^2;

q_{whole} - whole sun energy, W/m^2;

q_{conv} - warm stream passed from the surface of convection, W/m^2;

q_{rad} - warm stream passed by radiation, W/m^2;

α_B - the coefficient of radiant heat exchange of concrete and environment, $W/(m^2*{}^0C)$;

α_{conv} - coefficient of convective heat exchange on fence surface, $W(m^2 * {}^0C)$;

τ_1 - ability to handle film or sheet material of the first layer;

τ_2 ability to handle film or sheet material of the second layer;

φ_d - diffusional reflection ability of fence;

φ_1 - reflectional ability of the first layer of fence;

v - coefficient of kinematic viscidity, m^2/sec;

λ - warm throughput of air, $W(m * {}^0C)$;

g - gravity acceleration, m/sec^2;

θ_B - black degree of concrete surface;

σ - Stephen- Boltsman constant, $5{,}67 * 10^{-8}$ $W/(m * {}^0C)$;

E_0 - electrical field stress, W/m;

f - microwave radiation frequency, Hs;

ε - concrete dielectric perspicacity;

$tg\sigma$ - tangent of angle of concrete dielectric loses.

CURING OF CONCRETE.
EFFECT ON STRENGTH AND DURABILITY

Per-Erik Petersson

SP Building Technology

Sweden

ABSTRACT. The influence of curing conditions on compressive strength and carbonation rate has been studied in a field exposure test on two OPC concrete qualities. The influence of parameters such as the water-cement ratio, the time until form removal and the exposure conditions were studied. The results from the tests show that the difference between different curing conditions becomes more pronounced for dry than for humid climatic conditions. This means that laboratory tests cannot always be used to predict concrete behaviour in real structures. Field exposure tests are to be preferred. The results also indicate that wet curing produces somewhat higher strength than curing with plastic film while the two curing methods seems equally efficient for producing good durability. As expected, no protection at all after casting gives much poorer results.

The results presented in this report are primarily relevant for Swedish climatic condition, and the conclusions may be different for field exposure tests carried out in other countries.

Keywords: Concrete, Curing, Compressive strength, Durability, Carbonation, Field exposure test

Dr Per-Erik Petersson is head of the Department of Building Technology at the National Testing and Research Institute in Sweden. His main interest today is in durability of concrete, especially within the two areas of frost resistance and curing. He has also published a number of reports and papers in the field of fracture mechanics of concrete.

Radical Concrete Technology. Edited by R K Dhir and P C Hewlett. Published in 1996 by E & FN Spon, 2–6 Boundary Row, London SE1 8HN, UK. ISBN 0 419 21480 1.

INTRODUCTION

The influence of curing conditions on the properties of concrete has been studied in many investigations [1,2,3,4,5,6,7]. However, most of the investigations have been performed under closely-controlled laboratory conditions and very few results have been reported for field exposure tests on specimens that have been cured under different curing conditions. This means that most tests have been performed under relatively dry and warm conditions, at least when compared with the climatic conditions in Sweden and other parts of northern Europe.

In order to study the influence of real climatic conditions on the results, extensive field exposure tests have been performed at the Swedish National Testing and Research Institute. Specimens cured by different methods were placed in three different climates; indoors, outdoors protected from rain or outdoors exposed to rain. This paper presents some of the results from the investigation: a more detailed presentation is given in [8].

EXPERIMENTAL DETAILS

Concrete qualities

This paper presents results for two concrete qualities with water/cement ratios of 0.34 and 0.50. A Swedish Ordinary Portland cement (Degerhamn std P, CEM I 42.5) was used. This cement has a low alkali content (eq 0.5% Na_2O), a low content of C_3A (about 2%) and a low value of heat of hydration [9]. Natural gravel was used with a maximum aggregate particle size of 16 mm, the slump was 100 ± 10 mm and the air content was $5.5\pm0.5\%$ for both mixes. The air entraining agent used was a neutralised Vinsol resin and a melamine-based plasticizing agent was used for the concrete with the lowest water-cement ratio.

Mixing and casting

The aggregate and the cement were first mixed for one minute in a paddle mixer with a capacity of 350 litres. The air-entraining agent was mixed with some of the water and then added to the mixer with the first mixing water. The concrete was then mixed for two minutes and the slump was determined. Mixing then continued for another minute and the slump was checked again.

The specimens were cast in steel moulds. The moulds were filled in two layers and each layer was compacted on a vibrating table for 15 seconds. Finally, the top surface was smoothed using a steel rod.

A surface in contact with the mould was used as the test surface for some specimens. For these, plastic-coated plywood moulds were used and the bottom surface of the specimen was used as the test surface.

Curing

After casting, the specimens were cured in various ways, as shown in Table 1. All curing was carried out at a temperature of 20±1°C. During the first day, the specimens were kept in the moulds and sealed with plastic film so that evaporation was prevented. For the water-cured specimens, wet sponges were then placed under the plastic film during the first day, in order to obtain a high humidity, after which the specimens were immersed in water.

Table 1 The different curing conditions used in the tests
(W=water, PF=plastic film, A=air)

Type of curing	Time until form removal[1], days	Water curing, days	Covered with plastic film, days	In the air (RH=50%), days
W0	0	5	0	23
PF0	0	0	5	23
A0	0	0	0	28
W1	1	5	0	22
PF1	1	0	5	22
A1	1	0	0	27
W3	3	5	0	20
PF3	3	0	5	20
A3	3	0	0	25

[1]   0 days before form removal applies only for the horizontal top surface of the specimens while 1 and 3 days before form removal apply for surfaces cast against the mould.

Field exposure conditions

The specimens were subjected to two different outdoor field exposure conditions - exposed to (EXP) or protected from (PROT) rainfall. The test surface was turned upwards for the exposed specimens and positioned vertically for the protected specimens. In parallel with the field test, reference specimens were stored in a climate chamber at 65% RH and a temperature of +20°C (LAB). The specimens were placed in the different exposure conditions at an age of 28 days.

The field exposure station is located close to the National Testing and Research Institute in Borås, Sweden. The annual mean temperature in the area is normally close to +5°C with the highest value in July, +16-18°C, and the lowest in February, -2--4°C. The total amount of rainfall is about 800 mm/year. The relative humidity varies between about 90% in January and 65% in June.

Test methods

Compressive strength

The compressive strength was determined in accordance with Swedish Standard test method SS 13 72 10, with the exception of the curing conditions. The Standard requires the tests to be performed on 150 mm cubes.

Carbonation

For the carbonation depth, two 100x100x400 mm specimens were made for each combination of concrete quality, curing method and exposure condition. Testing involved splitting the specimens about 50 mm from one end and determining the carbonation depth by spraying 3% phenolphthalein solution on the surface immediately after splitting. The results presented here are for up to five years' exposure.

TEST RESULTS

Compressive strength

Laboratory tests

The influence of the curing conditions on compressive strength has been investigated in a number of projects: see, for example, [2,7]. The results indicate among other things that poor curing has a very adverse effect on strength. Normally, no curing at all leads to a drop in strength of about 15-25% at 28 days, relative to that of good wet curing, although sometimes the drop in strength can reach 50-60%.

Figure 1 presents results from this investigation, showing how the curing conditions influence the compressive strength for the concrete with a 0.50 water/cement ratio. When the concrete is cured in water for five days, the strength is 45 MPa. The strength of specimens stored in air for the entire period of time from casting to testing is about 13-14 MPa less, i.e. a drop of almost 30%. Curing using plastic film also results in reduced strength, although the drop is only 4-5 MPa or 10%.

The influence of delayed curing was also investigated, which in this case meant that the specimens were stored in air for two days directly after casting (in the mould during the first day, not protected against evaporation), before the start of curing in water for five days. As can be seen from Figure 1, the result was almost as good as when water curing started immediately after casting, and the strength is higher than in the case where curing was performed using plastic film. This indicates that poor curing conditions during the first days after casting can be largely compensated for by use of subsequent wet curing. This is in good agreement with results presented in [4], which indicate that the strength drop seems to be small (<4%) when wet curing is delayed by two days.

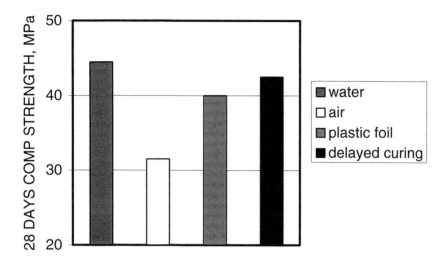

Figure 1   Compressive strength at 28 days as a function of curing conditions for an
OPC concrete (CEM I 42.5, W/C=0.50, air content=5.5%)

Field exposure tests

Figures 2 and 3 show how the compressive strength develops in different climatic
conditions for the two concrete qualities. The specimens were wet-cured for five days
and then stored in air (+20±2°C, 65% RH) until day 28, at which time they were
placed in the different climates. Each result represents the mean value for three
specimens. The results at the age "0" years represent the 28-day values.

The strength increases with increasing age for all curing conditions. The greatest
strength increase between 0 and 3 years (30-55%) was obtained for the specimens
exposed to rain, while the increase in strength of the specimens stored in the laboratory was limited to 5-15%. For the specimens stored outdoors, but protected from
rain, the strength increased by 25-35%.

The strength increase for the 0.34 water/cement ratio seems to be the same for the
protected and for the exposed specimens, while there is a clear difference for the
concrete with a 0.50 water/cement ratio. This means that the outdoor relative humidity at this field exposure station seems to be high enough for continued hydration
even after the curing period - at least, for low water/cement ratios. The explanation
for the fact that low water/cement ratios favour continued hydration is probably that a
low water/cement ratio leads to a finer pore structure, which gives a higher degree of
water saturation at a defined relative humidity. This ought to favour continued hydration.

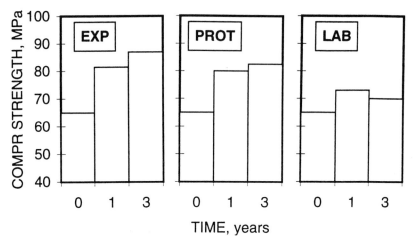

Figure 2   Compressive strength at different ages for concrete exposed to different climatic conditions. The results are relevant for OPC concrete (CEM I 42.5,air content=5.5%) with W/C=0.34. EXP=exposed to rain, PROT= protected from rain, LAB=laboratory (65% RH, +20°C)

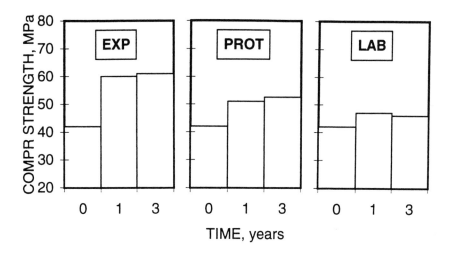

Figure 3   Compressive strength at different ages for concrete exposed to different climatic conditions. The results are relevant for OPC concrete (CEM I 42.5, air content=5.5%) with W/C=0.50. EXP=exposed to rain, PROT=protected from rain, LAB=laboratory (65% RH, +20°C)

Carbonation

The values for the carbonation depth after five years of exposure are presented in Table 2. Corresponding values after two years of exposure are published in [8].

The results show, as expected, that the carbonation rate increases as the environment becomes drier, i.e. the carbonation rate is highest for the specimens stored in the laboratory and lowest for the specimens exposed to rainfall. It can also be observed that the carbonation depth decreases when the time until form stripping increases. This effect is most pronounced where no further curing is performed after the removal of the formwork.

For the specimens stored in the laboratory, a significant difference was found between the three curing methods. Water curing results in the least carbonation depths, protection with plastic film gives higher values and no protection at all gives the poorest results. The results are relevant for top surfaces as well as for surfaces cast against the mould.

For the field-exposed specimens protected from rainfall, there does not seem to be any difference between water curing and protection with plastic film. However, the results for the "air-cured" specimens are much poorer than for other curing methods.

The difference between the three curing methods seems to be very small for the specimens exposed to rain. For the 0.34 water/cement ratio and for the case where the form is kept in place for three days, there is no difference at all. This means that, according to these test results, rain water is often able to cure the defects caused by insufficient curing immediately after casting.

The carbonation rate is normally assumed to follow [10]:

$$X = k\sqrt{t} \tag{1}$$

where X is the carbonation depth, t is the time and k is a constant depending on the environmental conditions and material properties.

Using Equation 1 and the results in Table 2, it is possible to make a rough estimate of the carbonation depth at an arbitrary age. Such estimates are presented in Figure 4, where the carbonation depth after 50 years' exposure is shown as a function of the curing time at 20°C. Results are given for three different exposure conditions and are valid for an OPC concrete with a water/cement ratio of 0.50, cured by protection against evaporation using plastic film.

Figure 4 shows that, for concrete structures exposed to rain, the carbonation depth after 50 years is also shallow for poor curing conditions. When the structure is protected from rain, the carbonation rate increases considerably. In order to limit the carbonation to, say, 15 mm, the curing time at 20°C must exceed about 24 hours. The corresponding value for the specimens stored in the laboratory is more than 200 hours.

Table 2 The carbonation depth (mm) after five years of exposure in different climates (EXP=exposed to rain, PROT=protected from rain, LAB= laboratory. The curing methods are W=water, PF=plastic film and A=air)

EXPOSURE AND CURING CONDITIONS		TOP SURFACE (not against mould)		SURFACE AGAINST MOULD	
		Time until form removal[1]		Time until form removal	
		0 day	0 day	1 day	3 days
		W/C = 0.35	W/C = 0.50	W/C = 0.50	W/C = 0.50
EXP	W	1.0	1.0	1.0	1.0
	PF	1.0	2.0	1.0	1.0
	A	2.0	2.0	1.5	1.0
PROT	W	1.0	3.5	2.5	2.0
	PF	1.0	3.0	2.0	2.0
	A	3.0	8.5	6.5	4.5
LAB	W	2.0	8.0	6.0	5.5
	PF	4.0	8.0	7.5	7.5
	A	6.0	19.0	11.5	7.5

[1] 0 day before form removal is relevant only for the horizontal top surfaces of the specimens
while 1 and 3 days before form removal are relevant for surfaces cast against the mould.

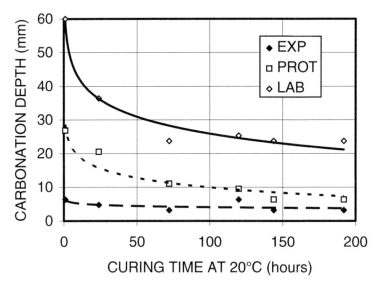

Figure 4 Calculated values of carbonation depth after 50 years' exposure as a
function of curing time at 20°C for different exposure conditions.
The curves are valid for an OPC concrete with a water/cement ratio
of 0.50, and the curing method is protection against evaporation.

CONCLUSIONS

On the basis of the results presented in this report, the following conclusions can be drawn:

- For the concrete qualities studied in this investigation, good wet curing results in a compressive strength that is about 10% higher than when the curing is performed using plastic film. No protection at all after casting leads to a strength drop of almost 30%.

- The compressive strength increases with increasing age, with the greatest strength gain being obtained under humid climatic conditions. The strength gain between 28 days and 3 years was 30-35% for the specimens stored outdoors exposed to rain, 25-35% for specimens stored outdoors protected from rain and only 5-10% for the specimens stored indoors under relatively dry laboratory conditions.

- The carbonation rate increases as the exposure conditions become drier. Storage in the laboratory (+20°C, 65% RH) gave the highest values, field-exposed specimens protected from rainfall gave lower values and specimens exposed to rain gave the best results.

- The drier the climate, the more pronounced the difference between different curing conditions. For the specimens exposed to rain, the differences are very small, while they are significant for the specimens stored under dry laboratory conditions. This means that laboratory tests cannot always be used for predicting the behaviour of concrete in a real structure. Field exposure tests are preferred.

- The test results from the carbonation tests indicate that wet curing and covering with plastic film are equally effective, while "curing" in air gives poorer results.

The results of the field exposure tests presented in this paper are primarily relevant for Swedish climatic conditions, and the conclusions may be different for field exposure tests carried out in other countries. The results imply that a drier, warmer climate than Sweden's probably results in a greater carbonation risk and also a greater sensitivity to the choice of curing method.

REFERENCES

1. MEYER, A. Investigations on the carbonation of concrete. Proc Chem Cem, Tokyo, 1968, pp 394.

2. POMEROY, C D. The effect of curing conditions and cube size on the crushing strength of concrete. Cement and Concrete Association, Technical report 42.470, London 1972.

3. KURZ, M. The influence of extremely short-term curing on carbonation in con-
    crete. Proceedings of RILEM Seminar on Durability of Concrete Structures Under
    Normal Outdoor Exposure, Hannover, 1984. pp 250.

4. BYFORS, J. Plain concrete at early ages. Swedish Cement and Concrete Research
    Institute, Report fo 3:80, Stockholm, 1980.

5. GOWRIPALAN , N et al. Effect of curing on durability. Concrete International,
    Vol 12, 1990. pp 47-54.

6. DHIR, R K, HEWLETT, P C, CHAN, Y N. Near surface characteristics of con-
    crete: intrinsic permeability. Magazine of Concrete Research, Vol 41, 1989:147.
    pp 87-97.

7. ANDERSSON, C, PETERSSON, P-E. The influence of curing conditions on the
    permeability and durability of concrete. Swedish National Testing and Research
    Institute, SP-Report 1987:7, Borås, 1987.

8. EWERTSON, C, PETERSSON, P-E. The influence of curing conditions on the
    permeability and durability of concrete. Part 2. Swedish National Testing and
    Research Institute, SP-Report1992:51, Borås, 1992.

9. MALMSTRÖM, K. The importance of cement composition on the salt-frost resis-
    tance of concrete. Swedish National Testing and Research Institute, SP-Report
    1990:7, Borås, 1990.

10. TUUTTI, K. Corrosion of steel in concrete. Cement and Concrete Research Insti-
    tute, Report fo 4.82, Stockholm, 1982.

THE INFLUENCE OF CEMENT ON SHAPING THE STRUCTURE AND STRENGTH OF HPC

K Flaga

J Mierzwa

Cracow Technical University

Poland

ABSTRACT.   The study on HPC properties are giving more and more valuable information about their outstanding qualities. Strength as the main feature of concrete, in its new generation which is HPC, is formatted by structure modification but especially by cement - superplasticizer - silica fume system. The quantitative character of interaction of that three components is not up to now explicitly identified. In the paper the attempt to evaluate this influence basing on author's research and Larrada's model has been carried out. The results can be used as the base for forecasting quantitative - qualitative composition of HPC concrete.

Keywords: microsilica, silica fume, plasticizer, Portland cement, High Performance Cement HPC, compressive strength.

Professor dr. hab. eng. Kazimierz Flaga is Director of Institute of Building Material and Structure of Cracow Technical University. He specialises in the use of concrete in Civil Engineering. Author of over 100 publications, Head of Concrete Construction Section of Polish Academy of Science.

Professor dr. hab. eng. Janusz Mierzwa is Head of Concrete Technology Laboratory at Cracow Technical University. He specialises in   technology and use of HPC concrete, structure durability, repairs and corrosion.

Radical Concrete Technology. Edited by R K Dhir and P C Hewlett. Published in 1996 by E & FN Spon, 2–6 Boundary Row, London SE1 8HN, UK. ISBN 0 419 21480 1.

INTRODUCTION

Since the time of application of concrete as the construction material the basic indicator of its technical utility has been compressive strength.
Gradually with time this feature has also become synthetic synonym of other properties, sometimes even those that are actually rather weakly correlated with compressive strength.
Result of acceptance of that concept of formatting composite qualities was that development research were focused mainly on achieving the highest possible compressive strength. So vital problem like concrete structure durability only later was to be recognised and expose its fundamental meaning. Two groups of concrete, used till now as the basic in designing and constructing are:

- simple concrete for structures with average strength $f_{cc} = (f_c cube) \approx 10$ to 30 MPa
- concrete of higher strength $f_{cc} = 35$ to 70 MPa used in prestressed and special structures.

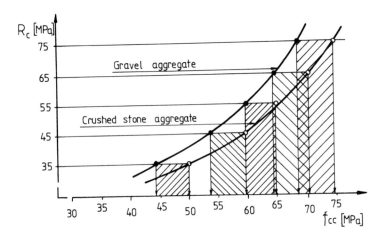

Figure 1   The ultimate strength of concrete from rounded and crushed aggregate and different class of cement.

In practice, because of lack or underestimation of the part played by admixtures, the highest level of strength of that second group of concrete was gained by minimisation of w/c coefficient and use of selected aggregate of high strength and granulation assuring minimum water demand. Often these activities were connected with special technological treatment.
Particular part were taken there by cement quality as its strength was the key factor deciding about achieving the highest strength. However, as further analysis shows, even application of particularly high strength cement up to $R_c = 75$ MPa , production of which is at the verge of profitability, is not able to assure conventionally produced concrete possibility to surpass $f_{cc} >$

75 MPa. Basing on Bolomey's equation , with verified accuracy, and expanded up to the value c/w = 3.2 (w/c = 0.312) and in alternative shape :

$$f_{cc} = \begin{cases} A_t(c/w - 0.5) \Rightarrow 1.2 \leq c/w \leq 2.5 \\ A_t(c/w + 0.5) \Rightarrow 2.5 < c/w \leq 3.2 \end{cases}$$

upper ranges of concrete strength made with rounded and crushed aggregate that can be obtained with application of cement of different class of strength beginning from 35 up to 75 are shown in Figure 1.

In practice to surpass only slightly the level of concrete strength f_{cc} = 70 MPa the class 75 of cement and crushed aggregate are necessary.

This limit was finally crossed not before introduction of highly efficient superplasticizer and microsilica opening entirely new concept of structural formatting concrete strength.

It is based generally on interference of high quality admixtures and additions which causes, as a result of reacting portlandite phase with microsilica, transformation of limiting layer grout-aggregate. The effect is quantitative increase of C-S-H phase and significant densification and strengthening of this weakest hitherto link of concrete.

The nature of these transformations is well described by Bentur-Gutman model [1].

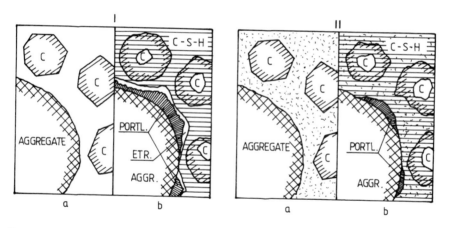

Figure 2   The draft of hydration transformation in contact grout-aggregate layer of concrete without addition (I) and with addition of microsilica (II) in the initial (a) and advanced (b) stage of hardening process.

The general concept of strength increase of concrete by way of increasing density of grout matrix structure and by modification of grout-aggregate contact layer by eliminating from its microporous structure $Ca(OH)_2$ is in a way new concept of hydraulic composite low formulated in 1895 by Ferret. Pertinence of that idea was verified as early as in seventies by Powers [2] and Neville [4] as well as Cottin [5] for which quantitative evidence is shown in Figure 3.

RESEARCH ON CEMENT INFLUENCE ON HPC STRENGTH

Generalisation of research which have been carried on to the present on cement influence on HPC concrete strength properties does not present itself so explicitly like in the case of typical structural concrete [5,6]. When the type of cement is taken into account world literature comprises almost all kinds of cement produced and conclusion which can be drawn from these publications is so that HPC concrete with higher or lower level of strength can be actually produced from every type of cement of class not lower than 35 [6]. For these applications as more effective cement which have mineralogical composition with high contents of silicate phase (C_3S+C_2S) and the level of amount of aluminate and ferrite phase (C_3A and C_4AF) [6,7,8] reduced to minimum are recognised. Value of total surface of higher quality cement proves to be quite good and taking into account distribution of function which takes place between microsilica and cement in BWK from binder particularly high milling (over 4.0 m^2/g) is not really demanded.

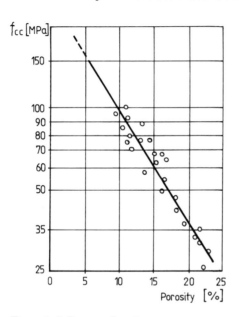

Figure 3 Influence of total porosity on paste strength.

Problem of cement quantity in HPC mixes and its limiting results more from physical issues than durability problems. Suggestions of some researchers [9,10,12] recommending not to exceed the value of 500 kg/m^3 are connected with typical in this concrete very high kinetics of hydration heat emanation and as a result serious problem of thermal and shrinkage stresses in construction.

From the point of view of HPC concrete strength influence of cement may be divided into three component effects:

- ability to create after hydration durable gel structures in beyond contact domain among grains of composite filling phase additionally compacted by microsilica,
- compliance to liquefaction under influence of certain rheological admixture which will decide about possibility of ultimate reduction of w/c coefficient, resulting from predisposition of mineral and granulometric composition of binder,
- effectiveness of transformation process of portlandite phase into CSH phase with parallel increase of microstructure density in created neoformations as a result of cement mineralogical composition and introduced into mix microsilica.

Therefore, taking into consideration cement influence, the final level of gained strength will be the result of interaction in cement - addition - admixture system. The attempt of quantitative estimation of this influence has been undertaken in the research presented below.

RESEARCH

In the carried out programme variability of cement strength, amount of microsilica and superplasticizer were taken into account. Three kinds of cement, with the following compressive strength : CP35 - 42.4 MPa, CP55 - 56.3 MPa and MPz III - 80.8 MPa, were applied. The amount of microsilica addition was 0.10 and 20 % of cement mass. As the admixture superplasticizer based on condensed naphtalene sulfonated formaldehyde (NFS) was used in the amount 0 or 2.5% of cement mass. Concrete mixtures had consistence 8.0 to 14.0 cm of slump. Chemical and mineral composition of 35 and 55 cement and microsilica are given in Table 1. Chemical composition of MPz III cement was compatible with production standard of „gypsum-less" binder while its specific surface was 5.4 m²/g by Blaine.

Concrete were made of aggregate composition containing natural sand (P-0/2) and fine granite (G1-2/8 mm) and coarse granite (G2-8/16 mm). Mixing proportions were P:G1:G2 = 1:0.74:0.89. Strength sampling were done after 28 days of curing in standard conditions on cube samples 15 cm (f_{cc}).

Table 1 Properties of cement and silica fume used for testing.

Component	Portland Cement		Silica
	PC 45	PC 55	fume
SiO_2 %	20.4	21.5	91.3
Al_2O_3 %	5.7	4.2	0.7
Fe_2O_3 %	2.6	4.7	1.2
CaO %	64.5	65.0	0.3
MgO %	1.5	0.8	0.2
$Na_2O + 0.658\ K_2O$ %	0.75	0.3	0.6
SO_3 %	2.6	2.4	-
C_3S %	58.0	59.4	-
C_2S %	14.7	16.9	-
C_3A %	10.7	3.2	-
C_4AF %	7.9	14.3	-
Fineness m²/g	2.6	3.2	19.0
Compr. Strength 28 days RcMPa	41.4	58.1	-

Table 2   Concrete mixtures kg/m³ and material properties

Mixture No	Series 1			Series 2			Series 3	
	1	2	3	4	5	6	7	8
Cement CP 35	515	606	608					-
Cement P 55				587	569	593	-	-
Cement MPz III	-	-	-	-	-	-	468	372
Aggregate 0/2	670	580	474	581	618	577	667	725
Aggregate 2/16	1094	946	953	950	1010	942	1089	1182
Silica Fume	-	61	122	-	57	119	48	37
Superplasticizer	13.0	15.0	15.0	-	13.3	·14.8	-	9.0
w/c - ratio	0.285	0.294	0.298	0.398	0.308	0.324	0.333	0.336
f_{cc} (cube 28 d)	59.9	63.9	70.6	66.7	79.1	91.2	70.9	74.1

RESULTS AND DISCUSSION

The results of compressive strength test were all within the range of 60 to 90 MPa. Differentiated level of Portland concrete strength CP 35 and CP 55 with very close amount of Σ (C_3S+C_2S) and several times difference of C_3A phase clearly manifested its influence on f_{cc} results of Series 1 in relation to Series 2. Simultaneously there was also clearly visible effect of silica fume and superplasticiser which is however smaller in case of CP 35 and CP 55 cement.

The other problem is behaviour of special „gypsum-less" MPz III cement. In the process of production its composition is supplemented with retarding - activating and plastifying admixtures. Eventually it is characterised by very low specific water and particularly high increase of initial strength with moderate (rather low) growth of 28-day strength. It proved that addition to mix made from this cement of superplasticizer in a way doubles previous effects and does not supports effectively strength growth even in spite of using microsilica. Both in Series 1 and in Series 2 the amounts of used cement were within the range of 500 to 600 kg/m³ so they were rather large.

Specification of indicators of cement strength effectiveness CR as a ratio :

$$CR = \frac{C[kg/m^3]}{f_{cc}[MPa]} = [kg/MPa]$$

on the background of silica fume addition S/C manifests quite large however straight in both series values (Table 3). Significantly lower level of CR is shown by mix with MPz III.

Konig's research [11] proved that when using high class cement and optimum selection of particularly effective superplasticiser it was possible to reduce CR even to 4.0 - 4.6 for B85-B105 class [11] (amount of microsilica S/C= 0.1).

Cement	Mix	Indicators	
	No	CR[kg/MPa]	S/C
CP 35	1	8.6	0
	2	9.5	0.1
	3	8.6	0.2
CP 55	4	8.8	0
	5	7.2	0.1
	6	6.5	0.2
MPz III	7	6.6	0.1
	8	5.0	0.1

To make quantitative estimation of influence of factorial group of cement -microsilica-superplasticizer strength, Larrada's analytical formula as the expansion of Ferret's low on high performance concrete relevant in the range w/c = 0.25-0.40 was used:

$$f_c = 4.91 \cdot R_c [1 + \frac{\omega}{A}]^{-2} \tag{1}$$

where:

$$\omega = G_c \frac{w}{c} = 3.1 \ \frac{w}{c} \ and$$

$$A = 1.4 - 0.4 \ exp \ (-11 \ s/c) \tag{2}$$

Accuracy verification of equation (1) + (2) based on Series 1 and 2 test results proved the value of average deviation of strength to be 7.1% with maximum individual values up to 14.0%. Values of analogical deviations for Series 3 (MPz cement) fluctuate from 36% to 61% so the formula is entirely inadequate.

Also Konig's investigation prove high accuracy of equations (1) + (2) for 1 and 2 Series, where average deviation is about 13%. So taking as positively verified formula (1) analysis of cement strength in the range from R_c = 35 to 65 MPa on concrete strength regarding variability of w/c and s/c was carried out. It proves that increase of cement strength of one class causes average strength increase of concrete:

with w/c = 0.25 for $\Delta f_{cc} \cong 15$ MPa when s/c = 0
to $\Delta f_{cc} \cong 20$ MPa when s/c = 0.30
and with w/c = 0.35 for $\Delta f_{cc} \cong 11$ MPa when s/c = 0
to $\Delta f_{cc} \cong 15$ MPa when s/c = 0.30

In the same time, assuming that R_c = constans, application of such effective superplasticizer enables, with constant concrete consistency, decrease of w/c from 0.35 to 0.25 and increase of concrete strength within the range from 30.5% with s/c = 30% to 38% for concrete without microsilica. The above given changes take place approximately homogenously in the whole range of analysed change of concrete strength from R_c = 35 to R_c = 65 MPa. Therefore simultaneous use of superplasticizer reducing w/c level to 0.25, cement R_c = 65 MPa and 30% of silica fume in comparison with concrete with w/c = 0.35 without silica fume and with cement $R_c = 35$ MPa can lead to change of concrete strength from $f_{cc} = 40.0$ MPa up to $f_{cc} = 130$ MPa.

The results of the above quantitative analysis based on verified Larrado's model give foundation to estimate the range of outcomes of material modification of concrete composition on account of combined concurrence of cement - superplasticizer - microsilica.

CONCLUSIONS

1. Carried out research and analysis proved that combined concurrence of cement - superplasticizer - microsilica system in durability, reductive and compactive range is sensitive on cement strength and in higher values R_c can cause significant effects of increase of HPC concrete strength.
2. Complex relations of reaction of three material parameters cement - superplasticizer - microsilica, as verifying research have proved, is quite precisely modelled by Larrada's analytical formula. So it enables, in the range of $w/c = 0.25$ to 0.45, to estimate influence of the parameters on the final strength of HPC.
3. As the analysis has proved, the influence of vast change of parameters of cement class ($R_c = 35 - 65$ MPa), superplasticizer and microsilica (from 0 to 30%) can cause change in concrete strength in approximate range from $f_{cc} = 40$ to 130 MPa.

REFERENCES

1.   GOLDMAN A., BENTUR A.: ACJ Material Journal, Vol. 86, No. 5/89, pp 440-470.

2.   POWERS T. 4th Int. Congress C.C. Washington 1960.

3.   NEVILLE A.M : Properties of concrete. 1973 Pit. Publ. London.

4.   COTTIN B. : Cement - Wapno - Gips no 6/1976 pp 157-159.

5.   GJORV E.O., LOLAND K.E.: Proceedings Nordic Res. Sem. on Condensed Silica Fume in Concrete; Feb. 1982 pp 293.

6.   CEB/FIP : High Strength Concrete - State of the Art; B-J No 197, Lausanne : CEB 1990.

7.   GAGNE R.; Durabilite au gel des betons de haute parformance mechaniques. Mat. el. Constr. 1990 No. 6 Vol. 87, pp 608-618.

8.   DIDERICKS U.; Material properties of high strength concrete at elevated temperatures. 13. IVB++ 1988 Helsinki pp 489-494.

9.   TACHIBANA D.; High - strength concrete, 2.Symp. Utiliz. HSC. Berkeley 5/1990.

10.   HELLAND S.; Temperature and strength development in concrete with less than 0.4. Proc. Utiliz. HSC. Stavanger 6/1987.

11.   KONIG G., GRIMM R., SIMSCH G.; Ductility of beams and columns made of HSC/HPC. Proc Int. Symp. „Brittle Matrix Composites 4" Warsaw Sept. 1994 pp 523-532.

THE EFFECTS OF TEMPERATURE ON THE PERFORMANCE OF HIGH STRENGTH CONCRETE WITH FLY ASH

T H Wenzel

T M Browne

Marquette University

Collins Engineers

K H Scherzer

Matsen Ford Design

USA

A Hassaballah

United Arab Emirates University

UAE

ABSTRACT. The effects of the temperature in which high strength concrete with fly ash is mixed, placed and cured is examined in this paper. A comprehensive experimental study was conducted to determine the compressive strength development of high strength concrete mixtures with fly ash (Class C and Class F) when subjected to different hot and cold weather mixing and curing regimes. Strength development and stress strain curves are plotted to reflect the data. These results are compared with data collected from specimens cured under standard laboratory conditions according to ASTM specifications. Conclusions are made regarding the effect temperature and fly ash replacement level have on the strength development of high strength concrete made with Class C or Class F fly ash.

Keywords: High strength concrete (HSC), Curing, Fly ash

Dr Thomas H. Wenzel is an Associate Professor and chairman of the Department of Civil and Environmental Engineering at Marquette University, Milwaukee, Wisconsin, USA. He is a past president of the Wisconsin Chapter of the American Concrete Institute and serves on several technical committees of the American Concrete Institute.

Mr Terence Browne is a Structural Engineer with Collins Engineers, Chicago, Illinois, USA. He was formerly a Research Assistant at Marquette University.

Mr Karl H. Scherzer is a Structural Engineer with Matsen Ford Design, Pewaukee, Wisconsin, USA. He was formerly a Research Assistant with Marquette University.

Dr Amr Hassaballah is an Assistant Professor in the Department of Civil Engineering at United Arab Emirates University, Al-Ain, United Arab Emirates. He completed his Ph.D. degree at Marquette University.

Radical Concrete Technology. Edited by R K Dhir and P C Hewlett. Published in 1996 by E & FN Spon, 2–6 Boundary Row, London SE1 8HN, UK. ISBN 0 419 21480 1.

INTRODUCTION

Advancements over the past twenty years have led to a substantial increase in the demand for high strength concrete (HSC). Structural engineers are using HSC more often because of a significant development in chemical admixture use and the availability of physical additives. Currently high strength concretes are available from the majority of concrete ready mix suppliers. The use of high strength concrete allows engineers to produce smaller structural members without a compromise in safety. These smaller structural members have economical benefits and can also lead to more innovative and attractive designs.

One additive commonly used in high strength concrete is fly ash. Fly ash is used in concrete to achieve economic, environmental, commercial and technical benefits. Improved concrete properties include workability, cohesiveness, pumpability, strength and durability.

Considering that concrete operations must be performed in many different climates with different curing regimes, it is important to study the influence these factors have on the concrete material properties. It is well documented that temperature has an effect on concrete material properties, durability, set time, and rheological characteristics. With the recent expansion of the HSC market, there is a need for information regarding the influence of temperature on HSC with fly ash (HSC-FA) in order to provide the most efficient concrete mix for a particular job site.

Previous research and studies clearly describe the effect mixing, placing and curing temperature has on ordinary portland cement (OPC) concrete. These studies have resulted in code guidelines from the American Concrete Institute including the Hot Weather Concreting [1] and Cold Weather Concreting [2] . These guidelines do not specifically address the effects temperature has on the development of HSC-FA material properties. Currently, specifications for hot and cold weather concreting with HSC-FA are the same as guidelines for OPC concrete placement. However, HSC-FA can exhibit different properties than OPC concrete under the influence of high and low temperatures.

Some researchers have shown that incorporating fly ash into a HSC mixture can enhance concrete properties in hot and cold weather, while other studies have concluded that HSC-FA mixtures are more sensitive to temperature than OPC concrete.

Currently, the two types of fly ash most commonly used are low calcium ASTM Class F and high calcium ASTM Class C. Until the early 1970's most fly ashes produced in the United States were ASTM Class F. Recently, high calcium fly ashes are being produced more readily and detailed research information is being sought.

The effects of the temperature in which high strength fly ash concrete is mixed and cured is examined in this paper. In 1993, a research program was initiated at Marquette University to determine the strength development of high strength concrete

(HSC) with and without fly ash when subjected to different mixing and curing temperature regimes. The study involved high strength concretes with both Class C and Class F fly ash replacement. This paper focuses on the influence of temperature during mixing, placing and curing on the compressive strength development of high strength concrete with fly ash.

EXPERIMENTAL PROGRAM

Objectives and Scope

The primary objective of this study was to investigate the effect of temperature on the strength development of HSC containing different replacement levels of Type C or Type F fly ash. This was accomplished through an extensive experimental research program. The experimental research program used a high strength concrete mixture with a W/CM Ratio of 0.3 and a 6" slump for workability. An empirical design approach was used to design a mix that met rigid strength and workability requirements. The average compressive strength of a 4" × 8" concrete cylinder for this mixture containing 0% fly ash replacement level and cured according to ASTM C192-90a [3] was 13,595 psi (93.8 MPa).

After the HSC mix was designed, cylindrical concrete specimens were cast at specified temperatures. These specimens were then subjected to similar environmental conditions during the curing process. Compressive tests and modulus of elasticity tests were completed after specified amounts of time (3, 7, 28, 56, and 91 days). After this was completed, the influence of the variables of environment and fly ash replacement level on compressive strength, modulus of elasticity and concrete plastic properties was examined and analyzed.

Concrete cylinders were cast and cured in various environments with varying fly ash replacement levels. The environments studied, (40°F - 60% R.H., 60°F - 51% R.H., 73°F - 44% R.H., 73°F - 100% R.H., 90°F 52% R.H. and 105°F - 31% R.H.), were chosen to simulate the following curing environments: cold weather, cool weather, temperate/partially humid weather, ASTM C 192-90a, warm/moderately humid weather and hot/arid weather, respectively. The fly ash replacement levels for both Class C and Class F included: 20%, 40% and 60%. The scope of this study included a total of forty-eight concrete mixtures from five different temperature regimes, four fly ash replacement levels, two types of fly ash and six curing conditions. All mixtures were subjected to compressive testing for strength and modulus of elasticity.

Materials

A low-lime content, Class F fly ash from the Oak Creek Power Plant in Oak Creek, Wisconsin was used in this study. The physical properties of the fly ash are in compliance with ASTM C618-93 [3].

For the Class C fly ash, a high-lime content fly ash from the Pleasant Prairie Power Plant in Kenosha, Wisconsin was used. The physical properties of the Class C fly ash were also in compliance with ASTM C618-93[3].

Type I Portland Cement was used in the research study. In addition, in order to develop a high strength concrete mixture, it was necessary to add silica fume to the mixture design. A constant silica fume replacement level of 15% was maintained for all mixtures throughout the study.

DISCUSSION OF RESULTS

Compressive Strength Discussion

A concrete mixture with a 28 day compressive strength between 6,000 and 12,000 psi (approximately 40 to 80 MPa) is classified as high strength concrete. The majority of the experimental mixtures reached this level, except for the mixtures that were mixed, placed and cured at 40°F and 60°F with 60% fly ash replacement levels. The mixtures that produced the highest compressive strengths were mixed, placed and cured at warmer temperatures (73°F, 90°F and 105°F) with lower fly ash replacement levels (0% to 20%). Figure 1 shows all of the compressive strength vs. age curves for the Class F fly ash mixtures. Only the Class F data is presented here.

Effects of Temperature on Compressive Strength

The experimental program produced HSC compressive strength results that agree with portions of previously published data. Klieger [4] stated that concretes mixed and cured at high temperatures would achieve high compressive strengths at 3 and 7 days, but the 28, 56 and 91 day strengths would be lower than the compressive strength of concrete mixed at normal temperatures. In agreement with Klieger [100], this study produced 3 and 7 day compressive strengths for the 90°F and 105°F mixtures that were always considerably higher than the other mixtures and the later age strengths were generally lower than the 73°F mixtures. Klieger also stated that concretes mixed at low temperatures reach maturity at a slower rate. The 40°F and 60°F mixtures clearly support this conclusion with extremely low early strengths. Later strengths (28 day, 56 day and 91 day) of the low temperature mixtures were also significantly lower than the higher temperature mixtures. This correlation is also supported by Klieger's data, but conflicts with research performed by Aitcin [5], Gardner, Sau and Cheung [6] and Gardner and Poon [7]. The conflicting studies stated that the low temperature had no long term detrimental effects on concrete compressive strength development.

The effects of temperature on compressive strength are best represented by the calculation of a strength ratio. The strength ratio is defined as the compressive strength of a concrete with a particular fly ash replacement level and mixing and curing regime, divided by the compressive strength of the concrete with the same fly

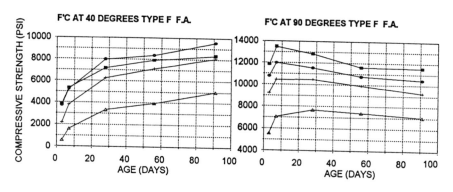

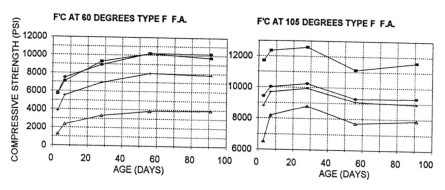

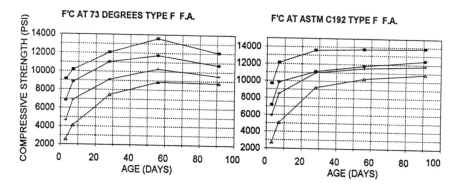

Temperature in degrees F,   1 psi = .0069 MPa

┌───┐
│ ―■― 0% F.A.   ―●― 20% F.A. ―+― 40% F.A. ―▲― 60% F.A. │
└───┘

Figure 1:   Compressive Strength Versus Age for Class F Fly Ash
              and Various Curing Environments

ash replacement level, mixed and cured according to ASTM C192 [3]. A number of previous studies have published specific correlations using this ratio. Strength Ratio vs. Age plots for each dry curing environment are shown in Figure 2 for Class F fly ash.

An analysis of the strength ratio plots do not yield a single optimum fly ash replacement level for all of the curing conditions, but the 20% fly ash replacement level seemed to produce the best results for both types of fly ash. Table 1 summarizes the optimum replacement level for each temperature range and fly ash type.

Table 1: Optimum Fly Ash Replacement Levels

Temperature and Fly Ash Type	Optimum Fly Ash Replacement Level
40°F Type C	20%
60°F Type C	20%, 40%
73° F Type C	20%, 40%
90° F Type C	20%, 40%
105°F Type C	20%, 40%, 60%
40°F Type F	20%
60°F Type F	20%
73°F Type F	20%
90°F Type F	20%
105°F Type F	20%, 40%, 60%

Effects of Fly Ash Replacement Level on Compressive Strength

The calculation of a fly ash replacement level strength ratio is an effective method to determine the effects the level of fly ash replacement has on compressive strength. The fly ash replacement level strength ratio is defined as the compressive strength of a concrete with particular fly ash replacement level mixed and cured at a particular temperature regime divided by the compressive strength of a concrete without fly ash mixed and cured at the same temperature regime. Plots of fly ash replacement level strength ratio vs. age for all curing environments are shown in Figure 3 for Class F fly ash.

The majority of the data suggests that an increased fly ash percentage directly causes a significant reduction in compressive strength. A few exceptions were evident in the data. For the 105°F Type C fly ash mixture, compressive strength values for all

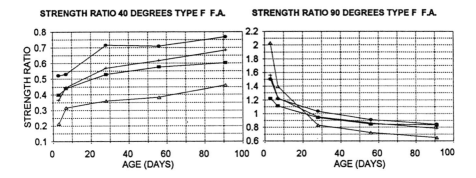

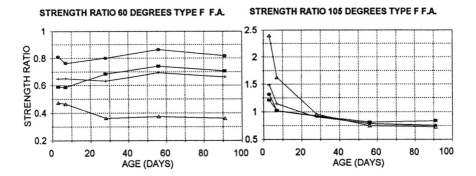

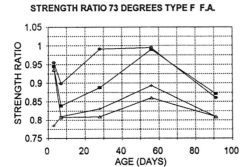

Temperature in degrees F,   1 psi = .0069 MPa

— 0% F.A.   — 20% F.A.   — 40% F.A.   — 60% F.A.

Figure 2:   Strength Ratio Versus Age for Class F Fly Ash
             and Various Curing Environments

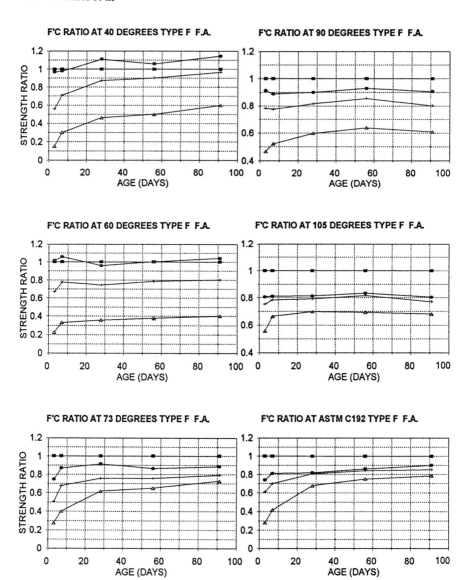

Temperature in degrees F, 1 psi = .0069 MPa

| ■ 0% F.A. | ● 20% F.A. | + 40% F.A. | ▲ 60% F.A. |

Figure 3:   Fly Ash Strength Ratio Versus Age for Class F Fly Ash
and Various Curing Environments

replacement percentages at 56 days or greater were higher than the 0% replacement values. The general results are similar to the findings of Rasoulian [8], but dissimilar to results found by Krishnamoorthy [9] and Hassaballah [10]. Hassaballah stated that "...incorporating Class C fly ash up to 30% by weight, will always increase the compressive strength of the concrete". This is also true for Class F fly ash except during the first two weeks.

A possible explanation as to why the compressive strength values of the HSC-FA did not reach the strength values of the 0% replacement mixture could be attributed to delayed pozzolanic effect. Since the mixtures with fly ash replacement contain less cement than 0% replacement mixtures, strength development will occur at a slower rate. When the concrete mixture develops, sufficient amounts of lime must be liberated to initiate the pozzolanic reaction. Once this reaction occurs, strength gain increases at a rapid rate. It may take between 1 and 3 months to complete the entire reaction. Researchers have stated that it is typical for fly ash concretes to reach and surpass 0% replacement strengths after 100 days of curing depending on the purity of the fly ash. Most of the data from this study suggests that the concrete was continuing to gain strength at 91 days. Without compressive strength data for 100 days and beyond it is difficult to predict how many of these mixtures would reach the strength of the 0% replacement mixtures, however, only the 20% replacement mixtures seem likely to reach this level of strength.

Modulus of Elasticity

The MOE data was very erratic and produced values for 2,350.000 psi to 16,950,000 psi. Ordinary portland cement concrete MOE values typically range from 2,000,000 to 5,000,000 psi. The majority of this study's MOE values are greater than the ordinary concrete MOE values which agrees with published data. Aside from this general statement, direct MOE comparisons cannot be made because of the limited amount of reliable data.

PRACTICAL APPLICATION OF RESULTS

The results of this study show the effects of curing temperature on strength gain for high strength concrete with different replacement levels of fly ash. In general, the cold weather HSC-FA mixtures experienced strength gains throughout the cycle of curing while the hot weather HSC-FA mixtures reached their maximum strength at an early age of curing.

These results need to be incorporated into specifications for hot weather concreting[1] and cold weather concreting[2] which currently do not address the effects of temperature on the development of HSC-FA material properties.

In addition, the study showed that the 20% fly ash replacement level produced the best results for all ranges of curing conditions. This is important in determining the cost and performance of a mixture in which fly ash replaces the cement and different curing conditions can be expected.

CONCLUSIONS

1   All of the high strength mixtures studied reached high strength f'c levels (6,000 psi- 12,000 psi or 40 MPa- 80 MPa) except for the 40°F and 60°F mixtures with 60% fly ash replacement levels.

2   Upon examination of compressive strength data, it can be concluded that an optimum fly ash replacement level cannot be chosen for all of the data, however, the 20% fly ash replacement level seemed to produce the best results for both types of fly ash in various curing environments. It is possible that a higher replacement percentage can be used for the 90°F and 105°F mixtures without a significant reduction in strength.

3   Concrete mixed and cured at 40°F and 60°F matured at a very slow rate and possessed extremely low early strengths. Typically these mixtures did not reach the long term strength of the other mixtures regardless of replacement percentage.

4   An increase in fly ash generally resulted in a reduction in concrete compressive strength.

ACKNOWLEDGEMENTS

The authors would like to express their appreciation to the Tews Company and Wisconsin Electric Power Company for providing some of the materials used in this study.

REFERENCES

1.   ACI Committee 305, "Hot Weather Concreting", American Concrete Institute, Detroit, Michigan, pp. 1-14.

2.   ACI Committee 306, Cold Weather Concreting, American Concrete Institute, Detroit, Michigan, 1978.

3.   ASTM, ASTM Annual Book of Standards, Concrete and Aggregate, Section 4, v. 4.02, ASTM, Philadelphia, 1991.

4.   Klieger, Paul, "Effect of Mixing and Curing Temperatures on Concrete Strength," *ACI Journal, Proceedings* v. 45, n. 12, June 1958, pp. 1063-1081.

5.   Aitcin, Pierre-Claude: Cheung, Moe S., and Shah, Vinay S., "Strength Development of Concrete Cured Under Arctic Sea Conditions," *Temperature Effects on Concrete,* STP-858, ASTM, Philadelphia, 1985, pp. 3-20.

6.   Gardner, N.J., Sau, P.L., and Cheung, M.S., "Strength Development and Durability of Concretes Cast and Cured At 0 Degrees C", *ACI Materials Journal,* v. 85, n. 6, November-December, 1988, pp. 529-536.

7.   Gardner, N.J., and Poon, S.M., "Time and Temperature Effects on Tensile Bond and Compressive Strength," *ACI Journal, Proceedings,* v. 73, n. 7, July 1976, pp. 405-409.

8.   Rasoulian, M., <u>Fly Ash in Concrete</u>, Louisiana Transportation Research Center Report No. 221, December 1991.

9.   Krishnamoorthy, S., "Performance of Fly-Ash Portland Cement Concrete Under Different Conditions of Curing", *Indian Concrete Journal,* v. 50, n. 12, December 1976, pp. 371-374.

10. Hassaballah, A.A., An Investigation of the Water to Cementitious Ratio of Fly Ash Concrete, Marquette University Thesis, Milwaukee, Wisconsin, 1993.

PERFORMANCE SPECIFICATIONS

Chairmen   **Mr J A Bickley**
Concrete Canada
Canada

Professor S Y N Chan
Hong Kong Polytechnic University
Hong Kong

Mr P Livesey
Castle Cement Limited
United Kingdom

Leader Paper

The Specification of Durability by Performance - When?

Dr T A Harrison
British Ready Mixed Concrete Association
United Kingdom

THE SPECIFICATION OF DURABILITY BY PERFORMANCE - WHEN?

T A Harrison

British Ready Mixed Concrete Association

UK

ABSTRACT. Durability performance tests can be used for research, a wide range of technical approvals varying from site approval to approval to include in a standard and as the basis for directly assessing specified performance requirements. Their evolution of use is likely to follow the sequence of research, technical approval then performance specification. Some requirements, such as cover to reinforcement and abrasion, are best placed directly on the structure while others, such as carbonation and chloride diffusion, can be adequately assessed on standardised concrete specimens made from fresh concrete. A number of tests take so long to complete that they only have a role as type approval tests while others are suitable for routine control. The current stage of development is summarised and an estimate given of the minimum time needed to standardise the various tests and criteria.

Keywords: Durability, Performance, Technical approvals, Specification, Cover, Carbonation, Corrosion due to chlorides, Freeze/thaw.

Dr Tom A Harrison is the Technical Director of the British Ready Mixed Concrete Association. He is chairman of a Concrete Society Working Party on durability design and performance specification. He serves on numerous CEN and BSI committees and is chairman of several of these committees, including the BSI committee on concrete.

Radical Concrete Technology. Edited by R K Dhir and P C Hewlett. Published in 1996 by E & FN Spon, 2–6 Boundary Row, London SE1 8HN, UK. ISBN 0 419 21480 1.

INTRODUCTION

In the closing address of Concrete 2000, comment was made on the fact that the many excellent papers on durability performance did not form a complete package which would lead to a system that could be used to specify durability by performance. This paper gives an industrial viewpoint on what is needed to have a workable system in the construction industry. Firstly, we are not all Luddites but we do oppose the contractual imposition of tests and criteria that are unproven, undeveloped and, in some cases, unworkable. We do support the controlled evolution of durability performance specifications, and such a process is described in this paper. An attempt has been made briefly to review and assess progress with the development of durability performance specifications and to predict, in broad terms, when they are likely to be widely available.

Within the context of this paper, mix limitations such as maximum water/cement ratio are not considered as a durability performance specification. Performance has to be established by using a durability performance test which, according to the Concrete Society[1], is:

> a) A test that directly assesses the resistance to a standard deterioration process, e.g. the freeze/thaw test, or

> b) a test that directly assesses the resistance of one of the phases of a deterioration process, e.g. carbonation tests, or

> c) a test that directly assesses a performance related parameter, e.g. cover or adiabatic temperature rise.

Any concrete test is a convention, and substantial background research is needed to relate the test and its associated conformity requirements to satisfactory performance in practice. This is not straightforward. For example, a low or high curing temperature may lead to inadequate performance which would not be shown by tests at 20°C, e.g. thaumasite formation and delayed ettringite formation. Durability performance testing has to be either relatively short term or accelerated and, relating this to the performance of the structure over its design life is an area fraught with difficulties. Passing the 'safe' test would be a reasonable prognosis for future performance. However a producer with a material that has a successful 50 year track record is unlikely to accept a 'safe' test and/or criteria that rejects this material.

Many of the durability performance tests currently under consideration take several months to complete. As such, while they are not suitable for routine control, they nevertheless have a function as an initial type approval test, which is:

> a) A test conducted prior to production to establish a mix that gives the specified performance, or

> b) a test used to establish equal or better relative performance from an unproven concrete or constituent material to a concrete or constituent material of established performance.

Tests which take months to complete may be suitable for ready-mixed concrete producers with continuity of supply of materials but they are not much use to a site wishing to start production quickly, using unknown materials. For each exposure class a number of options should be retained. These should comprise a traditional mix limitations approach and one or more performance options.

An initial type approval test would have to be supplemented with a routine control test. This is:

A test used for production/conformity/acceptance purposes to ensure that

a) the specified performance is achieved, or

b) the performance established by an initial type approval durability performance test is maintained, or

c) the performance translated to mix limitations is maintained.

A routine control test may not be a performance test but a simple indicator that the mix or constituent materials have not changed significantly, i.e. a concrete strength test.

Taylor[2] has stressed the importance of precision statements and estimates of uncertainty when specifying test methods and determining the criteria. The importance of these in a contractual situation cannot be over-emphasised as they form the basis by which industry determines the margin (the difference between the required value and the target value) and assesses its commercial risk.

Durability performance tests can be used for:

research;
technical approvals;
direct specification requirements.

As this is also likely to be the evolutionary process by which these tests are adopted, these uses are described in the following sections.

RESEARCH

A very high proportion of the research on concrete and its constituent materials is aimed at durability. In a few cases existing standardised tests are used and, in others, a test is developed to reflect the interaction between an environment and a specific product. However, in many cases, in-house tests are used or a test used elsewhere is modified to suit the equipment or views of the researcher. Rarely is the precision of such tests established (even for repeatability).

The use of different tests makes the comparison of results very difficult and delays or prevents industry benefiting from the considerable investment in research. A suite of standardised durability tests would be of benefit to users of concrete research. Equally important to industry is a clear record of the limitations of the test procedures and the relationship between performance in the test and performance in practice.

TECHNICAL APPROVAL

There are several levels at which technical approvals apply:

1. At the site level, to show that a specific concrete or constituent material performs as well as an approved concrete or material.

2. At the European Technical Approval level, to show that a specific source of a constituent material, combined with a wide range of other materials, gives concretes of equivalent performance to those with an established track record.

3. At the standardisation level, to show that a concrete or constituent material, at the limits of its draft product standard, performs as well as concretes or constituent materials already permitted in the standard.

In the short to medium term, these will be the most important applications of concrete durability performance tests. All current concrete standards are written in terms of material and mix limitations and this is likely to be true for the first revision of the European concrete standard, EN 206. Concrete specifications will take the same approach and they will reflect local materials and practices which both have established track records. However, the European Union is a single market with its standards reflecting a much wider range of materials and practices all of which can 'freely be placed on the market'. This places the public purchaser and many utilities in a difficult position if a constructor offers an alternative to the original specification. Whilst they are not obliged to accept the alternative, the reasons for rejection would have to be based on good technical or other grounds and not on protecting local industry. Durability performance testing will often provide a technical means by which the constructor's alternative can be accepted or rejected.

There will be problems. Durability performance tests for technical approval can be, and are likely to be, based on relative testing; the unproven concrete or constituent material being compared with a concrete or constituent material with a proven track record. However, unless the basis for comparison and the reference are clearly defined and agreed, there will be endless disputes. Taking as an example the application of the carbonation test to assess an unproven cement. If it is wished to minimise the differences between this unproven cement and other cements, the basis of comparison should be equal 28-day concrete strength and equal consistence. Conversely, if it is wished to emphasise the differences between cements, the basis of comparison should be equal water/cement ratio and equal cement content. Both approaches can be argued technically but they may well lead to different conclusions.

A further difficulty is the choice of reference cement. If the local cement is a high strength Portland cement, it can be argued that this should be the reference cement.

However, as these cements tend to give the lowest carbonation at equal water/cement ratio and cement content, by selecting this cement and basis for comparison, all other cements would be kept out of the local market.

At the other extreme are countries such as the UK, where none of the standardised cements are prohibited for use in carbonation conditions. The reference cement could be selected as the one that gives the highest level of carbonation in the test on the basis that it is a permitted cement. This could lead to some very questionable concretes passing the test.

The CEN draft test method leaves the users to select from three storage conditions:

a) 20°C ± 2°C and 65 ± 5% RH;

or

b) as above plus on the 28th day and every succeeding 28th day, the concrete test specimens are immersed in potable water for 6 hours;

or

c) as the first option plus on the 7th day and every succeeding 7th day, the concrete test specimens are immersed in potable water for 6 hours.

The object of having these three storage conditions is to allow the user to select a storage condition that approximates best to the intended conditions of exposure. The wetting cycles reduce the absolute level of carbonation and, when frequent enough, result in the concrete at a certain depth remaining wet and consequently prevent further carbonation. There is also the expectation that concretes stored in these different conditions will have different relative performances. There is a logic to having a choice of exposure conditions but this will not prevent the selection of the storage condition becoming a matter of dispute.

These types of practical problems, which occur when durability performance tests are used in a relative way, will increase pressure to find ways of specifying performance directly and in absolute terms.

DIRECTLY SPECIFIED REQUIREMENTS

This is the system where the specification contains absolute performance requirements linked to a specified test and procedure. The specification should contain clear rules on how, and by whom, conformity is to be established .

Absolute performance requirements can be applied directly to the structure or precast concrete unit, to concrete test specimens made and cured in a standardised way or to a constituent material. A recent Concrete Society Report, *Developments in durability design and performance based specification of concrete* [1], has taken a pragmatic approach to what should be the prime point of testing. Their conclusions with respect to in-situ testing are summarised in Table 1. Tests that take a long time to complete are only suitable as type approval tests and this is reflected in the table.

Table 1. Type of test and prime point of testing

Test	Constituent material	Concrete test specimen	In-situ test	Time to complete test
Cover			Routine	2 to 15 days
Carbonation		Type approval		1 year
Transport coefficient for chlorides		Type approval		49 days to 13 months(1)
Gas permeability		Routine		15 to 41 days
Freeze/thaw		Type approval when unknown materials used	Possible type approval	49 to 87 days (1)
Sulfate/sea water attack	Possible type approval	Type approval		13 months
Other chemical attack		Type approval		
Abrasion			Routine	$\geq$ 29 days(2)
Temperature rise and difference			Routine	14 days
Coefficient of thermal expansion		Type approval		5 days
Alkali-aggregate reaction	Type approval	Type approval if other guidance not followed		1 year
Delayed ettringite formation	Type approval			54 or 200 days (1)
Drying shrinkage	Type approval			11 days

Notes
1. Depending on test method   2. Depends on the in-situ development of maturity

A main difficulty with this approach is setting the criteria and conformity rules. Two basic approaches are possible:

i) test structures, concretes or constituent materials as appropriate that have performed satisfactorily over time, and develop criteria from these results:

ii) use an explicit design method [1] to establish the criteria.

Care is needed with both approaches. A concrete that is performing well in practice may be in excess of that specified and therefore the relationship between the specified and actual concrete supplied has to be determined. Constituent materials may have changed significantly over the years and simply batching the same proportions of the same notional materials may not give the correct indications. Explicit design methods are still under development and must be used with care and judgement. The two approaches to setting criteria can be combined to give greater confidence in the conclusions.

The tendency to set over-safe criteria must be resisted as this will hinder the continued use of materials, concretes and designs that have performed adequately over many years, and would adversely affect the competitive position of concrete vis-à-vis other competing materials.

The criteria must also take account of the test precision, uncertainty and the fact that the constructor or concrete producer will want to work with a margin that reduces their risk of rejection to acceptable levels.

The tests listed in Table 1 are at different stages of development.

STAGES OF DEVELOPING DURABILITY PERFORMANCE SPECIFICATION

Table 2 on the next page summarises the stages of developing performance specifications[3].

This table should be regarded as a general framework. Stage 2: Design methodology, is not required if the conformity criteria are to be based solely on previous experience.

An underlying assumption in the development of performance specifications is that they are an alternative to method, material and mix limitation specifications. Once a performance criteria is agreed, constructors and concrete producers will find the most cost effective way of passing the criteria. Stage 5: Limitations of test applicability, is therefore vital in that it will establish the bounds within which the performance criteria can operate. As technology develops, the implications for the durability performance criteria will have to be periodically reviewed.

Table 2 also illustrates the substantial effort needed before durability performance specification is a trusted and viable option alongside the traditional approach of material and mix limitations.

The next section reviews the progress with the development of durability performance specifications.

Table 2. Stages for developing a performance specification for concrete

Stage	Notes
1. Exposure class	Defining exposure classes with respect to a form of deterioration
2. Design methodology	Develop a quantitative design method and a definition for the end of the design working life
3. Test procedures	Develop test methods that relate to the output parameters of the design method.
	Develop sampling plans, precision statements and conformity procedures
4. Provisional conformity requirements	From stages 1 to 3 establish provisional conformity requirements. Check against traditional solutions.
5. Limitations of test applicability	Testing specimens that satisfy the performance requirement, but were produced at the limits of concrete variation and technology to confirm that adequate in service performance will be achieved. If not, limits on the materials that may be used will have to be introduced.
6. Production control / acceptance testing	Establishing effective systems for production control and acceptance testing
7. Trials	Full scale trials and long term monitoring of the resulting structure to confirm the conformity requirements

PROGRESS WITH DURABILITY PERFORMANCE SPECIFICATIONS

Cover

The lack of specified cover is the most common cause of corroding reinforcement. In broad terms, a 50% reduction in the cover to reinforcement results in a 75% reduction in the life. Cover can be measured with a sufficient degree of accuracy using a range of commercially available covermeters. To achieve an acceptable degree of accuracy, one of the more modern meters should be used. It should be calibrated for the specific concrete and reinforcement using a beam with an offset reinforcing bar. Reasonable precision is needed to minimise 'purchaser's risk'. The producer's risk is very low as, when an area of low cover is detected by the meter, a trial hole is cut to confirm that the cover is out of specification.

The conformity criteria are relatively simple; a minimum and possibly a maximum cover. However, there is scope for developing more sophisticated criteria using the concept of a characteristic cover and a lowest individual cover [1].

The cover shown on drawings, and to which the reinforcement should be fixed, is called the nominal cover. The minimum cover to which the conformity rules apply is some lower value. In the current UK design code[4] it is 5mm less than the nominal. However, various reports [5 to 11] have shown that even with good workmanship, achieving a margin of 5mm is not possible on site. A more achievable margin should be introduced in the UK codes and standards as a matter of urgency.

Given the importance of achieving the specified cover, and the fact that recent surveys show the specified cover is still not being achieved, the enforcement of the durability performance requirement for cover should have the highest priority. It can, and should be, applied immediately.

Carbonation

The concern of designers is with corrosion due to carbonation and not carbonation per se. Unless the internal relative humidity of the concrete cover reaches high levels, corrosion of reinforcement is sufficiently slow to give design lives of hundreds of years, even for carbonated concrete.

CEN TC51/WG12 is developing a carbonation test for use in a relative way. This should be regarded as a first step and its refinement into an absolute test is a relatively small step. What is more difficult is linking the output of the test to what is needed. Trial calculations [1] show that explicit design methods may provide a means for doing this but further research and development is needed. Given the time to complete a carbonation test, even an accelerated test, this type of testing only has a role in type approval testing where there is continuity in the supply of materials, e.g. a ready-mixed concrete plant or precast works.

The type approval test should establish the relationship between concrete strength and depth of carbonation. By interpolating, the strength necessary to achieve the specified depth of carbonation can be taken as the characteristic strength provided this is greater than that needed for structural purposes. Normal production control techniques can be used to ensure that this is achieved in at least 95% of cases, which will also result in the carbonation depth being less than specified in 95% of cases.

Corrosion due to carbonation is the only significant durability consideration for a substantial volume of concrete and therefore developing performance criteria could lead to significant economic benefit and as such warrants a high priority. The current stage of development of carbonation performance specification is given in Table 3.

Table 3. Progress with developing carbonation performance specifications

Stage	Notes
1. Exposure class	Basic system is developed. Research and guidance on correct selection of exposure needed
2. Design methodology	A number of analytical methods are available. Need calibration and refinement.
3. Test procedures	Relative test developed. Need to modify for use as an absolute test. Precision has to be established. Test procedures proposed.
4. Provisional conformity requirements	Still to be developed
5. Limitations of test applicability	No substantive work started
6. Production control / acceptance testing	Suitable system of routine control is likely to be strength
7. Trials	None in progress

Transport coefficient for chlorides

In saturated concrete, the chloride ion is transported into the concrete by diffusion. The relevant diffusion is based on the free chloride, but for simplicity the effective diffusion coefficient is often used. The change of diffusion coefficient with time, and the continuing debate over threshold values, complicate setting performance criteria.

However, in many structures where corrosion due to chlorides is a problem, the transport process is not just diffusion but a combination of processes including capillary suction, wick action, evaporation and hydraulic gradients. This is an area where further research is needed and it is likely to be a decade or more before these developments lead to general performance specifications for the concrete.

Gas permeability

One of the design methods for resisting corrosion due to carbonation[12] uses gas permeability as its main parameter in conjunction with a hydrate binding factor that depends on cement type. The significant advantage of such a system is the relative speed at which conformity can be established. Its disadvantage is that it is not directly related to corrosion due to carbonation and requires an adjustment for cement type.

How the specimens are to be conditioned is another key issue but research[13] has indicated possible methods. The precision of the test method is under investigation.

The advantage of speed of conformity applies equally to a range of other permeation tests. As permeability and permeation can be correlated to a number of aspects of durability(14, 15) and as the tests can be completed relatively quickly, there is interest in using such tests as a durability related measure of performance.

Freeze/thaw

CEN is developing two test methods for the assessment of the scaling resistance of concrete[1] and is working on test methods for assessing internal damage. RILEM is developing a different method[16] which may, in the future, be adopted by CEN.

The CEN test methods are relative tests due to indications of poor reproducibility. Work is in progress to improve the precision. Suitable performance criteria have to be established for the different exposure classes and the role of the test needs to be clarified. Views range from limiting testing to type approval where unproven materials or non-air entrained concretes are used to routine control testing on in-situ samples of all concretes subjected to freeze/thaw conditions.

Sulfate / sea water attack

CEN have developed a relative test method but, due to indications of poor reproducibility, it is only being progressed as a relative test. Its main role is likely to be in technical approvals, particularly for cements or combinations that are not recognised throughout the European Union as having sulfate resisting properties.

Evolution of the test into a test for the mortar fraction of concretes and the setting of absolute performance criteria for this mortar is likely to take at least a decade.

Abrasion

Providing the concrete has the potential to give an abrasion resistant surface, the finishing operations and curing will determine if this potential is achieved. In the case of abrasion, the structure should be the prime point of testing. Wear wheel testing[17] can be applied in-situ on flat or nearly flat surfaces, but the test method is not standardised nor is its precision established. Criteria have been proposed[17] and the test has a good track record of use in cases of disputes over abrasion resistance. Once the precision of the test method is established and shown to be reasonable, a direct performance requirement could be developed and standardised relatively quickly.

CONCLUSIONS

1. The use of durability performance tests will evolve from research, through technical approvals, to the basis by which directly specified performance requirements are assessed

2. Where execution is critical and poor execution not obvious, performance requirements are best placed directly on the structure or precast unit.

3. In most cases it will be adequate to place performance requirements from concrete on standardised concrete test specimens.

4. Industry is only likely to support standardisation of durability performance requirements if all the stages of evolution have been completed successfully.

5. The minimum time needed to develop standardised durability performance specification is estimated to be as follows:

Cover	Immediately
Corrosion due to carbonation	≤ 10 years
Corrosion due to chloride diffusion	≤ 10 years
Other corrosion due to chlorides	≥ 10 years
Freeze/thaw	≤ 5 years
Sulfate / sea water	$\geq 5, \ \leq 15$ years
Abrasion	≤ 5 years

REFERENCES

1. THE CONCRETE SOCIETY. *Development in durability design and performance based specification of concrete.* CS 109, Feb 1996. ISBN 0 946691 54 1

2. TAYLOR, H. *From research to practice: the design of specifications.* Magazine of Concrete Research, 47, No.170, March 1995

3. HARRISON, T A. *Framework for durability performance specification for concrete used in normal construction.* Proceedings XIth ERMCO Congress, Istanbul, 1995, pp216-224

4. BRITISH STANDARDS INSTITUTION. *Structural use of concrete. Part 1: Code of practice for design and construction.* BS 8110: Part 1: 1985

5. MORGAN, P R, NG, T E, SMITH, N H M and BASE, G D. *How accurately can reinforcing steel be placed? Field tolerance measurement compared to codes.* Concrete International, October 1982

6. CUR Report 113. The Dutch Centre for Civil Engineering research and Codes (Commissie Vor Uitvoering Van Research) Zoetmeer, Betonvereninging 1984. English translation: Report 113 Concrete Cover. C & CA

7. MAROSSEZEKY, M and CHEW, M. *Site investigation of the quality of rein-forcement placement on buildings and bridges.* New Zealand Concrete Construction, April 1990

8. MAROSSZEKY, M and CHEW, M. *Site investigation of reinforcement place-ment on building and bridges.* Concrete International, April 1990

9. CLEAR, C A. *A review of surveys of cover achieved on site.* BRE Occasional paper OP43, December 1990

10 BIRKLAND, P W and WESTHOFF, L J. *Dimensional tolerances in a tall con-crete building.* ACI Journal Proceedings V68, August 1971

11. DEUTSCHER BETON-VEREIN. *Recommendation Concrete Cover. Security of concrete cover in design, production and positioning of the reinforcement as well as of the concrete.* (Version March 1991). Betonwerk and Fertigteil - Technick (Concrete Precasting plant and Technology) Issue 5/1992

12. PARROTT, L J. *Design for avoiding damage due to carbonation-induced corrosion.* 3rd International Conference on Durability of Concrete, Nice 1994. ACI SP-145, 283-298

13. PARROTT, L J. *Moisture conditioning and transport properties of concrete test specimens.* Materials and Structures, 1994, Vol 27, 460-468

14. RILEM. *Performance criteria for concrete durability.* RILEM Report 12, 1995. ISBN 0 419 19880 6. Published by E & F N Spon

15. DHIR, R K, HEWLETT, P C , BYARS, E A and BAI, J P. *Estimating the durability of concrete structures.* Concrete, Vol 28, No.6, Nov/Dec 1994, pp25-30

16. SETZER, M J. *Freeze/thaw and de-icing resistance of concrete.* Proceedings XIth ERMCO Congress, Istanbul, 1995. pp225-238

17. CHAPLIN, R G. *The influence of ggbs and pfa additions and other factors on the abrasion resistance of industrial concrete floors.* British Cement Association 1990

BENCHMARKING THE DURABILITY
OF CONCRETE STRUCTURES

D C Spooner

British Cement Association

UK

ABSTRACT. A progress report is presented on a DOE, 'Partners in Technology' project, the purpose of which is to draw up guidelines for a future survey of the long-term durability performance of concrete structures. The interim conclusions are that best use of resources is achieved by visual surveys of a large number of structures, rather than more detailed study of fewer structures. Also that the survey should focus on the performance of structural elements in specific exposure situations, using the draft EN 206 classification of exposure. It is proposed that structures between 10 - 27 years old should be examined in the first survey.

Keywords: Concrete, Durability, Survey, Structures, Exposure.

Professor David C. Spooner is Director of Materials and Standards at the British Cement Association and visiting Professor at the University of Sheffield. His research interests include concrete properties and durability, test methods and the thermal performance of concrete buildings. He is active in both UK and European Committees concerned with standards for cement and concrete.

Radical Concrete Technology. Edited by R K Dhir and P C Hewlett. Published in 1996 by E & FN Spon, 2–6 Boundary Row, London SE1 8HN, UK. ISBN 0 419 21480 1.

INTRODUCTION

This paper presents a progress report on a project being undertaken under the UK Department of the Environment's 'Partners in Technology' scheme and entitled 'The Scope and methodology of a UK survey to provide the benchmark for the durability of concrete structures'.

A very considerable proportion of the UK annual investment in concrete research is targeted on durability but the relative importance of the various durability mechanisms - corrosion of reinforcement, freeze-thaw attack, sulfate attack, etc. - has not been established by survey, nor have the frequencies of occurrence of the mechanisms. As a result of this lack of guidance, it is not possible to be sure that research resources are being targeted in a way that ensures the greatest benefit from the work. Equally importantly, without a 'benchmark' of the national state of concrete structures regarding durability, it is very difficult to assess the effectiveness of changes in codes, standards and other specifications affecting materials, design and construction. The need for such a national survey has been recognised in the Department of Environment's strategy document 'Durability by Intent' [1].

Having established that, in principle, a UK national survey of concrete durability is necessary, it only takes brief reflection to realise what a large undertaking this could be and that there would be considerable demand on available resources. Clearly, as a first step, a study of the required scope and methodology of such a large undertaking is required in order to establish what needs to be achieved, how it might be done and at what cost. This latter study of the scope, methodology and cost is the subject of the DOE 'Partners in Technology' project and progress in the work is described here.

OBJECTIVES, PARTNERS AND MODUS OPERANDI

The main objective of the study is to produce a methodology for a future UK national survey which will establish the proportions which suffer specific durability problems. In order to devise that methodology, it was decided that the international literature would have to be examined to see what help could be found there. Since it was presumed that some of that literature would, in addition to methodology, contain data on the results of surveys, a further objective of this study was to provide a 'snapshot' of the current views of durability problems as far as was possible. The limitations of such a 'snapshot' were realised in terms of applicability of other countries' durability problems to UK conditions, where codes, standards and climate may well be different. None-the-less, the exercise was thought to be worth doing as it would provide at least some information fairly quickly.

The partners in the programme are:

> The British Cement Association
> Building Research Establishment
> Gifford and Partners, Consulting Engineers

University of Sheffield
DOE via the 'Partners in Technology' programme.

The project budget cost is £46K and will be completed in April 1996.

The work has been divided into tasks and these are:

Task 1   Obtain international literature

Task 2   Review scope, methodology and survey-
             results

Task 3   Produce recommendations and work sheets
             covering scope, scale and cost

Task 4   Report survey-results and recommendations
             for the future survey of UK concrete structures.

The tasks above have been shared among the partners according to their particular expertise. The project is controlled by reports on progress given at regular meetings of the partners and at which, the direction of the project is reviewed and target dates agreed for the actions.

REVIEW OF THE LITERATURE AND CONCLUSIONS DRAWN SO FAR

Results of the Review

Searches were undertaken of databases of the BCA (available from the BCA Library Service via 'Concquest' membership), of BRE and of Gifford and Partners and keywords such as 'durability', 'deterioration', 'survey' were used, among others and, not surprisingly, the lists of references obtained were very extensive. A list of some 500 references was finally obtained and the titles of these were assessed by each partner and 'scored' for likely value to the project. Based on this initial 'scoring', some 40 papers were read in detail and the conclusions put onto a database. One conclusion that was very clear was that the titles of papers frequently give no good indication of their contents! More usefully though, the papers divided into three general groups:

(a)   Methodology of surveys

(b)   Surveys of numbers of structures

(c)   Surveys of single structures or of only
      a very few structures.

Groups (a) and (b) were obviously directly relevant to the current project, whilst group (c) was less valuable. Unfortunately, it was clear that there was not much information available on the methodology of durability surveys. Frequently the guidance related specifically to structural (loading) assessments [2], sometimes contained extensive check lists of desirable information to be obtained but little on the basic principles of attempting a national survey. Too often it seemed that the guidance consisted of lists of everything it might be desirable to know about a structure from detailed drawings, through specifications, to results of visual, non-destructive and semi-destructive investigations. Whilst not denying the value of such information in limited surveys, such detail is clearly out of the question, from a timescale and resources point of view, when very many structures are to be examined.

Scope and Nature of the Survey

For us, a balance has to be struck between the number of 'structures' required to be examined to provide a 'significant' conclusion, the cost in time and staff resources of each 'structural' examination and the value of the information obtained in terms of level of detail on the durability problems.

After discussion, our view was that, at least initially, we required to examine a greater number of structures, albeit fairly superficially, rather than a few structures in great detail.   But would the results of a fairly superficial survey of a large number of structures provide useful information? It was evident from the literature that assessment of structural defects is probably not an exact science and that guidance would have to be given to 'assessors' on classification of durability problems. None-the-less, it was felt that a visual inspection, done by a trained and experienced engineer, would be able to provide useful information on the following factors:

Concrete quality:	-	compaction
	-	segregation
	-	plastic settlement
	-	other defects in the material
Concrete cracking:	-	differentiate between
		- thermal cracking
		- structural "
		- plastic "
Structural movements:	-	general expansion or contraction of elements
	-	foundation problems
Durability aspects:	-	frost attack
	-	rust stains indicating reinforcement corrosion (assess whether likely to be due to carbonation or corrosion)
	-	cracking due to expansive mechanisms
	-	cracking due to shrinkage mechanisms

- erosion due to chemical attack
- wear due to abrasion

The limitations of a purely visual survey have to be recognised:

- inaccessible parts of the structure can only be examined using, say, binoculars
- no direct evidence of the condition of, say, cover-to-steel can be obtained
- no direct evidence of the condition of sub-ground concrete can be obtained.

However, it is arguable that the resources needed to overcome these limitations of a visual survey are very great in comparison with the extra data which is obtained and are thus not really justified when establishing what proportion of structures suffer durability problems and categorising those problems into broad headings.

It may be that the results of the visual survey identify the need to conduct more detailed examinations on a limited number of structures. This might be, for example, to confirm the visual diagnoses or to investigate unfamiliar mechanisms of deterioration. Such detailed examinations would probably include the full range of NDT.

To Survey by Structure type or by Element Type?

It is clear from the literature that, not unnaturally, many surveys have been conducted by 'owners' of particular structural types (bridges, cark parks) [3,4] in order to help them plan future maintenance strategies and to manage resources. However, the question has to be asked whether, in any national survey, the division of the UK stock of concrete structures simply by structural type is appropriate or helpful. For example, a concrete column performs a similar function supporting a bridge deck as it does supporting an office-block floor. The reason, say, that more bridge columns currently suffer from reinforcement corrosion than do floor columns, is much more a function of the local exposure conditions than of the structural type.

The basis of current (and foreseeable) durability specifications for the materials side of concrete construction is to identify the likely severity of the exposure conditions and to select concrete properties accordingly. Thus the materials durability specifications are selected on the basis of the presumed local exposures, not on the basis of the structural types. The broad headings of durability exposure in the latest draft of the European concrete standard EN 206 [5] are:

No risk of corrosion or attack	Carbonation induced corrosion	Chloride induced corrosion	Aggressive chemical environments	Freeze-thaw attack

Internal chemical reactions leading to expansion, such as alkali-silica reaction, are not addressed in these headings but are referred to in the text, where national guidance is advised. Thus referring back to our example of columns, a bridge column over a motorway might be categorised in column (3) because of the risk from de-icing salts and an office-floor column in column (2).

As a result of this discussion, it seems more sensible to conduct the future survey from the point of view of the types of concrete elements rather than on the type of structure, and to register clearly in the survey the particular local exposure condition pertaining - for example, based on the EN 206 classification. When the question of numbers of concrete elements to be surveyed arises, then the type of structure would be relevant in providing a range of different exposure classes. Also, any database of the survey results would need to be addressable either on the basis of type of element or type of structure.

THE FIRST AND SUBSEQUENT SURVEYS

Age of Structures and Frequency of Surveys

An important question not addressed so far in this paper is that of the ages of 'structures' to be surveyed and, coupled with this, how frequently such national surveys should be conducted. The question of the age of the structures is one of most significance to the first survey, since it provides the 'benchmark' against which hopefully improving practices and procedures will be judged by the results of succeeding surveys. There seems little point either in surveying very young structures, such that no durability problems have had time to develop or in surveying very old structures, such that the causes of defects can now be expected to have been eliminated by historical changes in codes or 'state of the art'. It is felt that consideration of structures post 1970 is reasonable, and the date corresponds to the last major changes in the UK of the concrete codes governing durability. If it is assumed that the survey might be done in 1997, then this would mean that structures with ages up to 27 years could be considered. We probably don't know in general terms how long it takes for durability problems to become apparent and thus it is difficult to judge how soon after making some 'improvement' to practice, one might detect it by survey results. It seems sensible therefore in the first survey to cover structures with a wider age range than might at first seem appropriate, in order to try to determine when defects in durability become evident. Perhaps consider structures from 10 - 27 years of age? Depending on the results of the first survey, in terms of the age at which defects become evident, then the decision on the frequency of surveys can more reasonably be made.

Scope of Surveys and Costs

The present view is that the survey will be organised to report on the durability of concrete elements under a variety of exposure conditions and which are of ages of between, say, 10 - 27 years. Sufficient numbers of elements will be examined to provide some confidence that the results can be expected to apply to the general population of such elements in the UK, for the same exposure condition.

A classification of element types will be used which will describe the generic concrete type, non-reinforced, reinforced, in situ, precast, prestressed, etc. and the type of element, beam, column, panel, slab, etc.

The surveys will need to be conducted by specially trained engineers who will report on visual inspections, using methods to facilitate construction of a computer database. As part of their reports, the engineers will classify the environment local to the particular element.

The actual numbers of elements (structures) which need to be examined is a difficult question and logically should be a certain proportion of the total of such elements in the UK. As always, the number will be a compromise between what would ideally be required and what can reasonably be afforded. Certainly, several hundreds of structures will have to be examined, in order that the numbers of elements examined in each exposure classification are significant.

As to costs, they are proportional to the number of elements (structures) surveyed. If it were supposed that a visual examination and report of one structure could be done in two days by one qualified engineer, then we might estimate a cost of around £600-1,000 per structure. If 500 structures were thought to be needed statistically (there is no evidence yet to support such a figure), then the cost of the survey possibly approaches £0.3-0.5 million. However, this cost might reasonably be spread over two or three years, in order to limit the number of engineers involved in the assessments (and thus the subjective differences in those assessments).

REFERENCES

1. DEPARTMENT OF THE ENVIRONMENT. 'Durability by intent'. Strategy for DOE programme on concrete and reinforced concrete. Construction Sponsorship Directorate, May 1995.

2. INSTITUTION OF STRUCTURAL ENGINEERS. Appraisal of existing structures, 60 pp, London (1980).

3. THE DEPARTMENT OF TRANSPORT. The performance of concrete in bridges, 96 pp, Her Majesty's Stationery Office, April 1989.

4. LITVAN, G. AND BICKLEY, J. A. Durability of parking structures: Analysis of a field survey. Concrete Durability, American Concrete Institute, ACI SP-100, Vol. 2, 1503-1525 (1987).

5. CEN/TC104/SC1. Draft EN 206, Concrete, document N249, August 1995.

PERMEABILITY OF CONCRETE:
A PRACTICAL APPROACH

C Tait

A Kermani

F M Khalaf

Napier University

UK

ABSTRACT. The deterioration of concrete has caused concern for as long as concrete has been in use as a construction material. A review of literature shows that permeability is the governing factor of concrete durability. Virtually all current knowledge of concrete permeability has been obtained on unstressed and therefore uncracked laboratory specimens (together with some limited site data on structures of undefined stress history). This is in complete contrast with the concrete of real structural elements which contain microcracks caused not only by structural stresses but also by other environmental conditions. Testing of stressed samples show the importance of microcrack development and the influence that this parameter has over permeability.

The permeability of hardened PC concrete was studied in relation to the changes in the internal structure of concrete under various short term stress levels, ranging from zero to 70% of the ultimate load. The effects of an air-entraining agent and PFA as a part cement replacement were also investigated, as they effect the composition, strength and hence the propagation of microcracks in concrete. It was found that the permeability increased sharply with an increase in applied stress level of above 40%. The increase in test duration showed that the permeability increased to a maximum after 4 days eventually decreasing to a constant due to clogging and swelling of the concrete pores.

Keywords: Durability, Microcracks, Permeability, Porosity, Stress, Test duration.

Mr Colin J Tait is a research assistant in the Department of Civil and Transportation Engineering, Napier University, Edinburgh

Dr Abdy Kermani is a lecturer in the Department of Civil and Transportation Engineering, Napier University, Edinburgh.

Dr Fouad Khalaf is a lecturer in the Department of Civil and Transportation Engineering, Napier University, Edinburgh.

Radical Concrete Technology. Edited by R K Dhir and P C Hewlett. Published in 1996 by E & FN Spon, 2–6 Boundary Row, London SE1 8HN, UK. ISBN 0 419 21480 1.

INTRODUCTION

Since concrete was first used permeability, as its inherent property, has caused a variety of problems for engineers due to its ability to influence deterioration and hence durability. Permeability controls not only the ingress of water but of oxygen and chlorides which are necessary to start the electro-chemical reaction associated with carbonation and chloride attack.

Many test procedures have been developed over the last few decades, many now commercially available, and although they have shown that water/cement ratio, curing and various other factors are influential in governing permeability, they tend to focus on unstressed well cured laboratory specimens or in-situ samples of uncertain stress history.

Permeability of concrete is also directly related to its pore structure.   The existence of capillary and gel pores in cement paste, the air voids, capillary tubes and pre-existing microcracks govern permeability of the intact unstressed specimen. Stressing concrete causes microcracking to spread within the existing pore structure and in turn increase permeability.

By using pre-stressed and hence pre-cracked samples, this study shows the importance of the stress history of a structure and how influential this can be over the permeability.

REVIEW

Permeability can be defined as the rate of movement of a liquid or gas through concrete and is paramount in determining deterioration [1].   It has been realised that for an impermeable concrete, impermeable aggregate and paste once hardened as well as sufficiently workable mix to fill all the voids is required.   Concrete however, due to its constituent materials, their internal chemical reactions and the mixing and placing technique used in practice, combine to make this material inherently porous.

Durability and therefore permeability is related to the type and distribution of the pores which tend to concentrate on the smaller radii, hence the microstructure [2]. Voids initially form when water rests beneath aggregates [3], then follows a chemical reaction (hydration) between the water and cement [4] creating a dense, porous substance known as cement gel. This gel, although considered solid, consists of hydration products, water (combined or uncombined) and gel pores which are much smaller than capillary pores (about 1μm compared 2nm). These capillary pores are gradually reduced in diameter as the cement gel 'bridges' across them. Powers [5] suggests that the cement gel grows to 23 times greater than the original volume of the cement, depending upon cement fineness; Nyame [6] suggests that total elimination of capillary pores occurs at a water/cement ratio of less than 0.38. However, the cement gel also has an intrinsic strength based upon either physical attraction or chemical bonds and is the soul source of strength for cement paste.   Strength of concrete therefore is related to porosity with permeability more related to pore continuity [7].   Pore continuity can be indicated by calculating the pore size distribution (psd) using a variety of methods [1,6,8].

These methods include the Mercury Intrusion Porosimeter (MIP), Sorption methods to measure smaller pores and Helium or Methanol Pycnometers. Feldman and Beaudoin [9] and Day and Marsh [10] both suggest that it may be impossible to measure the psd of cement paste due to the stresses imposed by the MIP, with the later stating that a battery of testing methods should be used to measure pore continuity.

Permeability results, although notoriously difficult to compare, can be simply and adequately calculated using Darcy's equation, although must be applicable for laminar flows only. Viscous drag theories for measuring permeability have also been used [11] as well as complex relationships between pore size distribution and permeability as investigated and experimentally verified by Luping and Nilsson [12]. For calculation of gas permeabilities, mass flow theories may be used.

Various investigators have shown that the constituent materials can effect the concrete performance. Early work [13-15] shows that permeability was affected by grading and size of aggregate, end effects of specimens, test duration and applied pressure [16], temperature and most importantly, the water/cement ratio. For cement pastes that are at the same degree of hydration, permeability has been found to be lower at water/cement ratios of less than 0.6, where the capillary pores have become segmented [17].

Numerous subsequent work has also supported the idea that water/cement ratio is paramount with further work by Bamforth [18] showing that at very low temperatures, air-entraining agents and aggregate grading do effect the permeability of concrete. Nyame [19] suggests that the inherent permeability for mortar depends on both paste and then the aggregate. Aggregates, although impervious for all practical mixes, can effect the permeability of mortar by creating microcracks between themselves and the paste and also by making the ingress flow path more tortuous. The grading of aggregate has also been shown to control to a limited extent, porosity, drying shrinkage and workability of a mix [20-22].

Using a unique testing method Schonlin and Hilsdorf [23] were able to show that curing has a direct effect on the surface characteristics of concrete thus playing an important part in governing permeability. Permeability of steam cured concrete is also higher than that of wet-cured concrete. Dhir et al. [24] also showed that the permeability of cover concrete to reinforcement is also very sensitive to moist curing.

Test techniques for the measurement of permeability differ in two main ways. Direct methods allow test to reach steady state conditions, though this tends to be time consuming and may not occur until there is complete hydration of the cement. In-direct methods concentrate on the amount of water that can be forced into a specimen in a given time.

After initial simplistic test methods [25,26], there are now a wide variety of experimental apparatus commercially available for testing both in the laboratory and on site. Figg [27,28], Kasai [29] and Hansen et al. [30] all use semi-destructive test involving a hole being drilled into concrete samples and then pressurised, after which a measurement of permeability can be made. Hudd [31] used a non-destructive version of the Figg test.

A review of available in-situ concrete tests can be seen in refs. [32,33]. However the Initial Surface Absorption Test is the only test specified as a British Standard [34] and is simply a funnel being sealed to a specimen; water then added via a flexible tube with the inflow measured at set times. Problems arise with air locks in the tube, sealing of the cap and moisture content of the sample; (these are all synonymous with permeability tests).

Laboratory tests e.g. van der Muelen and van Dijk [35], Cabrera and Lynsdale [36], tend to concentrate on a fully enclosed and sealed permeability cells by which water or air can be applied to the specimen, in one direction only.

A study of the available data reveals that virtually all current knowledge of concrete permeability has been obtained from unstressed laboratory specimens with some limited site data on structures of some undefined stress history. This is in complete contrast with concrete of real structural elements which contain microcracks not only induced by stressing but also by various environmental conditions.

EXPERIMENTAL DETAILS

This investigation of the permeability of stressed concrete involved three differing mixes. Mix 'A' was an ordinary structural concrete using PC, mix 'B' was designed as a special structural concrete using PFA as a part cement replacement to supplement PC and mix 'C' used PC with an air-entraining agent as an admixture. A maximum coarse aggregate size of 20mm and a characteristic compressive strength of 30 N/mm^2 at 28 days was specified for all mixes.

The specimens were cast in cylindrical moulds, 200 x 100 mm dia., and cured in water at a constant temperature until time of test after which the ultimate compressive strength, f_u, was initially determined. The remaining specimens were then stressed to levels of 0, 0.3, 0.4, 0.5, 0.6 and 0.7 of the average cylinder crushing strength for a period of 5 mins. after which the load was removed.

A 50 mm section from each end of the specimens was then removed by a high speed cutter. This was due to doubts over the effects of end conditions caused during casting and stressing and also to gain information on the arrangement of the aggregates in the various mixes.

The permeability testing system contained three main parts. The pressure system operated from a low pressure air pump, which was further complimented by an air driven hydraulic pump to gain higher intensities (up to 10.5 N/mm^2, which corresponds to a head of 1070 m), which would then enable test times to be reduced, ensure sufficient outflow for accurate volume measurement and to allow for a more severe scenario than would occur in practical circumstances. The permeameters essentially consisted of a cylinder and two steel end plates; their interface sealed using neoprene 'O' rings. The gap between the specimen and the permeameters was filled with epoxy resin. Water entered the permeability cell via a centrally located steel tube in the top plate and then flowed through the specimen, eventually being caught by a funnel attached to the bottom plate of the permeameter. This funnel then discharged into a glass graduated tube which was loosely stoppered to prevent any evaporation.

RESULTS AND DISCUSSION

The effects of applied hydrostatic pressure, test duration and applied stress were considered. Applied hydrostatic pressure, Figure 2(a), shows that the coefficient of permeability only increases significantly with hydrostatic pressures of $0.4f_u$ or greater, with the worst case pressure of 10.5 N/mm^2 exaggerating the permeability by a factor of approximately 20.

The effect of test duration was studied for specimens pre-stressed to levels of 0, 0.4 and $0.7f_u$. Although the permeability increased to maximum after only a short period, there was continuous reduction in the rate of outflow with time thereafter. The permeability tended to decrease rapidly after the first 3-4 days, after which a levelling out was noticed, Figure 2(b).

1 : Compressed air
2 : Air filter and regulator
3 : Air hydropump (C.S. Madan)
4 : Reservoir
5 : Relief valve
6 : Permeameters

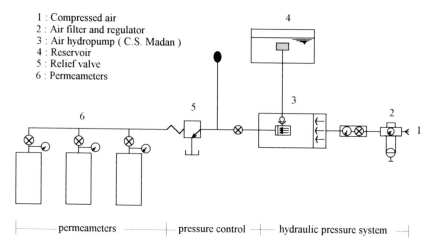

| permeameters | pressure control | hydraulic pressure system |

Figure 1 Hydraulic circuit for permeability apparatus

Clogging of particles due to 'wash-down' of fine particles from one area of concrete to another, swelling of the cement and further hydration are all thought to play their part in this phenomena, though it is felt that the amount attributable to each cause is dependant upon the condition of the test and constituents of the concrete. High hydrostatic pressures, with stressed specimens, clogging of pores was believed to cause reduction whereas with unstressed specimens at low hydrostatic pressures, swelling and further hydration were believed to be the main causes.

The effects on applied stress on mixes A, B and C under the maximum hydrostatic pressure are shown in Figure 2(c). Up to 30% of the ultimate load there were only modest increases in permeability for mixes B and C, mix A showing a slight decrease.

Above 40% stress level, the permeability increased rapidly for all mix types, and between 40% and 70% the specimens changed from 'low' to 'high' permeability concrete. Observations of the higher stressed specimens show a wash out of fine cement and silty particles that were produced at crack interfaces, indicating that as these were washed down they would be deposited in the narrower crack openings hence reducing the permeability.

CONCLUSIONS

This study has shown that the permeability of concrete can be greatly affected by applied stress and that additives, which influence the composition, strength and propagation of microcracks, are influential.

It was also found that applied hydrostatic pressure and test duration affect the permeability of the concrete specimens. Normally mixed PC concrete, at low stress levels , provided the best resistance to the penetration of water through its body compared to similarly stressed concrete containing fly ash. Air-entrained concrete permeability was shown to be greater at all stress levels, than PC concrete.

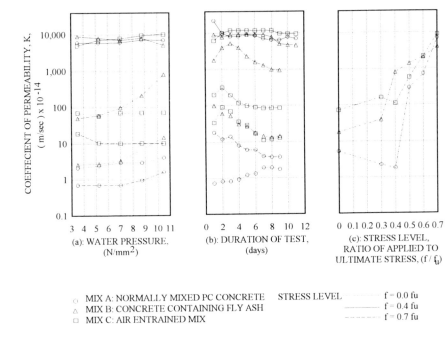

(a): WATER PRESSURE. (N/mm²)

(b): DURATION OF TEST, (days)

(c): STRESS LEVEL, RATIO OF APPLIED TO ULTIMATE STRESS, (f / f$_u$)

○ MIX A: NORMALLY MIXED PC CONCRETE
△ MIX B: CONCRETE CONTAINING FLY ASH
□ MIX C: AIR ENTRAINED MIX

STRESS LEVEL
———— f = 0.0 fu
———— f = 0.4 fu
- - - - f = 0.7 fu

Figure 2 Effects of applied hydrostatic pressure, test duration and applied stress

REFERENCES

1. NYAME, B K AND ILLSTON, S M. Relationship Between Permeability and Pore Structure of Hardened Cement Paste. Magazine of Concrete Research, Vol.33, No.116, Sept.1981, pp 139-145.
2. McMILLAN, F R AND LYSE, I.   Some Permeability Studies of Concrete. Journal of the American Concrete Institute - Proceedings, March 1929, pp 101-142.
3. MENG, B. Calculation of Moisture Transport Coefficients on the Basis of Relevant Pore Structure Parameters. Materials and Structures, Vol. 27, No. 167, April-May 1994, pp 125-134.
4. POWERS, T C. Structural and Physical Properties of Hardened Portland Cement Pastes. Journal of the American Ceramic Society, Vol. 41, No. 1, January 1958, pp 1-6.
5. POWERS, T C, COPELAND, L E, HAYES, J C AND MANN, H M. Permeability of Portland Cement Paste. Proceedings of the American Concrete Society, Vol.51, 1954-1955, pp 285-298.
6. NYAME, B K. Permeability and Pore Structure of Hardened Cement Paste and Mortar. Ph.D. Thesis, Department of Civil Engineering, University of London, 1979.
7. PEUTTALA, V E. Effects of Microporosity on the Compression Strength and Freezing Durability of High Strength Concretes. Magazine of Concrete Research, Vol.41, No.148, Sept.1989, pp171-181.
8. REINHARDT, H W AND GABER, K. From Pore Size Distribution to an Equivalent Pore Size of Cement Mortar. Materials and Structures, Vol.23, No.133, January-February 1990, pp 3-15.
9. FELDMAN, R F AND BEAUDOIN, J J. Pre-treatment of Hardened Cement Pastes for Mercury Intrusion Measurements. Cement and Concrete Research, Vol.21, No.2, July 1991, pp 297-308.
10. DAY R L AND MARSH, B K. Measurement of Porosity in Blended Cement Pastes. Cement and Concrete Research, Vol. 18, No. 1, January 1988, pp 63-73.
11. POWERS, T C, MANN, H M AND COPELAND, L E. Flow of Water in Hardened Portland Cement Paste. Highway Research Board, Report No.40, Washington, July 1959, pp 308-323.

12. LUPING, T AND NILSSON, L. A Study of the Quantitative Relationship between Permeability and Pore Size Distribution. Cement and Concrete Research, Vol.18, No.1, Jan. 1988, pp 63-73.

13. RUETTGERS, A, VIDAL, E N AND WING, S P. An Investigation of the Permeability of Mass Concrete with Particular Reference to the Boulder Dam. Journal of the American Concrete Institute Proceedings, March - April 1935, pp 382-416.

14. BUILDING RESEARCH TECHNICAL PAPER No. 3. The Permeability of Portland Cement (Revised Edition). Department of Science and Industrial Research, 1931, pp 62.

15. WILEY, G AND COULSON, D C. A Simple Test for Water Permeability. Journal of the American Concrete Institute, Vol. 34, September - October 1938, pp 65-75.

16. TYLER, I L AND ERLIN, B. A Proposed Simple Test for Determining the Permeability of Concrete. Journal of the Portland Cement Association and Development, September 1961, pp 2-7.

17. NEVILLE, A M AND BROOKS, J J. Concrete Technology - Revised. Longman Scientific and Technical, Harlow, England, 1993, pp 438.

18. BAMFORTH, P B. The Water Permeability of Concrete and its Relationship with Strength. Magazine of Concrete Research, Vol. 43, No. 157, December 1991, pp 233-241.

19. NYAME, B K. Permeability of Normal and Lightweight Mortars. Magazine of Concrete Research, Vol. 37, No.130, March 1985, pp 44-48.

20. HARRISON, W H. Durability Tests on Building Mortars - Effect of Sand Grading. Magazine of Concrete Research, Vol. 38, No. 135, June 1986, pp 95-107.

21. HARRISON, W H. Aspects of Mortar Durability. British Ceramics Transactions Journal, Vol. 89, No. 3, May - June 1990, pp 93-101.

22. BLOEM, D L. Effects of Aggregate on Properties of Masonry Mortar. Symposium on Masonry Testing, ASTM STP 320, American Society for Testing and Materials, 1962, pp 67-92.

23. SCHONLIN, K AND HILSDORF, H. Evaluation and Effectiveness of Curing of Concrete Structures. ACI Special Publication No. 100: Concrete Durability, Vol. 1, American Concrete Institute, Detroit, 1987, pp 207-225.

24. DHIR, R.K., HEWLETT, P.C. AND CHAN, Y.N. Near Surface Characteristics of Concrete: Intrinsic Permeability. Magazine of Concrete Research, Vol. 41, No. 147, June 1989, pp 87-97.

25. MERCER, L B. Permeability of Concrete-1. The Commonwealth Engineer, July 1945, pp349-357.

26. LINDSAY, J D. Illinois Develops High Pressure Air Meter for Determination of Hardened Concrete. Proceedings of the Highway Research Board No. 35, 1956, pp 424-435.

27. FIGG, J W. Methods of Measuring the Air and Water Permeability of Concrete. Magazine of Concrete Research, Vol. 25, No. 85, December 1973, pp 213-219.

28. CATHER, R, FIGG, J W, MARSDEN, A F AND O'BRIEN, T P. Improvements to the Figg Method for Determining the Air Permeability of Concrete. Magazine of Concrete Research, Vol. 36, No. 129, December 1984, pp 241-245.

29. KASAI, Y, MATSUI, I AND NAGANO, M. On Site Rapid Air Permeability Test for Concrete. SPT - 108, V.M. Malhotra, Ed., American Concrete Institute, Detroit, 1984, pp 525-541.

30. HANSEN, A J, OTTOSEN, N S AND PETERSON, C G. Gas Permeability of Concrete In Situ: Theory and Practice. SPT-108, V.M. Malhotra, Ed., ACI, Detroit, 1984, pp 543-556.

31. HUDD, R W. Measurement of the In - Situ Concrete Permeability. Ph.D. Thesis, Department of Civil Engineering, Loughborough University of Technology, 1989.

32. THE CONCRETE SOCIETY. Permeability of Concrete and its Control. Report of a Concrete Society Working Party on ' Permeability Testing of Site Concrete - A Review of Methods and Experience. Conference, London, December 1985, pp 130.

33. BASHEER P A M. A Brief Review of Methods for Measuring the Permeation Properties of Concrete In-Situ. Proceedings of the Institution of Civil Engineers. Structures and Buildings, Vol. 9, Issue 1, February 1993, pp 74-83.

34. BRITISH STANDARDS INSTITUTION. Methods of Testing Hardened Concrete for Other than Strength. London, BS 1881: Part 5: 1970.

35. VAN DER MEULEN, G J R AND VAN DIJK, J. Permeability - Testing Apparatus for Concrete. Magazine of Concrete Research, Vol. 21, No. 67, June 1969 , pp 121-123.

36. CABRERA, J G AND LYNSDALE, C J. A New Gas Permeameter for Measuring the Permeability of Mortar and Concrete. Magazine of Concrete Research, Vol.40, No.144, Sept. 1988, pp 177-182.

LIFETIME FACTOR METHOD IN DURABILITY DESIGN OF CONCRETE STRUCTURES

E Vesikari

VTT Building Technology

Finland

ABSTRACT. Traditionally two methods have been used in the mathematical formulation of durability design problems: the deterministic and the stochastic. A third method - the lifetime factor method - is introduced in this paper. The method has been applied and is discussed in the final report of the RILEM Committee TC 130 CSL "Durability design of concrete structures".

The lifetime factor method is based on the same safety principles as those applied in the stochastic method. However, with the aid of the lifetime safety factor (or the lifetime factor) the design problem is returned to the form of deterministic design. Thus the formulation of design is as easy as in deterministic design, allowing, however, effective control of safety over the service life.

The lifetime factor method is especially suitable for the structural design of concrete structures, as application of normal stochastic methods would be too complex. Another reason is that the method resembles the way in which the stochasticity of loads and material strengths are taken into account in the traditional mechanical design of structures.

Lifetime safety factors must be 'calibrated' with stochastic methods. The calculation methods for determination of lifetime factors are presented in this paper.

Keywords: Durability design, Service life, Concrete structures, Lifetime factor.

Mr Erkki Vesikari is Senior Research Scientist at VTT Building Technology, Finland. For more than 15 years he has studied the durability and service life problems of concrete structures.

Radical Concrete Technology. Edited by R K Dhir and P C Hewlett. Published in 1996 by E & FN Spon, 2–6 Boundary Row, London SE1 8HN, UK. ISBN 0 419 21480 1.

INTRODUCTION

Traditionally, problems of durability design have been formulated using two methods: the deterministic method and the stochastic method [4, 6].

In deterministic design the distributions of load S and resistance R are ignored. The design formula is written as follows:

$$R(t_g) - S(t_g) > 0 \qquad (1)$$

where   S is   the load
        R   the resistance of the structure and
        t_g the target service life.

The load is not necessarily a mechanical load, as is normally the case in the mechanical design of structures. It can also be a physical or chemical load imposed by the environment. The resistance of the structure is of the same character as the load and can therefore also be mechanical, physical or chemical.

S and R may represent the mean, characteristic or design values of the load and the resistance. Usually one or more durability models are incorporated in either S or R, or both.

Durability models usually show the loss of effective cross-section of materials or some material property with time. They may contain environmental, structural and material parameters. The designer gives these parameters such values that the condition of Formula 1 is fulfilled.

In stochastic design the distributions of load and resistance are taken into account. A requirement for the maximum allowable failure probability is added to the final condition.

Mathematically the requirement is expressed as: The probability of the resistance of the structure being smaller than the load within the service period is smaller than a certain allowable failure probability.

$$P\{failure\}_{t_g} = P\{R(t_g) - S(t_g) < 0\} < P_{fmax} \qquad (2)$$

where $P\{failure\}_{t_g}$ is   the probability of failure of the structure within t_g, and
        P_{fmax} the maximum allowable failure probability.

The problem can be solved when the distributions of the load and the resistance are known. If they are not known they can be assumed to follow some known type of distribution.

When the formulae for load and resistance are complex and many degradation factors affect the performance of structures, application of the stochastic design method may be difficult. In such cases it may be reasonable to apply the lifetime factor method [5, 7].

In the following the lifetime factor method is introduced mainly for the purposes of structural durability design. Application of the method is especially justified in mechanical design, as in traditional mechanical design the stochasticity of loads and strengths is also taken into account by safety factors, the method thus being in principle familiar to structural designers.

BASICS OF THE LIFETIME FACTOR METHOD

The lifetime safety factor method in the design of structures is based on the same safety principles as those applied in the stochastic methods. However, the calculations are easier, as with the aid lifetime factor the design problem is returned to the form of deterministic design. Compared with the deterministic method, the lifetime factor method provides the structure with controlled safety against falling short of the target service life.

The safety margin $\Theta(t)$ is defined as:

$$\Theta(t) = R(t) - S(t) \tag{3}$$

where Θ is   the safety margin
       R   the resistance (performance capacity of the structure)
       S   the load.

'Failure' takes place when the function yields a negative value, which also means the end of service life.

Service life has a distribution. The moment at which the mean function $\Theta(t)$ crosses the 0 line determines the mean of service life, $\mu(t_L)$. If structures are dimensioned according to the mean of service life, some 50% of them will fall short of the expected (target) service life (Figure 1).

Generally the maximum allowable failure probability must be smaller than 50%. Thus the mean service life must be longer than the target service life. The smaller the allowable failure probability, the greater is the difference between the mean and the target service life.

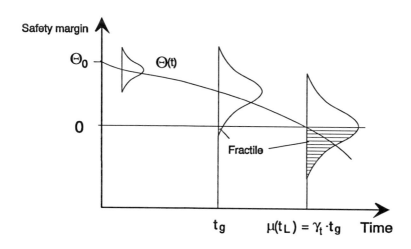

Figure 1   Mean service life and target service life.

Using the lifetime safety factor, the requirement of target service life is converted to the requirement of mean service life. The reason is that durability models available to designers show only the mean performance, or mean degradation. As designers operate with mean functions, every requirement of target service life must first be interpreted in terms of the corresponding mean service life.

The lifetime safety factor, γ_t, is the relation of mean service life to the target service life:

$$\gamma_t = \frac{\mu(t_L)}{t_g} \tag{4}$$

where γ_t is the lifetime safety factor,
$\mu(t_L)$ the mean service life and
t_g target service life.

The determining time in durability design is the design service life, t_d, which equals the mean service life. It is obtained by multiplying the target service life by the lifetime safety factor:

$$t_d = \gamma_t \cdot t_g \tag{5}$$

where t_d is the design service life.

The design formula can then be written as:

$$R(t_d) - S(t_d) > 0 \tag{6}$$

Example

To illustrate the design method, let us dimension the thickness of the concrete cover with respect to carbonation. In this case the thickness of the concrete cover represents the performance capacity R and the depth of carbonation represents the load S. 'Failure' takes place when carbonation reaches the steel bars.

The depth of carbonation (mean) is calculated using Formula 7:

$$\mu(s) = K_c \sqrt{t} \tag{7}$$

where $\mu(s)$ is the mean depth of carbonation (mm),
K_c the coefficient of carbonation (mm/√year), and
t time (or age of the structure in years).

The coefficient of carbonation depends on the quality of concrete and the environment. It can be determined with Formula 8 [3] as follows:

$$K_c = c_{env} \cdot c_{air} \cdot a \cdot (f_{ck} + 8)^b \tag{8}$$

where c_{env} is the coefficient of the environment, here $c_{env} = 1$,
c_{air} the coefficient related to air-entrainment, here $c_{air} = 1$,

f_{ck} the characteristic (cubic) compressive strength of concrete (MPa), here $f_{ck} = 30$ MPa and

a, b   coefficients depending on the binder of concrete, here a=1800 and b=-1.7.

Substituting the above values into Formula 8 gives:

$$K_c = 3.71 \ mm / \sqrt{year}$$

The design formula (Formula 6) is then

$$C - 3.71\sqrt{t_d} > 0 \qquad (9)$$

where C   is the thickness of the concrete cover (mm).

Let us then assume that the target service life is 50 years and the lifetime factor is 2.2. The design service life is then according to Formula 5:

$$t_d = 2.2 \cdot 50 = 110 \ years$$

Substituting this into Formula 9 gives:

$$C_{required} = 3.71\sqrt{110} = 39 mm$$

Without the lifetime safety factor the required concrete cover would be 26 mm.

A prerequisite for using the lifetime factor method is, of course, that the correct values for lifetime factors are known. To know the relation between lifetime factors and failure probabilities, stochastic methods must be applied. In the following the values of lifetime factors have been derived mainly for the mechanical design, but the results can be applied to any structural design.

DETERMINATION OF LIFETIME FACTORS FOR MECHANICAL DESIGN

In mechanical design the design Formula 6 is replaced with Formula 10:

$$R(t_d) - S(t_d) > \Theta_{min} \qquad (10)$$

where Θ_{min} is the minimum value of the safety margin.

The structural capacity R and the load S are assumed to be given with characteristic values of material strengths and loads.

The reason for applying Formula 10 instead of Formula 6 is that the necessary safety margin provided for a structure is not only in view of material degradation. The minimum safety margin, Θ_{min}, in Formula 10 guarantees the basic mechanical safety of the structure (related to normal dispersion of R and S).

Accordingly, in mechanical durability design the safety of the structure is divided into two parts:

1) mechanical safety and
2) durability safety.

The mechanical safety of the structure is considered in normal structural design practice, i.e. through the material and load safety factors. The safety factors are 'calibrated' by stochastic methods to the required mechanical safety. They also define Θ_{min}.

The durability safety is taken into account by increasing the capacity of the structure at the start of service life, so that even at the end of service life the safety margin will fulfil the minimum safety margin Θ_{min} with a certain required probability. In practice the durability safety is controlled by the lifetime safety factor.

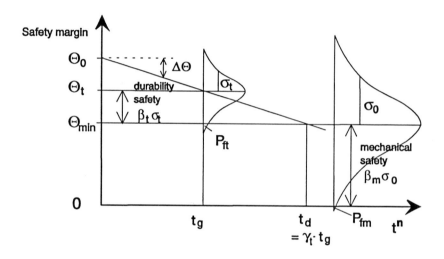

Figure 2   Mechanical safety and durability safety.

Mechanical safety

The mechanical safety is secured by Θ_{min}, which is defined as:

$$\Theta_{min} = \beta_m \cdot \sigma_o \tag{11}$$

where σ_o is   the normal standard deviation of the safety margin (assumed standard deviation of the safety margin in normal structural design) and

   β_m is the required mechanical safety index.

According to Eurocode 1 [2] the safety index β_m in the ultimate limit state must be at least 3.8, which corresponds to the failure probability $7.2 \cdot 10^{-5}$, and in the serviceability limit state 2.5, which corresponds to the failure probability $6.2 \cdot 10^{-3}$. These safety index requirements refer to the safety margin calculated by the characteristic values of loads and material properties. In traditional design the degradation of materials is not taken into account, in which case the safety after the service life will probably be smaller than stated by the above safety requirements.

For the above reason it has been considered appropriate to allow, in certain cases of durability design, a smaller mechanical safety index β_m ~~after the service life~~. However, if loss of bearing capacity of a structure can in any way put human lives at risk or otherwise have serious social, economic or ecological consequences, the safety index must be 3.8 even beyond the service life. Otherwise the safety index β_m may be 3.1, which corresponds to the failure probability $9.7 \cdot 10^{-4}$, provided it is ensured at the same time that the safety index ~~at the start of service life~~ fulfils the normal requirement of 3.8 [5].

In the serviceability limit state the β_m value of 2.5 is used if falling below the limit state could result in considerable economic loss or significant repair costs. It corresponds to the failure probability $6.2 \cdot 10^{-3}$. If projected economic losses are not notable the safety index beyond the service life may be 1.5, which corresponds the failure probability $6.7 \cdot 10^{-2}$.

In traditional structural design the mechanical safety is guaranteed in practice by the load and material safety factors. The characteristic material strengths are divided by the material safety factors and the characteristic loads are multiplied by the load safety factors. In general the minimum safety margin is the difference between the safety margins calculated by the characteristic and the design values of material properties and loads.

Durability safety

Durability safety is the safety against falling below the required mechanical safety at the end of service life due to degradation of materials. As normal dispersions of load and capacity are included in the mechanical safety, only the extra dispersion due to degradation is considered in durability safety.

In practical design durability safety is controlled by the lifetime (safety) factor. The relation between the durability safety requirement and the lifetime factor is derived as follows [5]:

Let us assume that the degradation-related loss of safety margin is proportional to t^n:

$$\Theta(t) = \Theta_0 (1 - k\, t^n) \tag{12}$$

where | Θ_0 is | the safety margin at moment $t = 0$ ($= R_0 - S_0$),
| k | a constant coefficient,
| t | time and
| n | an exponent (may vary between $-\infty$ and $+\infty$).

It is also assumed that the loss of safety margin $\Delta\Theta(t)$ ($= \Theta_0 - \Theta(t)$) is normally distributed around the mean and that the standard deviation of $\Delta\Theta(t)$ ($=$ degradation related standard deviation of Θ), is directly proportional to the mean of $\Delta\Theta(t)$. Thus the coefficient of variation, v_D, which expresses the relation of the standard deviation to the mean, is constant:

$$\mu(\Delta\Theta) = \Theta_0 k\, t^n \qquad \text{and} \tag{13}$$

$$\sigma(\Delta\Theta) = v_D\, \Theta_0 k\, t^n \tag{14}$$

where v_D is the coefficient of variation.

The lifetime factor γ_t is determined so that the probability of falling below the limit state Θ_{min} is the maximum allowable, taking into account both the decrease of the mean and the increase of standard deviation of safety margin with time.

In the following the lower index t refers to the target service life:
Θ_t is   the safety margin,
σ_t the standard deviation of $\Delta\Theta$ and
β_t the required safety index at moment $t = t_g$.

From Figure 2 we get:

$$\beta_t.\sigma_t = \Theta_t - \Theta_{min} \tag{15}$$

And further:

$$\beta_t = \frac{\Theta_t - \Theta_{min}}{v_D(\Theta_0 - \Theta_t)} = \frac{1}{v_D}\left(\frac{\Theta_0 - \Theta_{min}}{\Theta_0 - \Theta_t} - 1\right) \tag{16}$$

On the other hand from Figure 2 we get:

$$\frac{\Theta_0 - \Theta_{min}}{\Theta_0 - \Theta_t} = \frac{(\gamma_t t_g)^n}{t_g^n} = \gamma_t^n \tag{17}$$

Substituting this into Formula 16 gives:

$$\beta_t = \frac{(\gamma_t^n - 1)}{v_D} \tag{18}$$

and for γ_t:

$$\gamma_t = (\beta_t \cdot v_D + 1)^{\frac{1}{n}} \tag{19}$$

According to Formula 19 the value of the lifetime factor depends on β_t (the durability safety requirement), on the coefficient of variation of degradation, and on the exponent n. Note that the lifetime factor does not depend on the service life.

Nor does the lifetime factor depend on Θ_{min}. This means that Formulae 18 and 19 are valid also in the case $\Theta_{min}=0$ as in Formula 6. These formulae can therefore be used in any durability design, not only in mechanical durability design.

In principle, the requirement of durability safety can be set quite independently of the requirement of mechanical safety. However, the durability safety requirement should be somehow in line with the mechanical safety requirement.

The following study should give some guidelines as to what the durability safety index and the lifetime factor should be in the case of mechanical design. In the study the durability safety requirement is harmonised with the mechanical safety requirement.

Combined Mechanical Safety

Combined mechanical safety indicates mechanical safety when both the normal parameters of mechanical safety and the degradation-related parameters of safety are combined and treated together.

As before, we assume that the minimum safety index Θ_{min} is set according to traditional mechanical design, i.e. the safety factors of loads and materials are applied as usual. Also the formulae derived for β_t and γ_t are still valid (Formula 18 and 19).

The minimum mechanical failure probability P_{fm} defined by Eurocode 1 is described by that part of the distribution of Θ which has fallen below the 0 line (calculated by characteristic values of loads and material properties). As an extra condition for β_t and γ_t the combined mechanical safety at target service life t_g is said to be at least the same as the plain mechanical safety (when the degradation is not considered) at the design service life t_d. Using the notation of Figure 3 this means that:

$$P_{fm}{}' = P_{fm} \quad \text{and} \tag{20}$$

$$\beta_m{}' = \beta_m \tag{21}$$

From the above requirement it follows that when writing the design formulae for the design service life t_d ($=\gamma_t \cdot t_g$) the degradation-related parameters of safety need not be considered, as their effect on safety is included in the lifetime factor γ_t.

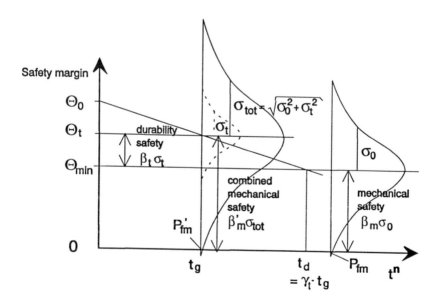

Figure 3   The principle of combined mechanical safety.

From Figure 3 we get the condition:

$$\beta_m \cdot \sqrt{\sigma_0^2 + \sigma_t^2} = \beta_m \cdot \sigma_0 + \beta_t \cdot \sigma_t \tag{22}$$

The standard deviations in Formula 22 and in Figure 3 are as follows:

σ_0 is the basic standard deviation of the safety margin (standard deviation of Θ, which is not dependent on degradation or time)

σ_t the extra standard deviation related to degradation and time (standard deviation of $\Delta\Theta$) at moment $t = t_g$.

σ_{tot} the combined standard deviation at moment $t = t_g$.

The combined standard deviation σ_{tot} is calculated as follows:

$$\sigma_{tot} = \sqrt{\sigma_0^2 + \sigma_t^2} \tag{23}$$

Formula 22 can then be worked out in the form:

$$\beta_t = \left(\frac{\sqrt{1 + \left(\frac{\sigma_t}{\sigma_0}\right)^2} - 1}{\frac{\sigma_t}{\sigma_0}} \right) \cdot \beta_m \tag{24}$$

The relation σ_t/σ_0 depends on the progress of degradation. In the following the relation σ_t/σ_0, which is rather difficult to manage in calculations, is replaced by the relation m, which denotes the relative diminishing of the safety margin in the interval $0 \rightarrow t_d$:

$$m = \frac{\Theta_0 - \Theta_{min}}{\Theta_0} \tag{25}$$

From Formulae 13 and 14 we get:

$$k = \frac{\Theta_0 - \Theta_{min}}{\Theta_0 t_d^n} = \frac{m}{\left(\gamma_t \cdot t_g\right)^n} \tag{26}$$

and

$$\sigma_t = \frac{m \cdot v_D \cdot \Theta_0}{\gamma_t^n} \tag{27}$$

For the basic standard deviation σ_0 we have:

$$\sigma_0 = \frac{\Theta_{min}}{\beta_m} \tag{28}$$

Still noting that $\Theta_0/\Theta_{min} = 1/(1-m)$ (from Formula 25) we get the relation between σ_t/σ_0 and m:

$$\frac{\sigma_t}{\sigma_0} = \frac{m}{1-m} \cdot \frac{v_D}{\gamma_t^n} \cdot \beta_m \qquad (29)$$

This can be substituted into Formula 24.

On the other hand we have Formula 18 for β_t. By setting the two formulae for β_t equal, the values of γ_t can be solved as a function of m.

The parameter m is used as a measure for σ_t/σ_0. Two extremes can be recognised: 1) when $m = 0$, $\Theta_{min} = \Theta_0$ and $\sigma_t/\sigma_0 = 0$ and 2) when $m = 1$, $\Theta_{min} = 0$ and $\sigma_t/\sigma_0 = \infty$.

In Table 1 the values of β_t and γ_t have been calculated with values of m varying between 0 and 1. The mechanical safety index β_m after service life is 3.8, 3.1, 2.5 or 1.5. The coefficient of variation is assumed to be 0.6 and the exponent n is 1 (linear degradation model).

In principle, γ_t could be any value in Table 1 if it is consistent with the real m. The real m depends on the amount of degradation during the service life and should be checked by calculations after the design. The real m must not exceed the value of m according to which the lifetime factor γ_t was applied in the design. In practical design a standard assumption for m is 0.7 corresponding to the values of γ_t 2.46 for $\beta_m = 3.8$ and 2.15 for $\beta_m = 3.1$ respectively. Following design of the structure the real m is checked to ensure that it does not exceed 0.7 [5].

The real value of m is determined using Formula 25. Θ is the safety margin (= R - S) determined using characteristic values of loads and material properties. The index 0 refers to the initial state of the structure and m to the final state, beyond the design service life t_d. Θ_m is determined from the final mechanical design solution by setting the load and material safety factors equal to 1. Θ_0 is obtained by setting also γ_t equal to 0. The calculations are easy when using a spread sheet application.

Table 1. Values of β_t and γ_t.

m	$\beta_m=3.8$		$\beta_m=3.1$		$\beta_m=2.5$		$\beta_m=1.5$	
	β_t	γ_t	β_t	γ_t	β_t	γ_t	β_t	γ_t
0.001	0.00	1.00	0.00	1.00	0.00	1.00	0.00	1.00
0.1	0.39	1.23	0.27	1.16	0.19	1.11	0.07	1.04
0.2	0.73	1.44	0.53	1.32	0.37	1.22	0.15	1.09
0.3	1.05	1.63	0.79	1.47	0.57	1.34	0.25	1.15
0.4	1.38	1.83	1.05	1.63	0.77	1.46	0.35	1.21
0.5	1.71	2.02	1.32	1.79	0.99	1.59	0.47	1.28
0.6	2.06	2.23	1.61	1.97	1.23	1.74	0.62	1.37
0.7	2.43	2.46	1.92	2.15	1.49	1.89	0.78	1.47
0.8	2.84	2.70	2.27	2.36	1.78	2.07	0.98	1.59
0.9	3.29	2.97	2.66	2.59	2.11	2.27	1.21	1.73
0.999	3.80	3.28	3.10	2.86	2.50	2.50	1.50	1.90

If m turns out to be greater than 0.7 the designer should check whether there is any way to reduce it by changing the dimensions or material specifications or by reducing the rates of degradation. If not, the required target service life may not be realistic in the environment concerned. The designer should then propose a shorter requirement for the target service life.

When $m = 1$ the requirement of β_t is the same as that of β_m (Table 1). The requirement for γ_t can be obtained directly from Formula 19. These values represent a case where σ_t totally governs over σ_0. It can be concluded from this that when applying these extreme values of γ_t no check for m is required following the design.

CONCLUSIONS

The lifetime factor method has been developed especially for the purposes of mechanical durability design. However, it can be applied in any structural design where the performance capacity of the structure is decreased due to degradation of materials. Compared with stochastic design the calculations are easier, as formulation of the design can be returned to the form of deterministic design.

The idea of the lifetime safety factor is similar to that of load safety factors and material safety factors in traditional design. It provides the necessary safety margin against falling short of the target service life. The values of lifetime factors are 'calibrated' using statistical methods.

REFERENCES

1. CEB-FIP MODEL CODE 1990. First predraft 1988. Lausanne: Comité Euro-International du Béton, 1988. U. S. (CEB Bulletin d'information 190a).

2. EUROCODE 1. Basis of Design and Actions on Structures. Part 1: Basis of Design. ENV 1991 - 1. CEN/TC250. Final project team document. 1993. 76 p.

3. HÄKKINEN, T. Influence of high slag content on the basic mechanical properties and carbonation of concrete. Espoo 1993. Technical Research Centre of Finland. Publications 141. 98 p. + app. 46 p.

4. KRAKER, A. DE TICHLER, J.W. & VROUWENVELDER, A.C.W.M. Safety, reliability and service life of structures. Heron 27(1982)1, 85 p.

5. SARJA, A. & VESIKARI, E. (ed.) Durability Design of Concrete Structures. Chapman & Hall 1996. Final report of RILEM Technical Committee 130-CSL. RILEM Report 14. 163 p.

6. SIEMES, A., VROUWENVELDER, A. & BEUKEL, A. VAN DEN. Durability of buildings: a reliability analysis. Heron 30(1985)3, pp. 3 - 48.

7. VESIKARI, E. Betonirakenteiden käyttöikämitoitus (Service life design of concrete structures). Helsinki 1995. Association of Finnish Civil Engineers RIL. RIL 183-4.9. 120 p. (in Finnish).

PREDICTION OF TIME TO REBAR CORROSION INITIATION BY CHLORIDES: NON-FICKIAN BEHAVIOUR OF CONCRETE COVER

C Andrade

J M Diez

C Alonso

J L Sagrera

Institute of Construction Sciences

Spain

ABSTRACT. The need to quantify the time taken by the chlorides to depassivate the steel reinforcement is recognized of interest as part of refined methods for service life prediction. This need is reflected in the increasing number of papers applying the solution of Fick's second law to chloride profiles obtained in cores taken from old structures, or in specimens submitted to laboratory short term experiments. In addition to define a chloride penetration model it is always necessary to fix the threshold value to depassivation. In present paper the mechanism of chloride penetration in marine atmospheres is analysed as a non-fickian process composed by several sideprocesses occurring simultaneous and consecutively. Concerning the chloride threshold, it is emphasized the need to dealt with it from a statistical point of view, as Haussman did in his original work.

Keywords: Chlorides, cover, modelling, corrosion, threshold, diffusion.

Professor Carmen Andrade is Director of the Eduardo Torroja Institute of Construction Science, CSIC, Madrid. She specialises in reinforced corrosion, and protection, concrete microstructure and durability. She has published more than 150 paper. She is involved in Technical Committees (CEN, RILEM, AENOR, etc).

Mr J M Díez is Mine Engineer actually a post-Doc student at the Eduardo Torroja Institute.

Dr Cruz Alonso is a Researcher of the Eduardo Torroja Institute of Construction Science, CSIC, Madrid. She is specialises in reinforced corrosion and protection. She has published more than 70 paper on the subject.

Dr J L Sagrera is a Researcher at the Eduardo Torroja Institute specialized on concrete durability. He has published many paper on the subject.

Radical Concrete Technology. Edited by R K Dhir and P C Hewlett. Published in 1996 by E & FN Spon, 2–6 Boundary Row, London SE1 8HN, UK. ISBN 0 419 21480 1.

INTRODUCTION

The chloride ingress into concrete may progress by following two types of processes: pure diffusion in water saturated concrete or by absorption/desorption phenomena due to dry/wet dayly cycles. From Collepardi et al [1] publications, the diffusion of chlorides in concrete is modelled following second Fick's law [2] irrespective of whether the process is pure diffusion or absortion/desortion, as both may be fitted to an square root of time evolution. What is quantitatively different is the diffusivity rate, but both may be solved by means of second Fick's law.

The assumptions to solve it are that: a) D is a constant, b) the surface concentration, C_s, is constant and c) the medium is semiinfinite. Then, the solution arrives to the expression known as "error function equation":

$$C_s = C_x (1 - erf \frac{x}{2\sqrt{Dt}}) \tag{1}$$

This equation (1) is widely accepted and used at present to calculate chloride diffusion coefficients, from fitting it to actual chloride profiles found in real structures submitted to marine ambients. From the mathematical fitting to the chloride gradients, Cs and D values are obtained.

The same procedure is applied to laboratory experiments in the tests named "ponding tests", where a salt solution is placed on the top of a cylindrical specimen and after a certain time, generally between 30 to 90 days, the chloride profile is analyzed and the D value calculated.

Following this mode of operation, OPC cement concretes present D values between 0.5-10 cm^2/s and blended cements between 0.1-5 cm^2/s.

LIMITATIONS TO THE USE OF THE ERROR FUNCTION EQUATION

In Figure 1 is shown the range of D values that using equation (1), would lead to avoid the presence of chlorides at the rebar level, in a period of 75 years for a cover of 3 cm. It may be deduced that relatively very low values of D would assure enough protection for the reinforcements.

These so low theoretical values of D have aimed in the past to restrict the use of certain concretes, or to recommend blending materials or high performance concretes, with very low w/c ratios. However, traditional concretes have experienced good performance in marine structures, which cannot be easily explained by means of short term experiments [3][4].

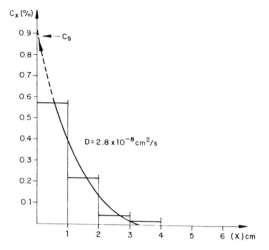

Figure 1 Range of chloride diffusion coefficients in concrete.

This apparent contradiction, may be justified due to: 1) short term experiments do not reproduce real behaviour, as D value may decrease along time and 2) the chloride threshold (Cx in equation (1)), inducing steel depassivation, is much higher than that at present considered (0.4% of cement weight or a Cl⁻/OH⁻ ratio of 0.6).

Variation of D Along Time

Concerning the variation of D value along time, it has been already confirmed by several authors [4][5][6]. Thus, Figures 2 and 3 depict the Cs and D values obtained, from chloride profile fitting in equation (1), in concrete blocks exposed to the tidal zone of a beach in the south of Spain. It can be deduced that Cs varies and that D values tend to decrease along the seven years of the experiment.

This example of real behaviour enables to illustrate the limitations in the application of equation (1) to real structures. In short, these limitations are:

The D coefficient depends upon the Cs value. Thus if Cs is smaller, maintaining the rest constant, the D value results higher (see Figure 4). That is, for the same Cx, D depends on the Cs value, which may randomly change with time in real conditions. This is illustrated in figure 4 were real chloride profiles are plotted and the D values are calculated by fitting equation (1). For the same Cx at the rebar level, different values of D are obtained in function of the Cx/Cs ratio.

If the chloride profile does not evolve with time (no more chlorides ingress along the time), the D value will numerically change with $\sqrt{t}$ [7]. That is, D changes mathematically although the chloride profile is "blocked". This may explain part of the lowering of the D values in old structures.

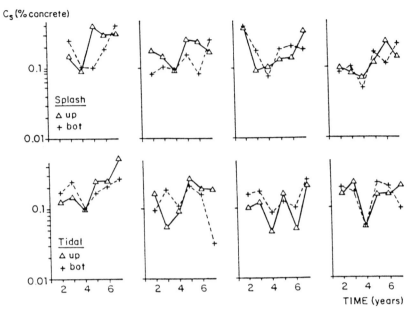

Figure 2 Chloride Surface Concentration evolution, Cs, along 7 years exposure of concrete block in tidal zone.

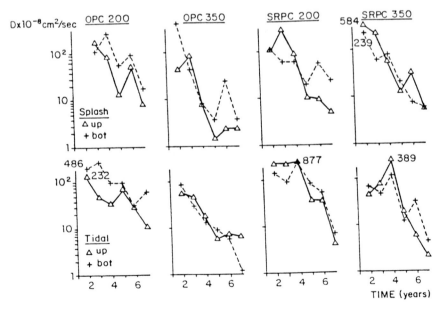

Figure 3 Diffusion coefficient, D, evolution along 7 years exposure of concrete block is a tidal zone.

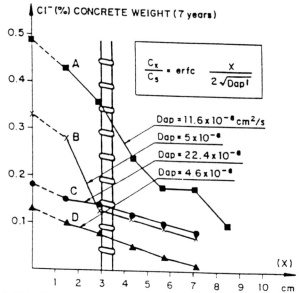

Figure 4 Chloride profiles examples. The value of D may lead to erroneous conclusions for rebar corrosion

In consequence, it seems not recommendable an indiscriminate use of equation (1), as the D coefficient has only a relative, instead of an absolute, meaning. It has mainly a comparative nature in standarized (in order to provide equal Cs values) tests. Also, these comments may justify why the D values obtained in short term experiments may fail in predicting long term behaviour.

Chloride Threshold

The complexity of this subject does not allow a comprehensive overview in present paper. However, it is important at least to emphasize, that it is another important factor if D values with comparative purposes are desired. It is known, that the amount of chlorides able to depassivate depends on parameters such as: alcalinity (OH⁻) cement amount and type, steel surface roughness, concrete moisture content and electrical potential of the rebar [8].

All these factors induce variations of the threshold which may vary one order of magnitude, making very difficult to deduce a reliable time period to steel depassivation, which is the final aim of the calculation of the D values.

Regarding the variability of the chloride threshold, it seems recommendable to dealt with it statistically. As in the case of mechanical strength, a mean value with a variation coefficient may be a much better way to approach this problem. At this respect, in recent studies of the authors, a mean value of 0.16% of Cl⁻ by mortar weight (Figure 5), with a variation coeficient of 0.23, has been found.

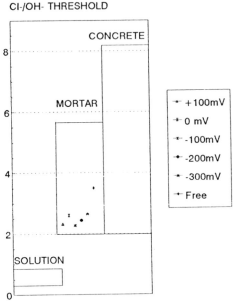

Figure 5 Chloride/hidroxide threshold frequency distribution in mortar immersed in sodium chloride solution. The critical value of Cl⁻ was 0.16 % by weight of cement, with a variation coefficient of 0.23.

REFINED MODELS OF CHLORIDE INGRESS

In order to develop more reliable models of chloride penetration rate, it is necessary to understand with higher precision the mechanisms of chloride ingress and to find out which are the laws of variation along time of Cs, Cx and D values [10]. The complexity of such processes makes unachievable their modelling at present. An approach step by step, seems at present the only mode of progress.

As a preliminar contribution, the following phenomena may be identified to happen in real structures suffering from marine environments:

A)   "Skin" effect - the first mm beyond the concrete surface carbonate either in aerial than in sea water submerged structures. In the case of submerged parts the carbonation appears as a consequence of the leaching plus the action of the bicarbonates present in the sea water. As well these concrete zones, closer to the concrete surface, may have a different composition (paste proportion) due to lack of compactation or of proper aggregate proportioning. Irrespective of the cause, the concrete usually may develop a "skin" with different transport characteristics than the concrete bulk. In consequence, the concrete cover should be modelled as a layered composite, having different D values in the several layers. At least two (skin and bulk) ones should be considered.

B) $Cs = f(t)$ - It was already shown in Figure 2 that Cs evolves along time, either in aerial or in submerged structures. It seems reasonable to think that the increase reaches a maximum (saturation).

C) $D = f(C)$ - Although assumed constant, D is a function of the chloride concentration itself. Its value increases as the solution is more diluted.

D) $D = f(t)$ - Progression of hydration and increase in mechanical resistance may be the reasons why D values decrease along time.

E) $D = f(binding)$ - Cement phases bind chlorides in function of the amount and cement tye (C_3A and mineral admixtures content), as well as of the chloride concentration itself [11].

All these phenomena difficults the rigourous modelling of the chloride ingress. However, they may be individually modelled and after, the additive effects of more than one can be tried. An example of this way of procedure is attempted in [12], where a numerical model is proposed.

Process of Chloride Ingress into Concrete in Marine Structures

In addition to identify the individual phenomena affecting the global process, its sequence of developing has to be tried.

Assuming aerial or submerged structures and considering the results shown in Figures 3 and 4, the following stepping periods of the process are proposed:

Period I   until 2-3 years after the contact of concrete with the chloride ambient. During this period all the phenomena (A to E) may be developing simultaneously, and therefore the chloride profile would be the result of several mechanisms occuring at the same time. A maximum in Cs may be achieved during this period (phenomenon B).

Period II   until 5-10 years. The skin (A) effect reaches estabilization. The rest of phenomena (C, to E) continue their evolution.

Period III   from about 10 years to ahead. The chloride profiles evolve more slowly (only phenomena C to D are active).

The global process is then composed of various phenomena which happens simultanea and consecutively. Such a behaviour is defined as "Non-Fickian" in [2], that is, it cannot be easily modelled by analytical procedures. Perhaps it has to be then tried numerically, unless some simplifications could be established.

In any case, a rigourous solution seems not possible at present, and therefore equation (1) should not be used for predictive purposes regarding time to steel depassivation. It, however, may be used for comparing concretes submitted to standarized conditions, although the D values so obtained are only of relative validity.

Simplified Model for Cover Calculation

For practical purposes, more simplified models have to be studied. In parallel to the development of sophisticated models, other simpler expressions, more familiar for the engineers have to be considered.

A suitable simple formula is that used in carbonation of the "square root law". This law considers the existence of a constant (K_{Cl} in present case):

$$x = K_{Cl}\sqrt{t}$$

x= penetration depth (cm)
t= time (years)   (2)
$K_{Cl}=$ cm/year$^{0.5}$

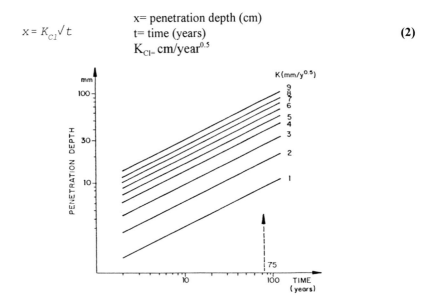

Figure 6   Representation of a square root law in log-log diagrams. The numbers in the parallel lines of slope 0.5 represent the values of the constant K

The constant K_{Cl} encounters all parameters related to concrete characteristics (type and amount of cement, w/c ratio, etc) and to the environment. It is interesting to realize that the units of the K_{Cl} may enable to equal it to the square root of the Diffusion Coefficient:

$$K_{Cl}[cm.year^{0.5}] = \sqrt{D} \, [\sqrt{cm^2.year}] \tag{3}$$

Therefore equation (2) may be expressed as:

$$x = \sqrt{D} \sqrt{t} \tag{4}$$

Plotting this expression in a log-log graph (Figure 6), in order to avoid steel depassivation in 50 years, the value of $K_{Cl} = \sqrt{D}$, should be $\leq 0.43 cm/year^{0.5}$ if the rebar is placed at a nominal cover of 3cm. That means a D value of $0.18 cm^2/year$ $(0.5 \times 10^{-8} cm^2/s)$, which agrees reasonably well with an apparently more rigourous calculation made form equation (1).

CONCLUSIONS

The conclusions that may be summed up are:

1)   The ingress of chlorides in concrete is a complex process composed by several phenomena acting simultanea and consecutively. It cannot be, therefore, only modelled by the simple application of the equation named of the error function.

2)   This equation should be restricted to laboratory or standarized experiments in order to compare concrete types with respect to a reference or in order to compare the behaviour of different elements of the same structure.

3)   Chloride threshold, which is the crucial parameter regarding steeel depassivation, has to be statistically treated due its variability and therefore, service life, should be quantified in probability terms or, at least, in a similar manner to mechanical strength.

4)   More advanced models of chloride ingress are needed, in order to better reproduce the performance of concrete structures submitted to real environments. Simultaneously, very simple engineering models have to be developed for practical purposes.

ACKNOWDLEGEMENTS

This work has been developed partially within the Brite-Euram Project 4062: "The Residual Life of Concrete Structures". The authors thank the partners K. Tuutti (Cementa) K. Petterson (CBI) and G. Fagerlund (Lund University) of Sweden for the useful discussions regarding present research. They also are grateful to the ERASMUS EU-Program for providing the grants of H. Didier and A. Pascal (Belgium) who carried out the work on the chloride threshold.

REFERENCES

1. COLLEPARDI, M., MARCIALIS, A., TURRIZIANI, R. Penetration of chloride ions into cement, pastes and concretes. J.A.C.I, vol. 55, 1972, pp.534-535.

2. CRANK, J. The mathematics of diffusion. Ed. Oxford Unviersity, 1975.

3. TUUTTI, K. Corrosion of steel in concrete. Swedish Cement and Concrete Institute (CBI) no. 4-82, Stockholm, 1982.

4. BAMFORTH, P.P., CHAPMAN- ANDREWS, J.F. Long term performance of R.C. elements under U.K. Coastal exposure conditions. Int. Congress on "Corrosion and Corrosion Protection of steel in cocrete" R.N. Swamy Editor, Sheffield, July 1994, pp. 139-156.

5. MANGAT, P.S., MOLLOY, B.T. Prediction of long term chloride concentration in concrete. Materials and Structures, vol. 27, 1994, pp. 338-346.

6. SANDBERG, P., TANG, L. A field study of the penetration of chlorides and other ions into a high quality concrete marine bridge column. 3rd Int. Conference on Durability of Concrete, ACI-SP-145, Nice (France), May 1994, pp. 557-571.

7. BAMFORTH, P.P. Concrete classifications for R.C. Structures exposed to marine and other salt-laden environments. Conference on "Structural Faults and Repair", 93, Edinburg, June 1993.

8. POURBAIX, M. Lectures on electrochemical Corrosion, Plenum Press, New York, 1973.

9. DHIR, R.K. JONES, M.R.and AHMED, H.E.H. "Concrete durability: estimation of chloride concentration during design life, Magazine of concrete, Res., 43, n° 154, 1991, pp. 557-571.

10. JONES, M.R., McCARTHY, M.J. and DHIR, R.K. "Chloride resistant Concrete" Concrete 2000. Ed. R.K. Dhir and M.R. Jones. 1993.

11. TANG, L., NILSSON, L.O. Chloride binding capacity and binding isotherms of OPC pastes and mortars, Cement and Concrete Res. 23, 1993, pp. 247-253.

12. SAETTA, A.V., SCOHA, R.V., VITALIANI, R.V. Analysis of chloride diffusion into partially saturated concrete, ACI Materials Journal, Sep-Oct 1993, pp.441-451.

STUDIES OF CONCRETE SILOS, BUNKERS, RESERVOIRS: DAMAGE ASSESSMENT MODELLING AND SERVICE LIFE PREDICTION

L M Poukhonto

Moscow State University of Civil Engineering

Russia

ABSTRACT. Concrete construction of silo, bunkers and reservoirs in many countries shows an increasing degree of damages and deterioration within a short of span 1 - 3 to 15 years.The decay of concrete due to cracking, spalling and rusting the steel reinforcement is now a common throughout the world.The low durability of concrete structures is due to several interactive factors: climatic condition together with construction practices and inadequate specification. The latter includes the deterioration in strength and stiffness, when structure is subjected to a limited number of low-cyclic load application.

The subject of this study is the development of a damage model of structures that permits the prediction in the service life span for concrete silo, bunkers, reservoirs, retaining structures, off-shore tanks. This paper presents also the experimental data obtained from the test of concrete specimens under low-cyclic long -term compressive loading and select the specific features of damage cumulation.

Key words: Concrete silo, reservoirs, retaining structures, deterioration, low-cyclic loading, damages cumulation, service life, prediction.

Dr Leonid M. Poukhonto is an Associate Professor of the Department of Reinforced Concrete Structures, Moscow State University of Civil Engineering. His research interests include: deterioration and damages of concrete structures, deteriorate assessment, inelastic behaviour, computer design, service -life prediction and durability of concrete structures: silos, bunkers and reservoirs; probabilistic characteristics of low-cyclic loading, cumulation damage analysis and evaluation. Last 10 years he has been involved in research on durability and service life prediction for reinforced concrete structures under low-cyclic load. His D Sc.(Eng) thesis is now in the closing stage.

Radical Concrete Technology. Edited by R K Dhir and P C Hewlett. Published in 1996 by E & FN Spon, 2–6 Boundary Row, London SE1 8HN, UK. ISBN 0 419 21480 1.

INTRODUCTION

The last two decades has witnessed on a global level a growing concern for problems of reinforced concrete structure deterioration. Similar to the observed trend of bridge deck deterioration, corrosion damage to parking garages, deterioration of coastal structures and corrosion in many other structures, concrete construction and maintenance show alarming degree of deterioration of silos, reservoirs, bunkers and retaining structures. Various case histories, effects and remedies for concrete wear and failure is detailed in [1,2,3,4,5]. The classification of the detected and the fixed damages of concrete silos, bunkers and reservoirs, including specific types of deterioration, is given in [14]. According to the Classification in Environmental classes relevant structures is mainly related to severely aggressive environment[8,9].

Experience gained over a number of years, construction practices and condition surveys carried out on corresponding concrete structures strongly suggest that deterioration in strength and stiffness is one the primary durability problems.

Wear resistance is, in this case, connected with the following additional features of concrete structures considered:

- this is mainly thin-wall elements having plane, revolution (cylindrical) or double curvature surfaces;
- stress-strain state is characterised by periodicity due to rising and falling levels of live load, that is low-cyclic loading;
- intensive interaction between anisotropic and nonhomogeneous material, such as reinforced concrete, and liquid or dry bulk stored (temperature, humidity, abrasion).

The purpose of this paper is to present the result of study of methods for predicting the service life of new concrete 'structures with the same examples of their application under low-cyclic loading.

This paper develops an approach prediction of service life of concrete structures with regard to deterioration mainly in strength and stiffness under long-term load repetitions at the initial stage by using calculation based on knowledge of degradation mechanisms and rate of degradative reaction.

RESEARCH SIGNIFICANCE

For the continued use of existing deteriorated concrete structures and for the prediction of long-life span of new structures it is important to evaluate the degree of deterioration. However, the common method to evaluate such deterioration in the case of long-time low-cyclic loading is yet to be established, because the same studies are needed. With this background experimental work has been conducted to examine the degree and the relationship of the strength and stiffness deterioration.

European and Russian Standards [8,9] accepted the approach to predict service life of relevant structures limited only by fatigue behaviour under high-cyclic loading.

The method of prediction of a remaining life span of existing deteriorated structures
in Russia is based on empirical assessment of damages [13,14]. In this case the damage definition is tied up with the degree of physical and chemical
deterioration and structural peculiarities with consequences regarding the

ability of concrete structures to resist further load and limitation of the acting forces and cracks.
The method used in this paper for prediction the service lives of new concrete structures is based on mathematical modelling of deterioration process together with accelerating testing and application of some stochastic concepts for description of load histories.

DEGRADATION MODELLING: DESIGN PHILOSOPHY

The approach accepted in this paper is connected with the deterioration relationship that describs the above mentioned degradation process, which may be suggested under following assumption.
Practical aspects of the theory of damage cumulation were applied because the real loading process can be represented as an nonstationary time series.
A block loading scheme was adopted as an ordinary model of the series by which the uncertainty and the complexity of applied load can be reduced. From this point of view starting period of the life span is the most important.

The behaviour of concrete structures considered is characterised by U-shaped function of wear intensity and cumulation damage relationship (Figure 1).

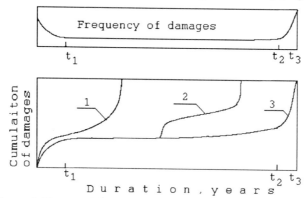

Figure 1. Types of long -term damage cumulation processes:
1- premature wear ; 2 - combined dererioration;
3 - deterioration due to low-cyclic loading

The function of damage cumulation D can be expressed as a discrete function of time, the general form being:

$$D = \begin{cases} a_1 t^{a_2} & ; t_0 \leq t \leq t_1 \\ a_3 t + a_4 & ; t_1 \leq t \leq t_2 \\ a_5 e^{a_6 t} & ; t_2 \leq t \leq t_3 \end{cases}$$

where
$t_1 - t_0$ = initial stage; $t_2 - t_1$ = middle stage ; $t_3 - t_2$ = final failure stage.
$t_3 - t_0$ =duration of loading, $a_1 - a_5$ =wear's coefficients.

Developing [15], the velocity change of structural resistance for the design sections considered may be written as

$$v_r(t) = v_t(t) + v_c(t) + v_e(t)$$

where : $v_t(t)$; $v_c(t)$; $v_e(t)$ - components depending on ages of materials; low-cyclic wearing; harsh environment influence.

The process of structural resistance deterioration due to low-cyclic loading can be expressed as: **Applied low -cyclic loads →Mikrocracking →Strength and stiffness deterioration.**

Strength and stiffness degradation velocity $v_r(t)$ is defined as the ratio between residual strain-strength increment for one cycle of operation $\Delta \varepsilon_{res}$ to the cycle duraiton Δt, when $t \to \infty$ i.e. $v_r(t) = d\varepsilon_{res} / dt$, and this ratio is proportional to the ratio $E_{cyc,1}$ - energy dissipation of one cycle to the total energy dissipation E_{tot}. Then energy increment due to N_i cycles is

$$\Delta E_i = \Delta E_o \exp - (N_i - 2) / A$$

where

$\Delta E_o = energy$ increment corresponding to first and second cycles ; $N_i =$ actual number of stress (load) cycles ; $A =$ coefficient depending on properties of materials , stress level , duration of loading.

The residual concrete strains cumulated to the end of $i-1$ cycles of load have been determined by the following expression:

$$\varepsilon_{res}^1(\tau_i, \tau_1) = \varepsilon_{res}^{i-1}(t_{i-1}, \tau_1) + ((\varepsilon_{el}^{i-1} + \varepsilon_c^{i-1}) - (\varepsilon_{ir}^{i-1} + \varepsilon_{cr}^{i-1})),$$

where ε_{el}^{i-1} – elastic strain at the time of application of load of $i-1$ cycle, ε_c^{i-1} – creep strain under sustained load of the same cycle, ε_{ir}^{i-1} – instantaneous recovery strain, ε_{cr}^{i-1} – creep recovery strain.

The effect of action combination shall be calculated by using the following load and action coefficient γ_{cb}. With this assumption life span of relevant concrete structures then is defined as $/ v_r(t) \gamma_{cb}$.

EXPERIMENTAL DETAILS

Taking into consideration the above mentioned practical aspect of the theory of damage cumulation , 3 series of plain and steel reinforced concrete specimens were tested using spring block type loading equipment and 2000 kN testing machine.

A present study was curried out to establish how significant the effect of long-term cyclic load was on concrete integrity, to explore the behaviour of specimens

and residual stresses and strains of concrete supporting silo columns under
low-cyclic compression and to control the degree of strength and stiffness
deterioration.

Test specimens and materials

The schematic showing for cyclic test including loading history and specimens
geometry is shown in Figure 2.

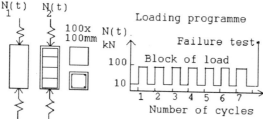

Figure 2. Schematic view of specimen geometry and loading history.

The cross-section of the specimens was 100 x 100mm and the total
length was 400mm. The specimens both for cyclic test and companion
monotonic tests were geometrically similar.
The manufacturing, curing and testing an additional series of plain concrete
cubes of 100 mm. and 150 mm. size were done to control the main concrete
properties during experimental program.
Materials used in the experiment are as follows: ordinary Portland cement;
coarse aggregate used crushed limestone passing through a sieve of a maximum
size 5 mm; fine aggregates: clean river sand passing through 0.8-1.0 mm.
Concrete mix with proportion 1 : 3.4 : 1.9 by weight and water cement ratio of
0.65 were used for all bathers of mix. Four longitudinal 6 mm and 10 mm hot
rolled high yield bars have been enclosed by 4 mm cold-worked mild steel
stirrups with   links spacing 100 mm.

Loading programme

These specimens were tested using two-stages loading programme: the first
stage-long-term block cyclic loading; the second stage-short -term failure testing.

Loading   block   was characterised by the ratio of variable load to dead load
8.3; cycle duration 10   days; total number of cycles was   equal 7;
frequency 1/864000 Hz was used; load asymmetry coefficient
$N=Nmin/Nmax=0.1; f_{ck,cube}=17.3$

MPa; $f_{ck,prizm} =12.9$ MPa , $E_b =1.84 \ 10^4 \ MPa \ \sigma_{b,min} / \sigma_{b,max}=0.06$--$0.15$;

$\sigma_{s,min} / \sigma_{s,max}=0.54$-$0.45$; $\sigma_{b,max} / f_{ck,prizm}=0.63$-$0.27$. The load minima were 10 kN
and load maxima were constant 93 kN and equal approximately to 50 % of
monotonic peak load.
When tested the first cycle, the age of concrete specimens were 21 days for the
first series and 240 days for the second series.

Consequently, ultrasonic monitoring and method of differential coefficient of transversal strain $\Delta v = \Delta\varepsilon_{tr} / \Delta\varepsilon_{ln}$ were used to examine microcracking process, which has been initiated by variable loading.

Test results and discussion

Figure 1 depicts three stages of damage growth. The strain increases very rapidly at the initial stage, and the rate of increase of strain becomes essentially constant at the intermediate stage and is suddenly increased at the final failure stage.

The test data obtained from the present study were applied to damage cumulation theory to investigate the validity of the model.

The increase of strain indicates the damage cumulation due to low-cyclic loading.

Since the distribution of strain components due to a long-term loading is unclear, cyclic behaviour was considered in terms of the residual strains in the reinforced or plain concrete as a relation of the creep and recovery creep to total strain.

The variations in stress-strain curves of the concrete specimens having strength and stiffness deterioration due to a cyclic load programme were compared to the similar curves of the specimens for companion monotonic tests without cyclic actions. Figure 3 shows the effect of great influence of starting variable loading on concrete resistance.

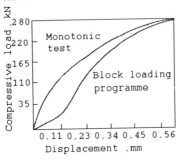

Figure 3. Effect of block loading programme

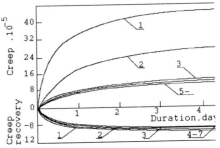

Figure 4. Creep curves and corresponding creep recovery curves 1-7 - number of loading cycles

Therefore the distribution of strain characteristics for every cycle has been determined at variable load level applied with maximum approximately 50-60 % of compressive strength. The results are represented in Fignure 4. The main creep and creep recovery (about 60-75 %) were reached after 2-3 cycles of loading. After 5-7 cycles the creep was practically stabilised for all specimens.

Present theoretical prediction based on previously reported energy increment,

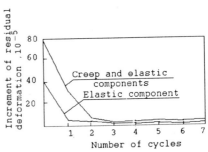

Figure 5. Residual deformation relationship

appears to be substantial by empirical results (Figure 5) The intensity of residual strains was found to be closely linked with the number of cycle of load applied. Comparisons with test results have shown that the calculated values of residual strains $\Delta\varepsilon_{res}$ are in agreement with measured values (5-12 %) .

CONCLUSIONS

The following conclusions may be drawn from this study.

1. Approximate methods using theory of damage cumulation was developed for the silo concrete structures for the case of nonstationary loading series. Cumulation degradation process for plain and reinforced concrete has been modelled using block scheme of low-cyclic long-term loading. Practical application of this method can be shown on the example of undersilo column that is the most loaded part of the silo.
2. Based on the experimental data reported herein an exponential expression is proposed to describe velocity change of strength and stiffness degradation during starting part of life span. The velocity change of residual strains was found to be closely linked with the number of applied cycles. It can be determined that the first cycle of loading presented the major portion of the increment of residual deformation; influence of others cycles was of little importance. Succeeding degradation is cumulated according to the exponential relation.
The result of these analysis correlates satisfactory with the result of the compression loaded specimens. The experimental results confirmed the approach adopted in this paper for given loading condition.Further work is required to establish the effects of the sequencing of variations in exposure condition and low-cyclic load parameters.
3. The work carried out is reduced to the investigation of the behaviour of under silo columns. However, the boundary conditions have been varied so that this method has wider application than just for the concrete column tested.

REFERENCES

1.   THEIMER, O F. Failures of reinforced concrete grain silos. ASME Publications. 1968, № 68-MN-36 .

2.   FUCHS,G. Collapse of a silo for fodder and grain and reconstruction. Paper of the International Symposium on silos, 3-rd Working Session of the IASS Committee of pipes and tanks. Some research and construction of silos. roclow, June 1974. pp 97.

3.   RUFFERT,G. Schaden an betonbauwerken.Ursachen-analysen-beispiele mit abbildungen. Verlagsgesellschaft Rudolf Muller Gmbh.Koln-Braunsfeld, 1982. pp 115.

4.   SATARIAN,S S. and HARRIS C E. Design and construction of silos and bunkers, Van Nostrand Reinhold Company. Chapter 10, Failures and repairs of solos and bunkers, pp 346-386.

5. DONTEN, K., KNAUFF, M. and SADOVSKI, A. Wzmocnienie przez sprezenie zbiornika na wode przemyslowa. Konferencja naukovo-techniczna: AWARIE BUDOWLANE, Szczecin miedzyzdroje. 26-28 Maja.1994,Tom2. Referaty. Politechnika szezinska. Wydzial udownictwa i architecture. pp 395-402.

6. BEEBY, A W. Design for life. Proceedings of the International Congress: Concrete 2000. Economical and durable constructions through excellence. 1994. Dundee. Scotland. pp 37-51

7. BOYLE, J T, SPENCE, J. Stress analysis for creep. Department of Univ. materials of Strathclyde, Glasgow. Scotland.; Butterworths. pp 360.

8. NORWEGIAN COUNCIL FOR BUILDING (NBR), Norwegian Standard NS 3473E. Standardisation concrete structures. Design rules., 4-th edition, Nov.1992 . pp 68.

9. EUROPEAN COMMITTEE FOR STANDARTIZATIOIN, European prestandard env. 1992-1-1. Eurocode 2:Design of concrete structures. Part 1. General rules and rules for buildings.,CEN, December 1991, p.p.250.

10. CLIFTON R J. Predicting the service life of concrete. ACI Materials Journal, November-December 1993, № 6, pp 611-617.

11. GRZYBOVSKI ,M. and MEYER,C. Damage accumulation in concrete with and without fiber reinforcement. ACI Materials Journal, V.90, №6, November-December 1993, pp 594-604.

12. BJEGOVIE,D., KRSTIC ,V., MICULIC, D. and UKRAINCZYK V. C-D-c-t diagrams for practical of concrete durability parameters. Cem. and Concr. Res. January 1995,Vol.25. №1, pp 187-189.

13. POUKHONTO, L M. and RASTORGUEV, B S. Damage cumulation model of concrete silo.Teoretyczne podsawy budownictwa. Referaty. Wydzial inzynierii ladovej politechniki Warszawsciej. Warszawa. 1994.p.p.160-163

14. POUKHONTO, L M. Damages and design problems of concrete silo and storage reservoirs subject to low-cyclic loading. 3-rd International Conference on the concrete future: concrete quality. Kuala-Lumpur March 1994, pp 167-175

15. KUDZYS,A. Reliability estimation of reinforced concrete structures. Vilnus Mokslas Publishers. 1985, pp 105-110

PREMATURE DAMAGE OF REINFORCED CONCRETE POLES. A PROBLEM RELATIVE TO THE DURABILITY

A G Sakellariou

Public Power Corporation

Greece

ABSTRACT. Centrifugal concrete poles are used to a great extent by the Public Power Corporation (P.P.C.) to support overhead distribution lines. A considerable number of these poles have shown signs of serious distress with a direct bearing on their strength. A serious effort has been undertaken by PPC to survey, record and classify the damage of approximately 1500 concrete poles in various parts of Greece. The damages recorded were mainly longitudinal cracks and spaling of concrete, as well as corrosion of the longitudinal steel reinforcement. From the recorded damages, it became clear that by far the overwhelming majority of cracking was associated with cracks parallel to reinforcement (80% of the damages). It is worth noting that more than half of these, were located at the seams of the molding forms. Below these cracks, the steel reinforcement was corroded. Laboratory tests carried out on 50 poles removed from the network, and on samples taken from 70 poles still in use have shown that cracks were always associated with carbonation of concrete and corrosion of reinforcement. In areas adjacent to sea, the contents in Cl-, at the locations of cracks, were critical. The results of this study helped in formulating the new P.P.C. specifications for the manufacturing of centrifugal poles. The new specification has encompassed our present day knowledge regarding durability (cover thickness, requirements for a denser and less permeable concrete) and incorporated improvements such as: increase of spiral reinforcement, controlled methods of accellerated curing.

Keywords: Concrete, Poles, Reinforcement, Corrosion, Cracks.

Dr Antonis Sakellariou is in charge, of the Concrete Laboratory, at the PPC Testing, Research and Standards Center (T.R.S.'C./PPC), Athens Greece. His main research interests include properties of concrete regarding durability.

Radical Concrete Technology. Edited by R K Dhir and P C Hewlett. Published in 1996 by E & FN Spon, 2–6 Boundary Row, London SE1 8HN, UK. ISBN 0 419 21480 1.

INTRODUCTION

In 1989, the Testing, Research and Standards Center (T.R.S.C.) started an ambitious research program [1] concerning the reinforced concrete poles used by P.P.C. to support Overhead Lines. This program lasted for six years and dealt with recording and classification of damage that these poles have shown, assessment of the extend of damage, and helped in establishing the new specifications for the manufacture of such poles. The sample used for this investigation consisted of 1500 poles installed at various areas of the country. The inspection, classification and the statistical evaluation revealed that the poles in the overhead lines had serious problems [2]. One, out of two poles in the sample, showed damage, in spite of the fact that the age of the poles was less than 40 years, and that the statistical sample, by 80%, had an age less than 20 years.

The main form of damage (77%) was associated with longitudinal cracking and corrosion of the underlying reinforcement. It became clear from the outset that the damage was due to development of internal stresses, associated with durability, and not to failure caused by external force. In order to determine the cause of damage, an extensive laboratory program was carried out. Tests were carried out on 50 poles brought to the laboratory, and on 70 poles in situ. These tests aimed at the quantitative evaluation of parameters related to durability. From the test results it became apparent, that the longitudinal reinforcement was not sufficiently protected and was corroded.

This program helped in preparing, a new PPC Specification in 1992, with guide lines for the design and construction of poles. The specification requirements, which were based on the conclusions of this research, helped to construct poles with design and techniques incorporating the present day knowledge on durability [3].

CLASSIFICATION OF DAMAGE

The reinforced concrete poles used by P.P.C. [4] were manufactured centrifugally, in the form of a truncated cone, with a ring-shaped section. Reinforcement was placed longitudinaly to a depth of 15mm from the outside surface. The reinforcement was surrounded by 2mm thick wire, placed helicoidally, while ring shaped steel rods placed at 1.00 m intervals kept it in place. The cross-section and the number of the reinforcing rods (ST I or/and ST III) varied according to the type of pole, and decreased from the base of each pole to the top. The concrete was of class C20/25, with a water cement ratio W/C<0,55 while the thickness of cover concrete was of the order of 15mm. The total cross-section of the longitudinal reinforcement was greater or equal to $16^o/_{oo}$ of the concrete cross-section. The length of the poles varied from 8 to 15 meters. The maximum load that a pole could carry, in any horizontal direction, applied at points located 20 cm underneath the top, varied between 4 and 65kN.

The poles inspected were 1500, out of a total number of approximately 200.000 poles, installed at the network by 1989 (percentage of 1%, approximately). This percentage was considered satisfactory, for this study. The poles came from different manufacturers and their use in connection with the overhead lines extended to 40 years back (1955 - 1995). Consequently, the age of the poles was less than 40 years, while more than one half of them had an age of less than 20 years. For the purpose of damage classification the age of poles is taken at five years intervals starting with the poles manufactured between 1955 and 1960. The poles examined were in operation at the network in eight different locations. Each location was selected with environmental criteria [5].

The classification of damage in relation to severity has shown that approximately half of the poles have a problem, ranging from small to serious. Damage was in the form of spalling and corrosion of reinforcement. The basic conclusion arising from the classification of damage was that approximately 80% of it concerned longitudinal cracks related to durability and only 20% of damage concerned cracks on planes at a 45° angle to the pole axis, which is due to lack of suitable reinforcement. Thus, the causes that created the majority of the damage, are related to poor design and construction when considered in relation to today's knowledge.

INVESTIGATION OF DAMAGE

Following the recording and classification of damage, tests were carried out on 50 poles which have been removed and also on 70 poles still in operation, with the object of getting an insight into the damage. The tests and measurements carried out on the removed poles, were important, because these poles could be destroyed, for the purpose of testing, if required. Thus, it was possible to check the adherence to specifications, to locate possible bad workmanship and perhaps, poor design. At the same time, it was possible to draw conclusions with regard to environmental effects, an experience that could be useful for subsequent examination of poles, in situ, at the network.

Inspection and Tests on Poles Removed from the Lines

Fifty-three (53) poles, selected chronologically, and classified according to the damage they presented, were cut transversally, in three sections. Measurements were taken on these sections, and out of them cored samples were retrieved, for testing. Measurements regarded the thicknesses of the cover of reinforcement, at the ring-shaped cross-section of the pole, both outside and inside, as well as the thickness of the concrete cross-section. Further, the type of the longitudinal reinforcement, the diameters of the reinforcing rods, the percentage of the reinforcement, in relation to the cross-section, and the minimum distance between the rods, were all recorded. It was found that the maximum diameter of the aggregate, was 20-25mm, and, also, it was found that the poles had no spiral reinforcement, what so ever. The results have been tabulated in Table 1, below.

Table 1: Parameters related to design of Poles

s/n	Design Parameters	Symbols-Units	Values
1	Concrete cover thickness, outside surface	Cout (mm)	9 -15
2	Concrete cover thickness, inside surface	Cin (mm)	20 - 30
3	Cross-section Thickness	S (mm)	45 - 65
4	Type of longitudinal Reinforcement	ST	I, III
5	Percentage of longitudinal Reinforcement	Fe (%)	25 - 60
6	Diameters of longitudinal Reinforcement	Φ (mm)	8, 10, 12, 14, 18
7	Minimum distance between longitudinal Rods	e (mm)	1 - 15
8	Spiral Reinforcement and pitch of the spiral	Φ (mm)/mm	Φ2/200

From the above, we may conclude, that:
a) The spiral reinforcement is practically non-existent. Instead, a 2mm diameter wire, was found to encircle the longitudinal rods, every 15 to 20 cm. The non-existence of the spiral reinforcement was, partly expected, as it had not been included in the design or the specifications. The specifications dealt only with bending loads taken

by the longitudinal reinforcement. Torsional loads or internal stressing, as reported here in detail, had not been considered and for this reason no torsional reinforcement had been included in the design. This finding alone can explain damage due to any form of radial stresses (e.g. due to the swelling of the corrosion products or due to expansion associated with humidity or temperature gradient) resulting to longitudinal cracking of the poles. Considering this, it is fortunate that serious damage due to external torsional loads, was found in only 2% of the poles examined.

b) The percentage of the longitudinal reinforcement is a lot higher that $16^o/_{oo}$, that the specifications required. Also, in all poles, rods with different diameters were found. Thus, the high percentage of the longitudinal reinforcement, the varying cross section of the steel rods, and their non-symmetrical installation, created an asymmetry to the rod spacing which was so reduced that big size aggregates were unable to pass through them.

c) The thicknesses of the reinforcement cover concrete, was adequate at the inner surface but inadequate at the outer surface. Approximately half of the removed poles had minimum cover thickness less than the 15mm which was required from the specifications. Normally, thickness varied between 9 to 15mm.

d) The outside surface of the poles, was not as it owed to be, a uniform outside surface of a truncated cone formed by revolution, but had two seams, along the joints of the mould, where concrete looked very porous. This concrete, with loose tissue (it consisted of poorly graded aggregates) was found in 50% of the poles, and was not of low strength only but also of high permeability, making an easy entrance for harmful substances, like humidity, carbon dioxide, chloride ions, etc.

On the samples retrieved from the removed poles laboratory tests were carried out to determine the material properties and to investigate the existence of substances which favour reinforcement corrosion. Finally, two fundamental characteristics were determined, which played important role on damage mechanisms, namely, the concrete permeability and the thicknesses of the corrosion oxides. For the permeability tests, the technique developed is described in ref. [6]. The results of these measurements appear in Table 2, below:

Table 2: Measured Properties affecting Durability.

s/n	Property	Symbols-Units	Healthy Position	Damaged Position
1	Concrete strength	R_c(MPa)	30 ± 10	---
2	Carbonation depth measured with phenolphthaleine	C_d(mm)	5 - 15	Up to the reinforcement
3	pH of pore water	pH	11,0 - 11,5	
4	Concentration of Chloride ions to a depth of 15mm	[Cl] % concrete W/W	<0,01	---
5	Concentration of sulphates to a depth of 15mm		2,2 - 3,4	
6	Coefficient of water permeability	K_{H_2O}(m/s)	$10^{-10} - 10^{-12}$	---
7	Thicknesses of oxides at the positions of damage.			
	(a) crack width: w<0,2mm	μm	---	10
	(b) crack width: 0,2<w<2mm	μm	---	200 - 300

From the above results, the following conclusions may be made:

a) The concrete strength of the poles shows a high dispersion. This picture is troublesome, since the average strength is by far below the value corresponding to durable concrete by today's standards and, also with regard to the magnitude of the dispersion (Coefficient of variation c:31%). This implies that the concrete used is not only of low strength, but also of bad quality.

b) Regarding carbonation depth in healthy positions, we remark that they are very small. Very rarely, they exceed 15mm, a fact that proves the very good compaction obtained by the centrifugal method. No differentiation of carbonation depth, was found in connection with the age of the poles. Regarding carbonation depths at the damage positions, we remark that carbonation depth extends up to the reinforcement, independently of the depth the reinforcement is located.

c) All pH values, at the healthy positions, vary around the value pH=11, values by far below the value pH=13 of the fresh concrete, a thing that shows that the concrete does not provide anymore that strong alcalic environment obtained during its preparation.

d) The conentrations of Cl- in areas not under the effect of sea breeze were found to be by far below the critical limit of 0,05% w/w of sample, and they lead to the conclusion that at least for the above poles, the chloride ions did not consist a dangerous factor for corrosion of steel reinforcement.

e) The results of the sulphates concentrations, prove that the quantities found are within the limits accepted by the specifications for the SO_3 content in the cement, a thing that shows that sulphates too, do not present a serious danger to account for the observed damage.

f) The order of magnitude of the water permeability coefficient that was found (10^{-11} m/s), is not particularly satisfactory, if we take into consideration that, today, the practice of durable concretes, requires coefficients by far smaller, of the order 10^{-13} m/s up to 10^{-16} m/s.

g) In order to determine the thickness of corrosion of the reinforcement where cracks were found, a metalographic microscope was used. It was found that the magnitude of oxidation, is a function of crack width. Capillary cracks, corresponded to thickness of corrosion between 10μm and 20μm, while cracks, 0,2<w<2mm corresponding to thickness between 200μm and 300μm. It appears that in order to have a crack (even a capillary one) corrosion with thicknesses of the order of 10μm is sufficient.

Tests on Poles in situ

The next step, of the research was to compare the above findings with the performance of poles in the network. The examination of poles in the network would also serve to evaluate the environmental influence on the development of damage, from the point of view of quantitative parameters, related to durability. It is to be noted that it has already been found that the poles are damaged more quickly and more drastically, when found at an unfavourable environment, from the point of view of physico-chemical effects (areas under the sea splash, areas with great humidity changes). Consequently, it was decided to select also poles from four areas of the Attica county, that represented different environments. The poles examined were all damaged. The selection was made so as to have poles of all ages, and of all degrees of damage. In these poles, one characteristic damage was located, at an accessible position (usually a crack). At the position of this damage, measurements were taken of the concrete cover thickness and, the depth of carbonation using phenolphthaleine, while the condition of the steel reinforcement, where there was one, underneath the damaged position, was checked. Small quantities of concrete taken from damaged locations were brought to the laboratory where the concentration of chloride ions was determined. The results of all these tests are shown in the Table 3 below.

Table 3: Measurements on Damaged Poles used in the Network

s/n	Area	Pole type (date)	Kind of Damage	Thickness of Cover (mm)	Carbonation Depth (mm)	%Cl-w/w Concrete	Reinforcement under location of damage
1	EVOIKOS	P 12 (1970)	Medium crack in the seam	27	>27	0,02	Corroded
2	EVOIKOS	P 40 (1975)	Longitudinal hair-like crack	17	0	0,14	Corroded
3	EVOIKOS	B 15 (1984)	Longitudinal hair-like crack	20	>20	1,05	Corroded
4	EVOIKOS	B 13 (1986)	Medium crack in the seam	10	>10	0,05	Corroded
5	ELEFSINA	E 95 (1966)	Large crack in the seam	20	>20	0,01	Corroded
6	ELEFSINA	E 121 (1974)	Longitudinal hair-like crack	15	>15	0,01	Corroded
7	PENTELI	NΠ10 4 (1961)	Medium crack in the seam	18	>18	0,04	Corroded
8	PENTELI	ΠΠ 17 (1967)	Longitudinal hair-like crack	28	>28	0,66	Corroded
9	PENTELI	ΠΠ 33 (1967)	Medium crack in the seam	20	>20	0,03	Corroded
10	MESOGIA	MM 64 (1968)	Longitudinal hair-like crack	27	>27	0,02	Corroded

From the above results the following conclusions were drawn:
a) The carbonation depth extends as far as the reinforcement, independently of the cover thickness.
b) Underneath every longitudinal crack, seam or not, there is always a corroded longitudinal reinforcement to be found.
c) The concentration of chloride ions is critical only in the area of the Evoikos Gulf, where the poles are in direct contact with the sea breeze.

On the above poles, measurements were also made to determine carbonation depth and concentration of chloride ions, at healthy positions. The aim of these tests was to compare quantitatively the substances that pollute concrete and reinforcement (carbon dioxide, chloride ions, rust) at the damage locations, and at locations where damage has started. The results are shown, on Tables 4 and 5 below:

Table 4: Tests on healthy and damaged locations of poles in situ.
Evoikos area, damaged locations.

s/n	Type of pole (date)	Degree of damage	Kind of damage	Thick- ness of cover (mm)	Carbo- nation depth (mm)	%Cl-w/w of concrete	Reinforce- ment under location damage
1	Λ 38 (1967)	MEDIUM	Longitudinal hair-like crack	10	>10	0,13	Corroded
2	P 12 (1970)	LARGE	Medium crack in the seam	27	>27	0,02	Corroded
3	P 40 (1975)	MEDIUM	Longitudinal hair-like crack	17	>17	0,14	Corroded
4	B 15 (1984)	MEDIUM	Longitudinal hair-like crack	20	>20	1,05	Corroded
5	B 13 (1986)	MEDIUM	Medium crack in the seam	10	>10	0,05	Corroded
6	Λ 122 (1986)	MEDIUM	Medium crack in the seam	13	>13	0,20	Corroded

Table 5: Tests on healthy and damaged locations of poles in situ.
Evoikos area, healthy locations.

s/n	Poles (date)	Degree of damage	Thickness of cover (mm)	Carbonation depth (mm)	%Cl- w/w of concrete	Condition of reinforcement
1	Λ 38 (1975)	MEDIUM	---	5	0,02	Non corroded
2	P 12 (1970)	LARGE	---	4	0,01	Non corroded
3	P 40 (1975)	MEDIUM	8	6	0,02	Non corroded
4	B 15 (1984)	MEDIUM	16	0	0,02	Non corroded
5	B 13 (1986)	MEDIUM	16	4	0,02	Non corroded
6	Λ 122 (1986)	MEDIUM	24	0	0,04	Non corroded

These Tables show that the differences in the carbonation depth, as well as in the concentration of chloride ions, in neighboring positions of the same pole, are enormous.

This finding justifies the term «pollution» that is used for the very high concentration of harmful substances that may appear at some locations of a healthy, in all other respects, concrete, resulting in local damage. Based on this difference, we may also explain the entirely different condition, in which the reinforcement underneath, is found.

Comments

The basic conclusion of this research, applying to the locations of damage, is the following:
(a) the concrete was carbonated,
(b) the reinforcement was corroded,
(c) in places near the sea, chloride ions concentration was very high.
The main cause of damage is considered to be due to the fact that longitudinal reinforcement was completely unprotected at the locations of seams, resulting in easy entrance of humidity and other harmful substances. In addition to seam locations longitudinal reinforcement was corroded at other locations, where either cover thickness was small, or where the environmental effects were unfavourable (e.g. salt contamination). Design and construction inadequaties such as the non-existence of any spiral reinforcement, and the high non-homogeneity have favoured stress concentration which caused a large number of cracks, reinforcement corrosion and spalling of concrete. As a result, one out of the two poles, in the statistical sample out of the 1500 poles examined, was damaged.

SUGGESTIONS FOR CONSTRUCTION IMPROVEMENT

The most usual cause of premature damage was found to be deffective construction. The main cause was the low quality of concrete at the seam locations (very porous), and the non-homogeneity. These lead to the creation of an entrance gate for harmful substances, humidity and carbon dioxide, and also to the development of internal differential stresses. At the same time, the non-existence of any spiral reinforcement left concrete underneath unprotected from the effect of stresses in a horizontal plane. Thus, under the effects of environmental exposure, reinforcement, sooner or later was corroded, resulting to cracking and subsequent progressive failure and consequent dramatic reduction of the life expectancy of the poles involved. It was, therefore considered necessary to establish new specifications with new design criteria and construction requirements, such, that would lead to renewal of construction equipment. In addition to the above, the new specifications would include clear recomendations and requirements regarding the materials (steel, concrete aggregates) in harmony with the existing regulations and instructions for durable structures [7].

The improvement of pole quality has to do with the following parameters: design, quality of materials, manufacturing equipment, construction process. Therefore, any attempt for improving future poles, should pass through reconsideration of the above parameters.

Design improvement had three targets: (a) Protection of poles from environmental effects, (b) standardization of materials and other pole characteristics, and, (c) designing for torsional in addition to bending stresses.
In order to increase the life span of the poles, the following steps were taken:
1. Spiral reinforcement was included with a minimum diameter of 3mm (for poles with a small load) and 4mm for the rest. The maximum spiral pitch was 5cm for the top 2 meters and 10 cm for the rest of the length of the pole.
2. It was specified that poles which presented loose tissue would be rejected and no

patching up was allowed. Also, it was specified that deffective steel forms would not be permitted.

3. The maximum diameter of the concrete aggregates (Φmax) was set to 12mm, and the minimum distance "e" between two longitudinal rods was specified to be not less than 20mm (>1,5Φmax). In this way, the risk of separation of fresh concrete, because of close spacing of the longitudinal rods, was avoided.

4. The minimum thickness of the concrete cover "c" outside the spiral reinforcement was increased from 15mm to 20mm (>1,5Φmax).

5. Clearly, it was specified that poles with accumulation of cement paste in their interior cavity would be rejected. Such accumulation would indicate non-homogeneity, originating from bad composition of concrete and compaction procedure.

6. Quality of concrete C40 (or B450 under the old regulations) was required.

7. Recommendations were made for waterproofing the heads of the poles, which are now shaped flat (without a cap).

8. The addition to the concrete of any accelerators that may contain harmful substances, favouring corrosion of reinforcement, such as those having $CaCl_2$, was prohibited.

9. Clear criteria were set, as to the methods of accelerated curing, that could be applied, in order to avoid problems of uncontrolled chemical reactions, quick drying, and development of differential temperature at the cross-section.

CONCLUSIONS

1. The reinforced concrete poles used by PPC for supporting the Distribution Overhead Lines, present damages, some of them less and some, more. Damage is mainly in the form of longitudinal cracks and corrosion of the underlying longitudinal reinforcement.

2. From the recorded damage it was found that the longitudinal cracks were located mainly (more than half of them) at the two seams of the concrete forms. It is to be noted that the poles are made by centrifuging in forms consisting of two halfs. This procedure results in two anti-diametric seams.

3. Laboratory tests and data taken from 50 poles removed from the network and 70 poles still in operation indicated that:
   - Corroded reinforcement was always found underneath longitudinal cracks independently of crack width.
   - Concrete along the crack was carbonated up to the underlying longitudinal reinforcement, even when reinforcement was at depths of 20 or 30mm.
   - At damage locations of poles installed near seaside, high concentrations of chloride ions were found, higher than the critical limit of 0,05% w/w of concrete.

ACKNOWLEDGEMENTS

This article is a part of a broader research program, on reinforced concrete poles, for overhead transmission lines carried out by PPC's Testing and Research Center.
The Author wants to express his thanks to Mrs A. Papathanassiou, Geologist, for the careful performance of most of the tests mentioned in this article.

REFERENCES

1. PUBLIC POWER CORPORATION. Research Programme of Concrete Poles. Internal PPC correspondance. CTRS/F.530: 2617/14.7.89 and 2780/28.7.89 (in Greek).

2. KARYSTINOS N. I., PAPATHANASSIOU A. G. AND SAKELLARIOU A. G. Damage Morphology of Reinforced Concrete Poles. 10th Greek Concrete Conference, Rhodes 1991, PP325-332 (in Greek).

3. P.P.C. Specification: EK 02.51/92. Poles of Reinforced Concrete. Athens, 1992, PP19 (in Greek).

4. PPC Specifiation: GR 51/71. Poles of Reinforced Concrete. Athens, 1971, PP18 (in Greek).

5. GRIESE, W AND MULLER, H. Erfassung von schaden an betonmasten und deren statische auswertung. Elektrizitatswirstschaft 85 (1986). H.3, S81-82.

6. PAVLAKIS G. A. AND SAKELLARIOU A. G. Air and Water Permeability on Concrete Cores CANMET/ACI International Conference on Advances in Concrete Technology May 1992. Athens PP1-21.

7. COMITE EUROINTERNATIONAL DU BETON. Durable Concrete Stuctures-CEB Design Guide. Bulletin d'information No 182, 1987.

CONCRETE TESTING DEVELOPMENT

O Sammal

Estonian Building Research Institute

Estonia

ABSTRACT. For better use of reinforced concrete it is necessary to study experimentally the real state of structures during assembly and service. The design of constructions and their members is based on an idealized and simplified pattern of stress, strain and force distribution, quite a bit of which in practice works out otherwise. The actual structural behaviour may, to a marked degree, differ from the rated one. Multiple reserves, sometimes also a near-collapse condition may exist. The determination of the technical resources and the reliability of structures, the verification of their actual state need examination of structures on site. Development of concrete oriented specific measuring means and test methods has become a vital necessity and the aim of the activity of the Estonian Building Research Institute. Three microprocessor complexes BCM, BPM and BTM have been developed for testing and direct measuring of strength, stress, and strain state in concrete structures. A new "negentropic prediction method for the determination of the serviceability of concrete structures" is proposed.

Keywords: Concrete, Strength, Stress, Strain, Measurement, Testing, In situ, Reliability, Method.

Olav Sammal, D.Ph., is Head of the Building Test Department, Estonian Building Research Institute, Estonia. He is specialized in the development of the new concrete oriented test methods and measuring devices BCM, BPM and BTM and using them in situ in building service. He is the author of the "impact-pulse (magnetoelastic) method" and the"negentropic prediction method" for the determination of the serviceability of concrete structures. He is individual member of I.A.B.S.E.

Radical Concrete Technology. Edited by R K Dhir and P C Hewlett. Published in 1996 by E & FN Spon, 2–6 Boundary Row, London SE1 8HN, UK. ISBN 0 419 21480 1.

INTRODUCTION

Concrete/reinforced concrete is a very important material for structural service of mankind nowadays and in the future. At the same time there exist great difficulties in testing or investigating the real behaviour and reliability of concrete structures. Heterogeneity and anisotropy of concrete, inconsistency of its physical and mechanical properties as well as discontinuity of the concrete medium differ to a considerable extent from the hypotheses accepted in calculations. The internal forces in precast and in-situ concrete structures as well are distributed according to the spatial force system. The role of the initial stresses induced in structural members during manufacturing and erection cannot be forgotten either. The actual structural behaviour may, to a marked degree, differ from the rated one. At the Estonian Building Research Institute microprocessor complexes BCM, BPM, BTM have been developed for testing and direct measuring of strength (BCM), stress (BPM) and strain (BTM) state in concrete structures. Negentropic prediction method for the determination of the serviceability of concrete structures is proposed.

CONCRETE STRENGTH MEASUREMENT (BCM)
Task

Nowadays the concrete strength checking system, based on destructing specimens (cubes, cylinders) made concurrently with the structural member of the same mortar, is quite doubtful because it is impossible to secure the uniformity of compacting (vibration) conditions and climatic factors during hardening of specimens and structures. In the world practice cases are known where buildings have collapsed due to poor strength of the concrete structures, although all specimens proved to meet the requirements.

The idea of the development of a new non-destructive, so-called "impact-pulse (magnetoelastic) method" for concrete strength and uniformity determination lies in using the magnetoelastic transducer developed at the Estonian Building Research Institute for force measurements in the hand-held impact device (sclerometer). The basic principle of this method is the following [1]:

At the moment of impact of the sclerometer to the surface of concrete, in the magnetoelastic transducer of the sclerometer a process of conversion takes place:

$$\frac{dP}{dt} \Rightarrow \frac{d\sigma}{dt} \Rightarrow \frac{d\mu}{dt} \Rightarrow \frac{dW_b}{dt} \Rightarrow U \tag{1}$$

where,

P and σ = force and mechanical stress in the magnetic core of the magnetoelastic transducer;

μ and W_b = magnetic permeability and magnetic flux in the same;

t and U = time and output signal of the sclerometer CK.

The process of converting the signal in the magnetoelastic transducer in the sclerometer CK by impact is shown in Figure 1.

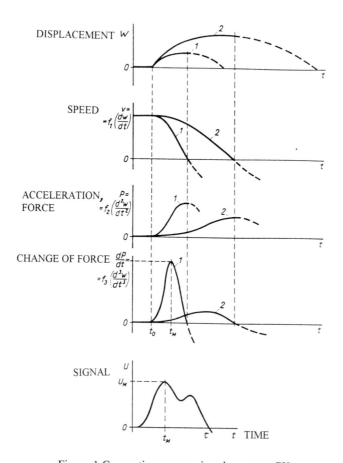

Figure 1 Converting process in sclerometer CK

W = displacement by indentation of the indenter into concrete surface;
$v = f_1(dW/dt)$ = indentation speed;
$P = f_2(d^2W/dt^2)$ = acceleration (or force) in the CK transducer at indentation;
$dP/dt = f_3(d^3W/dt^3)$ = change of force in the CK transducer by indentation;
$U = f(t)$ = output signal of the magnetoelastic transducer of CK;
U_M = first amplitude value of the output signal of the CK transducer at the moment t_M;
1 = process with a higher and 2 = with a lower strength of concrete.

Thus the principle of the method lies in recording the maximum speed of change in the contact force $dP/dt \Rightarrow U_M$ in the magnetoelastic transducer of the sclerometer CK (see Figure 1). The generated signal U_M is converted into digital code.

Complex BCM/CK

The complex BCM/CK has been developed to deliver multipoint impacts (for example, ten impacts to a test area within 10 seconds) and after that receive simultaneous indication on the display of the computed mean value and the statistical estimation as well as to memorize up to 1600 testing results in the memory of the recording unit BCM. The complex BCM/CK is calibrated by preliminary testing. There is a good correlation (> 0.90) between the electric signal and concrete strength in the range of 10...70 N/mm² (MPa). The impact energy of the sclerometer CK does not exceed 0.15 J. Set up routines are handled by menus and a 8 button keyboard. The instrument is fully operational immediately after turn-on. Memorized data are not lost even when the instrument is turned off. The instrument is powered by a built-in rechargeable battery with a capacity of 6 hours normal operation. The stored data can be offered to a printer or PC for further processing.

Concrete Strength Test Results

Determination of concrete strength and uniformity of four columns after erection in a building in Tallinn is shown in Figure 2 and 3.

	TEST RESULTS							ASSEMBLED COLUMNS
Col	Test No.	Rl, MPa	fcm, δ %, fck, grade	Col	Test No.	Rl, MPa	fcm, δ %, fck, grade	
K-1	1	36.5			21	33.1		13,26 ⟋ 40
	2	35.2			22	35.6		K-4
	3	35.4			23	30.3		
	4	34.4	37.0		24	34.1	35.0	31
	5	33.9	10.8%	K-3	25	40.2	8.0%	9,82 ⟋ 30
	6	38.5	30.5		26	37.9	30.4	K-3
	7	35.3			27	35.7		
	8	33.0	C 20		28	35.2	C 20	6,52 21
	9	44.6			29	32.2		20
	10	43.5			30	35.4		3,52 K-2
								11
K-2	11	38.2			31	52.6		10
	12	40.9			32	49.0		
	13	40.4			33	49.6		K-1
	14	43.2	41.5		34	48.6	49.5	
	15	43.3	5.7%	K-4	35	44.7	6.7%	
	16	40.1	37.6		36	46.1	44.1	
	17	40.1			37	48.8		2
	18	43.0	C 30		38	48.8	C 40	-3,08 1
	19	39.9			39	53.0		
	20	46.3			40	55.3		

Average value of concrete strength, fcm-cube, MPa: 40.8
Coefficient of variation, δ , %, at n=40: 15.7
Characteristic strength, fck-cube, MPa: 30
Concrete grade: C20...C40

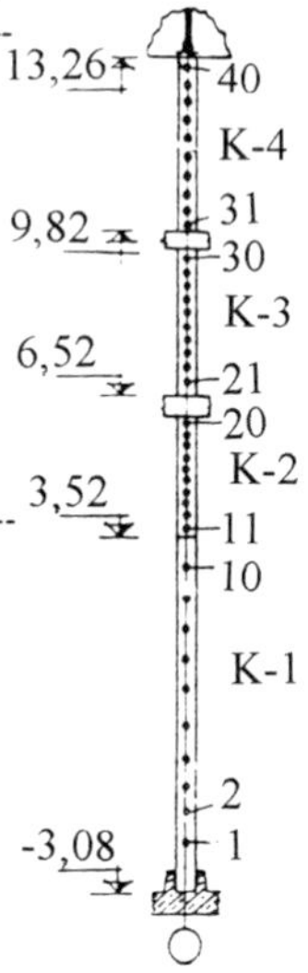

Figure 2 Test results of four columns after erection in a building

Explanation

° - 1...40 = local test areas of 10 impacts (test No);
R_i = concrete cubic strength in an area (mean value of test results of 10 impacts) measured by means of BCM/CK;
K-1...K-4 = four reinforced precast columns after erection in a building, designed $f_{cm\text{-}cube}$ = 40 N/mm^2 (40 MPa).

In reality the columns **K-1** and **K-3** do not meet the requirements, although the average strength of four columns does, $f_{cm\text{-}cube}$ = 40,8 N/mm^2 > 40 MPa.

Figure 3 Testing columns by means of BCM/CK

The complex BCM/CK is well qualified for operative discovering of concrete regions or zones with a low strength in prefabricated and in-situ manufactured structures.

CONCRETE STRESS MEASUREMENT (BPM)

Task

Many   concrete/reinforced concrete structures are calculated as statically
determinate but nearly always they are subjected to the spatial (hyperstatic) force
system. They depend basically on the actual rigidity of structural members and
joints. It is obvious that the emergence and introduction of new measuring devices
will allow to get a better insight into the state of stress in structures, perfect their
calculation and raise their reliability.
The ambiguity of the dependence between strains and stresses in concrete (due to
shrinkage, creep, microcracks, etc.) makes it impossible to precisely determine
stresses on the basis of strain measurement.
A possibility of direct measurement of compressive stress in concrete has been
shown by R.W.Carlson [2]. He determined the basic principle of stress
meter/transducer: it must be disc-shaped and more rigid than the medium (concrete)
surrounding the stressmeter. In that case the stressmeter is immune to inelastic
deformations (shrinkage, creep) of concrete. Figure 4 shows the test results of a
concrete prism with a stress transducer M moulded in the centre of it. We see the
typical relation between strain ε_σ and normal stress σ_N^c in the prism at loading (see
hysteresis loop), but the stress transducer M readings G show a very high correlation
$G=f(\sigma_N^c)$, and a very small hysteresis.

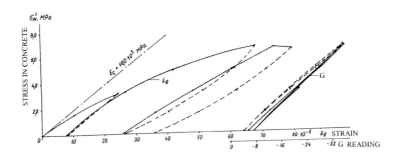

Figure 4 Strain and stress measurement in a concrete prism

It has been proved theoretically and by experiment [1] that the compressive stress
to be determined (along the normal axis **z** of M) may be determined by the formula

$$\sigma_z^c = K_1(\sigma_z^B - \sigma_0) \tag{2}$$

where, K_1 = correction factor (0.95...1.10) which depends on the type of the
stressmeter/transducer; σ_z^B = transducer reading after external loading of concrete
sample; σ_0 = initial reading.
As that, there is no need to know the modulus of elasticity and strains of concrete.
In case of two- or three-dimensional state of stress in concrete the stress transducer
records a small additional stress (~ 6 %) caused by transverse stresses.

Complex BPM/M

The new device BPM/M, developed at the Estonian Building Research Institute [1] consists of a small-sized concrete stress transducer M and a recording unit BPM. Stress transducer M (see Figure 5) consists of a metal (passive) disc in which a magnetoelastic transducer type AK is accommodated, actively responding to compressive stress along its normal axis, after moulding into concrete. There may be several separate transducers in a set to increase the accuracy of measurement and to receive the information about stress heterogeneity in concrete.

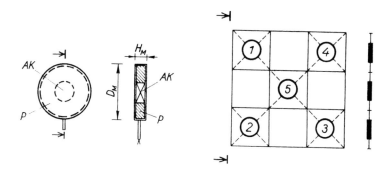

Figure 5   Stress transducer M20

AK = magnetoelastic transducer; P = metal (passive) disc; 1...5 = set with five transducers M20.

The main characteristics of the concrete stress transducers M20-B and M20-C are

Seq. No.	Parameters	M20-B	M20-C
1. Range of measurement of compressive stress, MPa		2 to 20	3 to 30
2. Basic instrumental error, MPa		1.0	1,5
3. Operating temperature, °C		up to 50	up to 50
4. Correction factor K_1		0.95	1.10
5. Dimensions, mm			
diameter		23	23
thickness		5	5

The modification of the recording unit BPM with a microprocessor system and connection with PC enables to carry out the necessary analysis of test results practically simultaneously with the measurement. BPM records and memorizes information from the transducers, calculates the stress state and internal forces in concrete sections tested (see test results).

Concrete stress measurement results

Figure 6 shows the test results of a fragment of a reinforced concrete column (cross section 20x20 cm) in which 12 stress transducers M20 were embedded.

TEST RESULTS

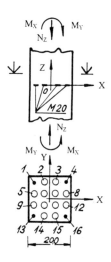

M20	Location		Meas.
No	X, cm	Y, cm	σ_Z, MPa
2	-2,5	7,5	7,7
3	2,5	7,5	5,2
5	-7,5	2,5	7,0
6	-2,5	2,5	11,6
7	2,5	2,5	8,7
8	7,5	2,5	3,6
9	-7,5	-2,5	8,0
10	-2,5	-2,5	13,4
11	2,5	-2,5	11,2
12	7,5	-2,5	6,1
14	-2,5	-7,5	6,1
15	2,5	-7,5	4,6

• 1, 4, 13, 16 = Reinforcement;
O 2...15 = Transducers M20

Figure 6 Determination of stresses and internal forces in a section of a fragment of a reinforced concrete column

Uniaxial compressive stresses σ_Z were approximated with a polynomial (plane)
$$\sigma_Z = 7.767 - 0.227x - 0.011y \qquad (3)$$
Approximation accuracy = residual standard deviation $s_Z = 3.07$ MPa
Internal forces in the concrete sections :

Normal force	$N_Z = \int_A \sigma_Z dA = 310.6$ kN	(4)
Moments	$M_X = \int_A \sigma_Z y dA = -14.5$ kNcm	(5)
	$M_Y = \int_A \sigma_Z x dA = -303.0$ kNcm	(6)
Eccentricities	$e_X = M_Y/N_Z = -0.975$ cm	(7)
	$e_Y = M_X/N_Z = -0.047$ cm	(8)

Stresses and forces in reinforcement are determinated by the reduced section rule:
$$\sigma_{s,i} = \sigma_{Z,i}\, E_s/E_c \qquad (9)$$
$$N_{s,i} = \sigma_{s,i}\, A_{s,i} \qquad (10)$$
where, E_s, E_c = module of elasticity of steel and concrete; x, y = co-ordinates; $A_{s,i}$ = cross section of reinforcement.
If tensile stresses (strains) are expected in concrete structures, concrete strain measurement using BTM/T is necessary.

CONCRETE STRAIN MEASUREMENT (BTM)

Task

In the last half century of the service of concrete/reinforced concrete structures there has become very important the problem of the dangerous cracks in many constructions.
15 years after the erection of the Tallinn TV tower there were noticed over 500 cracks on the surface of the concrete shaft and in the ceiling at the connection of the steel and reinforced concrete part at the level of +180m. To investigate how dangerous the radial cracks in the ceiling were the device BTM/T was developed at the Estonian Building Research Institute and used in situ.

Complex BTM/T

BTM/T is a two channel electronic device with microprocessor system, working with standard strain gages or special strain transducers, for example, type T moulded within concrete medium. Both channels operate with separate commutators. There is a special routine for differential measurement, memorizing the test results and working with PC.

Concrete strain test results

For the investigation of the changes in the radial cracks in the ceiling 64 standard strain gages were glued across the cracks. Analysis of the measurement results during two years shows that there is no correlation between the change in deformation of the cracks ε_C and time h, days;

$$\varepsilon_C = (5.6569 + 0.0004h)0.33.10^{-3} \qquad (11)$$

coefficient of correlation r = 0.08.
But there exists a correlation with the open air temperature t, °C (at the level of 200m) and ε_C:

$$\varepsilon_C = (6.2254 - 0.000483h - 0.06288t)0.33.10^{-3} \qquad (12)$$

coefficient of correlation r = 0.71.
It means that the changes in the radial cracks in the ceiling (+180m) have no progressive character and they do not represent any danger to the serviceability of the TV tower.

NEGENTROPIC PREDICTION METHOD

It is known that in a concrete medium heterogeneities (pieces of aggregate, microcracks, pores, etc.) are distributed stochastically. According to the classical probability theory the problem arises: which should be the probability level in deciding, particularly when the probability distribution differs considerably from the normal (Gaussian) law. With the increase of the probability level P the confidence interval of uncertainty widens respectively, and when P=1 it is equal to ± ∞ (seems to be useless).

In his information theory C.Shannon [3] proposed to determine the misleading effect by the formula

$$H(x) = - \int p(x)lnp(x)dx \qquad (13)$$

and call $H(x)$ noise entropy. Entropical bounds of uncertainty $\pm \Delta_e(x)$ are defined as the bounds of random events with sharply bounded uniform distribution with the equal value of noise entropy (or amount of misleading information) as by any other actual probability distribution

$$\pm \Delta e(x) = 1/2expH(x) \qquad (14)$$

To denote the concept "measure of ordering" L.Brillouin [4] proposed the term "negentropy" meaning negative entropy. Proceeding from the initial positions presented above, we propose the so-called "negentropic prediction method of the serviceability of concrete/reinforced concrete and other structures" [1]. According to this method, any loading of a structural member (e.g., formation of a stress field) may be regarded as certain ordering. In this way, any state of loading of a structural member or its section is juxtaposed by its negentropy. The maximum negentropy corresponds to some ultimate limit state, for instance, as that before failure of the structural member. This maximum negentropy we call "negentropic capacity Π_C ". The negentropy at the moment A of the structural examination which is determined by measurement results we call "negentropic charge Π_A". There are several variants possible for calculating the negentropy in the structures. For normal sections

$$\Pi = K*l*\int_A \sigma_z^2 dA \qquad (15)$$

where,

σ_z = normal stresses in the section A; K = constant depending on the material of the section; l = length of the section being examined (e.g., l = 1cm).

The negentropic capacity Π_C is calculated in accordance with the distribution of internal stresses/forces corresponding to the ultimate limit state (ULS) which is established by the standards and rules. The negentropic charge Π_A is calculated from the actual stress distribution determined by testing structures, for example, with BPM/M. The most probable failure of a structural member is predicted by comparing the negentropic capacity with the negentropic charge by their mean values $\overline{\Pi}_C$ and $\overline{\Pi}_A$.

The corresponding index is

$$\mathcal{H}^C = \sqrt{\overline{\Pi}_C / \overline{\Pi}_A} \qquad (16)$$

According to the information theory, every determination contains a certain lot of misleading information that is characterized by the entropical bounds of uncertainty $\pm \Delta_e(x)$. The prediction of the serviceability is possible according to the bounds of uncertainties, i.e., the loading of a structure can be raised until the upper bound $\hat{\Pi}_A$ does not exceed the lower bound $\check{\Pi}_C$. The corresponding index

$$\mathcal{H}^A = \sqrt{\check{\Pi}_C / \hat{\Pi}_A} > 1 \qquad (17)$$

means that there exist technical reserves regarding the service load. The values for additional load can be predicted or the structure can be tested by increasing the load until $\mathcal{H}^A \Rightarrow 1$. For example, testing results in Figure 6 show $\mathcal{H}^A = 1.41$ and $\mathcal{H}^C = 2.95$ [5],[6].

CONCLUSION

The practical application of the described methods and devices gives the researches, designers, inspectors and contractors some more confidence in erection and maintenance of new and more effective and reliable constructions and buildings.

© Estonian Building Research Institute. 7 Estonia Blvd, EE0100 Tallinn, ESTONIA

REFERENCES

1. SAMMAL, O. Stresses in Concrete and Prediction Method of the Technical Resources   of   Concrete   and   Reinforced   Concrete   Structures   and Constructions.Tallinn, Valgus, 1980 (Monograph in Russian).
2. CARLSON, R W and PIRTZ, D.   Development of a Device for the Direct Measurement of Compressive Stress. Journal of the American Concrete Institute,Vol. 24, No. 3, 1952, pp 201-215.
3.   SHANNON, C E. A Mathematical Theory of Communication. Bell System Tech. J., v. 27, July 1948, pp 379-423 (Part I); Okt. 1948, pp 623-656 (Part II).
4.   BRILLOUIN, L. J. Science and Information Theory. Academie Press. Inc. Publishers, New York, 1956.
5. SAMMAL, O. BCM/CK for Checking Concrete Strength and Uniformity Employing Impact Pulse (Magnetoelastic) Method. Quality Management in Building and Construction, Proceedings of EUREKA Conference Hamar/Lillehammer, June,1994, pp 431-436.
6. SAMMAL, O. and SAMMAL, K. Experimental Checking of Concrete Stress and Strength. FIP' 93 Symposium in Kyoto, Japan, Volume II, Oct., 1993, pp 1031-1038.

CORROSION OF STEEL FIBRES
IN CARBONATED CRACKED CONCRETE

P Schiessl

R Weydert

University of Technology

Aachen

Germany

ABSTRACT. In some applications of steel fibre reinforced concrete (SFRC), structural behaviour depends on the post-crack strain characteristic of the concrete. The durability of fibres in this case can be of great influence for the stability of the system. Whereas durability of fibres in cracks of concrete under chloride influence cannot be ensured, the behaviour of steel fibres in carbonated cracked concrete is hardly known. This paper presents results on the corrosion behaviour of different types of steel fibre in carbonated cracked SFRC-beams under outdoor conditions, showing that under unfavourable conditions, corrosion of fibres in the cracks can affect the durability of the SFRC.

Keywords: Carbonation, Cracks, Corrosion due to carbonation, Durability structural assessment, Steel fibre reinforced concrete (SFRC).

Professor Dr.-Ing. Peter Schiessl is Director of the Institute for Building Materials Research at the Technical University of Aachen. Besides other fields of activities he is working in the field of durability and repair of concrete structures with particular reference to corrosion of steel in concrete. He is active in CEN-, CEB-, RILEM- and various national committees.

Romain Weydert is a Research Engineer at the Institute for Building Materials Research at the Technical University of Aachen. He is specialised in corrosion of reinforcing steel and steel fibres in concrete and in corrosion protection.

Radical Concrete Technology. Edited by R K Dhir and P C Hewlett. Published in 1996 by E & FN Spon, 2–6 Boundary Row, London SE1 8HN, UK. ISBN 0 419 21480 1.

INTRODUCTION

A lot of research programs and publications deal with durability of steel fibre reinforced concrete (SFRC) under severe conditions (e.g. marine environments). Good durability is attested to steel fibres in uncracked concrete, even with high chloride contents of the concrete, whilst in cracked concrete, fibres corrode when chlorides can penetrate the cracks /3/. Sufficient durability can only be obtained when using stainless steel fibres.

Corrosion of steel in carbonated concrete occurs through the reduction of the pore fluid alkalinity caused by carbonation, thus allowing the loss of normal passivation of the steel /4/. In cracks of the concrete, steel is partly not protected by concrete so that the corrosion risk is unlike higher than in uncracked concrete. Corrosion can only be avoided if the high concrete alkalinity is transferred to the crack penetrating moisture.

The durability of steel fibres in carbonated cracked concrete has only been subject of few research programs /1/. For many application cases e. g. tunnel linings, precast garage-cells or other thin concrete members, the corrosion risk is mainly due to carbonation of concrete. Many attempts have been made to improve rules for design and construction of SFRC in guidelines and standards, with consideration of the structural behaviour in state II (cracked). The design of fibre reinforced concrete structures is mainly based on the determination of equivalent flexural strength values or toughness parameters of fibre reinforced bending test beams. The mostly unknown effect of fibre corrosion in cracks may be for example taken into account by reducing the structural depth of a certain structural member, thus allowing fibres to corrode in the section not used for loading design.

This paper presents the results of a research program dealing with corrosion of steel fibres in carbonated cracked and uncracked concrete, being of major interest for further guidelines and standards. The results are completed by investigations at 10 year old cracked reinforced fibre concrete members, stored under outdoor conditions.

EXPERIMENTAL DETAILS

Nine different types of fibre were used, the fibre contents varying between 30 and 120 kg/m³. The cement content (OPC) of the used mixes was 350 kg/m³ with w/c-ratios of 0.40 and 0.50. Additionally one shotcrete-mix was examined. An overview of the used types of fibre is given in Table 1. The fibre reinforced test beams had dimensions of 700 mm ×150 mm ×100 mm. To guarantee a uniform fibre orientation and eliminate formwork effects, the beams were sewed cut of a larger slab. At an age of 28 days one part of the beams were pre-cracked under third point loading with registration of the load-deflection behaviour. After the pre-cracking, one part of the beams has been exposed to quick-carbonation conditions under high carbon dioxide content of the air in special "carbonation chambers". After carbonation, the beams were exposed to normal outdoor conditions. The second part of the beams was

Table 1 Types of fibre used in the tests

FIBRE TYPE	PROFILE AND CROSS-SECTION SHAPE	SIZE l × d, mm	DESCRIPTION
1 2		60 × 0.8	drawn wire fibre, hooked ends/ type 2 galvanized
3		32 length	milled steel fibre
4 5		45 × 1.0 50 × 1.0	drawn straight wire fibre
6		60 × 1.0	drawn waved wire fibre
7 8		54 × 1.0 60 × 0.9	drawn wire fibre, enlarged/clinched ends
9		30 × 0.5	drawn wire fibre, hooked ends

exposed under the same conditions, being pre-cracked but not pre-carbonated. The crack widths of the beams varied between 0.2 and 0.4 mm, the crack opening being orientated upside. To adjust the crack width, the beams were put in special steel constructions as shown in Figure 1.

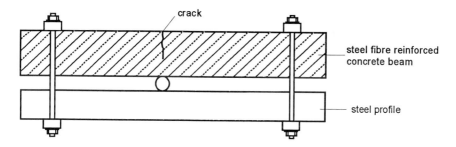

Figure 1 Schematic view of the loaded cracked test beam

RESULTS

After an exposure time of about 1.5 year the beams were loaded to failure under third point loading to determine the residual flexural strength. The crack zone was examined visually to determine the "area of corroded fibres" (see Figure 2), the failure mode of the fibres (percentage of broken and pulled out fibres) and the corrosion induced loss in cross section.

The carbonation depth at the end of the exposure time was between 8 and 10 mm in uncracked concrete and between 1 and 3 mm within the cracks for the pre-carbonated beams. Without pre-carbonation, the determined carbonation depths were 2 to 3 mm in uncracked concrete and below 1 mm within the crack.

The residual strength after 1.5 year for most of the beams was higher than the initial strength after 28 days. This was also true for beams where most of the fibres showed corrosion. As most fibres have been pulled out during rupture of the beam, irrespective of corrosion, the load increase of the beams is due to higher bond stress of the fibres. This strength gain is commonly observed, even when subjected to severe chloride application, and is probably due to a strength gain of the concrete as a consequence of ongoing concrete hydration processes under humid outdoor conditions.

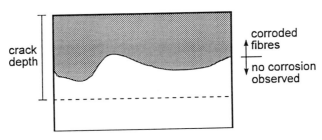

Figure 2 cross section of a test beam in the crack (example)

For further examinations, some corroded fibres were taken out to determine the reduction of the cross section due to corrosion in percent of the initial cross section. In some cases the reduction of cross section exceeded 20 %, the reduction being smaller with increasing crack depth. In the uncracked carbonated concrete, fibres were taken out and examined visually. Even with carbonation depth up to 10 mm, corrosion at mild steel fibres was only found for fibres with less than 2 mm concrete cover. No spalling of the concrete due to corrosion of the fibres was found. Apart from some visual affection of the concrete surface, fibre corrosion had no negative effect for uncracked carbonated concrete. The galvanized steel fibres showed less corrosion at the concrete surface.

In the crack, corrosion was observed for nearly all fibres at the upper half, surface near area of the crack. No major difference was found for pre-carbonated and other beams, although the corrosion induced loss in cross section and the percentage of broken fibres was slightly higher for pre-carbonated beams. Corrosion was less for smaller crack widths and was diminishing with increasing crack depths, but no critical crack width where no corrosion occurred could be found. In this context, further beams with smaller crack-widths are actually exposed.

Galvanized steel fibres did not show sufficient corrosion resistance in the crack zone. Compared with mild steel fibres, the galvanization leads to a time delayed start of corrosion. As expected, the corrosion behaviour of the different mild steel fibres was similar, whereas a loss in cross section is more serious for small fibre diametres than for bigger ones. This is shown by the fact that the percentage of broken fibres was higher for the smaller fibre diametres.

The fact that no major difference was found between pre-carbonated and other test beams shows that depassivation of fibres in concrete cracks cannot be avoided if air and humidity can penetrate the crack. After short time, the water penetrating the crack is no more influenced by the high alkalinity of the surrounding concrete, so that the free steel surface of the fibres can depassivate. In how far this process is followed by corrosion mainly depends on the "water history" of the crack, i. e. how often and how many water penetrates the crack. So, the results of this research program are only valid for the given test conditions with a horizontal surface being exposed to rain and cannot be transferred to other micro environmental conditions.

The examined cracks of the 10 year old cracked reinforced steel fibre concrete slabs were orientated vertically, not being exposed to direct rain. Fibres were taken out of the cracked zones after removing the surrounding concrete. Corrosion was found on the crack-crossing part of the fibres, this for crack-widths between 0.2 and 0.4 mm and up to 40 mm from the concrete surface. The corrosion induced reduction of cross section reached up to 25 %, being not much higher than for the fibres in the laboratory tests after 1.5 year of exposure.

DISCUSSION OF RESULTS AND PRACTICAL CONSEQUENCES

The research carried out shows, that fibre corrosion in cracks with crack-widths larger than 0.2 mm cannot be excluded under normal outdoor conditions. Until now, no critical crack-width can be given below which no corrosion is to be expected. Due to the great influence of the environmental conditions of the concrete (wet-dry cycles) on fibre corrosion, a critical crack-width in any case needs to be related to micro environmental conditions. The micro environment chosen in the test programme can be considered to be the worst for carbonation induced corrosion. This can be confirmed by the fact that corrosion of the fibres of the 10 year old concrete members is similar to the test beams after only 1.5 years of exposure. As the corrosion process is supposed to continue, a loss in strength with time has to be expected.

The consequence of possible fibre corrosion in cracked SFRC has to be assessed thoroughly. For most applications of SFRC however, post crack strain behaviour of the fibre reinforced concrete is not essential and corrosion of steel fibres can be considered to be harmless. This is true for all cases where fibres are used for crack limitation and crack control at early age concrete, for an increase of freeze-thaw-resistance of concrete or an increase of abrasion resistance of concrete.

In cases where steel fibres are expected to replace partly or totally the usual reinforcement in the state II (cracked), fibres are needed to ensure the equilibrium of forces and fibre durability must be guaranteed for the service life of the construction. Some questions remain for the application of SFRC in hyperstatic systems where fibres are expected to transmit forces. In the example of SFRC tunnel linings, the rotation capacity of the hyperstatic system leads to redistribution of the loads and plays an important role to ensure the stability of the tunnel system. For such hyperstatic structural systems, the effect of fibre corrosion until now has not been subject of investigations. It can be expected that the weakening through corrosion of fibres leads to a redistribution of the loads to neighbouring parts, thus not necessarily affecting the stability of the system. Further results are expected from research being actually done on SFRC-beams with smaller crack-widths and different environmental conditions.

CONCLUSIONS

Depassivation of steel fibres in cracks can occur even if there is no measurable carbonation at the inner crack surfaces. This means that under unfavourable conditions (crack wettings), corrosion cannot be excluded, even for very small crack-widths. Corrosion intensity strongly depends on the environmental conditions. Galvanizing of the fibres provides no sufficient corrosion protection.

Durability of steel fibres in cracks depends mainly on crack-width, fibre type, diametre and content, environmental conditions and water ingress in the crack. These factors must be taken into consideration when evaluating the durability of steel fibres in cracked concrete.

In most application cases fibre corrosion must not be overrated as it does not affect the stability of the construction.

REFERENCES

1. HANNANT, D.J. ; EDGINGTON, J.: Durability of Steel Fibre Concrete. Lancaster : The Construction Press Ltd, 1975. - In: Fibre Reinforced Cement and Concrete. Rilem Symposium, September 1975, (Neville, A. (Ed.)), pp 159-169

2. HOFF, G.C.: Durability of Fiber Reinforced Concrete in a Severe Marine Environment. Detroit : American Concrete Institute, 1987. - In: Concrete Durability - Katharine and Bryan Mather International Conference, Atlanta/Georgia, 1987, ACI SP-100, Vol. 1, pp 997-1041

3. MANGAT, P.S. ; GURUSAMY, K.: Permissible Crack Widths in Steel Fibre Reinforced Marine Concrete. In: Materiaux et Constructions (Rilem) 20, 1987, Nr. 119, pp 338-347

4. SCHIESSL, P.: Zur Frage der zulässigen Rißbreite und der erforderlichen Betondeckung im Stahlbetonbau unter besonderer Berücksichtigung der Karbonatisierung des Betons. Berlin : Ernst & Sohn. - In: Schriftenreihe des Deutschen Ausschusses für Stahlbeton, Nr. 255, 1976

CONCRETE IN USE:
VERIFICATION AND PROOF OF COMPLIANCE

W G Shearer

North Lanarkshire Council

UK

ABSTRACT. This paper outlines the sequence of events that have to be followed on a large roadworks construction site in order that the Engineer acting on behalf of the Employer can demonstrate compliance of the concrete supplied with the relevant Specification. It sets out the information to be supplied by the Contractor to demonstrate compliance of the constituent materials, the workability, durability and finally the strength of the hardened concrete. It explains how potential problems such as Alkali - Silica Reaction are dealt with. It compares the testing arrangements on a "conventional " contract with those on a Design and Build contract and suggests that irrespective of the type of contract that is used there will always be a need to demonstrate compliance of the concrete used with the Specification.

Keywords: Compliance, Specification, Constituent Materials, Workability, Durability, Strength, Alkali - Silica Reaction.

Mr W G Shearer BSc, MSc, FICE was the Resident Engineer on the St James Interchange project in Glasgow (value £27.5m) and the M8 Extension contract in Edinburgh (value £40m) - both were Design and Build Projects. Prior to that he worked as Depute Resident Engineer on the M80 Stepps Bypass Contact (value £36m) which was a 'conventional ' remeasurable contract based on the ICE 5th Edition Conditions of Contract.

Radical Concrete Technology. Edited by R K Dhir and P C Hewlett. Published in 1996 by E & FN Spon, 2–6 Boundary Row, London SE1 8HN, UK. ISBN 0 419 21480 1.

INTRODUCTION

Acceptance that concrete delivered to a construction site can be incorporated in a pour means it is deemed to have complied at that stage with all the requirements of the relevant Specification. Checks on the concrete's strength are subsequently made to ensure compliance with the specified characteristic strengths required by the design. Removal of defective concrete, after it has hardened, because the strength has failed to meet the specified requirements is however time consuming and costly.

It is therefore essential that prior to the final on site checks for workability and air entrainment being made there is a high degree of confidence that both the constituent materials comply with the Specification and that they have been mixed in the correct proportions. An examination of Figure 1 reveals the many varied factors which influence the strength of concrete and which therefore have to be checked prior to and during construction.

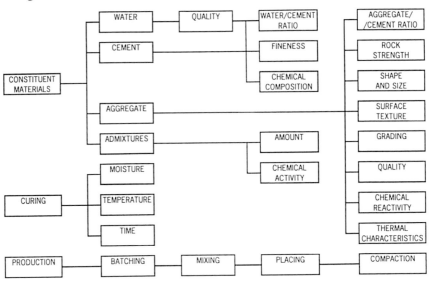

Figure 1   Factors Influencing Concrete Strength

Control of the maximum chloride content, the maximum sulphate content and of the alkali - silica reaction are particularly important in achieving sound concrete. The tighter too the standard of control exercised on site the lower the variation in test results will be and the higher the confidence level will be about the quality of the final product.

This paper examines how the process for checking compliance works on a large Design and Build roadworks construction project works and what records are kept to prove this compliance.

SPECIFICATION OF CONCRETE REQUIRED BASED ON PURPOSE

When deciding what concrete mixes will be specified the designer will have given consideration primarily to the conflicting demands of strength and workability so that what will result is a concrete of adequate durability. Consideration may also have been given to other factors such as the likelihood of chemical attack, the type of aggregate available locally, the means by which the concrete is to be mixed and placed, the degree of supervision and standard of quality control and the standard of finish to be achieved.

Having decided the requirements of the contract, a job specific list will probably be provided by the designer classifying the mixes to be used.   Such a list is shown in Table 1.

Table 1 A typical list of job specific concrete specifications

Mix Ref	Where Used	Class of Concrete	Min Cement Content	Max W/C Ratio	Admix.	Required Workability	RequiredType of Cement
1	precast bridge beams	50/20	325	0.45	N/A	Medium*	BS12
2	structural foundations	40/20	330	0.45	water reducing	Medium } } } }	mixture of BS 12 OPC and BS6699 GGBFS
3	abutment walls / deck slabs	40/20	330	0.40	water reducing and air entrainment	Medium } } } } } }	where slag ≯ 65% and ≮ 40% by weight of total amount
4	bearing plinths	50/14	345	0.40	water reducing	Medium }	}
5	general roadworks	30/20	300	0.55	N/A	N/A	N/A

* Medium is defined in BS 5328 as 25mm to 100mm.

Supplementary to the above, the Contract will dictate the Specification to be used.   This paper is based on the requirements of the Specification for Highway Works (hereafter referred to as the "Specification").   Table 1 is used in conjunction with the requirements of the Specification to enable the concrete manufacturer to design the mixes required.

CHECKING COMPLIANCE OF CONSTITUENT MATERIALS

Cement

The Specification or Table 1 will dictate the British Standard or Standards with which the cement has to comply. The Specification also dictates that the only accepted Quality Assurance Scheme for the manufacture of Portland Cement is that of the British Standards Institution. Part of the concrete supplier's Quality Plan should stipulate that the cement to be used in the Works has to be supplied by a company holding a National Accreditation Certificate. This Company will regularly sample and test their cement and provide the concrete manufacturer with Test Reports demonstrating compliance with the British Standards listed in the Specification. These Reports are in turn provided to the Engineer.

Aggregates

The criterion for a good aggregate is that it should produce the desired properties in both the fresh and hardened concrete. The properties of the aggregate known to have a significant effect on concrete behaviour have been shown in Figure 1.

The Specification requires that the type of aggregate to be used (ie as required by the design) shall comply with the following British Standards as appropriate:

normal aggregates

(i) BS 882 (Aggregates from natural sources for concrete).

(ii) BS 1047 (Air-cooled blast furnace slag coarse aggregate for concrete).

Lightweight aggregates

(iii) BS 877 (Foamed on expanded blast furnace slag coarse aggregate for concrete).

(iv) BS 3797 (Lightweight aggregates for concrete).

The engineer may specify or approve on request the use of aggregates other than those specified above, including gradings not covered by the appropriate British Standards, provided that there are satisfactory data on the properties of concrete made with them. If we assume that the aggregates to be used are to BS 882 then the Appendix in BS 882 sets down the information that the supplier should provide to demonstrate compliance viz:

    (a)   Source of supply
           1. Name and location of quarry.
           2. County or origin.
    (b)   Aggregate type.
    (c)   Typical Properties - see Table 2 below.

Table 2 Details of properties to be submitted as required by BS 882

Details of Properties to be submitted as required by BS 882	Relevant part of B.S. 812 (Testing Aggregates)	Approval Date
1. Shape		Test results shall be not
2. Surface Texture		more than 5 years old
3. Flakiness Index	Pt.105.1: 1989	for 1 - 5 properties.
4. 10% Fines value	Pt.111: 1990	
5. Aggregate Impact value	Pt.112: 1990	
6. Particle Density	Pt.2: 1975 }	Test results shall be not
7. Water absorption	Pt.2: 1975 }	more than 1 year old.
8. Grading	Pt.103.1: 1985	Test results shall be not
9. Fines	Pt 103: 1985	more than 3 yrs old for
10. Shell Content	Pt.106: 1985	8-11.
11. Acid soluble sulphate content	Pt.118: 1988 Pt.119: 1985	
12. Chloride ion content	Pt.117: 1988	If typical result less than 0.01% then yearly, otherwise most recent.
13. Drying shrinkage	Pt.120: 1989	If less than 0.05% then no more than 5 yrs old otherwise no more than 1 yr old.

The frequency of the tests carried out on site to check these properties will normally have been set down in the Contractor's Quality Plan and agreed with the Engineer. Failure to agree the frequency of testing prior to work commencing may lead to conflict later. Confidence based on consistent compliance may allow the frequency to be reviewed.

With the exception of alkali - silica reaction, the list of tests covers all the properties of the aggregates which are known to have a significant effect on concrete behaviour.

The submission of the above information in conjunction with that set out in Table 2 from the supplier per the Contractor will provide the above information as proof of compliance of the aggregates with the Specification.

Control of Alkali - Silica Reaction

Although rarely a problem in the United Kingdom, some cases have been reported since 1976 (BRE Digest 258).

The Specification allows the contractor two options to control alkali - silica reaction:

(i)   by using non reactive aggregates ie those which contain at least 95% by weight of one or more of the rock or artificial types listed in Table 17/3 of the Specification. BS 812 also gives details of how the petrographical examination of aggregates for alkali - silica reaction is to be carried out. When the coarse and fine aggregates are not accepted as wholly non reactive the amount of sodium oxide is not allowed to exceed 3.0kg per cubic metre of concrete.

(ii)   by restricting the content of equivalent sodium oxide in the mix.   The Specification sets out the method of determining the equivalent sodium oxide in the mix which accounts for not only the aggregates but also the concrete, slag and any admixtures.

Whatever method is used the Contractor has to demonstrate to the Engineer that alkali - silica reaction will not be a problem.   There are many documented cases of the deleterious effects of this concrete cancer.

Water

Most water used in concrete in this country will be taken from a Public Utility supply and under these circumstances it is generally taken as read that the water is acceptable for mixing concrete.   The location of the site however may require that the water needs checked. Seawater can cause efflorescence and staining and would therefore be avoided. Some impurities which are likely to have a detrimental effect when present in significant quantities include acids, alkalis, salts, silt, clay, organic matter and sewage.   If contamination is suspected then the Engineer should request that the water be tested in accordance with BS 3148.

Admixtures

The principal types of admixtures used in concrete are:

(i)   Air entraining agents
(ii)   Water reducing agents

and there are many proprietary brands of each available.

Air-entraining agents are by far the most commonly used workability admixtures because they also improve both the cohesiveness of the plastic concrete and the frost resistance of the resulting hardened concrete.

Water-reducing admixtures are used to increase workability while the mix proportions are kept constant or to reduce the water content while maintaining constant workability. The former may result in a slight reduction in concrete strength.

The Specification states that the quantity and method of using the admixtures shall be in accordance with the manufacturer's recommendations and in all cases shall be subject to the approval of the Engineer.   Normally they will have to comply with either BS 1014 (Pigments for Portland Cement and Portland Cement products) or BS 5075 (Concrete admixtures).

The Contractor will in any case have to provide the following information for the Engineer's approval:

(a)   the quantity to be used, in accordance with the manufacturer's requirements.
(b)   the detrimental effects caused by adding a greater or lesser quantity in kilograms per cubic metre of concrete;
(c)   the chemical name(s) of the main active ingredient(s);
(d)   whether or not the admixture leads to the entrainment of air.

If required by the Engineer the Contractor will also have to demonstrate the action of an admixture by means of trial mixes.   The use of calcium chloride in any form is not permitted.

CONCRETE MIX DESIGN

There are two basic types of concrete mix - prescribed mixes and design mixes.   The basic difference is that for prescribed mixes the mix proportions (cement, coarse and fine aggregate) are specified and tested for compliance whereas for designed mixes the concrete grade (characteristic concrete strength) is specified and tested for compliance.

The Specification requires however that all concrete shall be a designed mix.   Table 1 has already set down the mixes for the Contract.   It should be noted from Table 1 that with durability in mind the Engineer is responsible for specifying the minimum cement content and the maximum water / cement ratio.   In addition the Engineer per the Specification will set down:

        (i)   the Maximum Cement Content
        (ii)   the Maximum Chloride Content
        (iii)   the Maximum Sulphate Content
and   (iv)   how the control of Alkali - Silica Reaction is to be achieved.

Armed with the above information and the Specification requirement already discussed for the constituent materials, the contractor will be in a position to design the required mixes.

The concrete manufacturer will now provide the Engineer per the Contractor with evidence for each grade of concrete that the intended workability and mix proportions and manufacturing method which will produce concrete of the required quality. The target mean strength of the design mixes has to exceed the specified characteristic strength by at least the current margin. For the mix references 1 - 5 shown on Table 1. the current margin is 15N/mm² until statistical evidence provided by the supplier demonstrates otherwise.

The current margin is a statistical control introduced to reflect the normal distribution assumed for the spread in concrete strength typically found on road construction projects. This normal distribution (see Figure 2 under QUALITY CONTROL) forms the basis of mix design and statistical quality control procedures for satisfying the strength requirement.

It is possible that the concrete manufacturer will have used a similar mix design before. In this case he may alternatively submit to the Engineer full details of tests on trial mixes previously carried out in accordance with the Specification.

Once the mix designs have been approved in principle the concrete manufacturer will be asked to prepare trial mixes as follows: For each mix a set of six cubes is made from each of three consecutive batches. Three from each set of six is tested at an age of 28 days and three at an earlier age (normally 7 days) approved by the Engineer. The tests are carried out in a laboratory approved by the Engineer.

The Specification requires that the average strength of the nine cubes tested at 28 days has to exceed the specified characteristic strengths by the current margin minus 3.5N/mm². Once the trial mixes have demonstrated compliance any adjustments to the mix proportions can only be made with the Engineer's approval (minor changes in cement content up to 20kg/m³ are allowed dependent on the cement or cement mixes being used). This could be done for example to minimise variability of strength and to maintain the target mean strength.

CHECKING THE PRODUCTION

It is particularly important to check that the gradings of the aggregates used in the trial mixes are similar to those of the samples from the pit or quarry or other source of supply. Whilst the Specification calls for weekly testing it is more usual that they are tested daily - particularly at the start of concrete production.

It is also advisable to check the accuracy of the batching plant measuring equipment by periodical calibrations prior to and during production. The water content of each batch of concrete should be adjusted to produce a concrete of the workability consistent with the trial mixes.

CHECKING DELIVERY REQUIREMENTS

There are essentially four items to be checked when concrete is delivered to site ie:

(a)   timing
(b)   temperature
(c)   air entrainment
(d)   workability

(a)   The Specification requires that the concrete has to be compacted and in its final position within 2 hours of the introduction of the cement to the aggregate. Delivery slips with the batching time noted are handed over and checked by the testing staff and the Engineer's staff to ensure compliance.

(b)   The temperature of concrete is checked if a pour is being carried out under extremely cold conditions. If the air temperature is below $2°C$ then the surface temperature at the time of placing has to be $> 5°C$ and $\not> 30°C$ and be maintained at a minimum of $5°C$ until the concrete has reached $5N/mm^2$. If the temperature is below $2°C$ then checks are also made at the plant to ensure that the aggregates and water used in the mix are free from snow and frost.

(c)   The air content of fresh concrete is checked as described in BS 1881: Part 106. It is normally done for each batch unless sufficient confidence has been built up on a large pour to allow the Engineer to relax this requirement. The typical requirement for a 40/20 mix is $5\% \pm 1½\%$.

(d)   It is normal practice on roadwork construction sites to check the suitability of fresh concrete by carrying out a Slump Test as described in BS 1881. The Specification will set limits allowed for the slump to deviate from the designated value (based on those stated in BS 5328).

For a given water - cement ratio the principal factors affecting workability are the shape and grading of the coarse and fine aggregates and the aggregate cement ratio. Hence the emphasis that was noted earlier to check that the gradings are consistent with the trial mix gradings.

As with air entrainment it is normal to check the slump of every batch. If we define workability as the work required to produce full compaction then it could be argued that the Slump Test does not measure this property. It is useful however in detecting changes in workability.

If any of the above items do not comply with the Specification it may at best mean additional cost to the Supplier, but at worst could result in a major interruption to a concrete pour and serious cost implications for all concerned.

DURABILITY OF CONCRETE

Although no tests are set down to measure durability of concrete it is nevertheless at least as important as strength.   Mention has already been made at the Mix Design stage of those factors that the Engineer is responsible for specifying - all of which were concerned with combating the following internal factors which affect durability ie salts, alkali - silica reaction, volume changes, permeability and absorption.   The spray application of silane to fully cured concrete surfaces to protect against the ingress of road salts is now common practice.

The choice of aggregate is also important - aggregate having high shrinkage properties should be avoided in exposed concrete.   For roads where salt is likely to be used for de-icing purposes, the use of air entrained concrete is recommended.   Sulphate - resisting cement should be used if there is the likelihood of sulphate attack and so on.   A reduction in durability is accompanied by an increase in the water cement ratio and a corresponding reduction in concrete strength.   Hence the emphasis in control testing is on determining the concrete strength.

CONCRETE STRENGTH

If all the steps outlined so far have been followed and comply with the requirements of the Specification then providing proper compaction and curing is applied on site, the concrete should achieve its characteristic strength.

Compliance with the specified characteristic strength is almost always based on tests made on cubes at an age of 28 days.   The Specification sets down the rate of sampling required but on a large deck pour it would be normal (at least at the start of the pour) to take one cube from each ready mixed concrete wagon.

For compliance purposes as required by the Specification:

(a)   the average strength determined from any group of four consecutive test cubes has to exceed the specified characteristic strength by not less than 0.5 times the 'current margin'.

(b)   each individual test result has to be greater than 85% of the specified characteristic strength.

For mix references 1 - 4 shown on Table 1 the current margin is specified as 15N/mm² so that for mix reference 1 for example the above requirements become:

(a)   the average of 4 consecutive test cubes > 50 + 0.5 x 15 ie > 57.5N/mm².
(b)   each individual test result > 0.85 x 50 = 42.5N/mm².

Copies of all the test results are passed to Engineer to demonstrate compliance.

QUALITY CONTROL

Until fairly recently it would have been normal on large road construction projects for the Employer to engage a firm of Consulting Engineers to carry out the design (Local Authorities may have done it in house) and to have a Contractor build it under the supervision of the Engineer. The preferred option now in the United Kingdom is for Design & Build contracts where the design becomes that of the Contractor who will engage the services of a Consulting Engineer himself.

With regard to testing the change is a significant one. Before the Testing Consultant would have been under the control of the Engineer who in turn would be engaged by the Employer. Under Design & Build all testing is the responsibility of the Contractor. It is therefore essential that the Contractor is asked to operate a Quality System to demonstrate that all goods and services provided conform to the Specification. It is also prudent to specify in the contract documents that the laboratory used has NAMAS Accreditation (National Measurement Accreditation Service). Poor Quality Control by the concrete supplier for example will increase the mean strength required to produce the characteristic strength as shown in Figure 2.

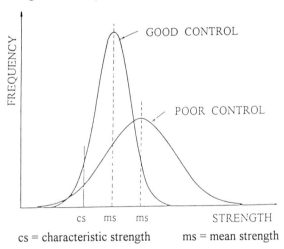

Figure 2 Effect of control on required mean strength

The poorer the control the higher the mean strength required and this will mean a lower water - cement ratio, this in turn will require a smaller aggregate - cement ratio for a given workability and a consequent increase in the total cost of materials. However there are also implications for the Employer or Purchaser; if poor quality control is being exercised by the Contractor or his sub-contractors then he may have to increase his own site supervision. Apart from the increased cost of such action there may be longer term concerns over the quality of the product and a maintenance burden on the Purchaser.

Whilst Design, Build, Finance and Operate contracts seek to place this long term maintenance obligation with the Contractor there will still be a duty on the Engineer to ensure that the Contractor discharges his new found duties under the Contract and that all the necessary records set down in this paper are properly provided to demonstrate the compliance of the concrete supplied with the Specification.

REFERENCES

1. BS 12: 1978 Specification for ordinary and rapid-hardening Portland cement.

2. BS 812: Testing aggregates (in several parts).

3. BS 877: Part 2: 1973 Specification for foamed or expanded blastfurnace slag lightweight aggregate for concrete.

4. BS 882: 1983 Specification for aggregates from natural sources for concrete.

5. BS 1014: 1975 Specification for pigment for Portland cement and Portland cement products.

6. BS 1047: 1983 Specification for air-cooled blastfurnace slag aggregate for use in construction.

7. BS 1881: Testing of concrete (in several parts).

8. BS 3148: 1980 Methods of test for water for making concrete (including notes on the suitability of the water).

9. BS 3797: Part 2: 1976 Specification for lightweight aggregates for concrete.

10. BS 3892: Pulverized-fuel ash, Part 1 and Part 2.

11. BS 5075: Concrete admixtures, Part 1: 1982 Specification for accelerating admixtures, retarding admixtures and water-reducing admixtures; Part 2: 1982 Specification for air-entraining admixtures; Part 3: 1985 Specification for superplasticizing admixtures.

12. BS 5328: 1981: Methods for specifying concrete, including ready-mixed concrete.

13. BS 6699: 1986 Specification for ground granulated blastfurnace slag for use with Portland cement.

14. Department of Transport: August 1986 Specification for Highway Works.

15.   Department of Transport: August 1986   Notes for Guidance on the Specification for Highway Works.

16.   JACKSON, N, and DHIR, R K. Civil Engineering Materials 1988.

17.   JOHNSON, R P. Structural Concrete (1967).

18.   McINTOSH, J D. Concrete Mix Design (Cement and Concrete Associations, London, 1964).

RISK REDUCING THROUGH QUALITY MANAGEMENT SYSTEMS FOR READY MIXED CONCRETE

J G Gunning

University of Ulster at Jordanstown

UK

ABSTRACT. This paper is based on the experiences of the author during 19 years of assessing Quality Management Systems within the N Ireland ready mixed concrete industry. It shows how improved supplier quality management has extended to the construction site by means of improved specifications and designs, by closer liaison between suppliers, contractors and designers, and by improved motivation and training of all concerned with production and supply of concrete.

Across the United Kingdom the development of an industry-wide product certification scheme based on ISO 9001 principles and on BS 5328 for Concrete [1] has culminated in the newest QSRMC Quality and Product Conformity Regulations [2]. These apply to all of the processes involved in the design, production and supply of concrete, and are the result of a major review of certification standards by a balanced group of producers, specifiers and customers. Essential requirements are identified and clarified for all parties. The costs and benefits of operating a third-party certified Quality Management System are examined in some detail, although these aspects can be difficult to quantify precisely.

The paper shows how partnership between producers and customers has resulted in a Quality Scheme which has reduced purchasers' costs yet increased the assurance to customers of the UK ready mixed concrete industry that their specified requirements will be achieved. A number of lessons learnt from the 12 year experience of QSRMC members are discussed and some international comparisons with regard to quality schemes for ready mixed concrete are drawn. This method of reducing the risk of defective concrete for all involved is put forward as a model for the entire construction process to follow.

Keywords: BS 5328, BS EN ISO 9000, Defects, Ready Mixed Concrete, Risk Reduction, Quality Management Systems.

Mr J G Gunning is a chartered Civil Engineer and a Fellow of the Institute of Quality Assurance and of the Chartered Institute of Building. He is a Senior Lecturer in the School of the Built Environment at the University of Ulster specialising in the teaching of construction management. Since 1977 he has also been involved in the assessment of quality management systems within concrete producers, and concrete testing laboratories. He has presented many papers on Quality Management at international conferences.

Radical Concrete Technology. Edited by R K Dhir and P C Hewlett. Published in 1996 by E & FN Spon, 2–6 Boundary Row, London SE1 8HN, UK. ISBN 0 419 21480 1.

INTRODUCTION

Quality assurance is a term with which most of the construction industry have become familiar in the past decade. Many have used it as the basis for rationalising their work practices, and there has been a rapid rise in the number of third party certification bodies. QA has mistakenly been expected to provide a cast iron guarantee of satisfactory quality, and the disappointment which results from failure to achieve full satisfaction has led to some scepticism about the benefits of a quality system. Some risk of defects still remains.

Many of these doubts have arisen because of the emphasis placed by quality assessors and auditors upon whether people have done what they said they would do, rather than whether what was being done was appropriate to the process and product in question. In other words many quality systems have failed to provide the necessary assurance because they are based on procedures which do not fully address the problem of achieving full product conformity. Instead they concentrate on a paper exercise which is dominated by form-filling and observation of documented procedures, regardless of whether these procedures actually ensure product conformity. This may be regarded as another example of the computer programmers' motto of "Garbage In, Garbage Out" (The GIGO Principle).

What is required in an effective quality system is that it is based on stringent regulations which specifically lead to increased consistency in achieving specified requirements. It must focus on the important criteria and not merely upon adherence to procedures for their own sake. In the case of the QSRMC Quality and Product Conformity Regulations [2], the important criteria have been clarified over the years by suppliers and consumers across the industry, resulting in a Quality Scheme which does provide real assurance of quality to the user. This is an example which might usefully be followed by other sectors of the construction industry who wish to minimise their quality risks.

THE LINK WITH QMS STANDARDS

Quality Management Systems have been based on BS 5750/ISO 9000 principles since 1979. Whilst many have queried the application of this Standard to construction, based as it is upon manufacturing philosophy, it has shown itself to be very durable throughout its transformation from BS 5750 [3] to the latest ISO 9000 series [4]. Research by Bentley of the Building Research Establishment [5] has shown that a relatively small proportion of construction defects (10-20%) are due to inadequate quality in the manufacture of materials and components. Other reasons for poor quality were ascribed to poor design information, poor supervision/management and poor site workmanship. In the case of ready mixed concrete, any of these three categories of weakness can undo all of the good work of a quality system for production.

Fletcher and Scivyer [6] have examined the performance of a range of building products (excluding concrete) in relation to published quality standards. They found that an ISO 9000 system for production was a prerequisite for, but not a guarantee of quality in the finished product installed on site. They emphasised the need for quality principles based on ISO 9000 to be applied to every stage of the construction process, from inception and design through to site construction by contractors and sub-contractors, and even to use of the completed structure. This was consistent with the findings of Bonshor and Harrison [7] that 50% of faults in traditional housing lay within the design sphere, and 40% were caused by site construction deficiencies. Faulty materials caused only 10%.

In reviewing quality management practices within the British Property Services Agency, Dalton [8] highlighted the need for QA to be a dynamic system with procedures for the regular review and updating of design, construction and maintenance procedures, based on integration and feedback. As with all standards, BS 5750 has been regularly reviewed and revised, to the point that it is now identical to the International Standard ISO 9000 and to the European EN 29000. Its latest incarnation is as BS EN ISO 9000 series, prepared under the direction of the BSI Quality Management and Statistics Standards Policy Committee. This was the outcome of a review process based on the Vision 2000 Strategy [9] for the long term development of international standards in the quality arena.

In 1994 the European Committee for Standardisation (CEN) accepted the ISO 9000 series, so that there now exists a Europe-wide, if not quite a world-wide, family of quality standards. These standards are generic, and independent of any specific industry or economic sector. They specify requirements which determine what elements have to be encompassed by a quality system, but they are not intended to enforce conformity.   In particular, the Introduction to each Standard within the series emphasises that their requirements are complementary to and not alternative to the technical requirements which are specified or appropriate. The design and implementation of a quality system are necessarily influenced by the particular objectives, products, processes and individual practices of the organisation.

BS EN ISO 9004 emphasises that a successful organisation should offer products which:

a. meet a well-defined need or purpose;
b. satisfy customers' requirements;
c. comply with applicable standards and specifications;
d. comply with the requirements of society;
e. reflect environmental needs;
f. are made available at competitive prices;
g. are provided economically.

It is precisely in order to achieve these objectives for the UK's ready mixed concrete industry that the QSRMC exists in its present form. The technical, administrative and human factors affecting the quality of its products are controlled within the QSRMC operations in satisfying the needs and interests of both customers and producers.   The objectives of QSRMC are therefore directly in accordance with the principles of BS EN ISO 9000 series, The new Regulations incorporate the requirements of BS EN ISO 9001: 1994 [Quality Systems; Model for quality assurance in design, development, production, installation and servicing], combined with the technical specifications of BS 5328; 1991 to provide assurance of product conformity of all concrete produced by QSRMC Member Companies. The Regulations now follow the BS EN ISO 9001 format in each of their 6 technical parts;

1. Quality System Requirements
2. Order Processing
3. Purchase and Control of Materials
4. Concrete Mix Design
5. Shipping, Production and Delivery
6. Control of Concrete

This consistent format enables clauses to be found in the same numbered sub-section in each of the 6 modules. The foundation of QSRMC Regulations as being upon the principles of the ISO 9000 series of Quality Management Standards is therefore quite transparent.

THE EVOLUTION OF QSRMC

The author (Gunning [10]) has set out the origins of QSRMC, which was established in 1984 to further the principles embodied within the BRMCA Authorization Scheme in 1968. BRMCA and BACMI merged their QA interests in ready mixed concrete in order to satisfy the emerging need for third party certification schemes to be independent of their members. QSRMC was therefore founded with an independent chairman of international reputation, and with a Governing Board made up with equal representation between suppliers and consumers. No single interest can dominate, and a wide-ranging view of concrete is embodied within these representatives. All Regulations must be approved by the Governing Board, having been drafted by working parties representing all interests.

In 1989, QSRMC joined the growing list of certification bodies holding accreditation from the National Accreditation Council for Certifying Bodies (NACB), which now forms part of the UK Accreditation Service (UKAS). It was the 9th third-party certifier to do so, and the list has now expanded into more than 50. Membership of NACB is the official Government seal of approval for QSRMC, which is supported by all major UK specifying authorities, including the Department of Transport, the Water Services Association, the National House-Building Council, the Ministry of Defence and the County Surveyors Society. NACB itself conducts annual assessments of all its accredited members, using as a basis its own regulations plus the procedures set out in the certifying body's own Quality Manual. QSRMC has a particularly detailed in-house management system, based on ISO 9000 principles and covering all aspects of its operations. Few comparable bodies have the same level of detail, or intensity of assessment of its own membership and staff.

QSRMC, as an independent, non-profit making organisation, challenges the concept of competition amongst certification bodies within a specific industry (except for general quality management systems). Courtier [11] argues strongly that one cannot compare the procedures and regulations of one quality scheme for a product with another for the same product. Hence a difference of assurance will arise - two schemes based on different requirements are unable to be identical. Therefore, in the United Kingdom, it is held by QSRMC that their own scheme, based as it is on industry-wide collaboration and covering at least 90% of the industry's production, is the only one capable of providing real assurance of product conformity. Widespread endorsement of this view by specifiers has reinforced the belief that to have more than one system of quality regulations covering a particular specialised industry is at best inefficient, and at worst self-deluding.

THE COSTS AND BENEFITS OF A QUALITY MANAGEMENT SYSTEM

It is relatively simple to quantify the costs of formulating and implementing QA in a construction organisation, although the costs of staff time are often overlooked. It is much more difficult to quantify the benefits, since they are primarily the absence of deficient work and an enhanced reputation. Juran [12] has categorised quality costs as;

1. Prevention Costs such as auditing and training.
2. Appraisal Costs, including inspection, checking, testing and control.
3. Internal Failure Costs resulting in reworking of defective items.
4. External Failure Costs such as replacement, complaint investigation and the consequential delays arising from failures.

Oakland [13] argues that all such costs should progressively reduce once a quality system is up and running. He has estimated the costs of failure as being up to 40% of turnover. Baden Hellard [14] has claimed that construction disputes in the UK come to a value of 7% of the industry's turnover, and if we assume that only 1/3 of the defective work becomes the subject of a formal dispute, we could estimate the costs of failure as around 20% of turnover. Sjoholt, of the Norwegian Building Research Institute [15], has identified some projects as being much worse, at a value of 25-30% of total project costs. Feigenbaum [16] has assessed that non-QA companies allocate 5% of total quality costs to Prevention, 25% to Appraisal and 70% to Internal and External Failure Costs. In the United Kingdom construction industry these costs could reach £10 billion per year. An estimate of the failure costs within the ready mixed concrete sector would certainly run into millions of pounds.

The costs of introducing a quality system into a small organisation has been estimated by Gunning [17] as £20,000 minimum. Others have estimated the costs as 2% of a firm's turnover. A common assessment is that initial costs are repaid within 3 years by the drop in failure costs, although there remains the ongoing expenses of auditing and reviewing, testing and quality control, maintenance of system documentation, operating QMS procedures and periodic reassessment/surveillance by the certifying body. The major single contribution to quality costs has been shown as the consequential costs of having to remove defective work. For example it might cost £50,000 to remove one load of defective concrete costing £300 from a high rise building several weeks after casting. It is important to have an appropriate method of measuring the costs of conformance, and of non-conformance, in order to validate the operation of the quality system. It is much more difficult to evaluate the benefits of a certified QMS. Marketing factors are seen as being by far the most important, but other benefits include;

- increased productivity, with less defective work;
- increased quality awareness, motivation and job satisfaction;
- equality of competition based on achieving quality standards;
- more effective planning, communication and problem solving;
- reduced costs of rectification of deficient work, and lower risk of its occurrence.

Obviously few of these benefits can easily be quantified, but companies with an effective costing system can at least compare on a year to year basis, the expenditure on repairs and replacement arising from failures. It may thus be possible to identify through cost-benefit analysis, the value to a company of a Quality Management System. Unfortunately the nature of some of the costs and of many of the benefits make this a very approximate exercise, but it is still worth doing.

Many companies have found that the discipline of a QMS makes them operate procedures which they should always have been following, and that staff become better motivated once clear lines of management responsibility are being observed. Gunning [18] reported that a major concern of concrete producers was a less than whole-hearted level of specification of a Quality Management System as a pre-condition of supply by some UK specifying authorities. Even today, the support by Government for QSRMC has not yet resulted in certification being made mandatory for the supply of ready mixed concrete to the Ministry of Agriculture, Foods and Fisheries. However, there is a growing requirement for QSRMC certification amongst specifying authorities, so the vast majority of the UK ready mixed concrete industry does indeed see positive benefits in being members of QSRMC. The cost-benefit comparison appears to justify any expense incurred, because of risk reduction.

THE REDUCTION OF DEFECTS WITHIN QSRMC MEMBERS

Statistics on the quantity of defective concrete are almost impossible to obtain. Producers are reluctant to admit any deficiencies which may have arisen, whilst contractors and designers are embarrassed at being connected, however slightly, with publicised failures. Nevertheless people associated with ready mixed concrete over the past twenty years have reported a steady reduction in the amount of defective concrete, together with a growing confidence in suppliers. This growth in confidence has been accompanied by some reduction in the quantity of compliance testing. Where this occurs the QSRMC scheme provides, through its regulatory processes (according to Barber [19]), assurance that there will be full compliance with specified requirements. Additional verification is possible through the checking of mix design certificates and the declaration of specified criteria on the delivery ticket. The strength of QSRMC lies in its regular and rigorous programme of assessments. Gunning [20] has outlined the procedures followed and the problems encountered in assessing quality systems for ready mixed concrete. Barber [21] has identified the following 3 major conditions required for assessment to be effective;

1. Assessors must be well-trained and fully conversant with Regulations (Regular staff audits ensure that standards are consistently maintained).
2. The assessment process must be properly structured, with due emphasis on the key factors affecting quantity and quality, (Well developed assessment schedules form the basis of each assessment).
3. There must be a system of ensuring prompt correction of any deficiencies identified during the assessment. (The method of discharge varies with the seriousness of the discrepancy, varying from written declaration, through detailed statements of corrective actions, to a full reassessment and possible eventual suspension)

Each assessment examines key aspects such as:

- the provision of clear data and work instructions;
- the adequacy of procedures, personnel and training;
- the setting of appropriate limits and targets;
- the effectiveness of action taken when limits are exceeded.

Essentially the system relies on in-house consistency, with regular in-house auditing to ensure continuing compliance with all Regulations. It should not be run simply with the objective of satisfying the assessor, but with quality and quantity (as ordered) being the key focus. The role of the third-party assessor should therefore be mainly to confirm and ensure that things are being done correctly rather than to try to score points by identifying minute deviations from the Regulations.

The experience of the past twelve years of QSRMC has seen a continuing decline in the number of deficiencies identified by assessors. The Central Records department keeps a running "score" of members' performance in the number and scale of breaches of Regulations, and the pattern shows a steady downward trend for almost every member, despite some tightening of the requirements over the years. It is generally felt, therefore, that use of an industry-wide Quality Scheme has resulted in a clear improvement in the performance of that industry, and to increased confidence in its ability to meet specified criteria, Product conformity means proper quality assurance through a disciplined management of all elements of supply. This in itself presents enormous benefits to suppliers and customers alike, and well repays any costs incurred in its achievement.

THE QSRMC EXPERIENCE

Many lessons can be learned from the history of QSRMC. The pattern of its evolution shows how technical requirements have finally been properly aligned to modern Quality Standards. Numerous Interpretations and Guidance Notes are now incorporated into the Regulations in clear sequential elements covering all aspects of the supply process.

The recent review process revealed that customers did not understand the control systems operated by producers across a family of related mixes with common characteristics. Most deficiencies were found to arise from human error, and can be prevented by good management practices and procedures. Hence there is an increased emphasis in the new Regulations upon effective systems of management at every stage of the production process from receipt of order to delivery on site. The 1995 Regulations incorporate many minor changes which increase efficiency without any deterioration in quality. There is a new emphasis upon positive action to avoid waste or increased costs. Some earlier concentration upon minor procedural and technical matters has been reduced, so that suppliers can focus on the economic achievement of full Product Conformity.

An improved Customer Complaints Procedure forms an important part in identifying and correcting any downward trends in quality. All complaints have to be documented, investigated and resolved in a way which prevents recurrence. Serious complaints can be referred by customers to QSRMC in writing, and an independent investigation made by an assessor. However most complaints can be resolved directory with the producer. Annual in-house reviews form another important "plank" in achieving Product Conformity. Full examination of systems and procedures, together with an analysis of the results of complaint and anomaly investigations, in-house audits and third-party assessments should enable a company to continue to improve its performance towards that elusive "Zero Defects" target.

One must distinguish between certification for Quality Systems only and certification which covers both Quality Systems and Product Conformity. When Product Conformity assurance is required, as it generally should be for concrete products, it is important to examine exactly what is meant by this term. The cliché definition of quality as "fitness for purpose" is inadequate because it is too vague. Most products have multiple purposes or requirements, and one needs to specify what is meant by "fitness". Many organisations are given Product Conformity certificates on the basis of an extremely sketchy specification of what will be deemed to conform. Customers, and others involved, should therefore carefully investigate what forms the basis of producers' claims to have Product Conformity certification for their products. In the case of the QSRMC, there is no debate as to what regulations or requirements are being satisfied - the QSRMC Quality and Product Conformity Regulations [2] lay down explicitly, over dozens of pages and hundreds of clauses, the basis upon which members achieve Quality System and Product Conformity compliance. Unlike many other criteria or Schedules of Requirements, these Regulations are widely publicised and open to comment and suggested improvements from any interested party. Not many Quality Schemes can match this degree of detail, or of openness to public scrutiny.

The development of QSRMC over the past decade has broadly mirrored the evolution of the ready mixed concrete industry in Denmark, Belgium and Netherlands where similar quality schemes operate. France has a certification scheme with partial coverage of the ready mixed concrete field, whilst the countries of southern Europe tend not to operate national schemes. Germany does not operate a national system, but has broad requirements in each of the Landers. North America seems less advanced with formal quality assurance for concrete.

CONCLUSIONS AND RECOMMENDATIONS

The development of QSRMC has shown how a Product Conformity scheme, fully reviewed and approved by representatives of all sectors of an industry, and operated through a professionally organised team of assessors, can provide great "added value" for purchasers, specifiers and their clients. The ordering, materials procurement, mix design, production, delivery and quality control are all performed through highly trained specialist staff working to appropriate procedures approved by QSRMC. Customers can have faith in suppliers' own quality control and records, even if they still feel a need for some compliance testing. This is reducing, as purchasers gain increasing confidence in the performance of the ready mixed concrete industry. Davies [22] feels that large sums of money are spent on compliance tests, the results of which often cause more concern about the quality of the testing regime than about the product itself. A responsible site quality plan for reducing deficiencies could more efficiently rely upon QSRMC as a basis, only carrying out compliance verification such as non-destructive testing on very critical structural elements.

The recent review of QSRMC has resulted in a detailed set of Regulations as a consensus of best practice. There have been extensive discussions with every sector of the construction industry, from manufacturers of cements, chemical admixtures and batching equipment, through calibration authorities and aggregate producers to contractors, specifying authorities and client bodies. The outcome of these discussions has been synthesised through various working parties into a clearly documented set of achievable requirements meeting the aspirations of all who wish to see high quality concrete, and who wish to minimise the risk of defects on their projects.

The DOE Quality Liaison Group [23] agrees that "many sector certification schemes have undoubtedly succeeded in providing added confidence in their users". Their report warns against using the Standard as a goal in itself rather than as a means to an end, and against allowing bureaucracy to dictate. It also recommends that "Procurement Policy should consider giving preference to tendering belonging to the accredited certification schemes for construction activities based upon common interpretation of the BS EN ISO 9000 series, to include continuous improvement, which should form the basis for industry sectoral schemes".

Forward thinking concrete companies are now adding Safety Management and Environmental Management Manuals to their Quality Manuals, based on similar format and on broad ISO 9000 principles. This epitomises how some producers have fully adopted and internalised quality assurance philosophy in every aspect of their business. Total Quality Management may be the next step for some of them. The author contends, that producers of ready mixed concrete and precast concrete products from most parts of the world could learn from the experience of QSRMC and develop their own quality certification schemes tailored to meet local requirements. The outcome of this would almost certainly be a reduction in the risk of low quality production, and enhancement of their own reputation and profitability.

REFERENCES

1.   BRITISH STANDARDS INSTITUTION. (1991) Concrete. BSI, London BS:5328; Parts 1-4.

2.   QSRMC. (1995) Quality and Product Conformity Regulations. QSRMC, Hampton, United Kingdom.

3.   BRITISH STANDARDS INSTITUTION. (1979) Quality Systems. BSI, London. BS:5750 Parts 1-6.
4.   BRITISH STANDARDS INSTITUTION. (1994) Quality Systems. BSI, London. BS EN ISO 9000 Series.
5.   BENTLEY, M.J.C. (1981) Quality Control on Building Sites. BRE Current Paper CP7/81, BRE, Watford, England.
6.   FLETCHER, K.E., SCIVYER, G.R. (1987) QA in the UK Building Industry: Its current status and future possibilities. CIB-W65 Symposium on Organisation & Management of Construction. E & FN Spon, London.
7.   BONSHOR, R.B., and HARRISON, H.W. (1982) Traditional Housing; a BRE Study of Quality. Paper 18/82. BRE, Watford.
8.   DALTON. (1987) Quest for Quality Developments in the Management of Quality by the UK's PSA. CIB-W65 Symposium on Organisation & Management of Construction. E & FN Spon, London.
9.   BRITISH STANDARDS INSTITUTION. (1993) A Vision for International Standards Implementation in the Quality Arena during the 1990's. BSI, London, PD 6538.
10. GUNNING, J.G. (1987) QA in the UK Construction Industry, with particular reference to ready mixed concrete production. CIB-W65 Symposium on Organisation and Management of Construction, E & FN Spon, London.
11. COURTIER, B. (1995) Changing the Rules, in New Civil Engineer Concrete Supplement, July, pp 16-17.
12. JURAN, J.M. (1988) Quality control Handbook (4th Ed.), McGraw Hill, New York.
13. OAKLAND, (1993) Total Quality Management (2nd Ed.), Butterworth Heinemann Ltd., Oxford.
14. HELLARD, R.Baden- (1993) Total Quality in Construction Projects. T.Telford Ltd, London.
15. SJOHOLT. (1988) Quality Management Systems in Construction, Norwegian Building Institute, Oslo.
16. FEIGENBAUM, A.V. (1991) Total Quality Control (3rd Ed.), McGraw Hill, New York.
17. GUNNING, J.G. (1995) The Financial Implications of Operating a Certified QMS within a Construction Organisation. International Conference on Financial Management of Property & Construction. University of Ulster, N.Ireland.
18. GUNNING, J.G. (1989) The Application of Quality Assurance to the Ready Mixed Concrete Industry in Northern Ireland. 6th EOQC Seminar - Implementation of Quality in Construction, Copenhagen.
19. BARBER, P.M. (1990) QSRMC - the key features. Concrete Journal, Vol. 24, No. 2. pp. 21-23.
20. GUNNING, J.G. (1993) The Audit and Assessment of Quality Management Systems for Ready Mixed Concrete Production. Concrete 2000 Conference on Economic and Durable Construction through Excellence. Dundee. E & FN Spon, London.
21. BARBER, P.M. (1995) The QSRMC Assessment and Certification Process. Construction Industry Briefing - Belfast.
22. DAVIES, N. (1995) The Requirements of Customers. Construction Industry Briefing - Belfast.
23. DOE QUALITY LIAISON GROUP. (1995) Constructing Quality - a strategy for quality in construction. Construction Document, DOE, London.

CONSTRUCTION TECHNIQUES

Chairmen **Professor S Besari**
Institut Teknologi Bandung
Indonesia

Professor J H Bungey
University of Liverpool
United Kingdom

Leader Paper

The Contribution of Fluidity Improving Technology to the Widespread Use of High-Strength Concrete

Dr T Yonezawa
Takenaka Corporation
Japan

THE CONTRIBUTION OF FLUIDITY IMPROVING TECHNOLOGY TO THE WIDESPREAD USE OF HIGH-STRENGTH CONCRETE

T Yonezawa

Takenaka Corporation

Japan

ABSTRACT. Over the past 20 years, high-strength concrete has come into widespread use throughout most of the world. This paper reviews the techniques that have been developed to improve the fluidity of high-strength concrete, and demonstrates that these fluidity-improving methods have contributed greatly to the popularity of high-strength concrete. The fluidity-improving techniques emphasized in this paper are: superplasticizers, long-life superplasticizers, silica fume, and super-superplasticizers. An outline of each technique is given here, together with examples of their application. Since they are prerequisites to understand these methods, pre-superplasticizer high-strength concrete technology and the rheological characteristics of high-strength concrete made with superplasticizers are reviewed.

Keywords: High-strength concrete, Superplasticizer (SP), Long-life superplasticizer (L-SP), Silica fume (SF), Super-superplasticizer (S-SP), Fluidity, Viscosity

Dr Toshio Yonezawa is a Chief Researcher, Research and Development Institute, Takenaka Corporation, Chiba, Japan. He specializes in the corrosion of steel reinforcement, pore solution chemistry, light-weight aggregate concrete and high-strength concrete. He is also a member of the committees for the Japan Concrete Institute, the Japan Society of Civil Engineers, and the Architectural Institute of Japan.

Radical Concrete Technology. Edited by R K Dhir and P C Hewlett. Published in 1996 by E & FN Spon, 2–6 Boundary Row, London SE1 8HN, UK. ISBN 0 419 21480 1.

INTRODUCTION

The use of high-strength concrete — that is, concrete with a compressive strength of 50 to 100 N/mm^2 — has now become common throughout the world in the construction of large structures such as super-high-rise buildings, bridges, and oil production platforms. Such very high compressive strengths are of great practical importance in the structural design of these structures.

Material costs for high-strength concrete is higher than for ordinary concrete due to high cement content and usage of extra materials such as superplasticizers and silica fume. However, the increase in concrete cost is more than compensated by the increase in concrete strength. The cost of a 60 N/mm^2 concrete is 1.5-2.0 times the cost of a ordinery 20 N/mm^2 concrete. This is why high-strength concrete is used for super-high-rise buildings and large span structures, to reduce construction costs.

High-strength concrete technology derives from a synthesis of research results related to concrete's basic characteristics, such as mechanical properties, durability, and fire resistance, with development results on concrete pumping techniques, high-speed formwork systems, and advanced quality control methods. The factor that has had most influence in promoting the widespread use of high-strength concrete, however, is the fluidity-improving techniques such as superplasticizers.

In this paper, a technical outline of four techniques developed to improve the fluidity of high-strength concrete are reviewed, together with examples of their application; the first-generation superplasticizer developed by Hattori et al., the long-life superplasticizer which increased fluidity retention, silica fume for decreasing viscosity, and the super-superplasticizer for ultra-high-strength concrete.

PRE-SUPERPLASTICIZER HIGH-STRENGTH CONCRETE TECHNOLOGY

Since concrete was first used as structural material in construction work, the development of stronger concrete has been the dream of concrete engineers and researchers and has been an important research field of concrete engineering.

In 1930, Yoshida [1] produced concrete with a compressive strength of 104 N/mm^2 by molding a concrete material with a water-cement ratio of 31% under a pressure of 10 N/mm^2. Around the same period, Menzel [2] produced concrete with a compressive strength of more than 100 N/mm^2 by steam curing. The high-strength concretes of this period differed substantially from today's high-strength concretes, because they had no fluidity whatsoever, and so could not be placed using ordinary forms.

In 1932, Scripture came up with a water-reducing agent in which lignosulfonic acid was the main component, and he patented this technology in 1939 [3]. Several other water-reducing agents based on organic acids were also reported around the same time. The basic aim of these water-reducing agents was to reduce the water content or increase the fluidity of ordinary concrete. At the same time, they made it possible to produce high-strength concrete with some fluidity at water-cement ratios of 35 to 40%.

The Willows Bridge, constructed in Toronto in 1966 [4], is a good example of this type of high-strength concrete in use. A water-reducing agent based on an organic acid was used to give the concrete a design strength of 42.2 N/mm^2 (6,000 psi) with a water-cement ratio of 33 to 37%, and a slump of 4 to 6 cm. This is a much lower strength than is possible with today's high-strength concrete, and the slump of 4 to 6 cm is very much lower than the present technology offers.

RHEOLOGICAL CHARACTERISTICS OF HIGH-STRENGTH CONCRETE WITH SUPERPLASTICIZERS

The superior characteristics of today's high-strength concrete, high fluidity at the same time as high strength, were made possible by the invention of the superplasticizer (SP). High-strength SP concrete has different rheological characteristics from ordinary concrete, and this difference is important in evaluating its fluidity.

Figure 1 (a) [5] shows the relation between the yield stress of cement paste and the dosage of SP, while Figure 1(b) gives the results of measuring the relation between plastic viscosity and SP dosage. As the water-cement ratio falls, both yield stress and plastic viscosity increase considerably, and the result is poor fluidity. However, while the addition of SP considerably reduces the yield stress, its effect on the plastic viscosity is weak.

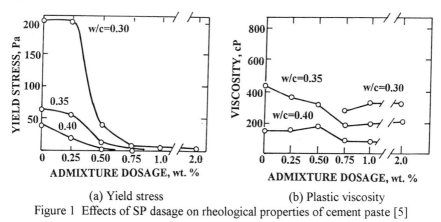

(a) Yield stress   (b) Plastic viscosity
Figure 1 Effects of SP dasage on rheological properties of cement paste [5]

Figure 2 [6] shows the results of measuring concrete viscosity with an L-flow meter [6] in the case of a slump value of 23 cm. The L-flow meter is an instrument for measuring fluidity, as shown in Figure 3. The larger the L-flow rate, the smaller the viscosity. It is clear that plastic viscosity increases with falling water-cement ratio at a particular slump value.

This is the major rheological characteristic of high-strength concrete. The slump and flow can be easily controlled by using SP, but the viscosity cannot be reduced easily. In other words, high-strength concrete is characterized by extremely high viscosity and poor flow rate.

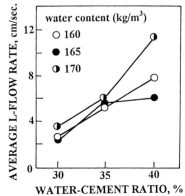

Figure 2 Effects of water-cement ratio on L-flow rate [6]

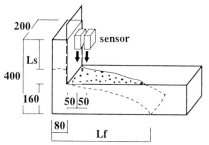

Figure 3 L-flow meter [6]

Figure 4 schematically compares the rheological characteristics of ordinary concrete and high-strength concrete. When the slump or flow of ordinary concrete is measured, the

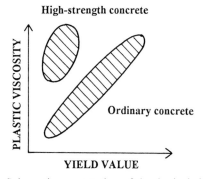

Figure 4 Schematic presentation of rheological characteristics
of high and ordinary strength concrete

fluidity of the concrete is sufficiently evaluated, because ordinary concrete has a relatively linear correlation between yield value and plastic viscosity. For this reason, slump or flow has been used as an index of fluidity of concrete. In high-strength concrete, however, there is only a weak relationship between these two parameters, making separate measurements of yield value and plastic viscosity necessary. These measurements are typically made with an L-flow meter, a two-point tester [7, 8], LCL apparatus [9], or other testers.

FIRST GENERATION SUPERPLASTICIZERS

The first SP in history came out of study by Hattori et al. [10] in the early 1960s, and it appeared on the market as early as 1964. The main component of this SP was a formaldehyde condensate of beta-naphthalene sulfonate (BNS) with the chemical structure shown in Figure 5 (a). Toward the end of the 1960s, Aignesberger et al. [11] developed another SP with a formaldehyde condensate of melamine sulfonate (MS) with the chemical structure shown in Figure 5 (b), and this came into practical use in the early 1970s in Germany.

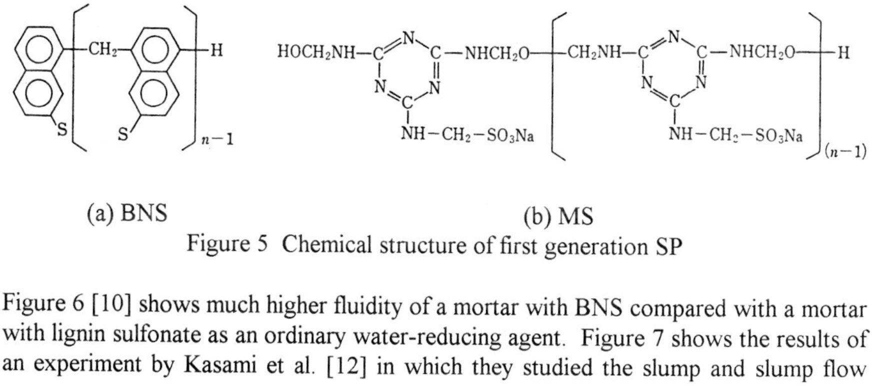

(a) BNS   (b) MS

Figure 5 Chemical structure of first generation SP

Figure 6 [10] shows much higher fluidity of a mortar with BNS compared with a mortar with lignin sulfonate as an ordinary water-reducing agent. Figure 7 shows the results of an experiment by Kasami et al. [12] in which they studied the slump and slump flow

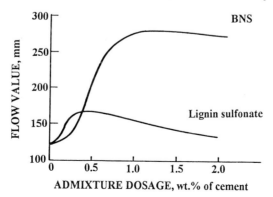

Figure 6 Effects of admixture dosage on the flow value of mortar [10]

increase of a concrete, containing water-reducing agent, by addition of SP. This demonstrates that the addition of BNS contributes considerably to improved fluidity.

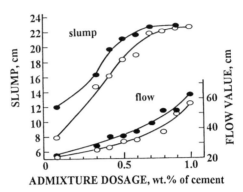

Figure 7 Effects of BNS dosage on slump and slump flow of concrete [12]

Figure 8 shows the results of an experiment by Aigensberger [13] in which the compressive strengths of concrete using MS and concrete without MS was measured for the same mix proportion. Concrete containing MS reached a strength of 64 N/mm^2 at a water-cement ratio of 37%, higher than that attained by the non-MS concrete.

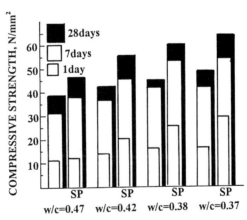

Figure 8 Effects of MS on compressive strength [13]

These first-generation superplasticizers thus realized high fluidity and high strength, but with a serious drawback in terms of concrete work. Figure 9 (a) shows data by Perenchio et al. [14] for the effects of an SP on slump loss in ordinary concrete having a water-cement ratio of 59 to 61%. Figure 9 (b) shows similar data but for high-strength concrete with a water-cement ratio of 33%. In both types of concrete, the use of the SP gave rise to a larger slump loss, but the loss was particularly noteworthy in the case of the high-strength concrete. This means that, when using high-strength concrete with an SP, the time between mixing and placing is greatly restricted.

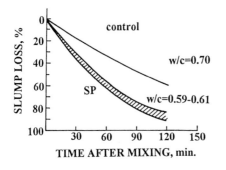

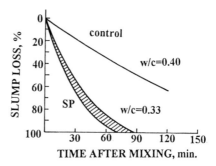

(a) Ordinary concrete   (b) High-strength concrete

Figure 9 Slump loss of concretes with and without SP [14]

The first Japanese concrete bridge [15] using BNS concrete of 60 N/mm^2 was constructed in 1974. The concrete for the bridge was produced taking into account the rapid slump loss; the concrete must be placed within 30 minutes after mixing. However, when temperature, transportation, etc., varied, it was difficult to place concrete before a serious loss of slump occurred.

The problem was solved by adopting a process in which the SP is added at the ready-mixed concrete plant, the mixed concrete is transported to the construction site, and then fluidity is recovered on site by re-dosage. The concrete is then placed before any harmful drop in fluidity takes place.

There are now many examples of construction around the world in which this SP re-dosage technique has been used. One is the Pacific First Center Building in Seattle (46 floors), where concrete with a slump of 15 cm was worked back to 25.5 cm at the site [16]. Another is the Melbourne Central Building in Melbourne (55 floors), where concrete of 6 or 7 cm was returned to 16 to 18 cm at the site [17].

For the construction of Melbourne Central Building [16], SP tank of 2.5 m3 and SP supply pump were necessary at the construction site. In addition, technicians for adding SP were neccesary. The SP re-dosage method had a drawback of necessitating extra facilities and manpower. There was an underlying need for a long-life superplasticizer which has a feature of a small slump loss over a period of 1 to 2 hours.

LONG-LIFE SUPERPLASTICIZERS

Long-Life superplasticizer (L-SP) is a particular superplasticizer that can provide concrete with extremely small slump loss. Most development work on L-SPs took place in Japan during the 1980s, and the admixtures that resulted from this work can be classified into two types according to the mechanism by which they maintain fluidity: one type compensates for the amount of SP taken up by the cement hydrates, while the other changes the structure of the SP such that is not easily taken by the cement hydrates.

The mechanism of a typical L-SP of the former type is shown in Figure 10 [18]. The reactive polymer shown in the figure is a non-water-soluble compound mainly comprising carboxylic acid salt, amide, and carboxylic anhydride. As a result of hydrolysis triggered by the alkaline condition in concrete, it releases a water-soluble dispersant little by little. This dispersant compensates for the decreased dispersing action of the SP. Figure 11 [18] shows changes in slump with time for concrete made with BNS and this reactive polymer. The figure clearly demonstrates that fluidity is maintained for a long period.

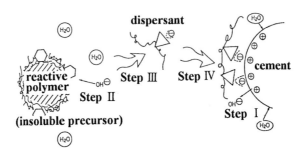

Figure 10 Mechanism of reactive polymer based L-SP [18]

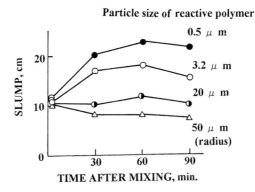

Figure 11 Slump loss of a reactive polymer based L-SP [18]

An example of the second type of L-SP is the graft copolymer. It has a steric structure which is easily adsorbed onto the surface of the cement, as shown schematically in Figure 12 [19]. This L-SP causes the cement to disperse as a result of steric hindrance effect and, unlike BNS which adheres to the cement surface two-dimensionally, it cannot easily be taken by the cement hydrates because of its steric structure.   This gives it a characteristically low slump loss [20]. There are many variations on this type of L-SP, but they are often known as "carboxylic L-SPs" in general. Figure 13 [19] shows the results of an experiment to investigate the change in slump of concrete made with this type of L-SP.

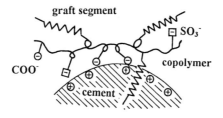

Figure 12 Steric structure of carboxylic type L-SP [19]

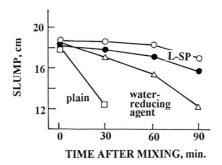

Figure 13 Slump loss of a carboxylic type L-SP [19]

Such long-life superplasticizers entered practical use in the mid-1980s. The majority of concrete super-high-rise buildings constructed in Japan since around 1986 were constructed with L-SP concrete. Figure 14 shows slump measurements of high-strength concrete with a water-cement ratio of 33 to 37% at the batcher plant and at the construction site during construction work [21]. Despite the 30 to 40 minutes taken to deliver the concrete, no loss in slump was measured, and it even rose slightly over time.

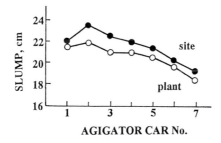

Figure 14 An example of the relation between the slump of
L-SP concrete at plant and site [21]

Figure 15 [22] shows the machine layout used to pump L-SP concrete up to the 45 floor of a building (about 160 m above the ground). A high-pressure (maximum: 22.5 N/mm^2)

pump, made by Putzmeister Corporation of Germany, and a concrete distributor was used for pumping and placing concrete.

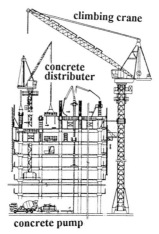

Figure 15 Arrangement of construction machines for concreting at site [21]

Placing high-strengh concrete using concrete pump and distributor enabled efficient concreting, compared with placing concrete using concrete buckets. According to the estimation for 311 South Wacker Drive Project [23], 9 minutes were required for placing the concrete at 40-50 story floors with an 1.5 m³ bucket; the placing rate was 10 m³/hour. By contrast, the placing rate was approximately 40 m³/hour, when concrete pump was used. This efficient concrete placing expedited construction time. The high-rise building shown in Figure 15 was constructed spending only 7 days at each floor. 311 South Wacker Drive Building was constructed at a rate of 5 days in each floor. The rapid construction of structures obviously reduces the construction costs.

SILICA FUME

Although SP and L-SP admixtures have made it possible to produce high-strength concrete with good fluidity, the concrete has extremely high viscosity when the water-cement ratio is below 30%, so workability and pumpability are very poor. The silica fume (SF) helped to solve the problems associated with high viscosity.

Silica fume comprises ultra-fine spherical particles of diameter of 0.1 to 0.5 micron and consisting mainly of amorphous silicon dioxide (SiO_2). It is a by-product of silicon metal or ferrosilicon manufacture and it came into practical use in Norway at the beginning of the 1970s [24].

It is a pozzolan and reacts with cement hydrates to fill the gaps between cement and cement hydrates. This action is known to make the concrete stronger. Figure 16 [24] shows an example of the effects of SF on the compressive strength of concrete. The early strength of the concrete is low, but over the long term it becomes 10 to 15% superior.

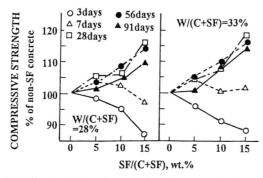

Figure 16 Effects of SF replacement on the strength of concrete [24]

This 10 to 15% gain in compressive strength when SF is used naturally has great importance as regards the efficiency of high-strength concrete. In the author's view, however, the improvement that SF brings to the fluidity of high-strength concrete is crucial. Not all types of SF on the market effectively improve fluidity, however, so careful selection is recommended. It is also worth giving the reminder that SP or L-SP are prerequisite to the use of SF.

Figure 17 [25] shows measurements of the L-flow rate [6] of high-strength concrete using an as-produced (undensified) SF. This clearly demonstrates that the addition of an SF leads to lower plastic viscosity since a higher L-flow rate means a lower viscosity.

Figures 18 (a) [26] shows the results of tests to determine the amount of L-SP required to obtain concrete with a slump of 23 cm, a water-binder ratio of 28%, and an SF/(C+SF) ratio of 10% for eight types of as-produced SF and six types of densified SF. Figure 18 (b) [26] shows the results of L-flow rate on the same concretes. The amount of L-SP needed is lower with the as-produced SF than with the densified SF. The L-flow rate also tends to be greater with the as-produced SF (that is, it has a lower plastic viscosity).

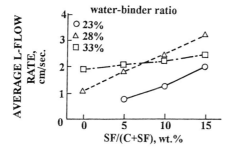

Figure 17 Effects of silica fume on the plastic viscosity of high-strength concrete [25]

Figure 19 [27] is an elevation of the Gullfaks C oil production platform and a layout of the concrete pipeline and batching plant used in its construction. Figure 20 [27] gives the pressure distribution along the pipeline as concrete was pumped. A very small amount of SF was used (2% replacement), but this agent clearly led to lower pressure in the pipeline.

This ties in with the experimental result [28, 29] that, with a fall in plastic viscosity, the friction loss per unit length inside a pipeline falls for a particular value of slump.

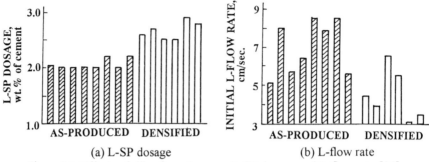

(a) L-SP dosage　　　　　　　　(b) L-flow rate

Figure 18 Effects of silica fume types on L-SP dosage and L-flow rate [26]

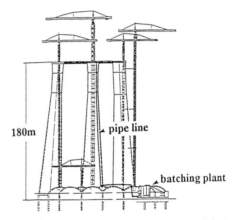

Figure 19 Arrangement of pipeline and batcher plant on Gullfacks C [27]

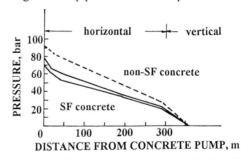

Figure 20 An example of pumping pressure reduction by silica fume [27]

The mechanism by which SF improves the fluidity of high-strength concrete has not yet been fully clarified. Bache [30] has proposed a DSP model to explain the improvement, according to which fine particles of SF uniformly disperse around cement particles which are in mutual contact, filling the gaps between them. This is shown in Figure 21 (a). A previous study by the authors on the cement-SF system [31] — in which a laser diffraction particle size analyzer and laser microscope were used — demonstrated that the fine SF particles agglomerate on the surfaces of cement particles, as shown in Figure 21 (b). This infers that the fluidity improvements arise both from the bearing action of the SF and electrical repulsive action of SP or SF.

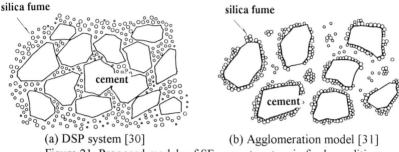

(a) DSP system [30]   (b) Agglomeration model [31]
Figure 21 Proposed models of SF-cement system in fresh condition

Silica fume is an ultra-fine powder and is essentially bulky. Accordingly, haulage is inefficient and it is difficult to store and handle in ready-mixed concrete plants. To overcome such handling problems, a plant as shown in Figure 22 [32] has been developed to supply SF in slurry form to batcher plants in Norway and North America. In Japan, the system shown in Figure 23 [33] has been developed for the supply of as-produced SF to batcher plants.

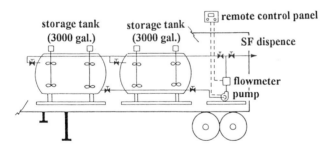

Figure 22 A slurry-SF supply plant to concrete batcher [32]

In North America, many super-high-rise buildings, such as Two Union Square Building [35], were constructed using the high-strength concretes made in combination with slurry-SF or densified-SF and SP.

Figure 24 [34] shows a super-high-rise building (39 floors) constructed with concrete from the Japanese plant. The concrete had a specified design strength of 60 N/mm2, a water-binder ratio of 28 %, a SF(C+SF) ratio of 10 %, and a slump flow of 45 to 50 cm. The 700 mm by 700 mm colums of this building consist of concrete-filled steel tube (CFT), into which concrete was filled in a single action to a height of 61.9 m by means of bottom-up concreting. The bottom-up concreting without compaction as shown in Figure 25 [33] is possible only by using high-strength concrete with very high fludity.

This concreting can be conducted with few workers under clean condition as shown in Figure 26 [33].

Figure 23 An as-produced SF supply plant to concrete batcher [33]

Figure 24 New Shinagawa Prince Hotel (39-story, CFT-structure) [34]

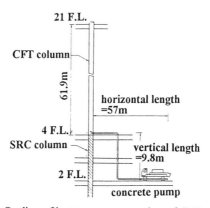

Figure 25 Outline of bottom-up concreting of CFT column [34]

Figure 26 Outlook of bottom-up concreting of CFT columns [34]

SUPER-SUPERPLASTICIZER

Use of SP or L-SP in combination with SF has made it possible to obtain high-strength concretes with good fluidity at water-binder ratios of 25 to 30%. However, when the water-binder ratio drops to 20 to 25% or below, the concrete once again becomes extremely viscous, thereby making construction work difficult.

This problem has been overcome through development of a super-superplasticizer (S-SP), which greatly reduces the viscosity of a cement-SF system when the water-binder ratio is 20 to 25%. This enables ultra-high-strength concrete to be produced with good fluidity [37, 38].

The compounds used as S-SP have the chemical structure shown in Figure 27 [36]. According to the results of tests to measure adsorption onto cement and zeta potential [36], S-SP appears to be characterized by both steric hindrance effects like carboxylic L-SPs and electrical repulsion effects like BNS.

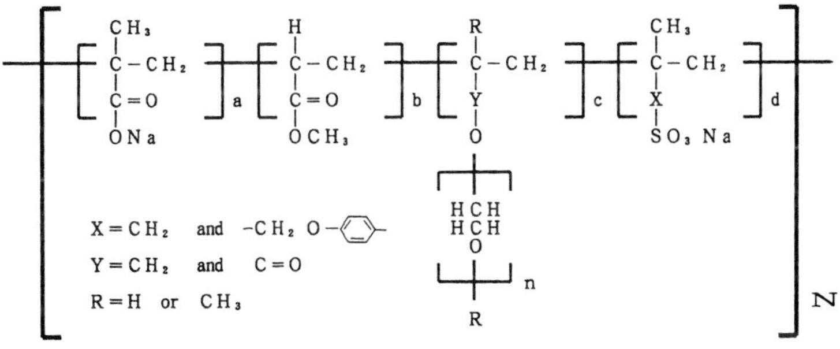

$$\left[\left[\begin{array}{c} CH_3 \\ C-CH_2 \\ C=O \\ ONa \end{array}\right]_a \left[\begin{array}{c} H \\ C-CH_2 \\ C=O \\ OCH_3 \end{array}\right]_b \left[\begin{array}{c} R \\ C-CH_2 \\ Y \\ O \\ \left[\begin{array}{c} HCH \\ HCH \\ O \end{array}\right]_n \\ R \end{array}\right]_c \left[\begin{array}{c} CH_3 \\ C-CH_2 \\ X \\ SO_3Na \end{array}\right]_d \right]_N$$

$X = CH_2$ and $-CH_2\,O-\langle\bigcirc\rangle-$

$Y = CH_2$ and $C = O$

$R = H$ or CH_3

Figure 27 Chemical structure of S-SP [36]

Figure 28 (a) [37] shows the level of dosage required to give concrete a slump flow of about 60 cm. Figure 28 (b) [37] compares the results of measuring the relation between L-flow rate and water-binder ratio. Compared with a BNS-based L-SP and a carboxylic L-SP, S-SP has a low admixture dosage and low viscosity in the region with a water-binder ratio of 25% or less. The S-SP has a steric structure and offers a long life — like L-SP — and extremely small slump loss.

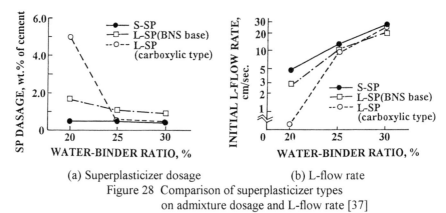

(a) Superplasticizer dosage   (b) L-flow rate

Figure 28 Comparison of superplasticizer types
on admixture dosage and L-flow rate [37]

Figure 29 [38] shows a bridge built with concrete specified to have a design strength of 100 N/mm^2. To achieve this strength, an ultra-high-strength concrete based on the S-SP and with a water-binder ratio of 20% and a slump flow of 65 cm was used. This concrete was made in the plant shown in Figure 23 and was transported spending 2 hours followed by 30 minutes of placing. Reports indicate that it retained its fluidity for a period of 2.5 hours [38].

A 38 stories high-rise building using the S-SP concrete of water-binder ratio of 23-25 % and specified design strength of 80 N/mm^2 is now under construction in Osaka, Japan. The high fluidity concrete with extremely low water-binder ratio is, again, used for bottom-up concreting of the concrete-filled steel column without compaction.

Figure 29 CNT Super Bridge [38]

CONCLUSION

Since the 1970s, structural engineers' demands for increased concrete strength have always been met, as concrete engineers and researchers endeavored to develop methods of securing both strength and fluidity. As this paper makes clear, first-generation superplasticizers, long-life superplasticizers, silica fume, and super-superplasticizer have each in their turn offered a successful response to the demand for concrete with increased strength and yet with adequate fluidity. Through these efforts, we are now able to make practical use of high-strength concrete of high fluidity at water-binder ratios as low as 20%.

REFERENCES

1.   YOSHIDA, T.: Making concrete with the highest strength, Journal of the Japan Society of Civil Engineers, Vol.26, No.11, pp.997-1016, 1930.
2.   MENZEL, C.A.: Strength and volume change of steam cured cement mortar and concrete, Proceedings of American Concrete Institute, Vol.31, Nov.-Dec., 1934.
3.   MIELENZ, R.C.: History of chemical admixtures for concrete, Concrete International, Vol.6, No.4, pp.40-53, 1984.
4.   BURGESS, A.J., RYELL, J., and BUTING, J.: High strength concrete for the Willows Bridge, ACI Journal, Vol.67, No.8, pp.611-619, 1970.
5.   ASAGA, K. and ROY, D.M.: Rheological properties of cement mixes, Cement and Concrete Research, Vol.10, pp.287-295, 1980.
6.   YONEZAWA, T., IZUMI, I., MITSUI, K., and OKUNO, T.: Measurement of the workability of high-strength concrete by L-flow meter, Proceedings of Japan Concrete Institute, Vol.11, No.1, pp.171-176, 1989.
7.   TATTERSALL, G.H.: The rationale of a two-point workability test, Magazine of Concrete Research, Vol.25, No.84, pp.169-172, 1973.
8.   TATTERSALL, G.H. and BAKER, P.H.: The effect of vibration on the rheological properties of fresh concrete, Magazine of Concrete Research, Vol.40, No.143, pp.79-89, 1988.
9.   LARRARD, F.D.: A method for proportioning high-strength concrete mixtures, Cement Concrete and Aggregate, Vol.12, No.2, pp.47-52, 1990.
10.   HATTORI, K.: Experiences with Mighty superplasticizer in Japan, ACI Publication, SP-62, PP.37-66, 1979.
11.   AIGNESBERGER, A. and KRIGER, H.: Zusatz von anionischen melaminharzlösungen baden, Zemenk-Kalk-Gips, Vol.21, No.10, pp.415-419, 1968.
12.   KASAMI, H., IKEDA, T., and YAMANE, S.: On workability and pumpability of superplastized concrete — experience in Japan, ACI Publication SP-62, pp.67-85, 1979.
13.   AIGNESBERGER, A. and KERN, A.: Use of melamine based superplasticizer as a water reducer. ACI Publication SP-68, pp.61-80, 1981.
14.   PERENCHIO, W.F., WHITING, D.A., and KANTRO, D.L.: Water reduction, slump loss and entrained air-void systems as influenced by superplasticizers, ACI Publication, SP-62, pp.137-155, 1979.
15.   MACHIDA, F., NAKAHARA, S., HIROSE, T., KOMADA, K., MIYASAKA, Y., ICHIKAWA, H., and KITSUTA, T.: Design and construction of a PC bridge-using high-strength concrete, Prestressed Concrete (Japan), Vol.16, No.5, pp.36-45, 1974
16.   RANDALL, V. and FOOT, K.: High-strength concrete for Pacific First Center, Concrete International, Vol.10, No.4, pp.14-16, 1989.
17.   BURNETT, I.: High-strength concrete in Melbourne, Australia, Concrete International, Vol.10, No.4, pp.17-21, 1989.
18.   IZUMI, T., MIZUNUMA, T., IIZUKA, M., and FUKUDA, M.: Slump control with reactive polymeric dispersant, ACI Publication, SP-119, pp.243-262, 1989.

19. KINOSHITA, M., YAMAGUCHI, S., YAMAMOTO, T., and TOMOSAWA, F.: Chemical structure and performance of new type high range water reducing AE agent, Cement Association of Japan, Proceedings of Cement & Concrete, No.44, pp.222-227, 1990.

20. KODAMA, K. and OKAZAWA, T.: Development of high-range water-reducing AE agent for high-strength concrete, Cement Association of Japan, Cement & Concrete, No. 546, Aug., pp.24-32, 1992.

21. YAGI, S., FUJII, T., TABUCHI, H. and SENDA, T.: Design and construction of a 41-story reinforced tube structure, Concrete Journal, Japan Concrete Institute, Vol.29, No.8, pp.35-47, 1991.

22. TSURUOKA, M.: Development of a construction system for super high-rise RC buildings, Mechanization of Construction, JCMA, No.6, pp.23-29, 1995.

23. Concrete giant rises on Chicago skyline; Concrete Construction, January, pp.5-10, 1990.

24. JAHREN, P.: Use of silica fume in concrete, ACI Publication, SP-79, pp.625-642, 1983.

25. TAKADA, M., UMEZAWA, K., OKUNO, T., and YONEZAWA, T.: An experimental study on mix design of high-strength concrete using silica fume, Proceedings of Annual Meeting of Architectural Institute of Japan, pp.621-622, 1990.

26. MITSUI, K., YONEZAWA, T., NAKASHIMA, M., and SUGIMOTO, M.: A study on the effects of physical properties and chemical composition of silica fumes on the properties of high-strength concrete, Proceedings of the Japan Concrete Institute, Vol.15, No.1, pp.63-68, 1993.

27. SANDVIK, M., HANG, A.K., HUNSBEDT, O.S., and MOKSNES, J.: Condensed silica fume in high-strength concrete for offshore structures, ACI Publication, SP-114, pp.1117-1129, 1989.

28. IZUMI, I., YONEZAWA, T., KAKIZAWA, T., and KOGA, T.: Experimental study on the placing of super high-strength concrete containing silica fume, Proceedings of Annual Meeting of Architectural Institute of Japan, pp.615-616, 1990.

29. KODAKA, S., WAMI, H., SAKURAMOTO, F., SUZUKI, K., and YANAGIDA, K.: Experimental study on pumpability of high strength concrete, Proceedings of Annual Meeting of Architectural Institute of Japan, pp.261-262, 1994.

30. BACHE, H.H.: Densified cement/ultra-fine particle based materials, The Second International Conference on Superplasticizers in Concrete, Oral presentation, 1981.

31 YONEZAWA, T., YANAGIBASHI, K., IKEO Y., and ASAKURA, E.: A study on the dispersion condition of silica fume in high-strength concrete, Proceedings of the Japan Concrete Institute, Vol.15, No.1, pp.69-74, 1993.

32. ELKEM MATERIALS: Brochure of EMSAC admixtures.

33. SATO, M., OHURA, T., OKUNO, T., and YONEZAWA, T.: Plant system for silica fume high-strength concrete, Proceedings of the Japan Concrete Institute, Vo.15, No.1, pp.75-80, 1993.

34. YONEZAWA, T., OKUNO, T., MITSUI, K., NUMAKURA, N., OHURA, T., and SATO, M.: Bottom-up concreting into steel tube column filled with ultra high-strength concrete using silica fume, Concrete Journal, Japan Concrete Institute, Vo..31, No.12, pp.22-33, 1993.

35. GODFREY JR, K.A.: Concrete strength record jumps 36%, Civil Engineering, October, pp.84-88, 1987.

36. KINOSHITA, M., SUZUKI, T., YONEZAWA T., and MITSUI, K.: Properties of an acrylic graft copolymer-based new superplasticizer for ultra high-strength concrete, ACI Publication, SP-148, pp.281-299, 1994.

37. MITSUI, K., YONEZAWA, T., KINOSHITA, M., and SHIMONO, T.: Application of a new superplasticizer for ultra high-strength concrete, ACI Publication, SP-148, pp.27-45, 1994.

38. YONEZAWA, T., TSURUMAKI, H., ANDO, S., NAKASHIMA, T., and KINOSHITA, M.: Design and construction of a PC bridge with depth-span ratio of 1/40, using 100 KN/mm^2 ultra-high-strength concrete, Journal of Prestressed Concrete, Japan, Vol.36, No.3, pp.11-23, 1994.

RESEARCH INTO DESIGN AND CONSTRUCTION OF LASER SCREEDED INDUSTRIAL FLOORS

J Knapton

University of Newcastle upon Tyne

UK

ABSTRACT   This paper provides design guidance which allows a floor constructed by a laser screed machine to be constructed to meet loading, durability and surfacing requirements.   The design method presented has been developed as part of a research project conducted under the guidance of the Newcastle University Floors Research (NUFLOOR) Committee.   It permits ground bearing concrete floors to be designed.   A study has been undertaken of the movement of two floors over a nine months period and the data is presented.

Keywords : Concrete Floors, Laser Screed, Fast Track Floors, Steel Fibres, Polypropylene Fibres, Joints, Microsilica, Floor Design Charts, Field Monitoring.

John Knapton is Professor of Structural Engineering at the University of Newcastle upon Tyne.   He specialises in pavement and floor design both with conventional rigid concrete and with pavers.   His publications include the British Ports Association pavement design manual, the British Standard for paver surfaced roads, the Civil Aviation Authority design manual for aircraft pavements surfaced with pavers and the definitive guide on laser screeded industrial floors. He is Chairman of the International Small Element Pavement Technologists (SEPT) Council and is a past Chairman of the Concrete Society, Northern.

Radical Concrete Technology. Edited by R K Dhir and P C Hewlett. Published in 1996 by E & FN Spon, 2–6 Boundary Row, London SE1 8HN, UK. ISBN 0 419 21480 1.

INTRODUCTION

Automatically laser controlled screeding machines have been introduced into the UK in order to increase the quality of industrial floors and to speed up their construction. Their introduction has encouraged the use of fibre reinforced concrete since the placing of conventional mesh immediately in front of the laser screeder militates against the high levels of production achievable with laser screeders. Steel fibres not only govern joint spacings in the same way as mesh, but additionally increase the strength of the concrete so permitting thinner floors. Polypropylene fibres increase concrete toughness and durability but do not influence joint spacings or slab thickness. For these reasons, automatically laser controlled screeding of concrete floors has an impact upon design as well as construction. This paper describes a method for designing laser screeded industrial floors using design charts. The charts introduce the concept of Single Equivalent Wheel Load, whereby combinations of point loads of differing values and contact areas can be converted into a single load value and then used to choose a slab thickness. Data is presented showing how joints in slabs open over the initial life of the slabs. The data confirms that the recommendations presented in this paper lead to slabs in which joints form correctly then operate in a manner to reduce stresses in concrete.

CONCRETE

Unlike most structural applications of concrete, floor slab design is based on the flexural strength of the concrete. A relationship between the 28 day characteristic compressive strength and the flexural strength of concrete is given in Table 1[1]. From Table 1 the relevant flexural strength can be determined and factors of safety applied resulting in the design flexural strength used in this paper.

Table 1 Relationship between concrete grade, compressive strength and flexural strength

Concrete Grade	C30	C40	C50
Characteristic cube strength (N/mm^2)	30	40	50
Flexural strength (N/mm^2)	3.8	4.5	5.1

Table 2 shows mix design guidance[2] for various grades of concrete. To achieve a durable wearing surface, concrete with a minimum cement content of 325kg/m^3 and water/cement ratio not exceeding 0.55 is commonly specified, although stricter limits may be applied if the floor is to be subject to heavy industrial use. The workability of fresh concrete should be suitable for the conditions of

**Table 2 Relationship between free water/cement ratio
cement content and lowest grade of concrete**

Maximum free water/cement ratio	0.65	0.60	0.55	0.50	0.45
Minimum cement content (kg/m^3)	275	300	325	350	400
Grade of concrete N/mm^2	C30	C35	C40	C45	C50

handling and placing, so that after compaction and finishing the concrete surrounds all reinforcement and completely fills its formwork.   Micro-silica concrete is specified for concrete floor construction where a tough marble-like wear resistant surface is needed.   Tests commissioned by a manufacturer of micro-silica concrete concluded the results shown in Table 3[3].

**Table 3 Test results showing the increase in 28 day compressive and
flexural strength obtained from samples of micro-silica concrete as
opposed to conventional C40 concrete**

	Conventional C40 concrete	Micro-silica concrete
Cement content (kg/m^3)	330	300
Water/cement ratio	0.55	0.45
28 day compressive strength (N/mm^2)	55	83
28 day flexural strength (N/mm^2)	5.9	7.7

SUBGRADE AND SUB-BASE

Subgrade is the naturally occurring ground or imported fill at formation level. Homogeneity of the subgrade strength is particularly important and avoiding hard and soft spots is a priority in subgrade preparation. Any subgrade fill should be suitable material of such grading that it can be well compacted. On very good quality subgrades, such a firm sandy gravel and sub-base may be omitted. The sub-base is the foundation to the floor slab and usually consists of an inert, well graded granular material although a cement-treated sub-base such as lean concrete or cement-bound granular material may be employed. In the case of wheel and rack loading the sub-base assists in reducing the vertical stress transmitted to the subgrade. In considering the value of the stresses induced in a slab under loading, the influence of the subgrade i treated as that of an elastic medium with a modulus of subgrade reaction (K). Modulus of subgrade reaction characterises the deflexion of the ground and/or the foundation under the floor slab. California Bearing Ratio (CBR) tests and plate bearing tests can be used to establish values. Values of modulus of subgrade reaction (K) for use in design are shown in Table 4. Table 7 shows the complete range of values from which the design values in Table 4 are developed.

Table 4 Modulus of subgrade reaction (K) for typical British soils[1]

	Typical soil description	Subgrade classification	Assumed (K) (N/mm^3)
Coarse grained soils	Gravels, sands, clayey or silty gravels/sands	Good	0.054
Fine grained soils	Gravely or sandy silts/ clays, clays silts	Poor	0.027
		Very poor	0.013

Chandler and Neal[1] suggest that the sub-base can be taken into account by enhancing the effective modulus of subgrade reaction (K) as in Table 5.

When a lean concrete sub-base is specified, the value of K of the subgrade material is used to calculate the required thickness of the concrete slab. This calculated thickness is then apportioned between the structural slab thickness (the higher strength concrete) and the lean concrete sub-base thickness. This relationship is shown in Table 6 when a C40 concrete is used for the slab and a C20 lean concrete is used for the sub-base.

SLIP MEMBRANES

A slip membrane is used to reduce friction between a concrete slab and its sub-base. The coefficient of friction with the use of a membrane is 0.2, compared with values of up to 0.7 when the concrete slab and sub-base are in direct contact. Prevention of loss of moisture and fines from the fresh concrete into the sub-base does occur although a slip membrane is not intended or required to serve as a damp-proof membrane. When damp-proofing is to be provided, thicker sheets or more elaborate measures may be required.

Table 5 Enhanced value of K when a sub-base is used

K value of subgrade alone	Enhanced value of K when used in conjunction with:							
	granular sub-base of thickness (mm)				cement-bound sub-base of thickness (mm)			
	150	200	250	300	100	150	200	250
0.013	0.018	0.022	0.026	0.030	0.035	0.050	0.070	0.090
0.020	0.026	0.030	0.034	0.038	0.060	0.080	0.105	-
0.027	0.034	0.038	0.044	0.049	0.075	0.110	-	-
0.040	0.049	0.055	0.061	0.066	0.100	-	-	-
0.054	0.061	0.066	0.073	0.082	-	-	-	-
0.060	0.066	0.072	0.081	0.090				

Table 6 The modified thickness of a concrete slab with a C20 lean concrete sub-base

Calculated thickness of slab (mm)	Modified thickness of slab(mm) when used in conjunction with lean concrete sub-base of thickness :		
	100mm	130mm	150mm
250	190	180	-
275	215	200	-
300	235	225	210

FIBRES AS REINFORCEMENT

With the use of laser screeders, fibres are often specified instead of conventional mesh because of the inconvenience in positioning individual mats of mesh immediately in front of the laser screeding machine as the machine progresses.

Table 7   Modulus of subgrade reaction values for a number
common subgrade and sub-base materials

	CBR	Modulus of subgrade reaction $K(N/mm^3)$
Humus soil or peat	<2%	0.005 - 0.015
Recent embankment	2%	0.01 - 0.02
Fine or slightly compacted sand	3%	0.015 - 0.03
Well compacted sand	10 - 25%	0.05 - 0.10
Very well compacted sand	25 - 50%	0.10 - 0.15
Loam or clay (moist)	3 - 15%	0.03 - 0.06
Loam or clay (dry)	30 - 40%	0.08 - 0.10
Clay with sand	30 - 40%	0.08 - 0.10
Coarse crushed stone	80 - 100%	0.20 - 0.25
Well compacted crushed stone	80 - 100%	0.20 - 0.30

As a consequence, plain concrete or fibre reinforced concrete are often specified for laser screeded floors.

Polypropylene fibres can be in fibrillated or monofilament form manufactured in a continuous process by extrusion of polypropylene homopolymer resin. Polypropylene fibres are not a substitute for conventional structural reinforcement or normal good curing procedures, but they may be used as an alternative to non-structural mesh for crack control purposes acting as a secondary reinforcement. The main purpose of polypropylene fibres is to provide crack control by distributing and absorbing tensile stresses which may occur as a result of shrinkage and temperature movements, particularly in the early life of the slab when the concrete has yet to reach sufficient tensile strength.

Steel fibres may be used in place of mesh reinforcement. The stresses occurring in a floor slab are complex and depending on the type of load, tensile stresses can occur at the top and at the bottom of the slab. There are, in addition, stresses that are difficult to quantify, arising from a number of causes such as sharp turns from fork lift trucks, shrinkage and thermal effects and impact loads. The addition of steel wire fibres to a concrete slab results in a homogenously reinforced slab achieving a considerable increase in flexural strength and enhanced resistance to shock and fatigue. The most commonly used steel fibre is the 60mm long hooked fibre. Hooked fibres are usually glued together (collated) with a special water soluble glue to form fibre plates which readily disperse in the concrete mixer. It is usual to consider only fibres with enhanced adhesion for reinforcement in concrete industrial floors. A manufacturer of anchored steel fibres commissioned TNO, Delft[4] to undertake flexural strength tests using fibres embedded in C30 concrete. Partly from these results and partly from work undertaken at the UK Cement and Concrete Association[1] the flexural strength values shown in Table 8 have been developed.

Table 8 Concrete design flexural strengths with steel fibres present

Concrete grade and steel fibre dosage	Flexural strength (N/mm^2)
Plain C30 concrete	2.0
$20kg/m^3$ steel fibre C30 concrete	2.4
$30kg/m^3$ steel fibre C30 concrete	3.2
$40kg/m^3$ steel fibre C30 concrete	3.8
Plain C40 concrete	2.4
$20kg/m^3$ steel fibre C30 concrete	2.8
$30kg/m^3$ steel fibre C40 concrete	3.8
$40kg/m^3$ steel fibre C40 concrete	4.5

LASER SCREED CONSTRUCTION

The laser screed machine has four wheel drive, four wheel steer and is operated by one person seated at a point of maximum visibility. Mounted on its twin axles a circular fully slewing turntable carries a counter balanced telescopic book having a 6.1m reach on the end of which is attached a 3.66m wide screed carriage assembly which comprises a plough, an auger to spread the concrete accurately and a vibrating beam for compaction. Test results have shown that the machine can compact concrete to depths in excess of 300mm. A self levelling laser transmitter is fixed at a visible point close to the work so as to project a 360 degree rotating beam across the working area. Depending on the type of transmitter, various inclinations of floors can be achieved including level, single and dual grades. The level of the laser screed is controlled by the laser beam which activates receivers mounted on the screed carriage assembly. During concreting the signals are relayed continuously to an on-board control box which automatically controls the level of the working screed head by direct intervention on the machine's hydraulic system. Outputs of between $2000m^2$/day and $4000m^2$/day are normal and $5000m^2$/day has been reported[3].

Table 9 shows joint spacings for various types of concrete. These joint spacings have been used successfully for many years in the UK. Some consider that spacings can be greater than 10m for steel fibre reinforced floors but the additional joint movement might lead to loss of aggregate interlock.

Table 9   Common concretes and suggested joint spacings

Concrete Type	Joint Spacing
Plain C30 concrete	6
Microsilica C30 concrete	6
20kg/m^3 ZC 60/1.00 steel fibre C30 concrete	6
30kg/m^3 ZC 60/1.00 steel fibre C30 concrete	8
40kg/m^3 ZC 60/1.00 steel fibre C30 concrete	8
Plain C40 concrete	6
Microsilica C40 concrete	6
20kg/m^3 ZC 60/1.00 steel fibre C40 concrete	6
30kg/m^3 ZC 60/1.00 steel fibre C40 concrete	10
40kg/m^3 ZC 60/1.00 steel fibre C40 concrete	10

PERFORMANCE OF FLOORS INSTALLED BY LASER SCREED

A site investigation has been undertaken of two grounds supported concrete industrial floors. Each 1000m^2 floor was constructed using polypropylene fibre reinforced concrete constructed by laser screed. The two floors were Unit 114/14 and Unit 114/10 of the Boldon Business Park, Tyne & Wear. Once the floor slabs had been concreted and the saw cut joints made, an arrangement of measurement points shown in Figure 1 was established on the concrete surface to permit joint movements to be monitored. Using an extensometer with an accuracy of one hundredth of a millimetre weekly measurements were taken during a period of 9 months commencing December 1993. Figure 2 shows the movement of each of the measured joints from week to week. The larger movements at the beginning of the slabs' life represent joint cracking and the influence of cracked joints on neighbouring joints can be seen. When all the joints had cracked and were working, a more uniform pattern became evident

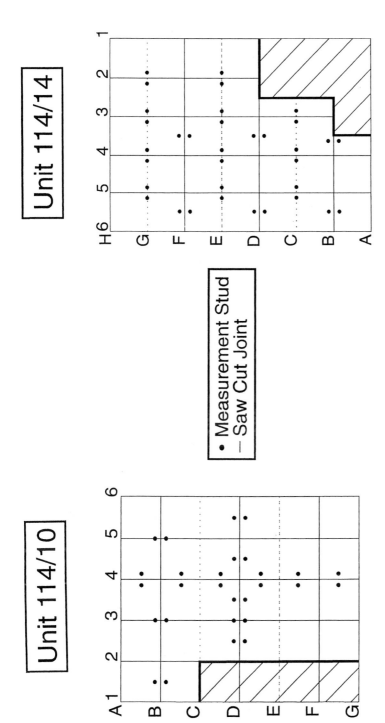

Fig. 1 Location of joint monitoring points in two floor slabs

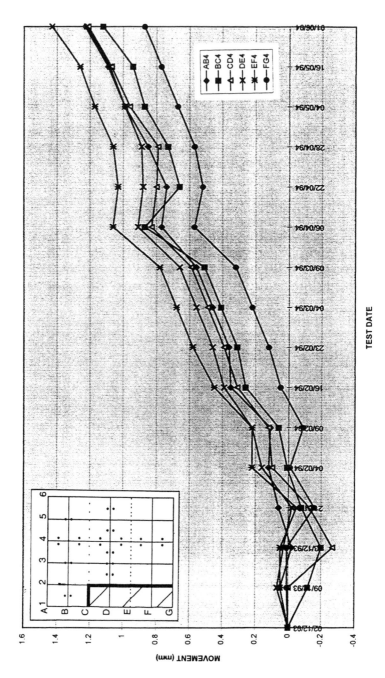

Fig. 2 Opening of monitored joints over a period of 9 months

throughout the slab. The data was used to calculate the cumulative movement of each joint with time in both floors. The cumulative movements are such that aggregate interlock can be assumed between cracked portions of slabs which means design can proceed on the basis of using mid-bay stresses. The cumulative movements are of the same order demonstrating uniform slab movement. This confirms that the laser screed machine and the design procedure set out in this paper together produce satisfactory durable industrial floors.

DESIGN

The following design procedure has been developed specifically for laser screeded floors taking into account the data presented in Figures 1 & 2. The design procedure comprises calculating stresses resulting from the loading regime and ground conditions and comparing those stresses with the strength of the concrete. Fatigue often leads to floor distress and a fatigue factor is built into the design method by the load safety factor. A load safety factor of 2.0 is recommended for an infinite number of load repetitions and has been used conservatively in all cases.

DESIGN METHOD FOR POINT LOADS

A set of design charts has been developed based upon calculating stresses resulting from imposed loads and comparing them with permissible stresses calculated by Westergaard equations and taking into account safety factors. The maximum flexural stress occurs at the bottom of the slab under the heaviest wheel load. The maximum stress under the wheel can be calculated by the Westergaard and Timoshenko[5] equation for mid bay conditions:

$$\sigma_{max} = \frac{0.275(1 + \mu)}{h^2} \cdot P.\log(0.36Eh^3/Kb^4) \qquad (1)$$

σ_{max} = flexural stress (N/mm^2)
P = point load (N i.e. characteristic wheel load x load factor
μ = Poisson's ratio, usually 0.15
h = slab thickness (mm)
E = elastic modulus, usually 20,000N/mm^2
K = modulus of subgrade reaction (N/mm^3)
b = radius of tyre contact zone (mm) = $(W/\pi.p)^{1/2}$
W = wheel load (N)
p = contact stress between wheel and floor (N/mm^2)

When two point loads act in close proximity, distance S apart, the stress beneath one point load will be increased because of the effect of the other point load. To calculate this stress magnification, the characteristic length (radius of relative stiffness, 1) has to be found from Equation 2. Solutions to Equation 2 are shown in Table 10.

$$I = (Eh^3/12(1 - \mu^2)K)^{1/4} \qquad (2)$$

Once the radius of relative stiffness has been evaluated, the ratio S/l can be determined so that Fig. 3 can be used to find M_t/P, where M_t is the tangential moment. The stress under the heavier wheel or point load is to be increased to account for the stress produced beneath the heavier load by the lesser applied load. The additional stress is:

$$\sigma_{add} = M_t/P.6/h^2.P_2 \qquad (3)$$

where P = greater point load, P_2 = lesser point load.

Sum the stresses and verify that the flexural strength has not been exceeded for the prescribed concrete mix. The design procedure is as follows.

(1)   Assess the Actual Point Load (APL) and Modulus of Subgrade Reaction (K) values to confirm the category of subgrade.

(2)   Determine distance between point loads.

    (i)   If the distance between loads is greater than 3m, the APL can be used directly (depending on the radius of contact zone, see stage (6)) calculate the thickness of the slab using the relevant design chart.

    (ii)   If the distance between loads is less than 3m, the radius of relative stiffness (1) has to be determined from Equation 2 or Table 10.

(3)   Calculate the additional stress σ_{add} resulting from the load nearby:

$$\sigma_{add} = \frac{M_t}{P} \cdot \frac{6}{h^2} \cdot P_2 \qquad (4)$$

where P = greater load (N)
P_2 = other point load (N)

(4)   From Table 2, a proposed concrete mix can be selected hence σ_{flex}. When two point loads are acting in close proximity (i.e. less than 3m apart), the greater point load (P) produces a flexural strength σ_{add} beneath the larger load. To calculate σ_{max},

$$\sigma_{max} = \sigma_{flex} - \sigma_{add} \qquad (5)$$

(5)   Calculate the Equivalent Single Point Load (ESPL) which when acting alone would generate the same flexural stress as the actual loading configuration.

$$ESPL = APL \cdot \sigma_{flex}/\sigma_{max} \qquad (6)$$

where APL = Actual Point Load

**Table 10 Radius of relative stiffness values for different
slab thicknesses and support conditions**

Slab thickness (mm)	Modulus of subgrade reaction (K) (N/mm³)			
	0.013	0.027	0.054	0.082
150	816	679	571	515
175	916	763	641	578
200	1012	843	709	639
225	1106	921	774	698
250	1196	997	838	755
275	1285	1071	900	811
300	1372	1143	961	865

E = Elastic modulus = 20000 N/mm²   *μ = Poissons Ratio = 0.15*

**Table 11 Point load multiplication factors for loads with
a radius of contact outside the range 150mm to 250mm**

Radius of contact(mm)	Modulus of Subgrade Reaction (K) N/mm³			
	0.013	0.027	0.054	0.082
50	1.5	1.6	1.7	1.7
100	1.2	1.2	1.3	1.3
150	1.0	1.0	1.0	1.0
200	1.0	1.0	1.0	1.0
250	1.0	1.0	1.0	1.0
300	0.9	0.9	0.9	0.9

(6)   Prior to using one of the design charts as shown in Figure 4, it is
necessary to modify the APL (actual point load) to account for contact
area as well as wheel proximity to obtain the ESP: (equivalent single point
load). The design charts apply directly when point loads have a radius of
contact between 150mm and 250mm. Some racking systems and pallet
transporters have a contact radius of less than 150mm and some vehicles
have a contact radius greater than 250mm. In these cases multiply the
point load by a factor in Table 12 prior to use in the design chart.

(7)   Use the design chart for the mix selected in (3) to determine slab
thickness.

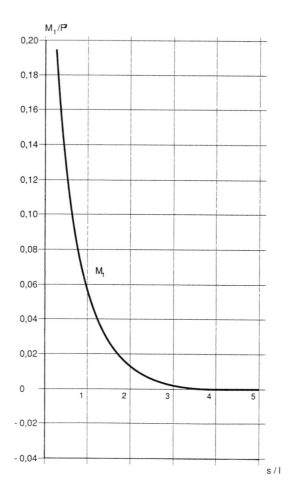

Fig. 3 Proximity factor chart example

POINT LOAD DESIGN EXAMPLE

Two point loads rest on a concrete floor. A 60kN point load is situated 1m away from a 50kN point load. The 60kN point load has a contact zone radius of 100mm and the 50kN point load has a 300mm radius. The ground conditions are poor (K = 0.027N/mm^3).

Assume thickness of slab = 200mm. Radius of relative stiffness, 1 is 843mm from Table 11.

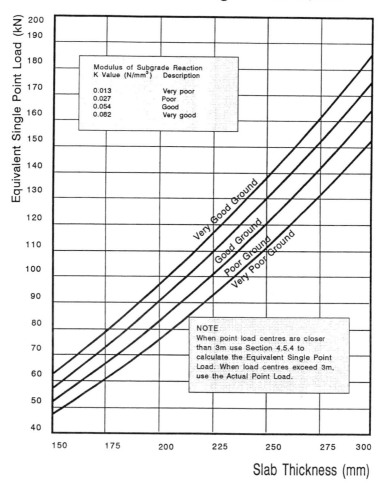

Fig. 4 **Typical design chart as used in point load**

Distance apart = 1m (i.e. S = 100mm) so, S/1 = 1000/843 = 1.186.

From Figure 3, M_t/P = 0.053

From Equation 4, σ_{add} = 0.053 $\cdot \dfrac{6}{200^2} \cdot$ 50,000 = 0.4N/mm^2

Try polypropylene fibre reinforced C40 concrete with a flexural strength of $2.4N/mm^2$

i.e. σ_{flex} = $2.4N/mm^2$

σ_{max} = $\sigma_{flex} - \sigma_{add}$ = 2.4 - 0.4 = $2.0N/mm^2$

This is the maximum flexural stress which the 60kN load can be allowed to develop. Calculate the Equivalent Single Point Load using Equation 6:

ESPL = $APL.\sigma_{flex}/\sigma_{max}$ = 60 x 2.4/2.0 = 72kN

From Table 12, the modified factor to be applied to the ESPL is 1.2 i.e. 72 x 1.2 = 86.4kN.

From the design chart shown in Figure 4, thickness of slab = 200mm.

A 200mm thick C40 concrete slab incorporating $0.9kg/m^3$ polypropylene fibre is adequate for this design.

REFERENCES

1.   Chandler, J.W.E. and Neal, F.E. (1988). The design of ground supported concrete industrial floor slabs. BCA Interim Technical Note 11, April 1988.

2.   BS8110 : 1985: Part 1. The structural use of concrete : Part 1 : Code of practice for design and construction.   British Standards Institution, London.

3.   BEKAERT Industrial floors with Dramix steel wire fibre reinforced concrete. N.V. Bekaert, S.A. 1990.

4.   Reports B88/607 and B88/751 (1988) TNO Institute for Building Materials and Structures, Delft.

5.   Westergaard, H.M. (1947) New formula for stress in concrete pavements of airfields. Proc. Am. Soc. Civ. Eng. vol. 73, No. 5, May 1947, pp 687-701.

RAPID CONSTRUCTION OF TALL BRIDGE PIER BY HYBRID-SLIP FORM METHOD

K Ukaji

Obayashi Corp

K Yamagata

Japan Highway Public Corporation

Japan

ABSTRACT Combining a new structural design concept and an unique construction scheme, the fast track construction of tall bridge pier is enabled by using a relatively simple site operation. The Hybrid-Slipform Method is intended to be applicable to a high pier structure exceeding 100m. The method has been applied to the piers of 31m and 34m high. The test project has been successful and shown greeter potential of the future deployment. Rapid construction, labor saving, cost competition and safety are major targets of the new method. The structure with hybrid of steel pipe and reinforced concrete also exhibit an advantage of the structural performance under destructive seismic loading, i.e., higher ultimate bending and shear strength as well as significant deformability. As we learned from Hanshin Earthquake Disaster(1995), the composite structure of the new method, which consists of steel pipe and concrete with prestressing strand spiral, is certainly a promising solution for an important tall structure to withstand the devastating dynamic forces.

Keywords: Rapid construction, Composite structure, Steel pipe-concrete hybrid, Slipforming, Prestressing strand spiral, Tall bridge pier, Labor saving, Ductility.

Dr Kenichi Ukaji is General Manager of Design Department, Obayashi Corporation, Tokyo, and Lecturer in Construction Engineering, University of Saitama and Musashi Institute of Technology.

Mr Keiji Yamagata is Manager of Structural Engineering Division, Japan Highway Public Corporation, and he is responsible of structural aspects in the corporation including research, development, design and construction.

Radical Concrete Technology. Edited by R K Dhir and P C Hewlett. Published in 1996 by E & FN Spon, 2–6 Boundary Row, London SE1 8HN, UK. ISBN 0 419 21480 1.

INTRODUCTION

The design maximum traffic speed of a new expressway between Tokyo and Kobe is up graded to 120 km/hr from the current design speed of 80km/hr. The construction of such expressway tends to be planned in rather mountainous area. The design standard of higher traffic speed and sever Japanese topography are typical reasons of large percentages of bridges and tunnels. Subsequently, many of those bridges require tall pier structures, i.e., 50-100m high. In order to speed up the construction of such expressways, a drastic improvement on the pier construction is essential. The improvement should subject not only construction speed, but also work force needed, construction cost, and safety.

Minor improvements of the traditional construction scheme as well as the conventional structural design of reinforced concrete structure can not meet the requirements. Therefore, a unique combination of construction scheme and structural design has been developed, namely, Hybrid-Slipform Method. The first part of the name, Hybrid, indicates a composite structure of steel pipe and reinforced concrete with horizontal reinforcement of prestressing strand spirals, while the second term, Slipform, represents slipforming construction.

A superstructure of tall piers usually have longer spans exceeding 100m because of overall economy. Subsequently, the piers must provide the higher load carrying capacities against vertical and lateral loading. Sometimes, the lateral loading during earthquake may exceed 1G. The structural stability under severe earthquake loadings requires that the pier structure should exhibit ductile behavior, such as the large deformability with sustaining the load carrying capacity. The composite structure of steel pipe and reinforced concrete is expected to meet such difficult requirements.

CONCEPT OF HYBRID-SLIPFORM METHOD

The design of a conventional reinforced concrete pier is an appropriate solution in terms of material efficiency and cost, while the overall construction cost as well as the construction period may not be optimum. The left hand side of Fig.1 shows a

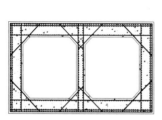
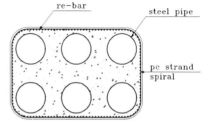

A. Conventional   B. Hybrid-Slipform Alternative
Figure 1   Conceptual Cross Section of Conventional and Alternative Designs

typical configuration of the conventional hollow type structure, which have one- or two-boxes, and the right hand side of Fig. 1 illustrates a typical cross section of the Hybrid- Slipform design, which is a composite of steel pipe and reinforced concrete [1].

A long term social trend in our society shows the shortage of skilled labor as well as the reduction of annual working hours. This trend enhances the need of drastic improvement on the construction schedule and labor force required for the bridge construction. The goal of Hybrid-Slipform is one half in construction period and one third in labor force compared to the conventional reinforced concrete construction. Figure 2 illustrates the equipment and structural arrangements of the new method.

Figure 3 shows the construction sequence of the method. The first step of construction is the erection of prefabricated pieces of steel pipe, which length is about 8m to 12m each. Those erected steel pipes should withstand wind load till concrete pour. At every 8m to 12m of the pipe erection, the group of steel pipes is strengthened by bracing, which connect each pipe together. When the erection reached to the pier top, a reaction frame is erected and fastened at the pipe head. The reaction frame supports the vertical slipforming load transmitted through the vertical pc strands.

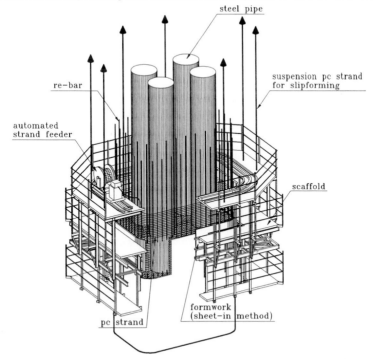

Figure 2   Illustration of Structural System and Equipment of
Hybrid-Slipform Method

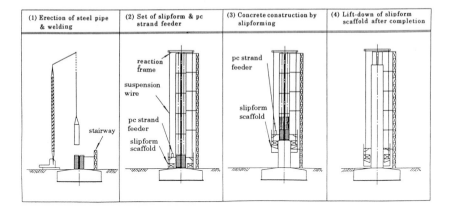

| (1) Erection of steel pipe & welding | (2) Set of slipform & pc strand feeder | (3) Concrete construction by slipforming | (4) Lift-down of slipform scaffold after completion |

Figure 3   Construction Procedure of Hybrid-Slipform Method

The second step follows the assembly of slipform, working decks, scaffolds and jacking system(center hole jack). The slipform system has three working decks. The top deck has a loop of rail track where the automated pc strand feeder runs on the track. The intermediate deck is used for form work, re-bar installation, concerning and jack operation. The bottom deck is used for removal of the sheet-in-form and concrete curing. The number of center hole jacks depends on the dimension of pier structure. A typical configuration is 8 jacks, two jacks at each side. The computer controlled jacking system provides the self-leveling mechanism of the slipform at every 2.5mm slip. The accuracy of the slipform positions on the horizontal and vertical planes is detected by the 3-dimensional positioning system.

Once completed the slipform installation, the highlight of Hybrid-Slipform Method begins with the pier construction by the daily repetition of cyclic jobs. The cycle consists of jacking up of the slipform system, sheet-in-form placement, concreting, curing and re-bar placement. The height of concrete placed by each daily cycle depends on the priority of construction schedule, the size and cost of the slipform system to be manufactured, the number of labor to be employed, the length of daily working hours, and the construction economy. It might be practical to chose the lift schedule of slipforming each day somewhere between 2m and 4m.

The jack up of the slipform from the previous old position to the new position is done by at once and takes one and half hours. The old sheet-in-form persists its old position and attached to the concrete poured one day before. As soon as the slipform is settled to the new position, the new sheet-in-form is placed for the next concreting. The advantages of utilization of the sheet-in-form are significant : a smooth surface finish of concrete as same as the fixed form, a drastic change of the slipforming procedure from continuous from lifting to at once lifting at any time, a better curing protection to the concrete surface, and more easier positioning of the slipform.

The horizontal reinforcement of pc strand is spiraled along the vertical reinforcement by using the automated strand feeder. The placement takes one and half hours by a few workers, which is needed for fastening the strands to the vertical re-bars.

When the slipform reached to the top of the pier at the completion, the working decks and scaffolds are easily lifted down on the ground by the jacking system. During the down process, any necessary finishing jobs along the pier surfaces can be done by using the working decks.

STRUCTURAL DESIGN

In order to assure the advantages of the rapid construction scheme, the high priority on the structural design is given to : (1) minimize the amount of vertical re-bar by steel pipe replacement, (2) utilization of pc strand for horizontal reinforcement with taking account its high strength, and (3) develop a simple and economical treatment of the steel pipe anchorages at the top and bottom of the pier. The conventional design of tall pier tends to require the mufti-layer arrangements of 51mm re-bar for the vertical reinforcement. If the vertical re-bar reinforcement could be a single row, it has a great impact on reducing the number of jobs on site and, subsequently, the construction time required for re-bar placement. On the other hand, if the number of pier is more than one, the erection of the steel pipes can be done in advance of the slipforming so that the steel pipe erection is off the critical path of construction schedule. The design procedure is as follows:

STEP 1 : assume pier dimension, size and spacing of vertical re-bar (a single raw only) , and diameter, thickness, spacing and number of steel pipe (see Fig. 1).
STEP 2 : calculate ultimate strength of the assumed pier cross section by replacing the steel pipe to an equivalent cross sectional area of reinforcing bar.
STEP 3 : check bending and bond stresses.
STEP 4 : calculate and check shear strength of the cross section in assumption of neglecting the shear resistance of steel pipe. If necessary, calculate an appropriate shear reinforcement.
STEP 5 : design structural details.

The bond stress and strength between steel pipe surface and concrete is one of the key factors of this composite structure to be monolithic. Most sever bending moment occurs at the both ends of the pier structure. Then, the pipe ends at the top and the bottom is required to be anchored to footing structure and / or superstructure by a simple and reliable manner. In order to increase the ultimate bond strength between the steel pipe and the surrounding concrete, an idea of deformed pipe surface is introduced to the steel pipe anchorage. This simple solution does not require any rigorous anchor bolt installation usually enforced by the conventional design. As an experimental test data shown in later, this solution tremendously   saves construction time, labor and cost.

To use of prestressing strand spiral in alteration of horizontal re-bar reinforcement is also an important decision on Hybrid-Slipform Method being successful. In terms of structural performance the spiral increases the ductility and deformability of the composite structure. Additionally, on-site construction activities become simple (see Fig. 2).

MODEL TEST OF DEFORMED PIPE END ANCHORAGE

The scaled model test of deformed pipe anchorage [2] has been carried out to investigate the load carrying capability of the new anchorage system. The test is also intended to verify the reliability of anchorage system as a vital part of rigid connection between the pier structure and the foundation and the superstructure. The steel pipe of the testing model is 60cm diameter and 9mm thick with the deformed outer surface of top 30cm. Non hoop reinforcement model is also tested for comparison purposes. The deformed pipe anchorage is strengthened by hoop reinforcement with 7cm clearance between the pipe surface. The concrete pedestal is jacked up by four jacks of 3000kN and the vertical force is applied at the target area of 30cm of deformed anchorage.

The test result is summarized in Fig. 4. The peak bond strength of the models is 5.6 N/mm² and 6.8 N/mm² for the non-hoop reinforcement model and the reinforced model, respectively. The design criteria of allowable bond strength 2.4 N/mm² is obviously too conservative assumption according to the test data. The factor of safety is 2.8. It is worthwhile to note that the deformed pipe surface exhibits higher bond strength even in the range of large slippage occurrence. Because of radiation tensile cracking appeared in the pedestal, which has a small dimension compared to the real footing or superstructure, the bond strength of a full size composite pier should be higher than the test results.

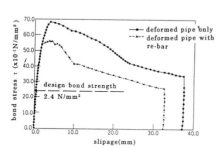

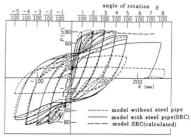

Figure 4   Test for Non-Anchor
Bolt Pipe Anchorage

Figure 5   Lateral Loading Test
of Composite Structure

MODEL TEST UNDER LATERAL LOADING

The model tests of Hybrid-Slipform structure is performed to evaluate the ultimate strength of composite structure. It is also investigated how large bending deformation of the structure can be attained with keeping concrete-steel pipe structure intact. The pier structure experiences most critical loading during earthquake shaking.

Therefore, the structural performance is most suitably represented by the cyclic lateral loading tests conducted herein. A typical experimental scaled model consists of a single steel pipe of 60cm diameter with 9mm thick and a rectangular reinforced concrete cover around the pipe[3]. The horizontal loading is applied at height of 340cm and the maximum bending moment and shear force are induced at the rigid connection of column and footing.

The testing models are designed to meet the following criteria in comparison of the full scale pier structure at Yokomichi Bridge project: (1) the ratio of ultimate bending strength of the steel pipe to the whole structure, (2) the ratio of shear span, and (3) the ultimate shear stress. The analytical computation of force-deformation relationship is performed for the reinforced concrete model without a steel pipe. Then, it is compared with the test results. Before applying horizontal loads, the vertical compressive stress of 1.5 N/mm^2 is given to the model. The column angle due to the loading is controlled to 1/200, 2/100, 3/100, 4/100, 5/100, and then increasing to failure.

Figure 5 shows the test results which explicitly indicate structural characteristics of the composite structure, i.e., the peak strength is equivalent in both model while the composite structure maintain significant strength even at large bending angle beyond 5/100. On the reinforced concrete model the bucking of re-bars occurred at the bending angle of 4/100, and then right after that, the collapse and falling down of concrete started, which follows to lose the structural strength. The ductile behavior of the composite model is well indicated by the parameter of energy absorption calculated by integration of hysterics loops. The composite pier structure maintains the energy absorption increase up to the angle of rotation 8/100, while the reinforced concrete model decreases it after reaching the peak at 4/100.

CONSTRUCTION OF YOKOMICHI BRIDGE

The first application of Hybrid-Slipform method was done at Yokomichi Bridge, Oita Expressway, Kyusyu, which is owned by Japan Highway Public Corporation. The bridge is a three span of continuous concrete bridge with two piers with the height of 34.5m and 37.5m. The pier has a typical cross section of 6.0m wide and 3.8m thick. Four steel pipes of 140cm in diameter and 14mm or 17mm thick are installed in each pier. A single row of vertical reinforcement by 38mm deformed re-bar and 12.7mm pc strand spiral with 5cm or 10cm spacing are arranged with 6cm concrete cover.

Following the construction sequence previously shown in Fig. 3, the pier construction commenced in September 1993 and completed in February 1994. A single lift of one day slipforming is chosen to be 1.8m high in working hours from 8am to 5pm each day. Figure 6 summarizes the actual construction time schedule of the two pier structures. The figure's left hand side illustrates the planned time schedule of the conventional reinforced concrete construction. Though the pier height is not tall enough as the Hybrid-Slipform method intended originally, the outcome seems to be successful that the total lobar force was reduced to 72%, and the time schedule was 60%, compared with the conventional method.

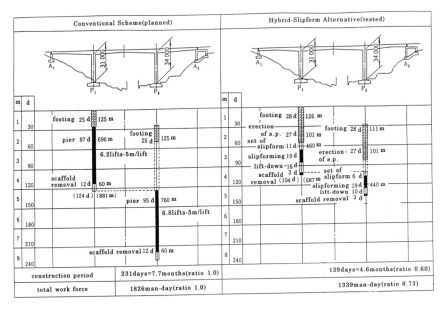

Conventional Scheme(planned)	Hybrid-Slipform Alternative(tested)

m	d					m	d				
1	30	footing 25 d	125 m			1	30	footing 28 d	126 m		
2	60	pier 87 d	696 m	footing 25 d	125 m	2	60	erection of s.p. 27 d	101 m		
3	90		6.2lifts-5m/lift			3	90	set of slipform 11 d	460 m	footing 28 d	111 m
4	120	scaffold removal 12 d	60 m			4	120	slipforming 19 d / lift-down 16 d		erection 27 d	101 m
5	150	(124 d)	(881 m)	pier 95 d	760 m	5	150	scaffold removal 3 d (104 d)	(687 m)	set of slipform 6 d	440 m
6	180			6.8lifts-5m/lift		6	180			slipforming 19 d / lift-down 10 d	
7	210					7	210			scaffold removal 3 d	
8	240	scaffold removal 12 d	60 m			8	240				

construction period	231days=7.7months(ratio 1.0)	139days=4.6months(ratio 0.60)
total work force	1826man-day(ratio 1.0)	1339man-day(ratio 0.73)

Figure 6   Comparison of Construction Period and Total Number of Work Force at Yokomichi Bridge

Photo 1   Ykomichi Bridge at Completion

Photo 2   Erection of Steel Pipe

Photo 3   Lift-up Slipform with
            Cast-in-Concrete Work

Photo 4   PC Strand Automated Feeder

Photo 5   Installation of Slipform

Photo 6   Lift-down of Slipform

CONCLUSIONS

The Hybrid-Slipform method provides an appropriate solution not only for rapid construction, but also for ductile structural performance suitable for the tall bridge pier. It is advantageous for a tall pier construction exceeding 40m high, i.e., a safe operation using even unskilled labor. As shown in the model tests, the composite structure of steel pipe and reinforced concrete is a promising way to strengthen any important superstructure built in highly seismic region.

ACKNOWLEDGMENTS

The authors owe to the advice made by Prof. H. Okamura and Associate Prof. K. Maekawa, University of Tokyo. The experimental model tests were conducted by Obayashi Research Institute under supervision of Japan Highway Public Corporation.

REFERENCES

1. Ichikawa, H, Hayakawa, S Sakai, H, Kato, T. Hashimoto, M,   Rapid Construction of High Pier by Using Steel Pipe-Concrete Hybrid Structure,   Annual Conference, Japan Society of Civil Engineers, 1993.

2. Mizuta, T, Mabuchi, K, Ouchi, H, Ichikawa, K,   Bond Strength of Deformed Steel Pipe,   Annual Conference, JSCE, 1994.

3. Ichikawa, H, Satuna, Y, Ouchi, H, Kobatake, K,   Lateral Loading Test of Hybrid-Slipfrom Pier Model,   JSCE, 1994

THE INNOVATIVE DESIGN AND CONSTRUCTION OF TRICKLING FILTER TOWERS

L G Mrazek

Sverdrup Facilities Inc

USA

ABSTRACT. This paper describes design concepts and construction methods associated with the construction of six (6) large trickling filter towers for the recent secondary wastewater treatment plant expansion at the Metropolitan St. Louis Sewer District's Bissell Point Wastewater Treatment Plant in St. Louis, Missouri, USA. These circular towers were designed using tilt-up concrete panels, vertical cast-in-place concrete pilasters and a cast-in-place concrete ring beam at the top of the open tower. The site is located in a moderate seismic region, requiring the computation of seismic forces and establishing a load path for these forces. Wastewater receives biological treatment while flowing through layers of corrugated polyvinyl chloride media, supported on precast concrete media supports; there is no liquid pressure on the wall panels.

Keywords: Environmental engineering concrete structures; tilt-up concrete; trickling filter towers; biological wastewater treatment; steel corrosion; concrete deterioration.

Larry G. Mrazek, P.E., S.E., is a Senior Engineer with Sverdrup Facilities, Inc., St. Louis, Missouri, USA. He specializes in the design and construction of environmental engineering concrete structures and the repair of concrete structures. He is an active member of ACI 350, "ENVIRONMENTAL ENGINEERING CONCRETE STRUCTURES", and he is a Fellow of the American Society of Civil Engineers, a member of the Prestressed Concrete Institute and a member of the International Concrete Repair Institute.

Radical Concrete Technology. Edited by R K Dhir and P C Hewlett. Published in 1996 by E & FN Spon, 2–6 Boundary Row, London SE1 8HN, UK. ISBN 0 419 21480 1.

INTRODUCTION

This paper describes the design and construction of trickling filter towers and precast media supports for the BISSELL POINT WASTEWATER TREATMENT PLANT, located in the City of St. Louis in the State of Missouri along the Mississippi River in the United States. (See Figure 1) Construction was accomplished in several contracts; the trickling filter towers project being one of the later contracts. Wastewater has received secondary treatment since October, 1992.

Six trickling filter towers, 41m in diameter by 13.7m high are packed with 76,455 cubic meters of corrugated polyvinyl chloride media which supports biofilm growth during the treatment process. Trickling filters at the Bissell Point Wastewater Treatment Plant biologically treat 568 million liters of high-strength wastewater per day. This treatment process occurs when the wastewater flows down through the packing media. Figure 2 indicates a partial site plan of the filters.

The site is located in a moderate seismic zone. Geotechnical investigations indicated the requirements for deep foundations and special provisions to protect foundation concrete from the effects of high sulfate concentration in the soils. Since the project is located adjacent to the Mississippi River, groundwater was an important factor in the design and construction of most of the structures at this plant. However, the Trickling Filter Towers are supported at grade, with minimal influence of groundwater effects. Fortunately, the site for the trickling filters was of adequate size to accommodate the sitecast/tilt-up design and construction concepts selected for these structures.

Construction materials in contact with this wastewater are subject to a highly corrosive environment. During conceptual and preliminary design, the design team investigated the effects of similar waste streams on concrete and reinforcing at other facilities. In addition, odor control was to be a future consideration by the Owner.

CONCEPTUAL STRUCTURAL DESIGNS

At the time of conceptual design, the size of trickling filter tower planned for this project was the largest in the United States (USA). Prior designs in the USA by the by the author and other engineers included precast, prestressed double tee walls, cast-in-place concrete walls and glass lined steel tanks. Due to architectural requirements by the Owner, the steel tank option was eliminated. Since the author was familiar with the tilt-up concrete concept, he pursued this approach from the beginning of the project. However, applying the tilt-up concept to this large a structure was new and an innovation to both the Owner and Sverdrup's project environmental engineer. The tilt-up approach satisfied process, architectural and structural requirements. A final selection was not made until a cost study was completed.

Table A shows alternative structural designs which were considered during the conceptual design phase of this project. Mat foundations were assumed to be pile supported for all alternatives. A rating system, shown in Table A, was developed for this study.

Figure 1 — Trickling Filter Towers at end of construction.

TABLE A - TRICKLING FILTER TOWER WALL SYSTEMS
ALTERNATIVE CONCEPTUAL DESIGNS

| WALL SYSTEM AESTHETICS | COST | | MAINTENANCE | FOUNDATION |
Description	Initial	Rating	Rating	Rating
Steel Tank Glass Lined	$350,000	4.5	8.5	3
P/C, P/S Wire Wound	$435,000	8	5	5.5
Cast-In-Place Reinf. Concrete	$240,000	7.2		6.5
Tilt-up Concrete	$155,000	6.5	5	8.0
Panels/Pilasters/ Ring Beam P/C, P/S	$142,000	6	6.5	7.5

NOTES:

1. Costs are based on 1985 U.S. Dollars.
2. Ratings are 1 (lowest-least acceptable) to 10 (highest).
3. Based on government guidelines, a minimum acceptable structural life of the filters is 40 years.

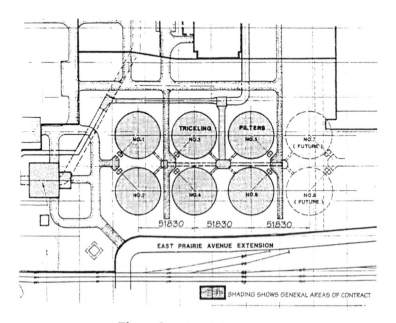

Figure 2 — Partial Site Plan

The coated steel frame with fiberglass panels was found to have several potential serious problems which included surface erosion of panels with exposure of fibers, susceptibility of panels to damage from ultra violet rays and possible deterioration of a translucent, fire retardant panel in three years.

Also, a design team member suggested using a flat precast concrete panel as an alternate to the tilt-up panel. This concept was not accepted due to the panel size and resulting shipping limitations.

A tilt-up concrete consultant provided guidance to the design team during this phase of the investigation and architectural concepts were developed for the higher rated alternatives.

The tilt-up panel concept with cast-in-place ring beam and vertical pilasters was selected, since it proved to be cost effective, involved less time to complete and also met architectural requirements.

BASIC DESIGN CRITERIA & STRUCTURAL MODEL OF SUPERSTRUCTURE

Design was based on BOCA 1984, Reference 1, which referenced ACI 318-83, "Building Code Requirements for Reinforced Concrete" and Commentary, Reference 2. In addition, guidelines presented in ACI 350R-83, Reference 3 were followed in order to comply with the state-of-the-art design for environmental engineering concrete structures. The following design criteria were assumed:

Design Method

"Alternate Strength Design" in Appendix B of ACI 318. This method is sometimes referred to as the Working Strength Design Method (WSD) in earlier ACI 318 and other codes throughout the world. The basic approach of ACI 350 is to maintain low reinforcing steel stresses such that crack widths are limited in order to prevent steel corrosion and subsequent concrete deterioration. Presently, ACI COMMITTEE 350 is developing a code for environmental engineering concrete structures. In this code, Sanitary Durability Factors are applied to load factors when designing with the Strength Design Method. These factors limit steel stresses to those computed when using the old WSD METHOD.

Compressive strength at 28 days = 27.6 MPa
Reinforcing Steel; f_y = 414 MPa
fc, allowable compressive strength in concrete = 9.32 MPa
fs, allowable tensile strength in flexure of reinforcing steel equals 124.2 MPa for structural elements in direct contact with wastewater or gases.
fs, allowable tensile strength when member is in direct tension for reinforcing steel equals 96.55 MPa.

It should be noted that the design of all structures, not just the subject trickling filters, was accomplished by a joint venture of two consulting firms; as a result, the Alternate Strength Design Method was selected for the entire plant addition, prior to conceptual design of the Trickling Filter Towers.

Other WSD design allowables are included in TABLES 2.6.4a and 2.6.4b of ACI 350R-83.

TYPE II CEMENT, in accordance with ASTM C150, was specified for concrete in contact with wastewater or gases, even though the concrete was specified to receive a coating or liner. Our research indicated the potential corrosiveness of the wastewater was severe enough to require additional measures to protect the concrete.

High sulfates in the soil resulted in specifying a TYPE II cement plus a Class F flyash for all concrete (primarily foundations) in contact with soils.

Loads

WIND - Velocity = 113 kilometers/hr. Importance Factor I = 1

FUTURE DOME LOAD: Snow = 958 Pa plus estimated dome load (aluminum)

SEISMIC - ZONE 1 BY BOCA and ZONE 2A by ANSI A58.1-1982, Reference 4. We used the more conservative value which results in approximately 2.5 times the seismic force specified in BOCA 1984. Presently, BOCA and the City of St. Louis base seismic design on Zone 2A, very close to the values which were assumed.

Concrete Weight - normal weight = 2320 kg/m^3

Plastic Filter Media = 719 Pa per 305mm of height.

Center Column Loads/Forces: As Furnished by Manufacturer. {Torque, Thrust and Overturning Moment}

Underground piping was designed for a temporary surcharge loading from heavy cranes operating near the tower during the erection of the tilt-up concrete panels.

Structural Model

Figure 3 shows a TYPICAL SECTION THROUGH THE "FILTER" WALL and Figures 4 and 5 the assumed MODELS which were applied to computer solutions for the lateral loading cases. There is no fluid pressure on the Trickling Filter Walls. A temperature change of 56°C was assumed for the structure in service.

The lateral load resisting system, which also provides lateral stability includes an assumed **load path** which transfers lateral loads from a cast-in-place concrete ring beam to pilasters and wall panels and then to the pile supported mat foundation. For conservatism, pilasters were initially assumed to transfer the total lateral loads, even though wall panels provided some shear wall action. At the time of design, conceptual connection details had already been developed in conjunction with state-of-the-art tilt-up concrete practices.

Process design indicated the possible need for a dome cover over each filter as a part of a future odor control system. Within several years after operation, odor problems became a reality and aluminum domes were installed on all six filters. See Figure 6.

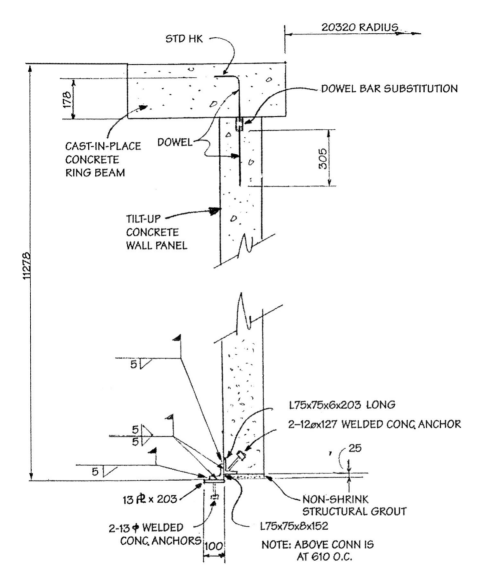

20320 RADIUS

STD HK

DOWEL BAR SUBSTITUTION

178

CAST-IN-PLACE
CONCRETE
RING BEAM

DOWEL

305

TILT-UP
CONCRETE
WALL PANEL

11278

5

L75x75x6x203 LONG

2–12ø x127 WELDED CONC. ANCHOR

5
5

25

5

13 𝆏 x 203

NON-SHRINK
STRUCTURAL GROUT

2–13 ⏀ WELDED
CONC. ANCHORS

L75x75x8x152

100

NOTE: ABOVE CONN IS
AT 610 O.C.

Figure 3 — Typical Wall Section

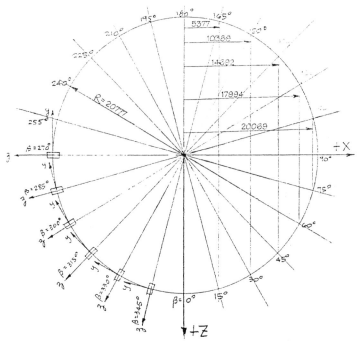

Figure 4 — Lateral Load Analysis - Structural Model
Geometry & Local Axis Orientation for Columns

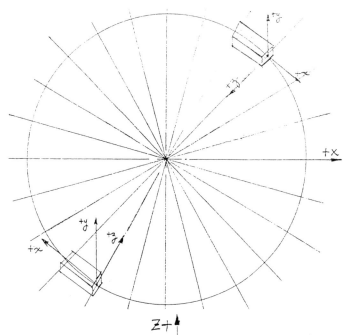

Figure 5 — Lateral Load Analysis - Structural Model
Local Axis Orientation - Ring Beam Segments

Figure 6 — Trickling Filter Towers in operation with aluminum domes

TYPICAL DETAILS AND CONSTRUCTIBILITY

During the conceptual design, alternate panel sizes (widths) were discussed with the Architect and the tilt-up consultant, in order to provide a cost effective design. Early in the planning, consideration was given to precasting the panels off site at a precast concrete plant. However, shipping constraints eliminated this approach. Pilasters served both structural and architectural purposes - these elements helped tie the panels together and minimized the typical problems of joint alignment and sealant performance experienced in precast concrete panel construction.

The cast-in-place concrete ring beam served as a structural member to transfer lateral loads and also served as a walkway to access the distributor mechanism supported on the center column. Mechanical threaded couplers were specified to splice the radial, curved ring beam reinforcing. This approach minimized the congestion which could occur with tension lap splices as specified in ACI 318. Ring beam and pilaster reinforcing were specified to be epoxy coated. Careful inspection by resident engineers was required to ensure damaged coating areas were repaired. This was especially true at the ring beam reinforcing splices with mechanical couplers. These couplers were specified to be shop coated before shipment to the site.

When planning the construction sequences, a true tilt-up panel construction as usually conceived was found to be difficult for the circular configuration. Moreover, the erection of precast concrete media supports proved to be a hindrance to the planned sequence of construction. After several meetings with local tilt-up contractors and our tilt-up consultant, it was decided to locate casting beds at pre-established locations, such that large cranes could lift and set a certain number of panels before moving (Refer to Fig. 7). Strongbacks were fastened to the panel and mat foundation before the panel was released by the crane. On the first series of panel erections, a strongback brace buckled; fortunately, the crane operator and other workers were prepared and the panel was stabilized from spinning in a revolving motion which could have resulted in injury to workers as well as damage the adjacent panels. That afternoon, the Engineer and Owner, along with tilt-up consultants and the tilt-up subcontractor met for several hours to discuss the morning's near accident and to develop further safety measures. There were no further problems with the remaining trickling filters during panel erection, when following additional safety guidelines and installing the heavier strongbacks recommended at the meeting.

PRECAST CONCRETE MEDIA SUPPORTS

During our research on similar facilities, environmental engineers discovered that quite a few media support systems at similar facilities had performed poorly. Hydrogen sulfide gases and the resulting mild acid attack led to reinforcing corrosion and concrete deterioration. Moreover, the frequently used precast concrete slat system was found to have minimal lateral support and was very labor intensive to install; moreover, this system was found to have resulted in several collapses, causing the plastic filter media to settle. Collapsing of the supports also occurred while the media, as high as 9.75m, was being placed, thus endangering the workers installing the modules of media. Cast-in-place concrete media supports would be very time consuming and costly. For example, a special drain trough is required in the top of the supports. For this project, a system of cast-in-place concrete piers with dowels extending into the precast concrete media support beams were designed for both this plant and another plant in the St. Louis area. Filter media

Figure 7 — Erection of tilt-up concrete panel.

manufacturers and installers of media concurred that this media support design was far superior to any installation for which they had been involved. Refer to Figure 7 which shows the erection of precast concrete media supports before most of the tilt-up panels were erected for the particular filter tower. Refer to Figures 8 and 9 for precast concrete media support details.

Epoxy coated reinforcing bars for media support piers were drilled and grouted into the mat foundation, using an epoxy resin grout. The original design had specified reinforcing dowel bar substitutions which proved to be costly from both a labor and material aspect. Piers were designed as vertical cantilever beam/columns to provide lateral stability for the precast media support modules which were seated over dowels extending up from the piers. After alignment and leveling, dowels were grouted. The Engineer required periodic pull-out testing of the dowels to ensure adequate bond to the foundation concrete. The following erection tolerances were specified for the precast concrete media supports.

> PLAN LOCATION OF SUPPORT PIERS: ± 12mm
> PLAN LOCATION - RADIUS OF SUPPORT LEDGE/FOUNDATION: +12mm, -6mm
> BEARING ELEVATION FROM NOMINAL ELEVATION AT SUPPORT
>   MAXIMUM LOW   12mm
>   MAXIMUM HIGH   6mm
> JOINT WIDTH BETWEEN UNITS: ± 10mm
> BEARING LENGTH: ± 12mm

DESIGNING AND DETAILING FOR DURABILITY

Considerable research and discussions with manufacturers of coatings and membranes preceded the final design of the filters. After consultation with an in-house specialist on coatings, the following approach was taken.

> Use TYPE II cement concrete and uncoated reinforcing bars for panels.

> Install an unbonded hypalon membrane attached to the panels and pilasters with stainless steel battens and stainless steel anchors.

Precast media supports received a coal tar epoxy coating after erection. It was necessary to specify an Architectural Grade B finish for the surfaces to receive the coating; i.e., bug hole sizes and honeycombed areas were limited. Before the painting contractor initiated work, the Engineer required both the painting contractor and coating manufacturer to meet at the site to verify the concrete finish was acceptable for the specified coating.

SUMMARY AND RECOMMENDATIONS

The design and construction of very large Trickling Filter Towers and media supports was discussed with an emphasis on practical considerations. Based on the writer's experience with both conceptual and final design as well as construction, the following recommendations are made.

1.   Develop a conceptual plan and study various alternative construction methods and materials.

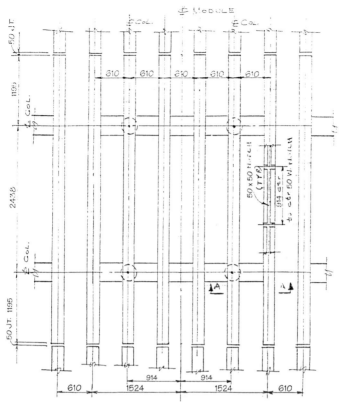

Figure 8 — Plan - Typical P/C Media Support

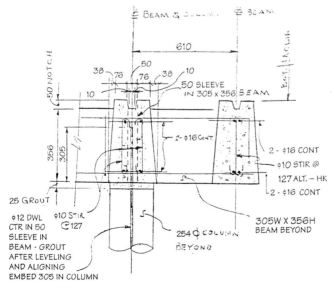

Figure 9 — Media Supports - Section A-A

2.   Understand the basic process design.

3.   Check the corrosiveness of both the fluids and gases.

4.   Follow design guides established for environmental engineering concrete structures, such as ACI 350.

5.   Watch the effect of vapor drive when specifying coatings or adhered linings. The trickling filter towers are a prime example of potential problems with coatings; this is the prime reason why an unbonded liner was selected.

6.   Always require the geotechnical investigation provide the following:

Soil pH
Soil Resistivity
Sulfate Concentrations in Soils
Chloride Concentrations in Soils

7.   Develop a structural model with an established load path early in the conceptual design phase. Also, look at connections during the early design phase. There have been occasions an entire design has been found to be faulty due to lack of feasible connections. This is especially true of precast or tilt-up construction. In the case of the Bissell Point Treatment Plant, seismic design was an important   design factor when developing connection details.

ACKNOWLEDGEMENTS

It should be noted that the Concrete Council of St. Louis awarded the Trickling Filter Project a Concrete Council Quality Award on January 13, 1992. The author extends his appreciation to Jon Bergenthal, Senior Project Manager, for providing process information contained in this paper.

REFERENCES

1.   BOCA Basic Building Code/1984, Building Officials & Code Administrators International, Inc.

2.   ACI 318-83, "Building Code Requirements for Reinforced Concrete", American Concrete Institute, Detroit, Michigan, USA.

3.   ACI 350R-83, "Concrete Sanitary Engineering Structures", American Concrete Institute, Detroit, Michigan, USA.

4.   ANSI A58.1-1982, American National Standards Institute.

MULTI-LEVEL FORMWORK FOR POST TENSIONED SLABS - IMPROVING THE CONSTRUCTION CYCLE

S L Kajewski

G Brameld

D Thambiratnam

K Hampson

Queensland University of Technology

Australia

ABSTRACT. Multi-level formworking is common practice in the concrete high rise construction industry whereby a limited number of formwork and shoring sets are cycled up the structure. This process allows freshly cast slabs to be supported on a number of lower level slabs rather than the building foundation. The primary purpose of this paper is to examine the effects of multi-level formworking for post-tensioned slabs with an emphasis on improving the safety and speed of the construction cycle. This paper demonstrates that post-tensioning is not necessarily as conservative as commonly believed but that it does enhance the safety during the multi-level formworking operation.

Keywords: Backpropping, Concrete Building Construction, Formwork, Multi-level Formwork, Post-tensioning, Reshoring, Shoring.

Mr Stephen L. Kajewski is the Post Graduate Course Coordinator for the School of Construction Management, Queensland University of Technology. His research interests include multi-level formworking processes; post-tensioned and prestressed concrete; and project management techniques in construction.

Dr Gerald Brameld is an Associate Professor in the School of Civil Engineering and is the Director of the Physical Infrastructure Centre Research Concentration.

Dr David Thambiratnam is an Associate Professor in the School of Civil Engineering.

Dr Keith Hampson is the Director of Research for the School of Construction Management.

Radical Concrete Technology. Edited by R K Dhir and P C Hewlett. Published in 1996 by E & FN Spon, 2–6 Boundary Row, London SE1 8HN, UK. ISBN 0 419 21480 1.

INTRODUCTION

Formworking represents a significant proportion of the costs associated with the construction of a multi-level concrete structure. To minimise these costs, formwork and shoring sets are cycled up the structure as construction progresses, eliminating the need for a new set of formwork and shoring with each new slab. This cycling of formwork and shoring sets leads to freshly poured concrete slabs being supported by other lower level slabs. These slabs may not necessarily have attained their 28 day ultimate strength prior to their being required to support dead loads and construction live loads from slabs above.

To determine the number of floors required to support a freshly cast slab during multi-level formworking, requires the analysis of a considerable range of time and non-time dependent factors. Simple methods do exist, that together with a number of simplifying assumptions, allow designers to quickly assess the load distribution that occurs when multi-level formworking reinforced concrete slabs.

It is a common practice when determining the multi-level formworking requirements for post-tensioned slabs to analyse the slab system as a conventionally reinforced system in the belief that this is a conservative approach due to the enhanced behaviour of a post-tensioned slab. While post-tensioning appears to improve the safety of the multi-level formworking operation at certain stages, to believe it a conservative approach may be unjustified.

FORMING MULTI-LEVEL STRUCTURES

In an ideal formworking situation for a multi-level structure, the formwork shores required to support the formwork and freshly placed concrete would continue to the foundation or ground level of the structure. In this scenario, the slabs are not required to carry their own weight, the weight of slabs above or other applied construction loads. Instead, all such loads are carried from the point of application, via the shores, to the ground.

Whilst maintaining a continuous load path to the foundations is certainly possible, and even common for low-rise structures, it is neither a practical nor a cost effective solution for the forming of most multi-level structures. Formworking for multi-level structures relies on a number of lower level slabs supporting the freshly placed slabs. In essence, this means that freshly placed slabs are often supported on lower level slabs that have yet to attain their full strength. There are three common shoring options: undisturbed supports, backpropping and reshoring.

With a system of undisturbed supports, the shores supporting the formwork remain undisturbed in their original position for the entire period over which the slab is required to be supported. As the slab remains fully supported whilst the shores are undisturbed, the slab is not required to carry any of its own load. The load from the freshly cast slab is fully transferred to the foundation level provided a continuous load path is available. If however load path to the foundation has been broken, the lower levels of slab interconnected by shoring support the new slab load in proportion to their relative stiffness.

When adopting a backpropping (backshoring) procedure, there are two common variations. The more rigorous process involves the installation of a secondary shore, adjacent to the original shores, directly supporting the formwork. When in position, the original shore, framing and formwork up to this secondary shore is removed and the weight of the slab is transferred to the secondary shore. A third shore is then installed snugly under the exposed slab soffit, in the approximately the same position as the original shore. The secondary shore and remaining formwork is then able to be removed.

A less rigorous but more practical process of backpropping involves stripping small areas of the slab, without the use of secondary shores and then backpropping the exposed area of the slab. This alternative allows the slab to deflect slightly but does not result in a significant redistribution of loads. As McAdam [1] explains:

> As the backprops have no load in them when first installed, there is a small amount of prop shortening and compatible slab deflection at the removal of the formwork supports. The result is a relaxation process with a small transfer of load up to upper and younger slabs.

The process of reshoring is similar to backpropping in that some portion of the slab is stripped prior to the installation of the shores (reshores). With reshoring however, larger areas of slab, often entire bays, are stripped prior to the installation of the shores. With the entire bay of the slab soffit exposed and unsupported, the slab is able to fully deflect. This forces the slab to support its self weight and some portion of the slab and construction loads over as a new load sharing equilibrium is reached amongst the interconnected slabs. Reshoring results in a load redistribution up through the structure of interconnected slabs, requiring the younger slabs to accept loads of greater magnitude earlier than that required with backpropping. Provided the shoring remains in contact with the foundations, the shoring loads are cumulative with a maximum shore load occurring at the lowest level (Nielsen [2], Grundy and Kabaila [3]). With reshoring, the only loads in the shores are due to the weight of the freshly cast slab and any applied construction loads. As Hurd [4] states:

> Such reshores are provided to transfer additional construction loads to other slabs or members and/or to impede deflection due to creep which might otherwise occur. Reshoring is done to facilitate maximum reuse of the formwork, making use of the strength of completed construction below as well as the partial development of strength in the member being reshored.

DETERMINING THE SHORING PROCEDURE

Designers, required to specify the number of floors to be used in the support of a fresh slab, are faced with a dilemma. As designers, they are primarily required to design for the safety and stability of the structure both during and after construction - the safety and stability of a structure is enhanced by increasing the number of interconnected floors.   Secondly, building contractors require a formwork and shoring procedure that is economical in both time and cost considerations - time and cost can both be reduced by increasing reuse of the formwork and reducing the number of interconnected slabs required. Clearly, the greater the number of floors required to be used in the support of a new slab, the greater the formworking cost (although it is not a linear relationship).

Many of the standards and codes of practice concerned with concrete construction and the formwork and shoring associated with such construction are vague and offer little advice to the design consultant with regard to shoring procedures for multi-level structures. It can be argued that these standards and codes of practice are deliberately vague due to the vast number of variables associated with multi-level concrete construction. ACI347R-88 cl. 2.5.3 [5] for example identifies the following variables:

- Structural design load of the slab or member,
- Dead load weight of concrete and formwork,
- Construction live loads, such as placing crews and equipment or stored materials,
- Design strength of concrete specified,
- Cycle time between placement of successive floors,
- Strength of concrete at time it is required to support shoring loads from above,
- Distribution of loads between floors, shores, and reshores or backshores at the time of placing concrete, stripping formwork, and removal of reshoring or backshoring,
- Span of slab or structural member between permanent supports,
- Type of formwork systems; and
- Minimum age where appropriate.

In addition to these construction-related variables, McAdam and Behan [6] examine a number of other construction and material property-related variables:

- The formwork system,
- Variations in the number of support sets and cycle times,
- Timing of the stripping/reshoring within the cycle,
- Formwork stiffness,
- The presence of stacked materials,
- The influence of the concrete frame including column and slab variations,
- Variations in the stiffness of the foundations,
- Slab shrinkage,
- Column shrinkage,

- Concrete creep, and
- Environmental effects including temperature and humidity.

With such a wide range of variables, many design consultants are unwilling (and sometimes unable) to properly analyse the structure during the multi-level shoring process. As such, these design consultants often resort to past experience or local practice. In an examination of the design practices, Hurd [4] identifies what is probably the most common design practice while at the same time, identifying the inherent dangers of such a practice:

> *Many multi-storey buildings have been built following a rule of thumb or local custom of using a certain number of stories of shores and reshores, often progressing at one storey per week. However, construction processes have become more mechanised and often significantly faster, at the same time that design practices are tending to produce buildings without the margin of extra strength once available. Under these conditions it is prudent to make a more rational judgement of how many stories of shores and reshores are needed for a given speed of construction.*

Despite the number of variables, it is possible to approximate the load distribution that occurs during multi-level shoring using a simple procedure first developed by Grundy and Kabaila [3]. While this method of analysis does not necessarily include the effects of many of the variables, and in fact makes a number of simplifying assumptions, numerous researchers (Blakey, Feld, Noble, Agarwal [7-10], Neilsen [2]) have indicated that reasonable approximations to actual shoring loads are obtained.

MULTI-LEVEL SHORING FOR POST-TENSIONED SLABS

Much of the literature to date concerning multi-level shoring has exclusively examined conventionally reinforced concrete slabs. This is certainly reasonable as most concrete structures are still constructed in this manner. Post-tensioning however is becoming more popular with designers as clients demand greater clear spans, thinner sections and improved short and long term serviceability. A number of immediate questions are raised with post-tensioned slabs that as yet remain unanswered:

- As the load on the majority of shores is actually lessened (if not eliminated) by the post-tensioning, can the number of fully shored floors be reduced?
- If the load balanced by the stressing of a lower level slab exceeds the dead load and construction loads applied, will this result in the lower level slab loading a higher level slab in reverse bending?
- Is it conservative to analyse the shoring procedure with the same assumptions as a reinforced slab? Are the factors of safety improved or worsened?
- Does the level of initial and final stressing play a significant role?

Determining Shoring Loads for Post-Tensioned Slabs

It is not the intention of this paper to provide a detailed, or even proven, method of analysis when examining the load distribution that occurs when post-tensioning multi-level formworked slabs. Certainly the research in this area is still in its infancy. The intention of this paper is to raise awareness of the sensitivities of multi-level formworking, particularly with post-tensioned slabs. It is also the intention of this paper to demonstrate some possibilities of multi-level post-tensioned work.

To examine the action of multi-level formwork load distribution with post-tensioned concrete slabs, the procedure developed by Grundy and Kabaila [3] is used. It should be noted at the outset that the method developed by Grundy and Kabaila [3] has a number of simplifying assumptions that limit the accuracy of the results obtained. As it is the intention of this paper to demonstrate some possibilities with multi-level post-tensioned work, the method of Grundy and Kabaila [3] will be sufficient to allow comparisons to be drawn between conventionally reinforced slabs and post-tensioned slabs.

Hurd and Courtois [11] and Hurd [4] provide a detailed description of the procedure together with a number of worked examples. Simply, the procedure expresses the loads in the interconnected system in multiples of typical slab loads and assumes that the loads are distributed between the interconnected slabs in direct proportion to their relative stiffness.

Shoring Scenarios

To demonstrate some possibilities for shoring procedures with post-tensioned flat plate slabs, four scenarios are examined:
- Conventionally reinforced - undisturbed support system (backpropping similar)
- Post-tensioned - undisturbed support system (backpropping similar)
- Conventionally reinforced - reshored system
- Post-tensioned - reshored system

To model the effects of the post-tensioning, the full slab dead load is assumed to be balanced. That is, at the level of full stress the slab carries its full self weight, thereby not contributing to the supporting shore or reshore loads. The validity of this assumption is yet to be proven but preliminary research by Kajewski et. al. [12] has indicated that this assumption may not be without foundation.

The concrete elasticity and compressive strength assumed in the scenarios are indicated in Figure 1.

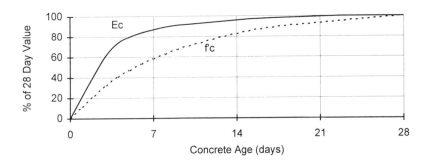

Figure 1 Assumed Concrete Properties

Conventionally reinforced - undisturbed support system (backpropping similar)

Reproduced from the classical example of Grundy and Kabaila [3], Figure 2 indicates a maximum slab ratio of 2.37 occurring on the level 3 slab at day 42. This slab has an age of 21 days. The level 3 slab was the last slab poured prior to the removal of the foundation level shores. Regardless of the number of shored levels, the last slab poured prior to the removal of the foundation level shores will be subjected to the largest slab load ratio.

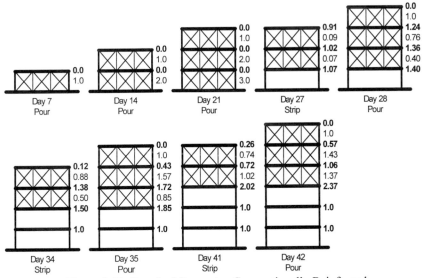

Figure 2 Undisturbed Supports - Conventionally Reinforced

Post-tensioned - undisturbed support system (backpropping similar)

Subject to the load balancing assumption outlined previously, when stressing a slab, the shores supporting the slab are assumed to be relieved of the slab load. This is indicated at day 12 in Figure 3 (Kajewski et. al. [12]). As indicated, the maximum slab load ratio in the post-tensioned system is reduced to 1.35. The age of the slab is also 21 days. This maximum ratio no longer occurs at the last slab poured prior to the removal of the foundation level shores but now occurs at the lowest slab of the interconnected system. The slab load ratios indicated at day 28 repeat up the structure at each pour date with the lowest level slab in the interconnected system being subjected to a load ratio of 1.35.

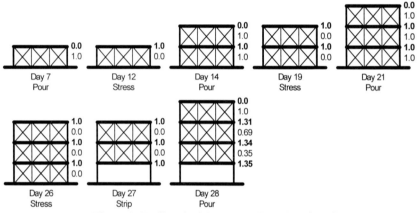

Figure 3 Undisturbed Supports - Post-Tensioned

Conventionally reinforced - reshored system

The perceived benefits of reshoring a conventionally reinforced slab is highlighted in Figure 4. Reshoring, by allowing slabs to carry loads earlier, reduces the maximum load ratio from 2.37 to 1.85. This reduction is not as significant as it first appears as the age of the slab is 14 days rather than 21 days as was the case in the undisturbed system.

Post-tensioned flat plate - reshored system

The reshored stressed system in Figure 5 indicates a maximum slab load ratio of 1.35. The slab load ratios produced in this system are identical to the undisturbed system. This indicates that the undisturbed system utilises an extra level of full shoring and formwork that is not warranted, at least when considering load distribution.

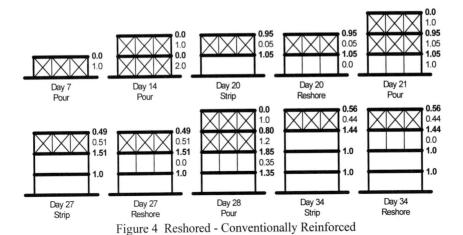

Figure 4 Reshored - Conventionally Reinforced

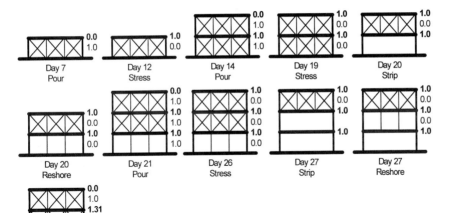

Figure 5 Reshored - Post-Tensioned

Comparison of Shoring Scenarios

Comparison of maximum slab load ratios is insufficient as a direct comparison. It is necessary to account for the relative age and hence strength of the slab in question. Hurd and Courtois [11] detail a method of determining factors of safety based on the assumption that the strength of a slab at a given age is proportional to the percentage of the 28 day (ultimate) strength.

Determining the factor of safety

Assuming a slab thickness of 150 mm, the slab dead load (G) is 3.53 kPa. The design live load (Q) assumed is 3 kPa which is equivalent to 0.85G. Concrete structures in Australia are designed according to AS3600 [13] for an ultimate load of:

$$W_u = 1.25G + 1.5Q$$

For the loads assumed, the 28 day ultimate capacity of a slab is therefore:
$$W_u = 1.25G + 1.5(0.85G) = 2.53G$$

AS3610 [14] indicates that during construction, the live loads to be assumed for the formwork system are 1 kPa on the uppermost slab and 0.25 kPa on lower level supporting slabs. As such, with 3 levels of supporting slabs and 1 freshly cast slab, the average live load during construction is 0.44 kPa or 0.12G. With 2 supporting levels and 1 freshly cast slab, the average live load during construction is 0.5 kPa or 0.14G. The actual load on a slab is the slab load ratio multiplied by the dead load plus average construction live load.

As an example calculation, consider the slab load ratio of 1.07 for the 20 day old slab indicated at day 27 in Figure 2. From Figure 1, at 20 days, the concrete is assumed to have 92% of the ultimate strength. Thus the factor of safety for this particular slab would be:

$$F.O.S = \frac{0.92 \times 2.53G}{(1G + 0.14G) \times 1.07} = 1.91$$

Factor of safety for scenarios

Table 1 indicates the lowest factor of safety for each of the four scenarios.

Table 1 Factors of Safety

SHORING SYSTEM	SLAB LEVEL	DAY	SLAB AGE (days)	SLAB LOAD RATIO	FACTOR OF SAFETY
Undisturbed - Reinforced	3	42	21	2.37	0.89
Undisturbed - Post-tensioned	2	28	7	1.31	1.00
Reshored - Reinforced	3	28	14	1.85	1.01
Reinforced - Post-tensioned	3	28	7	1.31	1.00

When comparing the undisturbed support systems, it is evident that post-tensioning, by relieving the shores of load, improves the factor of safety. In this case, from an overload situation of 0.89 to a factor of safety of 1.0. There is however negligible difference between the undisturbed post-tensioned systems and either of the reshored systems.

While it is evident that reshoring produces a better factor of safety than undisturbed supports, there is no evidence that post-tensioning is significantly safer than conventionally reinforced systems. The belief that analysing a post-tensioned slab system as a reinforced system is a conservative approach appears to be a misconception. There also appears to be no indication that the cycle can be improved significantly for multi-level post-tensioned slabs.

Examining the factors of safety throughout the cycle however indicates that the safety during the stripping and reshoring operation is enhanced for the post-tensioned systems. Examining level 2 at day 27 (strip) of the reshored systems for example indicates an improved factor of safety from 1.14 for the reinforced system to 2.04 for the post-tensioned system as indicated in Table 2.

Table 2 Factors of Safety During Stripping and Reshoring

SHORING SYSTEM	SLAB LEVEL	DAY	SLAB AGE (days)	SLAB LOAD RATIO	FACTOR OF SAFETY
Reshored - Reinforced	2	27	13	1.51	1.14
Reshored - Post-tensioned	2	27	13	1.00	2.04

REDUCING FORMWORK MATERIALS

Examining the shore loads for the reshored post-tensioned scenario indicates that the formwork shores are fully relieved of their load when the slab is tensioned and are not reloaded until the next slab is poured.

The scenario presented in Figure 6 details a situation in which the formwork is stripped and the slab reshored immediately following the tensioning of the slab. As the original formwork shores were not carrying load, there is no load redistribution as a result of this procedure.

This procedure allows for the slab formwork to be stripped at an earlier date than would normally be the case. In this particular scenario, the formwork from each slab is able to be removed 7 days earlier than the scenario presented in Figure 4 and Figure 5. This reduces the formwork material quantities from 2 sets of operational formwork with 1 set of reshores to 1 set of operational formwork with 2 sets of reshores. This is obviously subject to the development of suitable factors of safety.

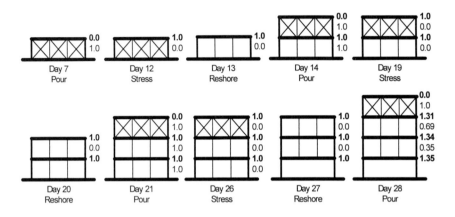

Figure 6 Early Reshoring Scenario

CONCLUSIONS

It was not the intention of this paper to present a fully detailed examination of the load distribution that occurs when multi-level formworking post-tensioned concrete slabs. The simple analysis method used however does serve to highlight the potential limitations of post-tensioned slabs.

It is evident from the examination of the four scenarios that while the load distribution is different for the conventionally reinforced and post-tensioned systems, the safety of the overall multi-level formworking process is not necessarily improved by the post-tensioning. As such, analysing a post-tensioned slab system as a conventionally reinforced system when considering multi-level formwork load distribution does not appear to be as conservative as one would first believe. Under certain circumstances (such as under-stressing) this may result in an overload situation. It appears however that post-tensioned slabs improve the safety of the formwork stripping and reshoring operation.

When using undisturbed supports on a multi-level project, it appears that the construction cycle could be improved marginally as there is an improvement in the factor of safety from the conventionally reinforced system to the post-tensioned system. This indicates that, to achieve a similar factor of safety for the post-tensioned system as would normally be achieved for a conventionally reinforced system, the construction cycle could be shortened, resulting in the slabs being loaded at an earlier age. This is not the case with a reshored system as the factor of safety was the same for both the conventionally reinforced and post-tensioned systems. Hence, an improvement in cycle time for a reshored system appears unlikely.

The overriding assumption in this paper is that the shores are completely relieved of the equivalent of 1 slab load when the slab is stressed to balance full dead load. This is currently the subject of research [12] but preliminary results are proving, at least to some extent, encouraging.

Further research is needed to fully assess the load distribution that occurs with multi-level formworking post-tensioned concrete slabs. Issues as the effect of the time dependent actions of creep, shrinkage, concrete elasticity and concrete strength need to be examined.

REFERENCES

1. McADAM, P., Modelling the Multi-Storey Formwork Process, Doctoral Dissertation, University of Queensland, Australia, 1989, pp 200.
2. NEILSEN, K., Loads on Reinforced Concrete Slabs and Their Deflections During Construction, Bulletin No. 15 Swedish Cement and Concrete Research Institute, Royal Institute of Technology, Stockholm.
3. GRUNDY, P. and KABAILA, A., Construction Loads on Slabs with Shored Formwork in Multistorey Buildings, ACI Journal, 1963, December, pp 1729-1738.
4. HURD, M., Formwork for Concrete, American Concrete Institute, Detroit, 1989.
5. AMERICAN CONCRETE INSTITUTE, ACI 347R-88 Guide to Formwork for Concrete, American Concrete Institute, 1988, pp 39.
6. McADAM, P. and BEHAN, J., Multi-Storey Formwork Loading, Concrete Institute of Australia, Brisbane, 1990, pp 18.
7. BLAKEY, F. and BERESFORD, F., Stripping of Formwork for Concrete Buildings in Relation to Structural Design, Transactions of the Institution of Engineers Australia, 1965, Vol CE7, No. 2, October, pp 92-96.
8. FELD, J., Reshoring of Multi-Storey Concrete Buildings, Concrete Construction, 1974, May, pp 243-248.
9. NOBLE, J., Stop Guessing at Reshore Loads - Measure Them, Concrete Construction 1975, July, pp 277-280.
10. AGARWAL, R. and GARDNER, N., Form and Shore Requirements for Multi-Storey Flat Slab Type Buildings, ACI Journal, 1974, Proceedings Vol 71, No. 11, November, pp 559-569.
11. HURD, M. and COURTOIS, P., Method of Analysis for Shoring and Reshoring in Multistorey Buildings, Forming Economical Concrete Buildings - Proceedings of the Second International Conference, American Concrete Institute, 1984, SP-90, pp 91-108.

12. KAJEWSKI, S., BRAMELD, G., HAMPSON, K. and THAMBIRATNAM, D., Multi-Level Formwork for Post-Tensioned Slabs, Proceedings - Concrete 95 Toward Better Concrete Structures, Brisbane, 1995.

13. STANDARDS ASSOCIATION OF AUSTRALIA, AS3600-1988 Concrete Structures, Australia, pp108.

14. STANDARDS ASSOCIATION OF AUSTRALIA, AS3610-1990 Formwork for Concrete, Australia, pp 42.

EFFECT OF WATER TEMPERATURE ON UNDERWATER STRENGTH PROPERTIES OF POLYMETHYL METHACRYLATE MORTARS

M A R Bhutta

Y Ohama

K Demura

Nihon University

N Maeda

Maeta Concrete Industry Ltd

Japan

ABSTRACT. Polymethyl methacrylate (PMMA) mortars are prepared with the proper binder formulations and mix proportions to have working lives of 30±5 minutes underwater and in air at temperatures of 0, 10, 20 and 30°C, and tested for flexural and compressive strengths and adhesion in flexure. From the test results, the effects of water temperatures in underwater placing or bonding and curing processes on the strength and adhesion of PMMA mortars are discussed. It is concluded from the test results that the use of PMMA mortars underwater in the water temperature range of 0 to 30°C is recommended for practical underwater construction work.

Keywords: Polymethyl Methacrylate Mortar (PMMA), Flexural Strength, Compressive Strength, Adhesion in Flexure, Water Temperature, Underwater Placing or Bonding.

Dr M.A.R.Bhutta received Ph.D. from Nihon University in Koriyama, Japan, and is a Research Engineer at Maeta Concrete Industry Ltd., Sakata, Japan.

Dr Y.Ohama is a Professor of Architecture at the College of Engineering, Nihon University, Koriyama, Japan. He has been involved in the research and development of concrete-polymer composites for more than thirty-five years.

Dr K.Demura is an Associate Professor of Architecture at the College of Engineering, Nihon University, Koriyama, Japan. His main interest has been in the research and development of concrete-polymer composites.

Mr N. Maeda is the President of Maeta Concrete Industry Ltd., Sakata, Japan. His research interests include precast concrete products, MDF cement and polymer concrete.

Radical Concrete Technology. Edited by R K Dhir and P C Hewlett. Published in 1996 by E & FN Spon, 2–6 Boundary Row, London SE1 8HN, UK. ISBN 0 419 21480 1.

INTRODUCTION

Polymethyl methacrylate (PMMA) mortar and concrete are normally made by mixing methyl methacrylate monomer-based binders with dried aggregates and polymerizing methyl methacrylate under proper curing conditions, and their properties are temperature-dependent because of the thermoplasticity of polymethyl methacrylate in the hardened mortar and concrete[1-3]. This fact also is a very important point to be considered in the underwater placing or bonding of PMMA mortar and concrete. Therefore, some investigations must be conducted to examine the effects of water temperatures in underwater placing or bonding and curing processes on the properties of PMMA mortar and concrete. The data obtained from such investigations will be effective to decide the allowable temperature range in the underwater applications of PMMA mortar and concrete.

PMMA mortars are prepared with the proper binder formulations and mix proportions to have working lives of 30±5 minutes underwater and in air at temperatures of 0, 10, 20 and 30°C, and tested for flexural and compressive strengths and adhesion in flexure. From the test results, the effects of water temperatures in underwater placing or bonding and curing processes on the strength and adhesion of PMMA mortars are discussed.

MATERIALS

Materials for Binder Systems

Binder systems were based on methyl methacrylate (MMA) monomer, together with trimethylolpropane trimethacrylate (TMPTMA) as a crosslinking agent, unsaturated polyester resin (UP) and polyisobutyl methacrylate (PIBMA) as shrinkage-reducing agents, benzoyl peroxide (BPO) as an initiator, and N, N-dimethyl-p-toluidine (DMT) as a promoter.

Filler and Fine Aggregates

Commercially available ground calcium carbonate (size; 2.5 µm or finer) was used as a filler, and silica sands (sizes; 0.04-0.30 mm and 0.21-1.19 mm) were done as fine aggregates. The water contents of the filler and silica sands were controlled to be less than 0.1% by heat drying.

TESTING PROCEDURES

Control of Working Life of PMMA Mortars
The working lives of PMMA mortars with the binder formulations shown in Table 1 and the mix proportions given in Table 2 were determined underwater and in air

Table 1 Formulations of binders for PMMA mortars

TEST TEMPERA-TURE, ℃	FORMULATIONS BY MASS					
	%				phr*	
	MMA	TMPTMA	UP	PIBMA	DMT	BPO
0	67.40	1.80	23.10	7.70	1.50	1.50
10	67.40	1.80	23.10	7.70	1.00	2.00
20	67.40	1.80	23.10	7.70	0.50	2.00
30	67.40	1.80	23.10	7.70	0.25	0.25

Note, * Parts per hundred parts of resin (MMA+TMPTMA+UP+PIBMA)

Table 2 Mix proportions of PMMA mortars

MIX PROPORTIONS BY MASS, %				Binder-filler ratio, B/F
Binder	Filler	Silica sand		
		No.4	No.7	
15.00	15.00	35.00	35.00	1.00

at temperatures of 0, 10, 20 and 30°C according to the Finger-Touching Methods specified in JIS A 1186 (Measuring Methods for Working Life of Polyester Resin Concrete), and controlled to be 30±5 minutes.

Preparation of PMMA Mortar Substrates

According to JIS A 1181 (Method of Making Polyester Resin Concrete Specimens), PMMA mortars were mixed with the binder formulations and mix proportions as shown in Tables 1 and 2 respectively. PMMA mortar substrates 40x40x80 mm for adhesion test in flexure were placed underwater and in air at temperatures of 0, 10, 20 and 30°C, and then cured underwater and in air at the respective temperatures for 1 day. After curing, the wet and dry bonding surfaces of PMMA mortar substrates were treated with the AA-150 abrasive papers specified in JIS R 6252 (Abrasive Papers). Then the wet and dry surfaces of PMMA mortar substrates were washed with pressurized water and blown by pressurized air respectively to remove all the dust particles.

Preparation of Specimens

In the preparation of specimens for flexural and compressive strength tests, PMMA mortars were placed into molds 40x40x160 mm underwater and in air at temperatures of 0, 10, 20 and 30°C, cured in the molds kept underwater and in air at the respective temperatures for 1 hour, and then demolded. After demolding, the specimens were cured underwater and in air at temperatures of 0, 10, 20 and 30°C for 0, 2, 5, 11, 23, 167 and 671 hours for strength development. In the preparation of specimens for adhesion test in flexure, PMMA mortars were bonded to PMMA mortar substrates 40x40x80 mm in the molds 40x40x160 mm underwater and in air at temperatures of 0, 10, 20 and 30°C, cured in the molds kept underwater and in air at the respective temperatures for 24 hours, and then demolded.

Strength and Adhesion Tests

According to JIS A 1172 (Method of Test for Strength of Polymer-Modified Mortar), the flexural strength test and adhesion test in flexure of cured specimens were conducted by use of the Amsler-type universal testing machine at the same temperatures as the curing temperatures. After flexural strength test, the broken portions of the specimens were tested for compressive strength by using the same testing machine according to JIS A 1172. After adhesion test, the failed crosssections of the specimens were observed for failure modes which were classified into the following three types: A: Adhesive failure (failure at the interface), M: Cohesive failure in the bonded PMMA mortar, and S: Cohesive failure in PMMA mortar substrates. The total area of the bonded surfaces was supposed to be 10, and the respective approximate rates of A, M and S areas on the failed crosssections were expressed as suffixes for A, M and S. The relative flexural strength, compressive strength or relative adhesion in flexure of the specimens was calculated by the following equation:

$$\text{Relative strength or adhesion } (\%) = (F_w / F_a) \times 100$$

where F_w and F_a are the strengths (MPa) or adhesions (MPa) of the specimens, placed or bonded and cured underwater and in air, respectively.

The ratio of adhesion in flexure to flexure strength of the specimens, bonded or placed and cured underwater and in air, was calculated by the following equation:

$$\text{Ratio of adhesion in flexure to flexural strength } (\%) = (\sigma_a / \sigma_f) \times 100$$

where σ_a and σ_f are the adhesion in flexure (MPa) and flexural strength (MPa) of the specimens, respectively.

TEST RESULTS AND DISCUSSION

Effects of Water Temperatures on Strengths

Figures 1 and 2 show the curing period vs. flexural and compressive strengths, and relative strengths of PMMA mortars placed and cured underwater and in air at different test temperatures. In general, the flexural and compressive strengths of PMMA mortars placed and cured underwater and in air increase sharply till a curing period of

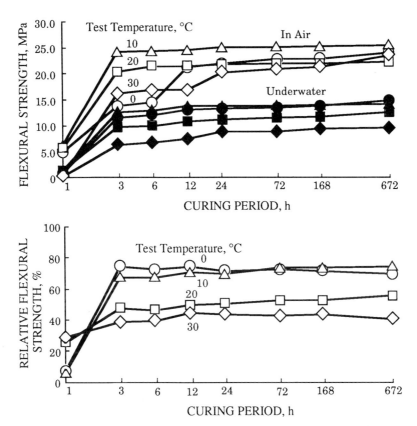

Figure 1 Curing period vs. flexural and relative flexural strengths of PMMA mortars placed underwater and in air at different test temperatures

3 hours, and become nearly constant at a curing period of 24 hours. Regardless of the test temperature, PMMA mortars placed and cured underwater and in air develop more than about 80% of 672-hour (28-day) flexural and compressive strengths at a curing period of 3 hours because of the quick setting of their binders. The relative flexural and compressive strengths of PMMA mortars placed and cured at different test temperatures range from about 40 to 75% at a curing period of 3 hours.

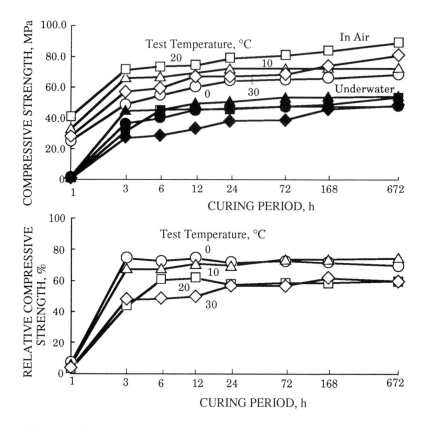

Figure 2   Curing period vs. compressive and relative compressive strengths
of PMMA mortars placed underwater and in air at different test
temperatures

Figure 3 represents the effects of test temperatures on the flexural and compressive strength developments of PMMA mortars placed and cured underwater and in air. In general, the flexural and compressive strengths of PMMA mortars placed and cured underwater reach maximums at a test temperature of 10°C, but the flexural and compressive strengths of PMMA mortars placed and cured in air do maximums at test temperatures of 10 and 20°C, respectively. Such strength development tendency due to the different test temperatures is attributed to the temperature dependence of polymerization reaction for the hardening of the binders and the themoplasticity of the polymeric binders after the hardening of PMMA mortars.

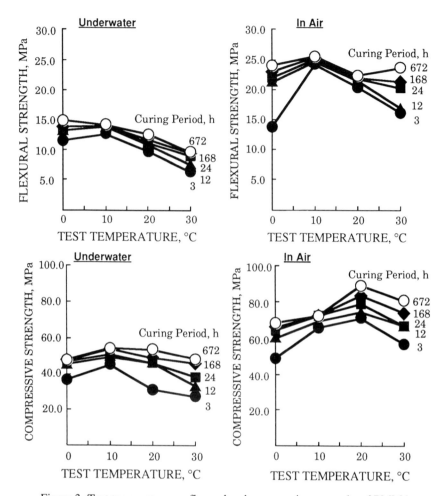

Figure 3 Test temperature vs. flexural and compressive strengths of PMMA
mortars placed and cured underwater and in air

Effects of Water Temperatures on Adhesion

Figure 4 exhibits the effects of test temperatures on the 24-hour adhesion in flexure of
PMMA mortars to PMMA mortar substrates, bonded and cured underwater and in air.
The 24-hour adhesion in flexure of PMMA mortars to PMMA mortar substrates,
bonded and cured underwater and in air, decreases almost linearly with a raise in the
test temperature. The failure modes in the adhesion test in flexure of PMMA mortars to
PMMA mortar substrates, bonded and cured underwater, are almost cohesive failures
in the bonded PMMA mortars and PMMA mortar substrates, which prove good
adhesion between the bonded PMMA mortars and PMMA mortar substrates
underwater.

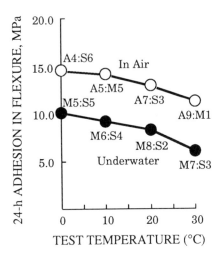

Figure 4   Test temperature vs. 24-hour adhesion in flexure of PMMA
mortars to PMMA mortar substrates, bonded and cured
underwater and in air.

Figure 5 illustrates the test temperature vs. 24-hour relative adhesion in flexure and
24-hour ratio of adhesion in flexure to flexural strength of PMMA mortars. Like the
adhesion in flexure as seen in Figure 3, the 24-hour relative adhesion in flexure of
PMMA mortars decreases almost linearly with a raise in the test temperature, and
ranges from about 60 to 70%. The 24-hour ratio of adhesion in flexure to flexural
strength of PMMA mortars bonded or placed and cured underwater and in air tends to
decrease with raising test temperature. The 24-hour ratio of adhesion in flexure to
flexural strength of PMMA mortars underwater is larger than that in air irrespective of
the test temperature.

CONCLUSIONS

The conclusions obtained from the above-mentioned test results are summarized as
follows:

1.   Generally, the flexural and compressive strengths of PMMA mortars placed and
cured underwater and in air increase with additional curing period, and become
nearly constant at a curing period of 24 hours. The flexural and compressive
strengths of PMMA mortars placed and cured underwater tend to reach
maximums at a test temperature of 10°C. By contrast, the flexural and
compressive strengths of PMMA mortars placed and cured in air are inclined to
reach maximums at test temperatures of 10 and 20°C, respectively. Such
strength development tendency due to the different test temperatures   is

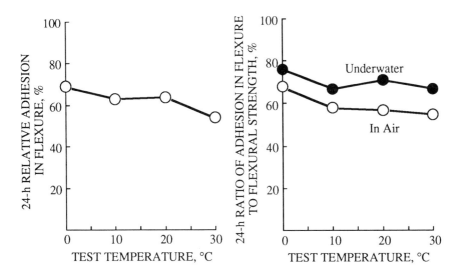

Figure 5   Test temperature vs. 24-hour relative adhesion in flexure and 24-hour
ratio of adhesion in flexure to flexural strength of PMMA mortars to
PMMA mortar substrates, placed or bonded and cured underwater and
in air

attributed to the temperature dependence of polymerization reaction for the
hardening of the binders and the themoplasticity of the polymeric binders after
the hardening of PMMA mortars. The flexural and compressive strengths of
PMMA mortars placed and cured underwater range from about 40 to 75% of
those in air.

2.   The 24-hour adhesion in flexure of PMMA mortars to PMMA mortar substrates,
bonded and cured underwater and in air, decreases almost linearly with raising
test temperature. The 24-hour adhesion in flexure of PMMA mortars to PMMA
mortar substrates, bonded and cured underwater, ranges from about 55 to 70%
of that in air. The failure modes in the adhesion test in flexure of PMMA mortars
to PMMA mortar substrates, bonded and cured underwater, are almost cohesive
failures. The ratio of adhesion in flexure to flexural strength of PMMA mortars
bonded or placed and cured underwater and in air tends to decrease with raising
test temperature. The 24-hour adhesion in flexure of PMMA mortars to PMMA
mortar substrates, bonded and cured underwater, ranges from about 70 to 75%
of the flexural strength of PMMA mortars, and is larger than that in air.

3.   It is concluded from the test results that the use of PMMA mortars underwater in
the water temperature range of 0 to 30°C is recommended for practical
underwater construction work.

REFERENCES

1. YAMASAKI, T AND MIYAKAWA, K. An estimation of the pot-life due to mix proportions of additives in REC. Transactions of the Japan Concrete Institute, Vol. 6, Mar. 1984, pp. 111-118.

2. OHAMA, Y, DEMURA, K, KOBAYASHI, T AND NAWATA, K. Temperature dependency of flexural behavior and water resistance of polymethyl methacrylate concretes. The Proceedings of the 30th Japan Congress on Materials Research, The Society of Materials Science, Japan, Kyoto, Mar. 1987, pp.157-161.

3. OHAMA, Y, DEMURA, K, KOBAYASHI, T AND DHOLAKIA, C G. Properties of polymer mortars using reclaimed methyl methacrylate. Materials Engineering, Vol.1, No.1, June 1989, pp. 97-104.

DURABILITY OF SPRAYED CONCRETE - A LITERATURE STUDY

E Nordström

Vattenfall Utveckling AB

Sweden

ABSTRACT. This study is aiming to compile the knowledge about durability of sprayed concrete and to elucidate the areas were further research is needed. The study is treating materials, additions, additives and their effect on durability. Further are factors like bond, cracking, freeze-thaw action, corrosion and leaching and their effect also included. In areas where little or no information specific for sprayed concrete is available, parallels with conventional concrete are drawn. The processes for durability should be fairly similar. Except from performance are the differences mainly the w/c-ratio (affects permeability) and the usage of accelerators for concrete sprayed with the wet-process.

The conclusion in this paper is that sprayed concrete with correct composition and good performance is a durable material. The durability in combination with advantages in performance makes sprayed concrete a very useful material.

Further research is needed about corrosion of steelfiber reinforcement in sprayed concrete, freeze-thaw action in combination with chlorides and methods to determine permeability in situ.

Keywords: Sprayed concrete, Durability, Cement, Aggregate, Reinforcement, Additions, Additives, Performance, Bond, Cracking, Freeze-thaw action, Corrosion, Leaching, Chemical attack

Mr Erik Nordström received his MSc in Civil Engineering from Luleå University of Technology in Sweden 1993. He is now working as a Research Engineer at the division of Concrete Technology, Vattenfall Utveckling AB, Sweden. His main research area is sprayed concrete durability with emphasis on corrosion of steelfiber reinforcement and freeze-thaw action. Other research areas are early thermal cracking in concrete and lightweight concrete with addition of flyash.

Radical Concrete Technology. Edited by R K Dhir and P C Hewlett. Published in 1996 by E & FN Spon, 2–6 Boundary Row, London SE1 8HN, UK. ISBN 0 419 21480 1.

MATERIALS

Cement

The chemical composition of the cement is the main factor affecting durability of sprayed concrete. Low C_3A-content increases the sulphate resistance and low content of alkali (K^+ & Na^+) reduces the risk for alkali aggregate reactions. The resistance against damage due to freeze-thaw action is also increased since the airpore structure is more fine and dense (Fagerlund [1]). Some authors claim that low content of C_3A would give a reduced chemical bonding of chlorides and therefore increased rate of corrosion. This is though refuted by Byfors & Tuutti [2]. Increased cement content reduces the rebound which gives an increased durability, but at the same time the risk for cracking due to shrinkage increases when more water is normally needed.

Aggregate

The aggregate should have sufficient strength, be resistant to freeze-thaw action and not be reactive to alkali. The gradation and the maximum aggregate size also affects the durability by affecting compaction, permeability, mechanical properties and rebound. Alkali-aggregate reactions arise from aggregate dissolving in the strongly alkaline environment the cement paste gives. Figure 1 shows the principle for the reaction, and if any of the components is missing the reaction will not take place.

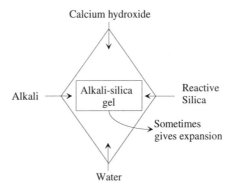

Figure 1.   Basic sketch for alkali aggregate reactions (Lagerblad & Trägårdh [3])

Reinforcement

Fibres are gradually replacing the conventional net reinforcement due to economy and performance criteria's. The fibres give effects like reduced plastic shrinkage, increased toughness at failure and limited crack widths. Glassfibres or other types of plastic fibres mainly affects the properties of the fresh sprayed concrete whereas steel fibres more affects the properties of the hardened concrete. According to Opsahl [4] the risk for debonding caused by shrinkage is reduced with addition of fibres. Other authors claims this is not the case (Holmgren [5]). Concerning durability performance is an advantage with fibres compared to conventional net reinforcement since it is easier to make a homogeneous concrete with fibres.

Additions

Commonly used additions are silica fume, fly ash and ground granulated blast furnace slag. They are used to achieve good workability, reduced permeability, higher resistance to leaching and chemical attack and reduced risk for alkali aggregate reactions. As the additions react with the free content of calciumhydroxide this could reduce the possibility of "self-repairing" small cracks (Fagerlund [6]). Material properties are shown in table 1.

Table 1.   Material data for Portland cement, silica and fly ash (Burge [7])

	Density of particles [kg/m³]	Bulk dens. [kg/m³]	Spec. surface [m²/kg]	SiO₂ content [%]
Portlandcem.	3.12-3.15	1.4	250-500	17-25
Silica	2.16	0.20-0.22	18 000-22 000	88-98
Fly ash	2.35	1.00	300-500	40-55

Silica

Silica is added either with the dry materials or as a slurry. Addition of silica in combination with a superplasticizer gives a cohesive mix which will give reduced rebound and better bonding to the sprayed surface (Durand, Mirza & Nguyen [8]). This also enables increased layer thickness and reduced usage of accelerators (Wolsiefer & Morgan [9], Fidjestøl [10]).

The hardened concrete will receive reduced permeability and therefore increase durability against freeze-thaw action (Burge [11], Glassgold [12], Mailvaganam & Samson [13], Morgan & al. [14], Wolsiefer & Morgan [9]). Addition of silica also gives increased durability against breaking down due to chemical attack and higher electrical resistivity which reduces the risk for corrosion (Wolsiefer & Morgan [9]).

If damages occur due to freeze-thaw action, in a concrete with silica fume but without entrained air, the breaking down is very fast compared to ordinary concrete (figure 2.) (Fagerlund [1]). Addition of silica also gives increased plastic shrinkage at early age and therefore higher demands on proper curing. Pettersson [15] points out the risk for a reaction between undispersed silica-gel and the alkaline cement paste. At normal dosage of well-dispersed silica is the risk for alkali aggregate reactions instead reduced (Lagerblad & Trägårdh [3]).

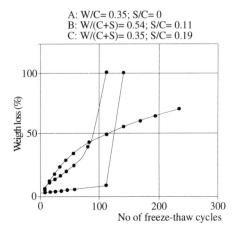

A: W/C= 0.35; S/C= 0
B: W/(C+S)= 0.54; S/C= 0.11
C: W/(C+S)= 0.35; S/C= 0.19

Figure 2.   Freeze-thaw test with salt solution of concrete with and without silica
fume and with no air entrainment. W= water, C= Portland cement, S=
silica fume (Fagerlund [1]).

Fly ash

In the fresh sprayed concrete fly ash has the effect of increased cohesion and reduced
risk for separation. The early strength growth is slower and this behaviour is
accentuated at low temperatures. Fly ash gives increased durability against sulphate
attack, reduced permeability (Mailvaganam & Samson [13]), increased freeze-thaw
resistance and reduced risk for alkali aggregate reactions. High content of fly ash also
gives reduced diffusivity for chloride ions. (Tuutti [16], Johansson, Sundbom &
Woltze [17]).

Variations in the remaining coal content can affect the water demand for the sprayed
concrete and possibly be the reason for the reduction of the air content of approx. 1%
(Fagerlund [1]). To high addition of fly ash gives an increased tixotropic effect,
reduced strength, increased shrinkage at desiccation (Mailvaganam & Samson [13])
and increased carbonation rate (Ljungkrantz, Möller & Petersons [18]).

Slag

GGBS makes the concrete less water demanding and reduces the risk for separation.
Other effects are increased resistance to sulphate attack and reduced chloride
permeability. A temporarily increased resistance to freeze-thaw attack is, in the long
run, replaced with reduced resistance. The reason is the increased water absorption in
the air-pore system. With freeze-thaw load in combination with water containing
chlorides the resistance is due to the amount of added GGBS. Luther, Mikols,
DeMaio & Whitlinger [19] claim that amounts of added GGBS between 35-65%
gives a reduction of the freeze-thaw resistance.

Additives

There are several various additives for sprayed concrete, both dry- and wet-mix. Their effect could be both on the fresh and the hardened concrete. The most commonly used additive for wet-mix sprayed concrete is the binder accelerator. Other types of additives are retarders, air-entraining agents, plasticizers, water reducers and polymers.

Accelerators

Accelerators are used to achieve early support in rock stabilisation and to make it possible to spray thicker layers. This gives reduced amount of discontinuities and a more homogeneous concrete with increased durability. The most commonly used binding accelerators consists water soluble salts of alkali metals or alkaline-earth metals. Most known are chlorides, silicates, carbonates and aluminates. Chlorides are used only to a limited extent because of the increased risk for corrosion. Many investigations (Burge [11], Manns & Neubert [20]) points out the effect of accelerators on durability. Alkaline accelerators give a reduction of the compressive strength at high dosages. Gebler & al. [21] also discovered a connection between reduced compressive strength and reduced resistance to freeze-thaw load. Probably is the phenomenon due to increased porosity and increased micro-cracking caused by the accelerated hydration. This is the reason to why Opsahl [22] observed increased permeability with the usage of accelerators. Other effects caused by alkaline accelerators are increased shrinkage with 20-50% both with wet- and dry-mix sprayed concrete (Manns & Neubert [20]). They also discovered doubled creep with normal addition of accelerators based on potassium carbonate/aluminate (wet-mix) and sodium carbonate/aluminate (dry-mix).

Air entraining agents

The purpose with air entraining agents is to create a concrete with small, well-distributed air pores. The air pores will act as expansion tanks for water pressed away by freezing. In dry-mix sprayed concrete the air entraining agent is added to the dry materials or in to the mixing water. This leads to difficulties in predicting the air pore content before spraying. There are many different opinions about the effect of air entraining agents in dry-mix sprayed concrete. According to ACI 506 R-90 there is no effect while Durand & Mirza [8] claims there is with reference to their tests.

When using wet-mix sprayed concrete the air entraining agent is added in advance and the air pore content could therefore be measured before spraying. Generally known is that the air pore content is reduced with 50% at spraying (Durand & Mirza [8], Seegebrecht & al. [23], Morgan & al. [14], Schrader & Kaden [24]). To high addition of air entrainers gives reduced strength.

PERFORMANCE

The performance is very important to achieve high durability and the skill of the personnel is essential. An optimally composed concrete could easily be wasted with erroneous performance.

Wet-mix sprayed concrete

High addition of accelerators at wet-mix spraying or variations of added water at dry-mix spraying reduces the possibility to achieve durable sprayed concrete. Non homogeneous concrete surrounding reinforcement could cause corrosion. Preparation and curing should also be correctly performed. To obtain an unobjectionable performance the staff should be competent and well experienced. The conditions when it concerns equipment, materials, planning of the site should also be satisfying.

For wet-mix sprayed concrete the advantage is that the concrete is ready mixed and can be controlled before spraying. Other advantages compared to dry-mix spraying is high capacity, less dust and reduced wastage. It is difficult to keep the dosage of accelerator at a constant level and uneven dosage can give differences in strength and discontinuities. An other disadvantage is that the equipment normally requires more water-rich concrete, but this can be solved with usage of water reducers or plasticizers.

Dry-mix sprayed concrete

The main advantage of dry-mix sprayed concrete is the possibility of making sprayed concrete with low w/c -ratio and that accelerators are not needed. It is also possible to use pre-bagged material and equipment with relatively low capacity, which can be an advantage at smaller repair jobs. When using dry-mix sprayed concrete the worker is adjusting the water content manually and this could make the properties vary considerably. Uneven distribution of the dry materials will give a variation of the water content in the sprayed concrete. The wastage is often greater with dry-mix sprayed concrete.

Preparation & Curing

Sufficient preparation of the surface that is to be sprayed is important. Anything that could reduce the bond has to be removed. When repairing concrete constructions all damaged concrete must be removed. Morgan & Neill [25] presents an investigation of bridges in Canada repaired with sprayed concrete. A common reason to failure was continued breaking down of the underlying repaired concrete while the repair it self was unaffected. Other factors reducing durability is reduced bond due to insufficient cleaning of rock surfaces with eg a film of oil from diesel engine exhaust or water containing soil or clay. The surface should be damp but with no free water on the surface for concrete and as dry as possible for rock. For rock surface strengthening it is therefore important with sufficient drainage of water from the surface to be sprayed.

When applying new layers of sprayed concrete it is important to remove old curing agents from the surface (should not be applied between layers) since some agents reduce bond effectively.

After spraying the young concrete must be protected against drying out to avoid cracking. Either by water or membrane curing. Some authors question the effect of some membrane curing agents. Particularly important is the curing of sprayed concrete containing silica while this gives increased plastic shrinkage (Opsahl [4]).

FACTORS AFFECTING DURABILITY

Bond

Interaction between the concrete/rock and the sprayed concrete is essential for the function of a thin sprayed concrete layer and it can be obtained by bond. Insufficient interaction can lead to failure or stability problems. Debonding can occur eg if the surface cleaning is insufficient or if damaged concrete is not removed sufficiently. If the reason for repair is an alkali aggregate reaction in concrete the reaction could proceed, if not removed sufficiently, and cause debonding. An underlying concrete with chlorides could cause a concentration of chlorides in the boundary between old and new concrete and by this cause debonding. The same phenomena with moist can cause debonding by water freezing in the boundary.

Shrinkage is another reason for debonding if the sprayed concrete does not crack. The risk for this type of damage is most common for thicker layers (Schrader [26]).

Cracking

Cracking can lead to reduced durability against eg corrosion, freeze-thaw action and leaching. Cracking is normally caused by movements in the underlying surface or in the sprayed concrete. The movements can be due to plastic shrinkage, desiccation shrinkage or load caused by temperature differentials.

Corrosion

Corrosion in sprayed concrete is activated by chlorides or by carbonisation. Chlorides could come from eg sea water in rock or from splashing of thaw salts from roads. Already a low chloride content can cause problems, since water that evaporates cause a chloride concentration on the surface. The penetration of chlorides can be both due to diffuse leaching and through cracks. The penetration is decreased with usage of fly ash (Johansson, Sundbom & Woltze [17]) but a negative effect is that the treshold value for corrosion is reached more quickly since the content of OH⁻ is reduced.

Carbonisation is a reaction between carbondioxide from the air and the calcium hydroxide in the sprayed concrete. By the decreasing pH the corrosion is initiated since the layer making the reinforcement passive is disappearing. The speed of carbonisation is ruled by relative humidity, carbon dioxide concentration, concrete permeability, amount of material to carbonise and the time after pouring. Resistance to carbonisation should be good for sprayed concrete since eg the permeability is low. Corroding fibres at a surface cause an unwanted aesthetical effect. Fibres inside the sprayed concrete are well protected and the corrosion of the fibre is usually limited. Mangat & Gurusamy [27] claim that this is due to the lime-rich coating that is protecting the fibre well since it has a large specific area. Fibres in small cracks also are generally well protected of the environment in the sprayed concrete, and the "self-repairing" effect is of great importance (Malmberg [28]). A field study by Malmberg [28] with investigations of structures in different environmental and climatic conditions show that in cases when corrosion should have appeared according to accepted theories, it did not always do.

Freeze-thaw action

Compared to conventional concrete sprayed concrete normally has lower permeability by usage of low w/c-ratios (mainly dry-mix), high binder content and low maximum aggregate size. This should lead to better freeze-thaw durability. The main difficulties are to maintain the air-pore system after spraying with the wet-mix method, and to predict the air-pore content with the dry-mix method (Morgan & al. [29]&[14], Schrader & Kaden [24], Seegebrecht, Litvin & Geblier [23]). There also are investigations indicating durability against freeze-thaw action without air entraining agents (Burge [11]). Fagerlund [30] points out the possibility of good resistance to freeze-thaw action due to enclosed air especially in dry-mix sprayed concrete. At standardised testing there is a possibility that the time for making the specimen saturated with water is not long enough for sprayed concrete. This would make the level for critical degree of saturation not being obtained. To be kept in mind is that already limited cracking will increase the permeability dramatically and by that the risk for freeze-thaw damages. An illustration of the theory with critical degree of saturation is shown in figure 3.

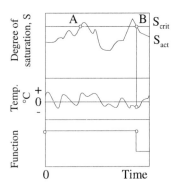

Figure 3.   Illustration of relations between degree of saturation, temperature and function. At point B damage occurs when the degree of saturation reaches the critical one at the same time as a sufficiently low temperature is ruling. (Fagerlund [30]).

Testing of freeze-thaw durability is divided in to two tests, one with water containing chlorides and one without, usually the resistance is lower to water with chlorides. The phenomena is probably due to osmotic pressures (pressure obtained by differences in chloride concentration) occurring between the surface and the pore water in the concrete. The damages are most commonly surface damages. According to the Norwegian Concrete Association [31] the most common damage on sprayed concrete due to freeze-thaw action is debonding from water-bearing rock. This points-out the importance of sufficient drainage behind sprayed concrete in zones with low temperatures.

Leaching

The low permeability of sprayed concrete normally makes it resistant to diffuse leaching of water. The ability of "self-repairing" thin cracks is reduced if one-sided water pressure is causing leaching by the running through of water. Especially if the water is soft the risk for leaching is great (Bodén [32]). Leaching in cracks will cause a local reduction of strength. Greater leakage of water will reduce the risk for leaching compared to conventional concrete since sprayed concrete has a lower permeability and a higher binder content. On the other hand the risk for leaching will increase since the cracks are more well distributed (especially with fibre reinforcement), the layers are thinner and the aggregate is less protecting.

Chemical attack

Chemical attack could be of eg acids, sulphates or salts. Acids cause surface damages in the sprayed concrete and reinforcement corrosion. The phenomena can occur when storing acids or in waste water pipes. Normally the sulphate contents in Swedish soils are too low to cause a reaction. A concentration at the surface of the sprayed concrete when water containing sulphate evaporates is though possible. Different kinds of salts could concentrate at the surface or at the boundary in the sprayed concrete and a growth of crystals could cause cracks and spalling.

FURTHER RESEARCH

From this investigation it can be said that more knowledge about corrosion of steel-fibre-reinforcement in sprayed concrete is needed. Especially the theories about fibre corrosion in cracked sprayed concrete, but also the effect on load-bearing capacity with corroded fibres. Not many investigations about freeze-thaw resistance for sprayed concrete with presence of water containing chlorides are found in this study and this can be an area for further research. This should be a common situation in road-tunnels with splashing of water containing thaw salts. To predict or measure the durability of a structure with sprayed concrete it could be convenient to be able to measure the permeability in situ. Available methods should be tested.

REFERENCES

1. FAGERLUND, G. "Betongkonstruktioners beständighet - en översikt", Cementa AB, 3:e upplagan, 1992 (in Swedish).

2. BYFORS, K. & TUUTTI, K. " Betonghandbok material ", AB Svensk Byggtjänst och Cementa AB, 2:a upplagan, 1994, pp 785-808 (in Swedish).

3. LAGERBLAD, B & TRÄGÅRDH, J. " Alkalisilikareaktioner i svensk betong ", CBI rapport, 4:92, 1992 (in Swedish).

4. OPSAHL, O.A. " Bruk av silika i sprøytebetong ", Norske Sivilingeniørers Foreningen, Fagernes Hotel, 21-23 april, Fagernes, 1982 (in Norwegian).

5. HOLMGREN, J. " Bergförstärkning med sprutbetong ", Vattenfall - handbok , 1992 (in Swedish).

6. FAGERLUND, G. " Vattenbyggnadsbetong ", Cementa AB, 1989 (in Swedish).

7. BURGE, T A. "Additives and mixtures for shotcrete" Tunnels & tunneling, Jan, 1993.

8. DURAND, B.; MIRZA, J. & NGUYEN, P. " ASTM C666 (A) Freeze-thaw durability of air-entrained wet- and dry-mix shotcrete ", Shotcrete for underground support VI, 1993, pp 188-196.

9. WOLSIEFER, J Sr. & MORGAN, D R. " Silica fume in shotcrete ", Concrete Int., pp 34-39, April, 1993.

10. FIDJESTØL, P. "Applied silica fume concrete", Concrete Int., Nov, 1993, pp 33-36.

11. BURGE, T A. "Fiber reinforced high-strength shotcrete with condensed silica fume" ACI report SP 91-57, 1991, pp 1153-1170.

12. GLASSGOLD, L I. "Shotcrete durability: an evaluation", Concrete Int., Aug., 1989, pp 79-85.

13. MAILVAGANAM, N P. & SAMSON, D. " The role of admixtures in the effective use of fly ash and silica fume in concrete mixes ", Proceedings from " Ash - A valuable resource ", 1987, pp 1-18.

14. MORGAN, D R.; KIRKNESS, A J.; McASKILL, N & DUKE, N. " Freeze-thaw durability of wet-mix and dry-mix shotcretes with silika fume and steel fibres ", Cement, concrete & aggregates, 1988, pp 96-102.

15. PETTERSSON, K. " Effects of silica fume on alkali-silica expansion in mortar specimens ", Cement & concrete research, vol 22, no 1, 1992, pp 15-22.

16. TUUTTI, K. " Korrosion på armering ", Marina betongkonstruktioners livslängd, seminariehandling, Uppsala, 1993, pp 85-101 (in Swedish).

17. JOHANSSON, L.; SUNDBOM, S. & WOLTZE, K. " Permeabilitet - provning och inverkan på betongs beständighet ", CBI rapport, 2:89, 1989 (in Swedish).

18. LJUNGKRANTZ, C. ; MÖLLER, G. & PETERSONS, N. " Betonghandbok material ", AB Svensk Byggtjänst och Cementa AB, 2:a upplagan, 1994 (in Swedish).

19. LUTHER, M.D.; MIKOLS, W.J.; DeMAIO, A J. & WHITLINGER, J E. " Scaling resistance of ground granulated blast furnace slag concretes ", Durability of Concrete, Third International Conference, Nice, France, 1994, pp 47-64 .

20. MANNS, W. & NEUBERT, B. " Mechanical-technological properties of shotcrete with accelerating admixtures ", Otto Graf Journal, 1992, pp 115-136.

21. GEBLER, S H.; LITVIN, A.; McLEAN, WILLIAM J. & SCHUTZ, R. " Durability of dry-mix shotcrete containing rapid -set accelerators " , ACI - Materials Journal, May-June, 1992, pp 259-262.

22. OPSAHL, O A. " A study of a wet-process shotcreting method-vol. I ", Norwegian Institute of Technology, Trondheim, BML report 85.101, 1985.

23. SEEGEBRECHT, G.W., LITVIN, A. & GEBLIER, S.H. " Durability of dry-mix shotcrete ", Concrete Int., Oct. 1989, pp 47-50.

24. SCHRADER, E. & KADEN, R. " Durability of shotcrete ", ACI report SP 100-57, 1987, pp 1071-1101.

25. MORGAN, D R. & NEILL, J. "Durability of shotcrete rehabilitation treatments of bridges in Canada", Paper TAC Annual conference, Winnipeg, Canada, 1991, pp c13-c51.

26. SCHRADER, E. " ʹMisconceptions about durability and bond in conventional and latex shotcrete ", ACI Fall Convention, Chicago, Oct., 1985.

27. MANGAT,P S. & GURUSAMY, K. "Corrosion resistance of steel fibres in concrete under marine exposure", Cement and concrete research, vol 18, 1988, pp 444-54.

28. MALMBERG, B. " Beständighet hos fibersprutbetong ", Föredrag vid Bergmekanikdagen 1994, SveBeFo, Stockholm, 1994, pp 79-92 (in Swedish).

29. MORGAN, D R.; McASKILL, N.; CARETTE, G G. & MALHOTRA, V.M. " Evaluation of polypropylene fiber reinforced high-volume fly ash shotcrete ", ACI - Materials Journal, March-April, 1992, pp 169-177.

30. FAGERLUND, G. " Betonghandbok material ", AB Svensk Byggtjänst och Cementa AB, 2:a upplagan, 1994, pp 711-726 & 727-783 (in Swedish).

31. NORSK BETONGFØRENING. "Sprøytebetong til fjellsikring", Norsk Betongforenings komite for sprøytebetong, Publikasjon nr.7, 1993 (in Norwegian).

32. BODÉN, ANDERS. "SFR Kontrollprogram, Bergkontroll - Sprutbetongens beständighet", Vattenfall Energisystem AB rapp. BEG PM 29/91, Oktober 1991(in Swedish).

APPLICATION OF HPC IN SHANGHAI

C Tianxia

Shanghai Construction Eng. Material Co

China

ABSTRACT. In recent years a large number of high-rising buildings, Nanpu and Yangpu bridges and the TV tower were constructed by using pumped high-strength concrete in Shanghai. Both Nanpu and Yangpu bridges are one-span, dual-pylon, dual-cable-stayed bridges over Huangpu river. with the main span length of 423m and 602m, pylon height (pumping height) of 150m and 208m, and strength of 40MPa and 50MPa respectively. The pumping height in one shot of the Shanghai TV tower is 350m. A series of measures had to be taken to ensure high-early strength, low creep and. optimum pumpability, such as optimization of concrete mixture, addition of fly ash, development of high performance pumping agent, application of advanced production know-how and strict quality control. In this paper, optimization of pumpability, properties of concrete mixture and characteristics of hardened concrete were illustrated.

Keywords: Pumping agent, Rheological characteristics

Miss Cao Tianxia is a member of ACI Committee 304, senior engineer of Shanghai Construction Engineering Material Company.

Radical Concrete Technology. Edited by R K Dhir and P C Hewlett. Published in 1996 by E & FN Spon, 2–6 Boundary Row, London SE1 8HN, UK. ISBN 0 419 21480 1.

INTRODUCTION

High performance concrete (HPC for short) overcomes the disadvantages of the original high strength concrete (HSC for short), which is usually stiff concrete that brings obstacles in the way of building engineering industrialization including the manufacture of premixing concrete and the application of pumping technique in construction. HPC has comprehensive superiority in the workability, stability and pumpability of the mixes and the strength, elastic modules and durability of the hardened concrete.

Since 1988, Shanghai Construction Engineering Material Company has been doing research on the mix design of the 40, 50, 60, and 80MPa HPC by using 425# slag cement and 525# ordinary cement. The 40-50MPa concrete have already been applied in construction. In 1994, all together 1.09 million cubic metres commercial concrete had been completed by the mixing plants of the company with an average strength of 45.5MPa, and 45% of them were 40, 50 and 60MPa concrete. The 80Mpa concrete was found on trial to be competent both in the laboratory and in the field, and will soon be taken into practical use.

In this paper, the slow-setting, high-early strength concrete used in Shanghai Nanpu and Yangpu bridges with the pumping height of 154m and 208m and the strength of 40 and 50MPa respectively, the 60MPa concrete used in Oriental-Pearl TV tower with the pumping height of 305m, the pumped concrete of 50MPa used in Shanghai highways and overflies construction, the foundation mass concrete continuously casted until completion in the construction of Xupu bridge, and so on are illustrated. The engineering practise has testified that the HPC inaugurates a new era in building engineering. Its application has played an important role in the concrete construction and improved the concrete properties.

CHARACTERISTICS AND MIX TECHNIQUE OF HPC

Characteristics

Unlike the traditional HSC, which is merely stressed on strength, the HPC is emphasized on the durability index as a primary technical parameter. For example, the hardened concrete should have high density, high elasticity, low creep, low permeability and wonderful corrosion resistance. So it is indispensable for HPC, to have high fluidity without segregation and bleeding during the plastic stage i.e. the slump can be effectively controlled so as to ensure HPC's high pumpability and high compatibility. In order to prevent the stress crack during the early-age of setting and hardening, HPC should have low heat of hydration, low dry shrinkage (these are especially important for the mass concrete) and some particular requirements (such as slow-setting and high-early strength concrete used in the main tower and crossbeam of a bridge). Only in this way can the hardened concrete gain sufficient strength together with high durability. Designed with a comprehensive consideration instead of a merely high-strength seeking, the concrete with high density, high durability, low heat of hydration and low creep is so called HPC.

Mix Technique

As discussed above, HPC still belong to the category of concrete. So its ingredients are just the same as those of the well-known ordinary pumping concrete. They are, water, cement, aggregate, addition and admixture. The key point here is that you should get the knack of mix design and thus you can achieve high performance in different cases.

The following points should be mastered based on the scientific research and the producing practise in the past eight years according to my opinion.

(1) Development of pumping agents, which can match with different cements and meet various requirements of the high performance concrete.

The adaptability of high performance pumping agent to cement is of particular importance in HPC. The slump loss with time is a basic reflection of this adaptability. In general, high range water reduction should be used to obtain a w/c ratio as low as possible in order to produce high density concrete. But the accompanying large slump loss could not meet the demand of rheological behaviour. In order to solve this contradiction, so as to keep a uniform rheological behaviour of the mixture during the course of mixing, transporting and pumping, a high quality pumping agent matching with various applied cements should be developed. And only after this work could the different technical based on different HPC indices and temperatures be attained. For example, during the construction of Nanpu and Yangpu bridges, the Nanpu-l, Nanpu-l(slow), Nanpu-2, Nanpu-2Ha, and Nanpu-2Hb pumping agents matching with Ningguo 525R# ordinary cement were developed so that 40 and 5OMPa HPC with various retarding time in four seasons (8-19h), fcu3d->32MPa and 154m and 208m pumping height respectively could be achieved. In the TV tower construction, FI'H-2D pumping agent was developed in order to achieve 60MPa concrete with 350 pumping. height in one shot. Recently during the design of 80MPa commercial pumping concrete, YJC-2 high. quality pumping agent was developed. It has a water reduction greater than 25%, a initial-setting time greater than 10h and a 10- 15mm slump loss after 1 hour.

In a word, it is very important to develop pumping agents in spite of the enormous manpower and material resources consumed in the hundreds, or even thousands, of experiments on scientific research and field trials.

(2) Mineral additives as an essential ingredient of HPC.

It is well known to use fly ash, zeolite powder, ground slag and silica powder as auxiliary binding materials or additions of cement for the purpose of cement clinker saving. And just because of this inherent view point, people regard the quality decrease in concrete with additions as certain to happen. On quite the contrary, the additions can improve the properties of concrete and thus become an important ingredient in HPC. Firstly, the high strength and high durability of HPC are partly due to the density of matrix. But to improve the density by raising the cement contents inevitably brings disadvantages. In fact, the cement heat of hydration increases with an increase in cement content. Then certainly a great deal of micro crack would come into being because of a rising temperature and shrinkage stress during the hardening process. All these do harm to concrete performance. Secondly, the alkalinity of the fly-ash concrete is lower; thus the alkali-aggregate reactions are restrained and the durability is improved. Thirdly, the substitution of less active admixture to cement can reduce the viscosity of flesh mixture. Thus the rheological characteristics can be more easily controlled to meet various requirements during construction. Fourthly, the superfine active addition or the Grade-I raw fly ash is better than the superfine high strength cement. Experiences coming from many great mass concrete construction undertaken by us. These have showed that Grade-1 fly ash, which replaces 20-30% cement, can largely reduce the heat of hydration. The foundation mass concrete of Yangpu and Xupu bridges with thickness of 5m and 6m, and concrete cast amount of 7600m3 and 10436m3 respectively both showed crackles. The 50MPa concrete with high elasticity, low shrinkage and a pumping height of 208m used in Nanpu bridge, and the 40, 60MPa concrete with a pumping height of 350m used in the Oriental Pearl construction are both achieved by using Grade-1 raw fly ash. Its characteristics of low water demand, high proportions of spherical ash and fine ash improved not only the concrete strength but also the easiness of casting.

(3) Optimization of aggregates and scientific mix design.

Aggregates used in HPC should be hard and well-grading. The suitable sand is medium or coarse, one with a fineness modulus greater than 2.6, a silt content less than 2%; the suitable coarse aggregate should have a silt content less than 1%, a needle and flat content less than 5% and a maximum size less than 25mm. The aggregate size effect is fairly important in HPC. According to the research of Pro.P.C.Aiticin in Canada, there is so strong a bonding between aggregate and cement matrix that the stress can be transmitted through the aggregate-paste interface in HPC. The smaller the aggregate size, the stronger the bond between aggregate and cement matrixes. Because of its higher intrinsical strength, coarse aggregate with a smaller size should be used in HPC.

Based on the demanded strength, durability and constructing requirements (such as the initial setting time, the slump before pumping), parameters of HPC mix are estimated after a full consideration about the conditions of construction, potation and environment temperature. Then the six ingredient mix design should be carried out according to GBJI 19,88 ((Applying Technical Specification of Concrete Addition)) GBJ 146-90 ((Applying Technical Specification of Fly Ash Concrete)) ((Guide for Structure Design and Construction of HSC)) edited by HSC committee of the Chinese Civil Engineering Institute and so on. But can the parameters be determined only after concrete on repeated trial both in the laboratory and the field have showed competence. Besides this, style and content of pumping agent together with the sand percentage should be adjusted promptly as the pumping height, temperature, and raw materials are changed. Be sure the coarse aggregate content is not over small so as to keep a high elasticity modulus of concrete.

(4) Advanced technical equipments and strict quality control.

There are difficulties in the construction of HPC. So advanced technical equipments and strict quality control are indispensable during the mixing, transporting, pumping, casting and curing of concrete.

Chart 1. Application of HCP in Shanghai

CATEGORY	NAME OF CONSTRUCTION	HEIGHT (m)	VOLUME OF CONCRETE (m³)	STRENGTH	SLUMP (cm)	Cement (C)	Mineral Additives (MA)	Admixture	Aggregate	C	MA	w/c	Bleeding	Air content (%)	mfcu 3d	mfcu 7d	Sfcu 28d	Cv %	Shrinkage (x10⁻⁶)	Creep	Frost resistance (%)	Permeability	Carbonation
Civic Construction	Nampu Bridge	154	15230	C40	16-20	525	II f.a.	Np-I	5-25	400	40	0.42	45	3.0	33	55	4.7	8.5	678	1.9	2.5	S>38	Good
	Yangpu Bridge	208	18550	C50	16-20	525	I,II f.a.	Np-II	5-25	440	44	0.38	79	3.0	35	60	3.9	6.5	519	1.1	5.5	S>30	Good
	East Pearl Tower	350	13500	C60	18-22	525	SM	FTH-2D	5-25	544	68	0.31	52	2.7	-	68	3.4	5.0	-	-	2.1	S>35	Good
	East Pearl Tower	350	20000	C40	18-22	525	I f.a.	JRC-2D	5-25	420	53	0.42	-	-	-	52	3.5	6.7	-	-	-	-	-
	East Pearl Tower	225	20000	C50	16-20	525	I f.a.	JRC-2D	5-25	440	55	0.40	-	-	-	65	3.6	5.5	-	-	-	-	-
	East Pearl Tower	180	20000	C60	16-20	525	I f.a.	JRC-2D	5-25	480	40	0.38	65-70	3.0	-	73	6.9	9.4	613	1.3	11.3	-	Good
High-Rise Building	Gongxiao Centre	90	2200	C60 56d	16-20	525	II f.a.	FTH-2D	5-25	460	35	0.37	70-84	2.0	-	68 (56D)	-	-	490	1.67	-	S>35	-
	New Century Blg.	98	5084	C40	12-16	525	II f.a.	EA-I	5-25	446	50	0.43	-	-	-	49	3.8	7.0	-	-	-	-	-
	New Century Blg.	21 flrs	2327	C50	12-16	525	II f.a.	Np-II	5-25	440	50	0.39	-	-	-	57	3.3	5.7	-	-	-	-	-
	Baoan Tower	138	4724	C40	16-20	525	II f.a.	EA-I	5-25	475	50	0.42	-	-	-	47	3.0	6.3	-	-	-	-	-
	Baoan Tower	38 flrs	3363	C55	10-14	525	II f.a.	Np-II	5-25	480	50	0.35	-	-	-	61	3.8	6.3	-	-	-	-	-
	Zhi Building	90	1914	C60	10-14	525	II f.a.	Np-II	5-25	440	50	0.37	-	-	-	66	4.8	7.2	-	-	-	-	-
	Orient Trade Blg.	150	5725	C60	12-16	525	II f.a.	Np-II	5-25	440	50	0.37	-	-	-	67	4.5	6.7	-	-	-	-	-
	Mingcheng Garden	105	2453	C50	14-18	525	II f.a.	Np-II	5-25	450	70	0.37	-	-	-	57	4.3	7.4	-	-	-	-	-
Continuously Pouring Large-Mass Concrete	East Pearl TV Tower	1.55 thick	3716	C40 S12	10-14	525	II f.a.	Np-II	5-40	360	70	0.41	-	-	-	48	5.2	10.8	-	-	-	-	-
	East Pearl TV Tower	1.55 thick	1796	C60	14-18	525	II f.a.	Np-II	5-25	490	50	0.36	-	-	-	67	2.2	3.3	-	-	-	-	-
	Yangpu Bridge	5 thick	7600	C30	10-14	mineral 425	II f.a.	WL-1	5-40	330	60	0.49	55	-	-	35	2.6	7.3	-	-	-	-	-
	Xupu Bridge	6 thick	10436	C30	10-14	mineral 425	II f.a.	EA-1	5-40	330	60	0.49	-	-	-	39	2.8	7.1	-	-	-	-	-
Foundation	Victory Tower	2.5 thick	6620	C40 S8	10-14	mineral 425	II f.a.	EA-1	5-40	420	70	0.40	-	-	-	47	2.8	5.0	-	-	-	-	-
	Meitan Tower	2 thick	21000	C35 S8	10-14	mineral 425	II f.a.	EA-1	5-40	440	70	0.42	-	-	-	45	5.3	11.9	-	-	-	-	-
	Jinmao Tower	4 thick	13500	C50	12-16	mineral 525	II f.a.	EA-1	5-40	-	-	-	-	-	-	58	4.4	7.7	-	-	-	-	-
	Sha World Trade Centre	1 thick	24000	C40 S8	-	-	-	-	-	-	-	-	-	-	-	-	-	-	-	-	-	-	-

The automatic single-shaft horizontal mixers are adopted in all mixing plants of our company. Advanced technique thus brings high efficiency. The fluidisation is increased as a result of slower feeding speed of admixtures. The automatic adjusting function on weighing error of the control system can give a precision better than state requirements under continuous mixing. The mixer vehicles are MR4500 mixer lorries with MITSUBISHI chassis. Most of the adopted pumps are the truck pumps or fixed pumps made in Germany or America. All these high quality equipments ensure the continuous production of concrete and the quality of construction. As a conclusion, the application of advanced technical equipments is an essential factor of the construction of HPC.

We also sign the producing and casting order strictly, strengthen the examination of raw materials, set up a technical re-check system in key-post of production. These give an important guarantee for the HPC quality.

CONCLUSION

HPC will imperatively substitute for HSC according to city development. It is advantageous to the improvement of concrete performance and the buildings service function, to the engineering construction speed and the city developing speed, thus do great benefit to both the economy and the society.

Practise has also showed that HPC is not a very hard achieved material. Provided that we optimize the raw materials, pay attention to the adaptability, design scientifically adjust promptly and control the qualities strictly, HPC can certainly be well-designed and well-constructed.

Of course our present HPC is far from the one with an eminent pack ability that can fill up every nook and cranny of the mould which is now being studied in advanced countries. And in the case of strength, not only 80MPa, 100MPa concrete have been widely used in advanced countries but 120MPa concrete has also begun its service in construction. But in our country, it is not until recently that 80MPa concrete is taken into practical use. Nevertheless we must realize that we are making progress. HPC technique is now on the list of the Ten New Techniques which should be popularize and adopted in the building industry in our country. Looking forward to our future, we are filled with boundless confidence that HPC would be a general trend of the concrete industry in our country.

NEW MATERIALS TECHNOLOGY

Chairmen **Dr R J Collins**
Building Research Establishment
United Kingdom

Professor F P Glasser
University of Aberdeen
United Kingdom

Leader Paper

Advanced Cement-Based Composites

Professor S P Shah and Mr S Marikunte
Northwestern University
USA

ADVANCED CEMENT-BASED COMPOSITES

S P Shah

S Marikunte

Northwestern University

USA

ABSTRACT. Plain, unreinforced cementitious composites in general, and concrete in particular, are characterized by low tensile strengths, and low tensile strain capacities; that is, they are brittle. The tensile strain capacity of cement-based materials can be dramatically improved by the addition of fibers. While small amount of fibers (<1%) have been successfully used in bulk concrete construction to overcome brittleness of cement, new processing techniques have helped in the manufacture of thin-sheet products with as high as 15% volume of continuous and discontinuous fibers. Fibers in such large quantities fundamentally alter the nature of the cement-based matrix, and the inherent tensile strength and strain capacity of the matrix itself are greatly enhanced. Such improvements have lead to a new breed of cement-based materials. This paper presents the recent advancements in processing technologies, mechanical behavior, toughening mechanisms and interface properties of fiber reinforced cement composites.

Keywords: Cement, Composite, Concrete, Fibers, Interface, Toughness, Strength

Surendra P. Shah is a Walter P. Murphy Professor of civil engineering at Northwestern University, Evanston, Illinois, and Director of the NSF Center for Science and Technology of Advanced Cement-Based Materials. He has published more than 300 technical articles and has edited 12 books. He has received the ACI Anderson Award, ASTM Thompson Award, the RILEM Gold Medal Award, and the Swedish Concrete Award. He has been awarded the Alexander von Humboldt Senior Visiting Scientist Award to Germany as well as the NATO Visiting Senior Scientist to France.

Shashidhara Marikunte is a Research Associate at the NSF Center for Science and Technology of Advanced Cement-Based Materials, Northwestern University, Evanston, Illinois. He has authored or co-authored more than 25 technical papers and his research interest is in concrete materials and structures.

Radical Concrete Technology. Edited by R K Dhir and P C Hewlett. Published in 1996 by E & FN Spon, 2–6 Boundary Row, London SE1 8HN, UK. ISBN 0 419 21480 1.

INTRODUCTION

Cement-based materials are characterized by low tensile strengths, and low tensile strain capacities. The tensile strain capacity of cement-based materials can be dramatically improved by the addition of fibers. While small amount of fibers (<1%) have been successfully used in bulk concrete construction to overcome brittleness of cement, new processing techniques have helped in the manufacture of thin-sheet products with as high as 15% volume of continuous and discontinuous fibers [1, 2]. Considerable advances in improvement of processing technologies and performance have resulted in the emergence of a new breed of cement-based materials. New processing technologies, reduced water-cement ratio and incorporation of organic and inorganic materials have resulted in less porous materials with improved performance.

FIBER REINFORCED CEMENT COMPOSITES (FRC)

Plain, unreinforced cementitious materials are characterized by low tensile strengths, and low tensile strain capacities; that is, they are brittle materials. They thus require reinforcement before they can be used extensively as construction materials. Historically this reinforcement has been in the form of continuous reinforcing bars, which could be placed in the structure at the appropriate locations to withstand the imposed tensile and shear stresses. Fibers, on the other hand, are generally, short discontinuous, and are randomly distributed throughout the cementitious matrix. Fibers used in cement-based materials are primarily made of steel, glass, polymer or derived from natural materials. Typical properties of fibers used for cement application are presented in Table 1. Since fibers tend to be more closely spaced than conventional reinforcing bars, they are better at controlling cracking. It is important to recognize that, in general, fiber reinforcement is not a substitute for conventional reinforcement. Fibers and steel bars have different roles to play in modern concrete technology, and there are many applications in which both fibers and continuous reinforcing bars should be used [1, 2].

Conventional Fiber Reinforced Concrete Exhibiting Strain Softening

Conventional fiber reinforced concrete is typically used for bulk field applications involving large volumes of concrete. For bulk construction and with conventional mixing techniques, it is practical to use short fibers (about 25 mm) with a relatively coarse diameter (0.25 mm) and a volume fraction of less than 1%. Addition of larger quantity of fibers tend to increase the viscosity of the cementitious matrix which makes it very difficult to handle and place. Typically, steel or polymeric fibers are used and the matrix is concrete (maximum aggregate size of 20 mm). In these composites since the ultimate strain capacity of the matrix is lower than the strain capacity of the fibers, the matrix fails before the full potential capacity of the fiber is achieved. The fibers that bridge the cracks formed in the matrix contribute to the energy dissipation through processes of debonding and pull-out. As a result the post peak response continues to remain strain softening, but with a less steep slope compared to plain concrete (Figure 1). Thus, the fibers improve the ductility of the material or more properly, its energy absorption capacity. In addition, there is often an improvement in impact resistance, fatigue properties and abrasion resistance. Fiber reinforced concrete is now widely used for reducing shrinkage cracking. The introduction of fibers is known to reduce considerably crack widths resulting from restrained shrinkage cracking [3, 4].

Table 1   Typical Properties of Fibers

Fiber	Diameter (μm)	Specific Gravity	Tensile Strength (GPa)	Elastic Modulus (GPa)	Strain to Failure (%)
Steel	5 - 500	7.84	0.5 - 2	210	0.5 - 3.5
Glass (Alkali Resistant)	9 - 15	2.60	2.0 - 4.0	70 - 80	2.0 - 3.5
Polypropylene (Fibrillated)	20 - 200	0.90	0.5 - 0.75	5 - 77	8.0
Cellulose (Kraft)	20 - 120	1.54	0.3 - 0.5	24 - 40	-
Carbon (High strength)	9	1.90	2.6	230	1.0
Asbestos (Chrysotile)	0.02 - 30	2.60	3.5	165	2 - 3
PVA (Polyvinyl alcohol)	15	1.30	0.9	29	-

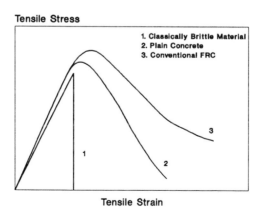

Figure 1   Typical tensile stress-strain curves for a classically brittle material, plain concrete and low volume strain softening FRC

Fiber Reinforced Cement Composites Exhibiting Strain Hardening

Recently with improvement in processing technology and in performance of cement-based materials it is possible to develop composites with a substantially higher tensile strength and with a strain hardening type of response rather than strain softening [5, 6, 7]. This is illustrated in Figure 2. There are several factors that can influence the response of the composites: geometrical and mechanical properties of fibers, properties of the matrix, interfacial characteristics and the spatial distribution of fibers. To develop cost-effective advanced cement-based composites, one needs to develop an accurate micromechanical understanding of how fibers reinforce quasi-brittle cement-based materials [1, 5].

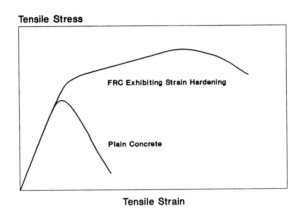

Figure 2   Performance of concrete with micro fibers exhibiting strain hardening

MECHANICAL PROPERTIES: TENSION

Experimental Evaluation of Conventional Fiber-Cement Composites

For bulk construction and with conventional mixing techniques, it is practical to use short fibers (about 25 mm) with a volume fraction of less than 1%. Typically, steel or polymeric fibers are used and the matrix is concrete (maximum aggregate size of about 20 mm). The stress-strain response of such composite can be approximated by three segments. The first segment, representing precracked region, can be defined by the modulus of elasticity of the matrix. The second segment can be taken as a zone of nonlinear deformation between the first cracking of the matrix and the ultimate tensile strength of the composite. For most cases, this segment is relatively small. The third segment of the curve representing the postpeak (reserve) strength can be attributed to the fiber pull-out resistance. Recent advances in instrumentation have made it possible to test tension specimens under closed-loop stable conditions. Using this type of test setup, postpeak responses can be obtained even for plain concrete and FRC.

strength and strain capacities of a matrix may improve the design procedures of high-performance cement-based composites.

The stress-strain curves of cement-based matrices reinforced with 8.7% and 13.4% volume fractions of polypropylene fiber are shown in Figure 4 [1, 5, 9]. In order to achieve such high volume fractions, special manufacturing techniques were used. The stress-strain curves of these composites can be divided into two parts, an initially stiff region followed by a substantially less stiff region. The endpoint of the initial, stiff part of the curve is often termed the bend-over point (BOP). The stress-strain curves were observed to be essentially linear up to this point. The stress-strain responses of the matrix calculated from the response of the composite specimen are shown in Figure 4 [5]. In order to calculate the matrix stress-strain curve, the contribution of the fibers subjected to an identical elongation was subtracted from that of the composite. The peak value of the matrix stress is observed to occur at about the BOP and depends on the volume fraction of fibers. This implied that macroscopically linear behavior of a matrix up to a stress level of 15 MPa may be the result of the suppression of localized cracking in the presence of fibers. Note that even at a strain level of 1.2%, the matrix capacity is almost 8 MPa. Similar results have also been observed in cement-based matrices reinforced by steel, glass, or carbon fibers [5]. With such a high fiber volume fraction of fibers, microcracking is stabilized, and a homogeneous distribution of microcracks can be found even at very high strain levels (1% strain, which is about 100 times the strain localization of plain concrete). Thus the mechanical behavior of the composite is fundamentally different from the mechanical behavior of the matrix, which typically is a cement paste or cement-based matrix.

The ACBM Center at Northwestern University has recently patented extrusion technology to incorporate discontinuous fibers in large quantities (in excess of 4%) and has achieved a level of performance comparable to that obtained using continuous fibers [6, 7]. In this continuous process, cement composites can be extruded through a die to obtain different size and shape of products. One needs to control the rheology of uncured cementitious paste to successfully accomplish processing. Techniques and organic and inorganic additives are needed to modify the rheology to produce the required thixotropic behavior. The advantage of introducing extrusion techniques into the cement product processing is that the materials are formed under high shear and high compressive forces. With this technology tensile and bending strengths comparable to those achieved by continuous fiber technology have been achieved. These sheets are substantially strong, tough and flexible. Extruded fiber reinforced cement products were made with portland cement and microfibers (e.g. carbon, PVA, cellulose, polypropylene and steel) up to 4% by volume. The most successfully extruded sheets have tensile strength of 7.5 MPa and exhibit strain hardening behavior with a strain of 1% (Figure 5). Multiple cracking associated with strain hardening behavior was also observed. With the properly designed die and the properly controlled material viscosity, the fibers can be controlled to align in the load-bearing direction, the matrix and the fiber packing densified to a low porosity and the interface between the fiber and matrix enhanced.

Figure 3 represents the response of plain mortar and steel fiber-reinforced concrete tested in uniaxial tension; the stress-displacement responses for various fiber volume fractions are also shown [8]. These responses were obtained using notched specimens loaded such that the notch mouth opening displacement increased monotonically. Plain mortar matrix exhibit linear elastic behavior up to about 50% of their tensile strengths. The onset of inelastic behavior prior to the peak load suggests the initiation of the microcracking process, although no cracks are detected prior to the peak loads. Almost immediately after peak loads, deformations become localized with the eventual widening of a single crack. The unreinforced matrix exhibits significant traction capacity at displacements several times those observed at the peak load.

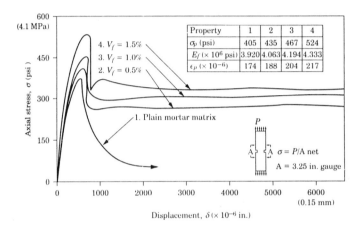

Figure 3   Tensile stress-strain response: Comparison of plain and steel fiber-reinforced concrete [8]

Steel fiber-reinforced composite behavior is linearly elastic up to about 80% of the matrix tensile strength. Nonlinear deformations take place beyond the linear elastic limit. Composite peak stress and corresponding displacements (or strains) are larger than the corresponding values for the unreinforced matrix. After the composite peak stress, the load-carrying capacity abruptly drops to a stress level termed by some investigators as the postcracking strength. One single crack becomes visible at the critical section. With further increases in displacement, the load-carrying capacity gradually drops with increases in displacement.

Experimental Evaluation of Fiber-Cement Composites Exhibiting Strain Hardening

A possible means of increasing the tensile-strength capacity and dramatically increasing the tensile-strain capacity of cementitious materials is by the addition of fibers sufficiently higher than the critical volume. Recently, with a variety of processing techniques, it has become possible to incorporate up to 15% fibers by volume into cement-based materials. As the fiber volume fraction increases and as fibers become more uniformly dispersed, they can hinder the growth of microcracks, suppress localization, and consequently substantially increase fracture strength and strain capacities of the matrix. Understanding how and why fibers alter fracture

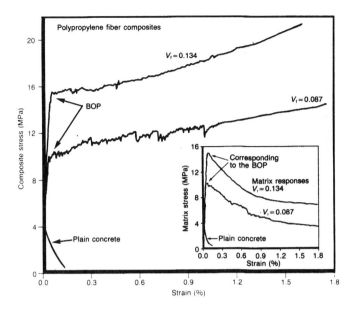

Figure 4   Stress-strain response of fibrillated polypropylene-fiber cement-based composites [5]

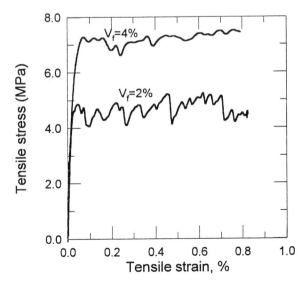

Figure 5   Tensile stress-strain behavior of extruded fiber-cement composites [6, 7]

MECHANICAL PROPERTIES: BENDING

In many applications fiber-reinforced concrete is subjected to a bending action. Hence, the behavior of composites in bending has been investigated extensively. One of the primary reasons for adding fibers to concrete is to improve the energy-absorbing capacity of the matrix, which can be evaluated by determining the area under the stress-strain curve or by the load-deformation behavior. In the case of bending, the area under the load-deflection curve is used to estimate the energy-absorbing capacity or toughness of the material. Increased toughness also means improved performance under fatigue, impact, and impulse loading. The toughening mechanism provides ductility. The composite's ability to undergo larger deformations before failure is often measured using a toughness index.

The contribution of fiber to toughness is well known. However, how to measure this toughness and express it as an index useful for design purposes is under debate. The major factors that affect the load-deflection performance and hence the evaluation of toughness are the following: fiber type, fiber geometry, fiber volume fraction, matrix composition, specimen size, loading configuration, loading rate, deflection-measuring accuracy, feed-back control, and stiffness of the machine compared with the stiffness of the specimens. After the load-deflection curve is obtained, how to evaluate the factors that contribute to the improved performance is also under debate. To overcome at least some of these difficulties, a different approach was developed in which toughness was characterized by a unitless value, termed as the toughness index, which is the area under the curve up to a specified deflection. There are at least three procedures (ACI Committee 544 of the American Concrete Institute, The Japan Concrete Institute and ASTM C 1018) that are suggested to quantify toughness using the load-deflection curve obtained in static bending [10, 11, 12].

The amount and type of fibers play a major role in influencing flexural strength and toughness of FRC. For a given fiber type a higher volume fraction provides more energy absorbing capacity or toughness. However, larger quantity of fibers tend to increase the viscosity of the cementitious matrix which makes it very difficult to handle and place. Higher amounts of fiber can be incorporated into the matrix with special manufacturing techniques (e.g., extrusion).

CONTROL OF RESTRAINED SHRINKAGE CRACKING WITH FIBERS

When a small quantity of coarse fibers are added there is no significant change in strength characteristics, but improved strain softening behavior (toughness). Toughness is important in design because quite often the failure is not by increased gravity load but by increased deformation (e.g. earthquake loading, differential thermal and shrinkage stresses, etc.). One way to assess the benefits of adding low-volume fibers for slabs is to compare their tensile response with a particular emphasis on the strain softening.

Shrinkage and temperature cracking is an important consideration for flat slabs, floors, parking garages and highway pavement. Concrete shrinks if it is exposed to drying environment. If concrete is allowed to shrink freely then there is no problem. However, in reality concrete is always restrained, which results in tensile stresses. Depending upon the relaxation (creep), age of exposure, elastic properties of concrete, degree of restraint, and potential free shrinkage crack develops. To simulate this complex phenomena in practical way ring test has been developed. In this test concrete is cast around a steel annulus (Figure 6) which provides a uniform restraint. After curing the ring exposed to a controlled drying environment. Drying is allowed to take place only from the surface to create axisymmetric situation and thus the dimensions of the specimen are not a factor. During drying the surface of the specimen is monitored for cracking and its development with time using a microscope.

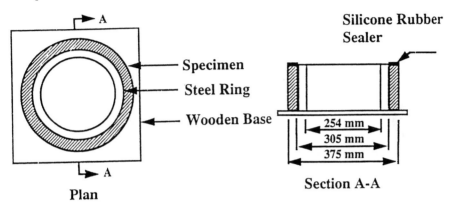

Figure 6   Ring test for restrained shrinkage cracking

Once a crack is developed in a brittle material with no strain softening, then that crack will be wide and the specimen now is without restraint. However, if there is some resistance at the crack-surface, then there is a possibility that a second crack forms. With fibers the possibilities of multiple cracks are increased. Thus with strain softening material, width of individual cracks are small. Theoretical models for the prediction of crack widths under restrained drying shrinkage conditions have been successfully developed. Figure 7 compares the theoretical and experimental values of average crack width of steel fiber reinforced concrete with time. Thus fiber reinforced concrete is now widely used for reducing shrinkage cracking [1, 3, 4].

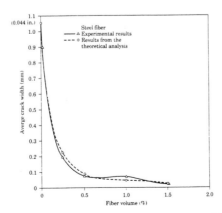

Figure 7   Average crack width versus fiber volume: steel fibers [1]

INFLUENCE OF FIBER PARAMETERS

Fiber parameters play a key role in improving the tensile behavior of cement-based materials. Fiber parameters which influence the behavior of the composite are (a) length and aspect ratio; (b) distribution; (c) interface bond: and (d) modulus of elasticity of the fiber.

Figure 8 presents the influence of fiber length and aspect ratio on the tensile behavior of cement-based matrix. Coarser enhance the crack resisting capacity of concrete. However, they do not significantly affect the tensile strength, but only slightly improve strain softening behavior (Figure 8 a). Coarse fibers (fibers with diameter in millimeter range) as used in conventional fiber reinforced concrete can not interact with microcracks but can bridge localized macrocracks. This means that they will influence primarily the post-peak response. Micro fibers with short length even though increase strength, are brittle (Figure 8 b). The length of these fibers are not sufficiently long to bridge the crack. When fibers have small diameter, are sufficiently long and closely dispersed (Figure 8 c), then it is possible to have composites with increased tensile strength as well as with strain hardening type of response.

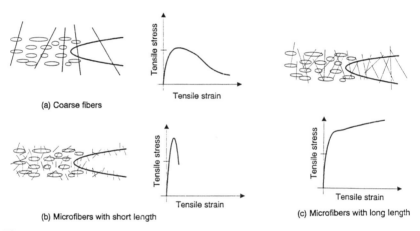

(a) Coarse fibers

(b) Microfibers with short length

(c) Microfibers with long length

Figure 8   Influence of fiber length and aspect ratio on tensile behavior

Fiber distribution is very important to take full advantage of fibers. Composites with smaller fiber-fiber distance show increased peak stress (Figure 9).

The fiber-matrix interface bond critically influence the ability of fibers to stabilize crack propagation in the matrix. Debonding and sliding at the interface have a significant influence on total energy consumption during crack propagation when a large crack is allowed to form in the matrix. For fiber reinforced cement-based composites a weak transition zone between ordinary portland cement paste and fiber has been observed (Figure 10). This weak zone includes a duplex film, a Calcium Hydroxide layer, and a porous layer consisting of Calcium Silicate Hydrates and Ettringite, and as a result, the interface behavior may primarily depend on this weak zone rather than on the bulk cement matrix. This weak zone can be densified when silica fume is added to the cement paste [2].

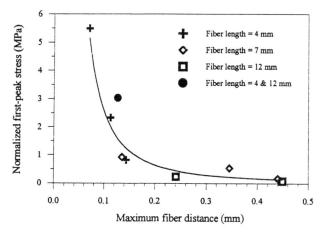

Figure 9   Influence of fiber distribution on the tensile behavior [13]

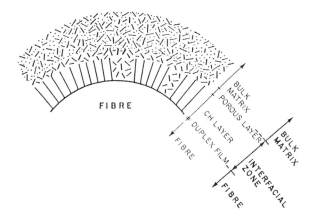

Figure 10   Interface bond between fiber and cement matrix [1]

INFLUENCE OF PROCESSING

There are several ways one can process a larger volume of fibers. Hatschek process (Figure 11 a) used for asbestos fibers is now slightly modified to accommodate other fibers such as wood fibers. Due to the shorter length of these fibers these composites are fairly brittle in nature [14, 15]. Pultrusion (Figure 11 c) and reticem process (Figure 11 b) are used for reinforcing continuous fibers or mats in cement matrix. Due to continuous fibers or mats these composites exhibit improved strength as well as strain hardening behavior. Recently, at Northwestern University's Center for Advanced Cement-Based Materials extrusion process (Figure 11 d) has been developed to incorporate large quantities of short fibers into cement matrix [6, 7]. The advantage of introducing extrusion techniques into the cement product processing is that the materials are formed under high shear and high compressive forces. In this processing technology it is not only possible to obtain flat shapes but also structural shapes (I sections, channels, tubes, hollow and solid tubes).

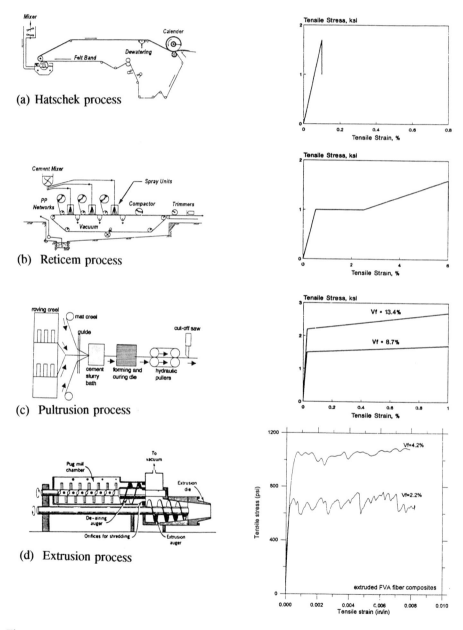

Figure 11   Influence of processing parameters on tensile behavior

TOUGHENING MECHANISMS

Toughening mechanism of the class of fiber reinforced cement-based composites in which the volume of short randomly distributed fibers is in the range of 1% to 2% is now well understood. It is now generally accepted that, for this class of fiber-reinforced concrete, the major contribution of the fibers is after the matrix strain localization, which occurs around the peak of the tensile stress-strain curve. The strain softening after the peak can be expressed in the form of a relationship between fiber-bridging closing pressure versus crack opening displacement [16]. Such a relationship can be determined experimentally as well as from micromechanical considerations.

Several attempts have been made to understand toughening mechanisms of cement-based matrices reinforced by a high volume fraction of fibers [17]. Concepts based on linear elastic fracture mechanics (LEFM) are not sufficient to understand the complex toughening mechanisms in composites with high fiber volume fraction. Quantitative optical microscopy to examine crack propagation in high-volume-fiber-reinforced cement-based composites. Recently laser holographic interferometry has been used to quantitatively measure surface displacement (strains) of fiber-reinforced cement-based specimens subjected to uniaxial tension [18]. Based on the crack development observed using these procedures, it appears that cracks initiate at a very low load level and are few and widely dispersed, no more than 1 mm in length, as schematically shown in Figure 12 a. Cracks begin to localize near the point when the matrix contribution is maximum; this point is referred to as the bend-over-point (BOP) (see Figure 12 b). Note that debonding and sliding at the fiber-matrix interface (which may occur) are omitted in Figure 12. With further straining, the localized bands seem to move closer together (Figure 12 c). This process eventually leads to the homogenization of microcracking (Figure 12 d). It appears that the suppression of the localization phenomenon may lead to the observed enhanced tensile strength of the matrix (BOP, point "B" in Figure 12). The stress at this point can be as high as 5 times the tensile strength of the plain matrix. Homogenous distribution of cracking observed at very high strain levels (point "D" in Figure 12) may explain the apparently high level of matrix contribution observed at this level.

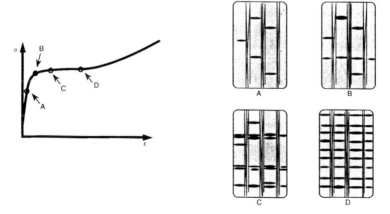

Figure 12   Stages of microcracking [18]

FUTURE

Recently, with a variety of processing techniques, it has become possible to incorporate up to 15% fibers by volume into cement-based materials [1, 2, 5, 6, 7]. Fibers in such large quantities fundamentally alter the nature of cementitious matrices. The responses clearly indicate the enormous increase in both strength and ductility. With such a high fiber volume concentration of fibers, microcracking is stabilized, and a homogeneous distribution of microcracks can be found even at very high strain levels (1% strain, which is about 100 times the strain localization of plain concrete). Thus the mechanical behavior of the composite is fundamentally different from the behavior of the matrix, which typically is a cement paste or cement-based matrix. Such beneficial interactions can lead to a new class of fiber-reinforced concrete structures as demonstrated in Figure 13 [5, 6]. The bending capacity of the compact, reinforced composite (made of cement, silica fume, and steel microfibers) is an order of magnitude higher than that of conventional reinforced concrete and, in fact, approximates that of structural steel.

APPLICATIONS

The applications of fiber reinforced composites are as varied as the types of fibers that have been used. Steel fibers have been used in pavements, in shotcrete (tunnel linings), in dams, and a variety of other structures. Increasingly, fibrillated polypropylene fibers are being used as secondary reinforcement, to control plastic shrinkage cracking. The use of thin-sheet fiber reinforced cement composites ranges from major components in industrial manufacturing to uses in commercial, residential and agricultural construction. Commercial and residential uses of thin-sheet fiber reinforced cement products are mainly for the production of flat and corrugated sheet roofing elements, exterior and interior wall panelling, equipment screens, fascial, facades and soffits, substrate for tiles, window sills and stools, stair treads and risers, substrate for applied coatings, and utility building cladding panels. Agricultural uses of thin-sheet fiber reinforced cement composites are mainly for farm buildings, sidings, stalls and walls, poultry houses and incubators, green house panels and work surfaces, and fencing and sunscreens.

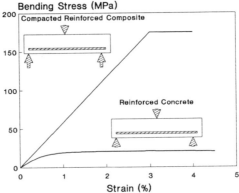

Figure 13   Compacted reinforced composite beam made of cement-based matrix and microfibers [5, 6]

Asbestos and wood fibers have long been used in pipes and corrugated or flat roofing sheets. Glass and carbon fibers are used primarily in precast panels (non-structural), as shown in Figure 19 [5, 6]. New processing technologies (e.g. extrusion) provides flexibility of manufacturing not only flat shapes but also structural shapes (I-section, channels, pipes, hollow and solid tubes).

Figure 14   Applications of FRC: Fiber reinforced wall panels [Ark Mori Building, photograph courtesy of Kajima Corporation, Tokyo, Japan]

REFERENCES

1.   BALAGURU, P.N. and SHAH, S.P., Fiber-Reinforced Cement Composites, McGraw Hill Publishers, New York, 1992, pp 530.
2.   BENTUR, A. and MINDESS, S., Fiber Reinforced Cementitious Composites, Elsevier Applied Science, New York, 1990, pp 449.
3.   GRZYBOWSKI, M. and SHAH, S.P., Shrinkage cracking of fiber reinforced concrete, ACI Materials Journal, Vol. 87, No. 2, 1990, pp 138-148.
4.   SARIGAPHUTI, S., SHAH, S.P. and VINSON, K.D., Shrinkage cracking and durability characteristics of cellulose fiber reinforced concrete, ACI Materials Journal, Vol. 90, No. 4, 1993, pp 309-318.
5.   SHAH, S.P. and OUYANG, C., Mechanical Behavior of Fiber-Reinforced Cement-Based Composites," Journal of American Ceramic Society, Vol. 74, No. 11, 1991, pp 2727-2953.

6. MARIKUNTE, S. and SHAH, S.P., Engineering of Cement-Based Composites, Concrete Technology: New Trends, Industrial Applications, Proceedings, The International RILEM Workshop, E & FN Spon, London, 1994, pp 83-102.

7. SHAO, Y., MARIKUNTE, S. and SHAH, S.P., Extruded Fiber Reinforced Composites, Concrete International, Vol. 17, No. 4, pp 48-52.

8. GOPALARATNAM, V.S. and SHAH, S.P., Softening Response of Plain Concrete in Direct Tension, Journal of American Concrete Institute, Vol. 82, No. 3, 1985, pp 310-323.

9. SHAH, S.P., Do Fibers Increase Tensile Strength of Cement-Based Matrices, ACI Materials Journal, Vol. 88, No. 6, 1991, pp 595-602.

10. ACI COMMITTEE 544, Measurement of Properties of Fiber Reinforced Concrete, ACI Materials Journal, Vol. 85, No. 8, 1988, pp 583-593.

11. JAPAN CONCRETE INSTITUTE, Method of Test for Flexural Strength and Toughness of Fiber Reinforced Concrete, Standard SF4, JCI Standards for Test Methods of Fiber Reinforced Concrete, 1983, pp 45-51.

12. ASTM, Standard Method of Test for Flexural Toughness of Fiber Reinforced Concrete, ASTM Standards for Concrete and Mineral Aggregates, Vol. 4.02, C 1018, 1989.

13. BETTERMAN, L.R., An Investigation of the Tensile Properties of PVA Fiber Reinforced Mortar, Ph.D. Theses, Northwestern University, Illinois, 1994.

14. MARIKUNTE, S. and SOROUSHIAN, P., "Statistical Evaluation of Long-Term Durability Characteristics of Cellulose Fiber Reinforced Cement Composites," ACI Materials Journal, Vol. 91, No. 6, 1994, pp 607-616.

15. SOROUSHIAN, P. and MARIKUNTE, S., Moisture Effects on the Flexural Performance of Wood Fiber-Cement Composites, Journal of Materials in Civil Engineering, ASCE, Vol. 4, No. 3, 1992, pp 275-291.

16. WECHARATANA, M. and SHAH, S.P., A model for predicting fracture resistance of fiber reinforced concrete, Cement and Concrete Research, Vol. 13, 1983, pp 819-829.

17. SHAH, S.P. and RANGAN, B.V., Fiber reinforced concrete properties, Journal of American Concrete Institute, Vol. 68, 1971, pp 126-135.

18. STANG, H., MOBASHER, B. and SHAH, S.P., Quantitative damage characterization in polypropylene fiber reinforced concrete, Cement and Concrete Research, Vol. 20, 1990, pp 540-558.

CEMENT AND SPECIAL COMPOSITES FOR HPC CONCRETE AND STRUCTURE RECONSTRUCTION

S Peukert

Institute of Mineral Building Materials

J Mierzwa

University of Technology

H Mroz

Institute of Mineral Building Materials

Poland

ABSTRACT. Numerous, ascertained during the last decades, cases of concrete structures decrease in durability or damage caused necessity of working out new cements and new technological - materials systems of repair which could ensure fast recovery of endangered buildings. To meet these problems special cement MPz, which is characterised by particularly beneficial features, was elaborated at IMMB Krakow. Thanks to very high kinetics of initial strength growth and high final strength it is exceptionally well suitable to repairs and reconstruction. In the paper its characteristic and examples of application have been presented.

Keywords: cement, concrete, durability, reconstruction, corrosion.

Stanislaw Peukert, Professor Dr hab. Eng., Scientific Director of Institute for Mineral Building Materials Cracow Division. Specialist in technology of special and ordinary cement, author of many patents and applications of new technologies in Polish cement plants, author of over 100 publications on production, properties and application of cement, co-author of MPz class cement.

Janusz Mierzwa, Dr hab. Eng., Professor of Cracow University of Technology, director of Concrete Technology Laboratory, author of tens of publications on concrete and concrete structures.

Henryk Mroz, MSc. Eng., Head of Concrete Investigation Laboratory at IMMB, specialist in technology and application of cement, co-author of MPz class cement.

Radical Concrete Technology. Edited by R K Dhir and P C Hewlett. Published in 1996 by E & FN Spon, 2–6 Boundary Row, London SE1 8HN, UK. ISBN 0 419 21480 1.

INTRODUCTION

Problem connected with exploitation of concrete buildings and structures during the last tens of years have demonstrated that, less perceived till now, issue of concrete durability should be included to the most important in concrete technology. Irrespective of environmental causes, conditioning maintenance of construction durability, there emerges a problem of selection of proper composition of concrete as a material of durable structure from the one side and from the other an issue of having adequate technological - materials systems of repair enabling quick repair of damage. In both cases the decisive factor in a set of material parameters becomes cement. Requirements which are made to it are connected with a quality ensuring cement high corrosion resistance but above all possibility of achieving high 28-days strength corroborating high tightness of concrete composite on the level of macro and microstructure as well as getting high increment of initial strength.

There are exactly such resulting features that distinguish now introduced in more and more range of applications type HPC concrete.

In this paper a slightly different MPz cement, which features in the face of above given requirements prove to be in many aspects even better then the highest quality Portland cement industrially produced, will be put forward.

THE GENERAL IDEA OF FORMATTING HIGH STRENGTH CEMENT AND MPz CEMENT COMPOSITION

The main factors which determine the strength of cement are [1]:

- phase composition and spatial net structure of clinker,
- degree of cement powdering,
- contents of subsidiary components,
- presence of additions which accelerate hardening,
- presence of additions controlling setting and modifying paste characteristic,
- thermal-humidity conditions in the phase of initial curing of paste.

Assuming that the strength of cement grout is a function of: contents, structure, strength of crystalline frame and tightness of the whole system, test of forming high strength and rapid hardening cements were performed in several directions. One of them was idea of reducing dimensions and modification of section of C_3S by proper running process of crystallisation and in practice through high saturation of clinker with lime and using in the same time post extraction sludge which have as the main component calcium orthosilicate [2,3]. Results of these investigations enabled to start in the seventies production of rapid hardening cements Super 40 and Super 50. They are characteristic by very high kinetics of initial strength growth and high strength after 28 days (Figure 1). These cements contained not only great amount of C_3S but also C_3A while specific surface F_B was about 3200 to 3600 cm^2/g. In later investigations special attention was focused on influence of structure defects and connected with it grout porosity. They resulted in placing both phases of calcium silicate properly in the process of building tight and dense CSH structure.

Function of alite and belite phases in creating of strength development in time of cement grouts is well illustrated by curves on Figure 2. Not lesser importance for strength growth has cement graining. In certain extent increase of fineness has very beneficial influence on intensification of initial strength growth. It was proved by investigations results [4] which are shown in Figure 3. Speed of hardening process can be also regulated by use of chemical

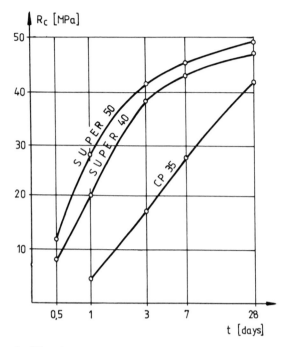

Figure 1   **Kinetics of strength growth of cement SUPER 40
and 50 and Portland Cement CP 35**

modifiers introduced as activators that change binder solubility and create with its components complex chemical compounds (e.g. Rozenberg T.J.[5], Joisel A [6], Roberts M.H. [6]) or as nucleus which create centres intensifying crystallisation of hydrated phases of hardening paste (e.g. J. Grzymek [2], Ratinow W.B. [5]). Within the range of the first group of modifiers specialists interest is focused on such compounds like $Ca(NO_3)_2$, $Al\ Cl_3$ $Al(NO_3)_3$, $Al_2(SO_4)_3$, Na_2CO_3, NaOH, KOH and particularly active K_2SO_4, K_2CO_3, $CaCl_2$ and Na_2CO_3. During the last ten years positive effects were also achieved by introduction of active micro-silica which effectively influences pozzolana transformation $Ca(OH)_2$ into low alkaline phase CSH and also by introducing to grout soluble polymers which leads to elimination of structure macro defects [3]. When elaborating technology of cement production especially high- performance MP type and modified MPz some of the above mentioned tendencies were taken into account particularly:

- high exposition of silicate module in clinker together with lessening dimensions of alite crystal,
- high contents of silicate phases at the cost of aluminate and ferrite phases,
- very fine cement milling,
- selection of proper combined additions strongly accelerating setting and hardening as well as retardants which cause significant lessening of structure porosity,
- elimination of gipsum addition as a setting process regulator („gypsum-less cement").

PROPERTIES OF TYPE MPz CEMENT

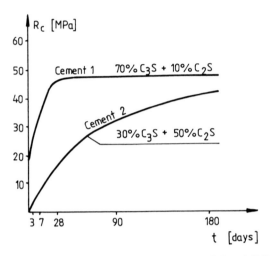

Figure 2 Consequence of silicate minerals C_3S and C_2S in
shaping the development of cement strength in time.

Type MPz cements were elaborated as a group of special binders characterised by particular
and regulated dynamics of initial strength growth and in the same time by achieving high 28th
days strength.
Produced in that group special cement MP type characterises itself by instant growth of
strength which can only be compared with the strength activity of polymer resins.
Evolution of its initial strength within the time of 24 hours against a background of other
cements is displayed in Table 1.

Table 1 Increase of strength of grouts (1:1) made from different cements
during first 24 hours of hardening.

Type of	Compressive strength Rc [MPa] after hours					
cement	1	2	3	4	12	24
CP35	0.0	0.0	0.0	0.0	0.5	4.5
SUPER 40	0.0	0.0	0.5	1.0	8.0	22.0
SUPER 50	0.0	0.0	1.0	2.0	14.0	29.0
MP	8.0	21.0	32.0	37.0	44.0	52.0

Table 2 Basic physical properties of MPz type cements

Type of cement	Blaine's spec.surf. [cm²/g]	Specific water [%]	Setting conditions		change of volume	
			start	finish	Le Chat. [mm]	Cake (-)
MPz - I	6600	22.0	2 min	6 min	0.0	norm.
MPz - II	6000	20.0	35 min	50 min	0.0	norm
MPz - IIIA	5600	19.0	1.05 h	1.20 h	0.0	norm
MPz - IIIB	5500	21.0	1.20 h	1.40 h	0.0	norm

To differentiate possibilities of technological utilisation in application to particular needs a group of cements, modified in comparison with MP type cement, were worked out and denoted as MPz. The four types of MPz cement differ by the rate of setting and hardening within the limit of time from 3 min. (MPz I) up to 80 min. (MPz III B). The last corresponds approximately with setting of ordinary cement used in building.

Basic physical properties of cements in that group are given in Table 2.

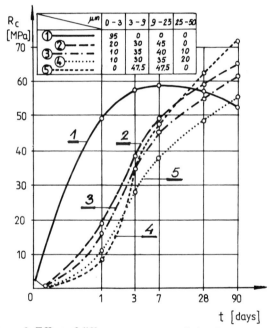

Figure 3 Effect of different cement granulation from the same clinker with 5% addition of gypsum on strength distribution in time.

Development of compressive strength of these cements beginning from 4th hour till 1 year in tests on standard grouts is shown in Table 3. In consideration of totally different conditions of liquidity of MPz cement grouts as a basis for determining the amount of water with standard proportion of standard sand, identity of consistency, for which applied water-cement ratio is given in the second column of Table 3, was assumed.

Table 3   Strength distribution for grouts (1:3) made from MPz cements

Cement	w/c	Compressive strength [MPa] after time							
type	ratio	4h	12h	1 day	3 days	7 days	28 d.	90 d.	360 d.
MPz-I	0.33	32.2	45.6	51.4	54.6	59.0	63.0	65.0	67.0
MPz-II	0.32	21.8	40.5	48.2	58.6	68.0	75.5	79.6	80.4
MPz-IIIA	0.30	11.0	22.5	36.0	58.0	69.0	80.0	84.5	86.0
MPz-IIIB	0.34	10.0	21.5	35.0	58.0	70.5	83.0	86.0	90.0

APPLICATION OF CEMENT MPz TYPE

There are different possibilities of using MP and MPz cement. For instance MP cement proved to be particularly effective in application for anchoring of lining of pit shafts and drifts to shaft cheek. Cement MPz I, because of its very short setting time and unusually high rate of strength growth, has characteristic of emergency binder indispensable for sealing up, repairs and reconstruction in civil engineering.

The group of MPz II, MPz III A and B, from the technical applications point of view, is treated as technologically normal binder, but with exceptionally useful properties. Concrete made from them, within the range to 200 cycles of freezing and defreezing, show practically 100% freeze resistance (Table 4).

Table 4   Concrete from MPz II cement tested

Number of	Loss of	Compressive strength [MPa]			Concrete
cycles	bulk	Comparative	Frozen	Loss	composition
freeze-defr.	$\Delta G[\%]$	series	series	$\Delta R[\%]$	(kg dm³/1000)
25	00.3	44.4	45.9	-3.37	
50	0.14	49.9	49.6	0.02	c=327.0
100	0.19	49.3	48.8	1.02	w=138.0
200	0.27	50.4	48.0	4.76	k=1915.0

Investigation on resistance of these cements to sulphate carried out on grouts, kept during two years in 1.0% solution of Na_2SO_4, replaced every 4 weeks, and comparatively in distilled water proved, considering both bending strength and length variation (Graff-Kaufman method), practical lack of corrosion changes.

When used to high strength concrete, MPz IIIA and B cements give very high increase of initial strength and significant value 28th days strength. The level of final strength explicitly places these concrete in a group of HSC concrete. Exemplary composition and compressive strength distribution of concrete with MPz III B cement gives Table 5 For comparison tests of concrete with the same aggregate (granite 2/16 mm; sand : granite = 1 : 2) and with Portland cement 45 to which superplastifier (3.0%) and micro-silica (15%) were added.

Table 5 Strength of concrete with MPz III B and Portland 45 cement.

Cement	Compressive strength (days) f_c [MPa]					Concrete mix composition				
	1	3	7	28	90	c	w	k	Sp	Mk
Mpz IIIB	54.7	68.9	74.6	86.5	90.4	490	156	1755	-	-
CP45	42.1	54.4	61.3	89.0	93.5	520	148	1670	16	78

Investigation results confirm very high dynamics of initial growth of strength of concrete with MPz III B especially noticeable in first and third day of curing. In this range of time MPz III B cement assures strength over 30% higher then that of Portland cement 45 modified by addition micro silica and 3% superplastificizer.

In 28th day and later strength of both concrete becomes equal. To assure possibility of quick use for different repair and reconstruction applications a series of mixes with differentiated granulation and technical parameters with MPz type cement was worked out. All these mixes are characterised by very good workability and very low shrinkage. Some of their qualities are given in Table 6

Table 6 Some properties of compositions with MPz type cement

Composition	Max. dim. of aggreg Dmax	Min. time consistence constant	Shrinkage after 90 d (air-dry)	Compressive strength after n days f_c [MPa]		
	[mm]	[min]	[mm/m]	1	3	28
M - 38	4.0	30	0.25	30	50	70
M - 38/1	0.5	25	0.65	14	32.	42
M - 38/2	1.0	25	0.40	18	38	53
M - 38/3	13.0	40	0.15	17	53	75
M - 38/4	22.0	40	0.06	16	60	80

Technical applications comprised numerous group of cases where it was necessary to repair construction very quickly and put a building to further operation, reconstruct extensive corrosion defect or construct in a very short time a new, heavy loaded, element. It concerned particularly bridge structures, hydrotechnical objects, chimneys and cooling towers as well as foundations and concrete pavements. These cements proved to be very advantageous in temperatures below zero. Because of necessity of very careful formatting of mineral composition of clinker and very fine milling MPz cements are more expensive than ordinary Portland cement. Their production and numerous applications have proved that direction of formatting gypsum-less cements should be recognised as developmental, though further investigation will surely lead to gradual reduce energy -consumption and decrease production cost.

CONCLUSIONS

1. The idea of formatting „gypsum-less" cement according to general rules given in the paper proved that it was possible to get binders of MPz type which were characterised by very advantageous physical and strength properties.

2. Investigations and application which have been done up to the present prove that grouts and concrete on MPz cement have high increase of initial and 28th days which places them in the HPC concrete group. The level of strength 80 to 90 MPa is achieved with moderate (as for these concrete) use of cement and without use of superplastifiers additions without silicate.

3. In the range of physical properties concrete with MPz type cement are distinguished by increased resistance to aggressive sulphate environment and very low shrinkage which allows to accept them as „shrinkless".

REFERENCES

1. Peukert S., Cement Wapno Gips. Nr 1/1994 pp 4-8
2. Grzymek J., Cement Wapno Gips. Nr 3/1963 pp 43-48
3. Kurdowski W., Chemia cementu. Wyd.Nauk.PWN 1991
4. Werynski B., Wpływ uziarnienia na własności fizyczne zaczynów cementowych. Dys.Doct. 1972
5. Ratinow W.B., Rozenberg T.J. Dobawki w beton. Moskwa 1973.
6. RILEM Int.Symp. on Admixtures for Mortar and Concrete. Brussels 1967.

HIGH EARLY STRENGTH PORTLAND CEMENT BINDERS FOR USING AT LOW AMBIENT AND ELEVATED TEMPERATURES

M A Sanitsky

V Melnyk

State University Lviv Polytechnic

P M Chaba

Academy of Civil Engineering of Ukraine

Ukraine

ABSTRACT. This paper presents results of studies on high-early strength binding materials made on the basis of special portland cements with complex chemical admixtures which include surface-active substances and alkali metals salts. These binders are characterized by reduced water consumption and regulated setting period, and accelerated hardening at low, ambient and elevated temperatures.

KEYWORDS: Alkali metals salts, Surface-active substances, Ordinary Portland Cement (OPC), Gypsum-free Portland Cement (GFPC).

Professor, Dr.Sc. Myroslav A Sanitsky is the head of the division "The Technology of Binding Materials (Cements)", State University "Lviv Polytechnic", Lviv, Ukraine. He specialises in the investigation of rapid-hardening portland cement compositions for using at low, ambient and elevated temperatures. Professor Sanitsky has published widely. He is an Academician of the Academy of Civil Engineering of Ukraine.

Dr Volodymyr M Melnyk is an engineer, Chair of The Chemical Technology of Silicates, State University "Lviv Polytechnic". He is Supervisor of the Test Laboratory of Building Materials.

Mr Petro M Chaba is General Director of PBO "Lvivmiskbud". He is an Academician of the Academy of Civil Engineering of Ukraine.

Radical Concrete Technology. Edited by R K Dhir and P C Hewlett. Published in 1996 by E & FN Spon, 2–6 Boundary Row, London SE1 8HN, UK. ISBN 0 419 21480 1.

INTRODUCTION

Production of high early strenghth cements with regulated setting period permitting to raise the rate of concrete set up providing its hardening at low and negative temperatures is one of the main directions of chemistry and technology of binding systems development. The most radical way of the cement systems effectiveness increase is their activation by complex chemical admixtures of polyfunctional action [1].

Alkali compounds are rather strong activation agents of cement systems hardening [2,3]. Because of synergy alkali admixtures significantly improve plastificating properties of such surface-active substances as lignosulphonates (LS) and the like. Hence number of alkali-bearing admixtures (soda, liquid glass and others) in amounts exceding 1 mass. percent leads to sharp stiffening of concrete mixes. At the same time other salts (nitrite and sodium chloride) may be used in increased amounts (up to 10 mass persent) and they do not exert destructive effect on concrete mixes. That is why it is necessary to investigate the reasons of different effect nature of alkali metals salts on portland cement structureforming processes at different temperatures and show the possibility of production of high early strength portland cement binders to be used at low, ambient and elevated temperatures.

EXPERIMENTAL DETAILS

Portland cements ordinary (OPC) and gypsum-free (GFPC) were used. Alkali metals salts (sodium and potassium chloride, nitrite, nitrate, sulphate, carbonate, silicate) in complex with admixtures- plasticizes (lignosulphonates etc) were included in mix of cement compositions. Such cements included active mineral additives, unconventional additives, fillers.

A number of physicochemical analysis methods were used for investigation of hydration processes of cement compositions (X-ray phase, thermal, IR spectroscopy, electron microscopy, X-ray microprobe spectroscopy, flame photometry, mercury porousmetry).

Physico-mechanical tests of cements and concretes were carried out according to usual procedures. Concrete prism strength and modulus of elasticity of concrete and its frost resistance, imperviousness to water, corrossion resistance during hardening at ambient, low and negative (down to -35° C) temperatures were defined.

RESULTS AND DISCUSSION

The results obtained from the time of setting of OPC with alkaly metals salts admixtures show that they can be divided into two groups [4]. The first group consists of sodium salts such as chloride, nitrate, nitrite not affecting strongly on time of setting. The second group consists of sodium carbonate and silicate, and potassium salts. Even small dosage of these compounds results in the sharp acceleration of grout setting and the loss of the system plastic properties.

The results of investigation of microstructure and phase composition formation regularities of the cement pastes with alkaline metal salts of second group admixtures showed the causes of destruction phenomena, frequantly appearing due to the alkaline admixtures effect.The following types of interaction between gypsum and these admixtures take place: sodium carbonate and silicate undergo exchange reactions with gypsum forming calcium carbonate or hydrosilicate and mirabilite; potassium sulphate and gypsum form less soluble compound syngenite, which is complex double salt $K_2Ca(SO_4)_2H_2O$; other potassium salts also result to the formation of syngenite instead of low-soluble part of gypsum dihydrate.As a result gypsum is removed out of the process of early structure formation and doesn't serve its main function - to be inhibitor of three-calcium aluminate setting in portland cement. Fibrous syngenite crystalls which are formed reduce the thixotropic properties of paste to a great extent and it causes cement stone destruction, lack of concrete strength and its structure-technical properties. Given results affirm the expediency of alkali metals salts of the second grope usage as admixtures to portland cement without gypsum dihydrate.Admixtures of the first group - $NaCl, NaNO_2, NaNO_3$ do not interact with gypsum dihydrate and do not effect paste setting time of OPC.

The admixtures of the second group together with surface-active substances permit to obtain GFPC with specified setting time. It is significant that potassium salts have greater effect than sodium salts so cations together with anion part of alkaline metals salts admixtures effect essentially early structure formation of fine grounded portland cement clinker [5].But the admixtures of the first group even in complex with enhanced content surface-active substances (lignosulphonate and the like) up to 1,0 mass percent do not delay the beginning of GFPC setting and system stiffens rapidly.

Comparative strength structural tests of OPC and GFPC with alkali-bearing anti-freezing admixtures showed that at negative temperatures NaCl, $NaNO_2$ (the admixtures of the first group) correspond to OPC but potash (admixture of the second group) correspond to GFPC to greater extent. In addition there is a possibility of introduction of alkali metals salts of second group in increased amounts which are the hardening accelerator and anti-freezing admixtures.

The properties of the alkali metal carbonates and sulphates (with plasticizing/inhibiting admixtures) as high-effective hardening activators are fully discovered in GFPC in the process of its utilization. They also serve as a setting terms regulator for binding material which makes it possible to obtain rapid-hardening portland cements binders. The properties of these cements can be determined based on the type and the amount of plasticizing, inhibiting, alkali-bearing admixtures and mineral additives.

Introduction of optimal amount of active mineral additives and fillers replacing clinker in the composition of portland cement binders with alkali activation provides the energy saving in their production and facilitates the growth of cement stone durability at the expense of alkali component binding into hydrate phases.GFPC with carbonate additives exhibit increased effectiveness when hardening at ambient, low and negative temperatures. It is so due to modifying effect of $CaCO_3$ on alumobearing phases in the presence of gypsum is exhibited not fully. The usage of unconventional additives (fly ash, burnt rock, ets.) for the high early strength portland cement binders is possible.It was established that for these processes activation it is worthwhile to use alkaline sulphate-containing product - wastes synthetic falty acids. These additives also intensify the portland cements milling.

In the processes of early structure-forming in portland cements incorporating alkali-bearing admixtures in amounts exceeding 2 mass. percent, the liquid phase has a lower concentration of calcium ions as compared to ordinary portland cement. According to X-ray spectral microanalysis data, the alkali metals salts admixtures are mainly concentrated on the outer hydrates composition. Therefore, scanning electron microscopy shows formation of more fine crystalline microstructure of stone, whereas the trimethylsilylation method shows considerable increase in the polymerization rate of silicon-oxygen anions in hydrosilicatecs. In the process of C_3A hydration with the said admixtures, crystals of hexagonal AF_m -phases are formed which can tightly fit in to each other. So, there are fever of cement grains moving apart in the early period of hydration than in the period of ettringite acicular crystals formation. Besides, breakdown effects which occur during the transition of the high-sulphate form of calcium hydrosulphoaluminate into the low-sulphate one and which are accompanied by change in crystal habit and density,do not occur with the increase of hardening time. Together, the factors contribute to the forming of the dense, strong but least stressed structure of the cement stone.

PRACTICAL APPLICATION OF RESULTS

Concrete based on rapid-hardening portland cements with alkaline activation attains 50-70 % of the branding strength in 16 hours at ambient temperatures and in twenty-four hours it attains 70-100 % of branding strength, when hardening at lower positive temperatures and at frost down to $-10°$ C the possibility of construction of monolithic structures with terms of loading which are close to summer loading is created. At temperatures $-15°$ C such concrete attains 80-100 % of branding strengh in 28 days and posses the capability of hardening at frost $-30°$ C.(Table 1)

Table 1 Compressive strength of concrete based on OPC and GFPC at ambient and negative temperatures

CONCRETE	TEMPERATURE °C	COMPRESSIVE STRENGTH, N/mm^2		
		7 day	28 day	90* day
OPC	+20	20,4	22,5	22,8
	-15	12,9	17,5	21,0
GFPC	+20	25,7	30,0	31,2
	-15	18,4	29,6	32,0

* 28 days at temperature -15° C, then 62 days at temperature +20° C

In case of using rapid-hardening portland cements prism strength, modulus of concrete elasticity and its adhesive strength with reinforcement is not less than similar indecies for ordinary concrete which hardens under normal conditions even if it was affected by negative temperatures. Besides, concrete is characterized by high value of frost resistivity (F 300 and greater) and increased water-proofness.

CONCLUSIONS

1. The establishment of the causes of destruction phenomena, frequantly appearing due to the alkaline additive effect, allowed to form a group of high early portland cement binders with regulated setting period and accelerated hardening at different temperatures.

2. The use of GFPC allowed to produce rapid-hardening binders.Technological methods of control over the binding systems properties presuppose a direct influence on the processes of their early structure formation and liquid phase composition. They are determined by selection of mineralogical composition of the portland cement clinker, type and member of retarders, plastificators, alkaline metals salts and mineral additives as well as temperature. As a result of synergism the alkaline admixtures significantly improve the plastisizing effect of the surfase-active substances like lignosulphonates, pentaeretrite production waste, etc.

3. The use of the high early strength portland cement binders has the following advantages: the terms of the objects commission reduse; the formwork turnover speeds up; it can be also used for making concrete and prefabricated reinforced concrete without its steam curing.

4. The use of GFPC with complex chemical admixtures for concrete in winter-time construction allows to effectively carry out construction and erection work without heating at negative temperatures (down to -30° C) which is material, energy and labour saving.

REFERENCES

1. RAMACHANDRAN, V S, FELDMAN, R F AND BEAUDOIN, Y Y. Concrete science. Tretise on Current Research, Heyden, 1981.pp 286.

2. KRIVENKO, P V. Special Slag-alkaline Cements, Kyiv, 1992, pp 192.

3. DAVIDOVITS, J. Properties of Geopolymer cements.First International Conference named after V.D.Glukhovsky,Kiev, 1994, V.1, pp 131-150.

4. PASHCHENKO,A A, SANITSKY, M A AND SHEVCHUK G Ya. Special Features of the Portland Cement Compositions with Alkaline Metals Salts Admixtures. Cement, No 7, 1990. pp 17-19.

5. SANITSKY, M A AND SOBOL, H S. Gypsum-free Rapid Hardening and Mixed Portland Cements. Proceedings of the 9th Internarional Congress on the Chemistry of Cement, New Delhi, 1992, Vol 3, pp 438-443.

CONTROLLED LOW STRENGTH MATERIALS (CLSM)

W E Brewer

Brewer & Associates

USA

ABSTRACT: Controlled Low Strength Material, referenced by the acronym CLSM, is defined by the American Concrete Institute as a material having a 28 day compressive strength less than 8.28 MPa (1200 psi). Its primary ingredients are portland cement, fly ash and filler aggregate.

Although CLSM's have been in use for a number of years, there still remains confusion as to CLSM's construction benefits and economical savings. The principal use of CLSM has been as a controlled density fill, referenced by the acronym CDF, in place of conventionally placed backfill. Other possible uses for CLSM mixtures consist of fills for: thermal, corrosion, pavement base, structural, and permeability.

This paper provides basic knowledge about CLSM mixtures: history, applications, material components, design, engineering properties, manufacturing, specifications, and cost. It also prescribes implementation methods for using CLSM mixtures.

Keywords: aggregate, backfill, economics, compressive strength, concrete, Controlled Density Fill (CDF), Controlled Low Strength Material (CLSM), corrosion, flowability, fly ash, K-Krete, pavement base, permeability, quality control, removability, specifications, thermal.

Mr William E. Brewer is President of Brewer & Associates, Toledo, Ohio. He has been involved in the study of Controlled Low Strength Material (CLSM) for over 30 years and was the Founder and original chairperson of ACI's Control Low Strength Material committee 229. He has over 45 years of experience in various phases of civil engineering and has written numerous papers and lectured on a variety of subjects.

Radical Concrete Technology. Edited by R K Dhir and P C Hewlett. Published in 1996 by E & FN Spon, 2–6 Boundary Row, London SE1 8HN, UK. ISBN 0 419 21480 1.

INTRODUCTION

The specified conventional backfill technique, for excavations supporting loads, has been the placement and tamping of granular material in layers. The granular material is placed in layers and tamped to achieve the desired compaction (density). In many cases, the material was dumped full depth into the trench; never tamped or adequately compacted.

In the early 1970's, engineers started looking at alternatives to conventional backfilling materials and methods [1, 2]. One alternative was the development of a material designated as K-Krete (CDF); where CDF stood for Controlled Density Fill. The material had a low strength, in terms of concrete, having a 28 day compressive strength in the range of 0.69 MPa (100 psi). K-Krete was a patented material process developed by the Detroit Edison Co., Detroit, Michigan and Kuhlman Corp., Toledo, Ohio [3, 4, 5, 6]. The material is still being sold under the trademark name of K-Krete.   Because of this material's success, other similar materials have been developed and sold with a variety of trade names: M-Crete, S-Crete, Flowable Fill, Flash Fill, Flowable Grout, Flowable Mortar, One-Sack Mix and so on.

By 1980, it was evident to the early developers of low strength materials that technical information about this product was not being developed. Some information was published in trade magazines but not on a consistent basis. An American Concrete Institute (ACI) committee 229 was formed to correct these deficiencies. The ACI 229 committee is designated as Controlled Low Strength Material with the acronym of CLSM. The committee defined low strength to be a material with a 28 day compressive strength of less than 8.28 MPa (1200 psi). The creation of the ACI 229 committee helped extend the news of CLSM. In recent years, ready mixed concrete trade associations have published numerous articles on CLSM uses [7, 8, 9]. Even with this extended publicity, CLSM uses were confined because of misunderstandings regarding construction applications and a realistic pricing structure.

USES AND APPLICATIONS

Although this paper will primarily address backfilling with CLSM controlled density fill (CDF), the reader should be aware that CLSM is really a family of possible mixtures for a variety of uses: pavement base, structural fill, thermal fill, anticorrosion fill, high or low permeability fill and so on. Each mixture is designated by a three letter acronym such as: CPB (controlled pavement base), CDF (controlled structural fill), CTF (controlled thermal fill), and so on. The CLSM mixtures can vary for each application and within each application to achieve the required end results. For example, when backfilling around a culvert, the initial backfill below the end wall elevation could be very fluid. Above the end wall elevation, the flowability of the mixture can be adjusted to avoid CLSM-CDF's flowing over the top of the end wall [10].

MATERIAL COMPONENTS

The basic components for CLSM-CDF are portland cement, fine aggregate, fly ash, and water. These components can be varied depending on a component's availability and quality. Any replacement, or adjustment, of the components requires laboratory testing prior to its use to insure that the specified end results can be met. In the case of CLSM-CDF's material components, when mixed and placed, must possess four end result properties: flowability, removability, strength, and a competitive price.

Portland Cement

Type I portland cement (*ASTM C150*) is usually used with CLSM mixtures. This is not to say that in some circumstances other types of cement could or should not be used. The amount of portland cement used, for CLSM-CDF, is in the range of 3% of the total mixture's weight. The purpose of the portland cement is to achieve cohesion and control strength in the mixture. For typical backfills, where future removability is anticipated, the compressive strength (C') should be less than 0.69 MPa (100 psi) in 28 days.

Fine Aggregate

The fine aggregate, known as filler, makes up the major portion (72%) of a typical CLSM-CDF mixture. The majority of fillers have initially been fine aggregate (*ASTM C33*), as used in the manufacture of concrete, because of its availability. While this material proved to be an excellent CLSM-CDF filler, it was not the must economical material to use. Materials not meeting ASTM standards have proven to be more economical. The filler should possess adequate gradation, similar to the requirement as set forth in *ASTM C33,* to insure proper flowability. Another consideration is the material's particle angularity; sharp edges reduce flowability.

Fly Ash

The major source for fly ash, for the manufacture of CLSM, is produced from the burning of coal in the generation of electricity. The fly ash is electrostatically collected. Fly ash used in the manufacture of portland cement concrete complies to *ASTM C618*. This same fly ash specification can be used for the manufacture of CLSM. Non-specification (*ASTM C618*) fly ash can also be used; prior testing should be conducted to insure end result acceptability. Fly ash makes up approximately 8% of a typical CLSM-CDF mixture.

Water

Water is used in a CLSM mixture for flowability and hydration. The large amount of water in the mixture, while affecting yield, does not necessarily affect the compressive strength (C'). A typical CLSM-CDF mixture would consist of approximately 17% water. The water used in CLSM mixtures, as with portland cement concrete mixtures, should be potable.

DESIGN CONSIDERATIONS

The design of a CLSM mixture depends on its desired end use and the availability of materials for manufacturing the mixture. To help identify possible end uses and related CLSM engineering properties see *Table 1*.

Table 1 CLSM uses and related engineering properties

Use	CLSM Acronym	Engineering Properties
Backfill	CDF	Flowability Removability Strength
Pavement Base	CPB	Flowability Strength
Structural Fill	CSF	Flowability Strength
Thermal Fill	CTF	Flowability "R" value Removability Strength
Anti Corrosion Fill	ACF	Corrosion Flowability Removability Strength
Permeability	CPF	Permeability Removability Strength

Each use requires an understanding of the intended application and the mixture properties of the CLSM. Following are examinations of each application and its engineering properties as referenced in *Table 1*. The examinations should help explain the relationship between the application and the CLSM mixture.

Backfill (CLSM-CDF)

This is for all types of backfills, but primarily excavations on which loads will be applied. The engineering properties are: flowability, removability, and strength.

Flowability

In the case of utility trenches, flowability consists of "pouring" the CLSM-CDF mixture in a trench and having it flow around and along the conduit. A test for determining flowability is found in the section on Specification and Standards of this paper.

Removability

This property must be considered if future excavation is a possibility. Laboratory and field research shows that a compressive strength (C') of less than 100 psi is required. This value is based on a number of construction factors: excavation equipment, cutting edge, impact, power, direction of excavation, CLSM-CDF's strength and density. CLSM-CDF field research (1990) conducted by the Cincinnati Gas & Electric Co., Cincinnati, Ohio has resulted in the development of a empirical equation.

Note: The length restriction of this paper does not allow for a complete discussion on CLSM removability. A complete discussion on the subject of removability can be found in reference [11].

Strength

The strength of CLSM-CDF mixtures are determined by unconfined compressive tests. In the early development of CLSM concepts, the primary investigators were concrete research oriented and they naturally referenced strength to megapascal (MPa) ((pounds per square inch (psi)). When using CLSM-CDF as an alternative to conventional backfilling, Newtons per square meters (N/mm² ((tons per square foot (Tsf)) is more appropriate. The major concern for CLSM's removability is to keep the strength low. A CLSM mixture with a compressive strength of 0.69 MPa (100 psi) possess a good bearing capacity 0.77 N/mm² (7.2 Tsf).

Pavement Base (CLSM-CPB)

The acronym, CLSM-CPB stands for Controlled Low Strength Material - Controlled Pavement Base. As with CLSM backfill research, pavement base research has also been conducted by the Cincinnati Gas & Electric Company. This research dealt with load transfer studies on backfilled trenches consisting of various CLSM-CDF strengths and properly placed conventional backfills [12]. In addition to this field research, previous laboratory work included California Bearing Ratio (CBR) tests for different CLSM strength ranges. This CBR testing was conducted according to *ASTM D1883*. The specimens were tested at various ages rather than after the 96 hours of soaking as used for soils. The results of this testing are shown in *Figure 1*. CLSM strength correlation with CBR values is necessary since pavement design methods incorporate CBR's or CBR correlation in their design [13].

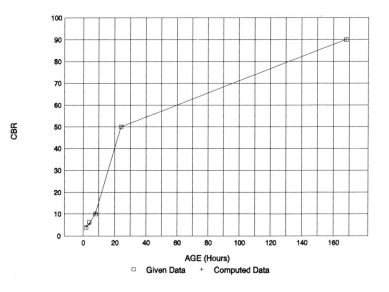

Figure 1   CBR values for CLSM-CPB

Flowability for CLSM-CPB's deals with the ability to regulate the CLSM mixture's flow on steep pavement grades. The CLSM mixture's flowability is regulated while maintaining the desired compressive strength. This is similar to controlling the flow for end slopes on culverts.

The AASHTO (American Association of State Highway and Transportation Officials) Road Test (1968-1972) developed a term known as *pavement coefficient* and is primarily used in the design of flexible pavements [13]. A mathematical correlation has been developed for determining pavement coefficients when the compressive strength is known. This correlation is shown in *Figure 2*.

This figure (*Figure 2*) was developed using the general AASHTO equations and laboratory investigation on CLSM mixtures. It shows that for a compressive strength of 6.9 MPa (1000 psi), the resulting coefficient (a_2) would be approximately 0.28. It is also interesting to note that 34.5 MPa (5000 psi) concrete would have a coefficient approximately equal to 0.60. These coefficients can be used in the design of a flexible pavement, but some engineers have used this coefficient approach to the design of rigid (concrete) pavements. Separate CLSM reports discuss CLSM-CPB applications when designing and constructing pavements [*10, 11, 12*].

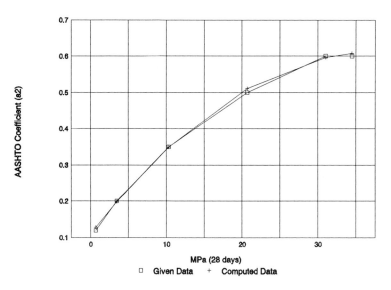

Figure 2   AASHTO coefficients based on compressive strength

Structural Fills (CLSM-CSF)

CLSM-CSF's are regulated by flowability and compressive strength.   The compressive strength will usually be higher than the strengths of CLSM mixtures for trench backfill.   In some cases the compressive strength has been as high as 8.28 MPa (1200 psi).   The design strength is based on foundation calculations and the bearing capacity of the supporting soil.   CLSM-CSF's have been used to support spread footers, grade beams, and load bearing walls.

The placement of CLSM-CDF requires a good flowable material.   The placement should not require any labor for placement other than someone to direct the placement from the ready mixed truck via its chute.   If truck chutes will not reach the placement location, CLSM-CSF can be pumped.

Thermal Fills (CLSM-CTF)

CLSM-CTF offers a number of opportunities for designers of structures. The CLSM-CTF mixture is modified to increase its insulation properties. While flowability and removability are important aspects of a CLSM-CTF mixture, thermal conductivity is the prime concern. CLSM-CTF mixtures can be designed to meet or surpass the thermal conductivity values of light weight concrete. This is achieved by using different fillers and/or admixtures.

Anti Corrosion Fills (CLSM-ACF)

Corrosion tests have shown a reduction of corrosion on metal pipes with the use of a CLSM-ACF backfill [14]. Laboratory and field research confirm that CLSM mixtures can be designed to reduce corrosion. Fly ash meeting *ASTM 618* standards has a Ph in the range of 9 to 11. The mixture's flowability and removability are important factors but controlling corrosion is the must important. Continued research on this subject is currently being conducted by the Cincinnati Gas and Electric Company.

It should be noted that the fly ash used in a CLSM mixtures does not cause corrosion. This is possible because fly ash's small particle size helps reduce moisture transmission through the mixture.

Permeability Fills (CLSM-CPF)

Permeability fills can be created by changing the CLSM's density through the filler material's gradation. Mixtures can be designed for a wide range of permeabilities. Laboratory tests have resulted in permeability values from .305 to 228.6 (10^5 m/min) ((1 to 750 (10^5 ft/min) for a standard CLSM mixture with varying filler types.

As with the other CLSM mixtures: flowability, strength and removability are also important. Each mixture has a prime governing factor, such as strength or permeability, that will be its major mix design consideration.

MANUFACTURING AND SUPPLYING

Ready mixed concrete equipment has generally been used to manufacture CLSM mixtures. This is not to say that other types of equipment and mixing procedures have not or could not be effective. Since the early CLSM concepts were developed by ready mixed concrete producers, it was natural for ready mixed concrete equipment to be used. The important thing to remember is proper mixing of the CLSM components. Without proper mixing flow, removability, and strength will not be achieved or controlled [7].

The usual method for transporting CLSM-CDF mixtures to the project has been with the use of ready mixed concrete trucks. With the advent of CLSM-CDF mixtures the ready mixed concrete trucks should now be designated as a material mixer and transporting truck. The CLSM-CDF mixture is usually placed by "pouring" directly from the truck into the trench or excavation. The material can also be pumped as previously referenced.

SPECIFICATIONS AND STANDARDS

CLSM specifications vary with each type of construction application. While all CLSM specifications are similar; the mixture design and placement requirements could be different. Many agencies have written specifications for CLSM-CDF. In the majority of the cases they spell out: intended use, materials, mixture design, placement, testing, and method of measurement and payment. While it is not the intent of this paper to write a generic specification for CLSM mixtures, a general "model" CLSM-CDF specification is presented in italics accompanied by notes.

Intended Use

In lieu of compacted granular backfill the contractor may use Controlled Low Strength Material - Controlled Density Fill (CLSM-CDF).

Materials

CLSM-CDF is a controlled density fill made of cementious materials; Portland Cement, water and selected filler material such as fly ash and aggregate. The combining of these materials shall result in a material that will achieve a compressive strength (C') of less than 0.69 MPa (100 psi) at 28 days and permit easy removability at some later date.

The CLSM-CDF material will have good flowability to allow for uniform placement in the trench. Good flowability means that the material will seek its own level while holding all mixture components in suspension.

NOTE: It is important to understand that the CLSM mixture must be able to flow into the trench thereby eliminating all labor requirements for placing. Several tests have been developed for laboratory determination of adequate flow. Early researchers developed the opened ended, 7.62 cm diameter x 15.24 cm long (3" diameter x 6" long), cylinder test. An open ended cylinder is placed on a level surface and the CLSM material is poured into the cylinder. The cylinder is then lifted, with a smooth vertical motion, to allow the material to flow out on the level surface. Good flowability requires no material segregation and a spread of approximately 20.32 cm (8") in diameter.

Mixture Design

Trial mixtures for CLSM-CDF should be made. For a CLSM-CDF the 28 day compressive strength should be less than 0.69 MPa (100 psi). A starting point for trial mixtures, to achieve the 0.69 MPa (100 psi) strength, could be:

Portland Cement (Type I)	*45.4 kg (100 lb.)*
Fly Ash (Type F)	*136.2 kg (300 lb.)*
Aggregate Filler	*1180.4 kg (2600 lb.)*
Water	*272.4 kg (600 lb.)*

NOTE: Some agencies will spell out mixture proportions; others just state they require a 28 day strength to be less than 0.69 MPa (100 psi).

Placement

The CLSM-CDF material may be placed directly from the truck chute into the trench. No vibration or compaction is required.

Testing

The suggested controls and field testing procedures are in regard to the final end use of the mixture. These tests consist of the following:

ASTM C138, Test for Unit Weight
ASTM C39, (modified) Test for Compressive Strength
Flow Test, no ASTM designation

The ASTM C39 Standard has been modified in regard to the cylinder's size and rodding requirements. The cylinder size can be either 10.16 cm x 20.32 cm (4" x 8") or 7.62 cm x 15.24 cm (3" x 6"). Naturally, 15.24 cm x 30.48 cm (6" x 12") cylinders can be used, but smaller cylinders yield satisfactory results. To simulate field placement, no rodding should be done after placing the mix in the cylinder. The cylinders should be allowed to stand, undisturbed for at least 48 hours.

NOTE: Some agencies require unit weight tests. This would be important if a specific density was the major design requirement. For general backfilling, compressive strength and removability have been the prime considerations.

Quality Control

Quality control for CLSM-CDF mixtures requires attention to the uniformity of materials, proper charging and mixing equipment, and material transportation to the project site.

NOTE: It cannot be over stressed that materials for CLSM mixtures must be uniform to provide control of flowability and strength. CLSM specifications have used American Society for Testing Materials (ASTM) standards as designated for concrete's components for quality control.

        ASTM C33   - Concrete Aggregate
        ASTM C150   - Portland Cement
        ASTM C618   - Fly Ash

These standards have helped to provide for uniformity in CLSM mixtures. Materials not meeting these standards have successfully been used. Use of non-ASTM materials require laboratory testing to determine flowability, removability, and strength of the mixture.

Method of Measurement and Payment

The quantities which constitute the completed and accepted structure will be measured for payment by the cubic meter (cubic yard). Payment will be made on the unit price basis for the measured cubic yard.

ECONOMICS

For estimating purposes transportation costs for concrete are used for the transporting cost of CLSM-CDF mixtures. Technically the CLSM-CDF supplier should consider less wear on equipment (blades) and faster placing times. Since the majority of the CLSM-CDF mixtures contain smaller sized aggregate than concrete, blade wear is greatly reduced. Because CLSM requires no vibration or work after placing, placement time is reduced from the usual 10 minutes per cubic yard for concrete, to 10 minutes or less for the entire CLSM-CDF load. Placing CLSM-CDF mixtures can significantly increase the "turn around" equipment time.

Construction Considerations and Backfilling

When considering total CLSM-CDF costs, related construction requirements must also be considered by the contractor. Related construction requirements include: trench width, OSHA, and speed of backfill placement. Trench widths can be reduced with the use of CLSM-CDF since a wider trench is not required to achieve adequate compaction around the conduit. This trench reduction width also reduces excavation costs and the amount of backfill material required.

Occupational Safety and Health Administration (OSHA) regulations require sloping sides for trench excavations [15]. For conduit placement, with a "steel box" and CLSM-CDF backfilling, sloping sides could be eliminated since no personnel are required to be in the trench during backfilling.

Backfilling is expedited with the use of CLSM-CDF because there are no delays for compaction testing in the trench. Backfilling can be as fast as the CLSM-CDF material can be poured into the trench. If the backfilling operation is expedited, it reduces total construction time.

Material Costs

To determine a proper pricing structure for a CLSM-CDF a knowledge of material costs is required. Material costs will generally vary with: geographical locations, time of year, and competition.

It cannot be over emphasized that the cost of the fill material has the greatest significance in determining the cost of a CLSM-CDF mixture. All available materials for possible aggregate filler should be investigated. The material can be a nonstandard material that could satisfy CLSM mixture requirements. The Hatfield Station project (Pennsylvania) is an example where bottom ash was used as the filler material [2].

Yield Of Mixture

There are two yield considerations for CLSM mixtures: the plastic or wet yield and the hardened or subsided yield. The absolute volume of the wet mixture is calculated in the same manner as for concrete. Because of the high water content, a significant amount of water will bleed off a placed CLSM mixture. Therefore, the hardened volume will be less than the initial wet volume. A typical subsided yield would have a reduced volume of approximately 6 to 8%. This reduced hardened volume must be reflected in both price and volume requirements. The specific gravity of the components, as with concrete, will affect the final hardened volume. A wet over yield could be adjusted by reducing the aggregate filler as long as proper flowability is maintained. Another consideration is that a higher wet yield will increase the subsided volume. A volume reduction of 6 to 8% would result in a final subsided volume of 0.736 m^3 to 0.722 m^3 (26.0 to 25.5 cu.ft.) respectively.

Competitive Price

The cost of using CLSM-CDF in place of conventional backfill has been debated between many CLSM-CDF producers and contractors. Favorable economics for using CLSM-CDF depends on the specification enforcement of the conventional backfill method and the cost of backfill materials [16, 17].

Consider a roadway trench with the following dimensions: 0.914 m (wide)(3'), 1.829 m (deep)(6'), 12.192 m (long) (40'). The total backfill material requirement would be 20.40 m^3 (26.67 cu.yd.) less the pipe's displaced volume. Using unit costs of: $37.62/m^3 ($28.76/cu.yd.) CLSM-CDF, $14.00/hr. labor and $10.46/m^3

($8.00/cu.yd) granular backfill, a cost comparison study shows that the CLSM-CDF cost to be $ 867.03 as compared to $ 836.74 for the conventional fill. It should be noted that for every one foot (.305 m) reduction in trench width, with CLSM-CDF use, results in a cost reduction of $255.64. While the $867.03 is larger than the $ 836.74, OSHA cribbing or side slope regulations need to be added to the conventional backfill's costs [15]. Lower CLSM-CDF costs could be achieved by considering less equipment wear and faster placement times. The estimated unit costs used in this example are from the Area Paving Council (APC), Toledo, Ohio.

IMPLEMENTATION PROCEDURES

To answer the question, "how to implement CLSM applications?", the following punch list is suggested:

- Investigate the availability of local aggregate filler.
- Perform laboratory tests on proposed CLSM mixtures.
- Review costs of CLSM mixtures with local ready mixed concrete producers. Assist in cost analysis of CLSM applications versus conventional applications.
- Consider demonstration project in lieu of conventional construction practices. The demonstration project allows for the monitoring of costs and construction activities.
- Write and adopt specifications for the CLSM application.

CLOSURE

The purpose of this paper was to furnish basic knowledge about CLSM materials: history, applications, material components, design, engineering properties, manufacturing, specifications, cost, and how to implement their use. The reader should be aware, as noted in the text, that detailed reports are available for the various topics referenced. If copies of these detailed reports are desired, the reader should contact Brewer & Associates, P.O. Box 8, Maumee, Ohio 43537-0008.

REFERENCES

1. BREWER, WM. E. The End of the Backfill Problem. Concrete Construction Magazine, Oct. 1975.

2. ENGINEERING NEWS RECORD (ENR). Slurry Fills Eases Big Pipe Setting, Nov. 11, 1976.

3. UNITED STATES PATENT - 4,050,258 - Brewer et al. Method of Building Embankments and Structure Supports of Backfilling, Sept. 27, 1977.

4. UNITED STATES PATENT - 4,050,261 - Brewer et al. Method of Backfilling, Sept. 27, 1977.

5. UNITED STATES PATENT - 4,050,950 - Brewer et al. Controlled Density Fill Material Containing Fly Ash, Dec. 13, 1977.

6. UNITED STATES PATENT - 4,062,195 - Brewer et al. Method of Bedding A Conduit Using Controlled Density Fill Material, Dec. 13, 1977.

7. BREWER, WM. E. Controlled Low Strength Material (CLSM) & the Ready Mixed Producer. Tennessee Concrete, Vol. 2, No. 2 Summer 1988.

8. HEWITT, GENE. If It Plays in Peoria It Will Do So Everywhere. Illinois Ready Mixed Concrete Newsletter, Feb. 1989.

9. OHIO PAVER, CDF Expedites Subdivision Development, Feb. 1988.

10. BREWER, WM. E. The Design And Construction Of Culverts Using Controlled Low Strength Material - Controlled Density Fill (CLSM-CDF) Backfill: Proceeding Of The First National Conference On Flexible Pipes, Columbus, Ohio, A.A. Balkema, Rotterdam, 1990.

11. BREWER, WM. E. Factors Governing The Removability of Controlled Low Strength Material - Controlled Density Fill (CLSM-CDF). The Cincinnati Gas & Electric Company, Cincinnati, Ohio, 1991.

12. BREWER, WM. E. Load Transfer Comparisons Between Conventionally Backfilled Roadway Trenches And Those Backfilled With Controlled Low Strength Material - Controlled Density Fill. The Cincinnati Gas & Electric Company, Cincinnati, Ohio, 1991.

13. AMERICAN ASSOCIATION OF STATE HIGHWAY AND TRANSPORT-ATION OFFICIALS (AASHTO). AASHTO Guide For Design Of Pavement Structures, Washington, D.C., 1986.

14. BREWER, WM. E. Controlled Low Strength Material - Controlled Density Fill (CLSM-CDF) Research - Corrosion Testing. The Cincinnati Gas & Electric Company, Cincinnati, Ohio, 1991.

15. OCCUPATIONAL SAFETY AND HEALTH ADMINISTRATION (OSHA). 29 CFR Ch. XVII, 1926.652 (7-1-89 Edition).

16. BREWER, W. and HURD, J. Economic Considerations When Using Controlled Low Strength Material (CLSM) As Backfill. Transportation Research Board, Washington, D.C., 1991.

17. BREWER, WM. E. The Economics of Using A Controlled Low Strength Material - Controlled Density Fill (CLSM-CDF) As Compared To Conventional Backfill. The Cincinnati Gas & Electric Company, Cincinnati, Ohio, 1991.

THEORETICAL FUNDAMENTALS OF THE APPLICATION OF EXPANSIVE CEMENT CONCRETES

I Chartschenko

J Stark

HAB Weimar - Universitat

Germany

ABSTRACT. The greatest stability of properties is found in expansive cements made on the basis of calcium aluminate sulphate and/or the formation of ettringite. The factors influencing the properties of ettringite-forming expansive cements are represented in a phenomenological model. The factors influencing the properties and the process of expansion are analysed. By controlling the process of hydration during the hardening of the expansive cement the linear change of length may be kept within the range of 0.1 to 200 mm/m and the properties of concrete and reinforced concrete structures may be improved. Expansive cements which are produced by mixing commercial portland cement with an expansive component offer considerable technological advantages and possibilities of further development.

Keywords: expansive cements, controllable formation of structures, ettringite, pH-value, pore structure, durability

Dr Igor Chartschenko is Ass. Prof. of the Institute of Building Materials at the University of Architecture and Building in Weimar, Federal Republic of Germany.

Professor Dr Joachim Stark is director of the Institute of Building Materials at the University of Architecture and Building in Weimar, Federal Republic of Germany

Radical Concrete Technology. Edited by R K Dhir and P C Hewlett. Published in 1996 by E & FN Spon, 2–6 Boundary Row, London SE1 8HN, UK. ISBN 0 419 21480 1.

PHENOMENOLOGICAL MODEL OF EXPANSIVE CEMENTS

Practical experience shows that the greatest stability of properties is found in expansive cements made on the basis of calcium aluminate sulphate and/or the formation of ettringite. The properties, i.e. first of all the process of expansion, which may be represented in the form of a tree ("expansive cement tree") (fig. 1), depend on "external" and "internal" factors of influence. By the influence of these factors on the mechanism of ettringite formation, on the morphology of ettringite phase, on the amount of ettringite phase formed, on the kinetics of ettringite formation and on the properties of the portland cement matrix, which is called the fifth root of the expansive cement tree, the process of the formation of microstructure during the hardening of expansive cement can be controlled. The "external" factors of influence are temperature (sun), moisture (rain or drying-out) and the extent of impediment to the expansion process.

The "internal" factors of influence are, first of all, the hydraulic activity of the Portland cement portion, the expansive components and the additives which determine the content of ions in the reaction solution in the process of hydration. In addition, there are, of course, the water-cement ratio and the additives which enable the desired coordination between the hardening of the Portland cement matrix and the formation of ettringite to be achieved. The desired properties of expansive cements and a control of the expansion process in time and space can only be achieved if the processes which all five "roots" have to go through under the "external" and "internal" factors of influence are under control. As in nature, it is difficult to decide which root is the most important for the life of the tree and the quality of the fruit (in this case the properties of expansive cements). All of them make a different but equally important contribution.

HYDRATION AND FORMATION OF THE MICROSTRUCTURE IN EXPANSIVE CEMENTS UNDER THE INFLUENCE OF "INTERNAL FACTORS"

The physical-mechanical characteristics of expansive cements and expansive cement concretes are closely correlated with the relationship between the speed of hardening of the expansive cement matrix (= Portland cement portion) and the speed of ettringite formation in the first 3 to 5days of hardening and/or with the active formation of the microstructure of the hydrated cement. This may, for instance, be achieved by deliberately changing the fineness of both components of the expansive cement, i.e. the expansive cement matrix and the expansive component / 1 /.

According to several authors / 1; 2; 3 / and our own test results / 4 / there is no universally valid relationship between the amount of ettringite formed and the amount of expansion. An interdependence does only exist if the quality of all components of expansive cement is constant and the external conditions do not change / 2 /.

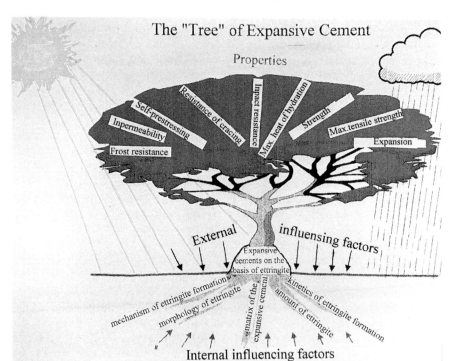

Fig. 1: Expansive cement "tree"

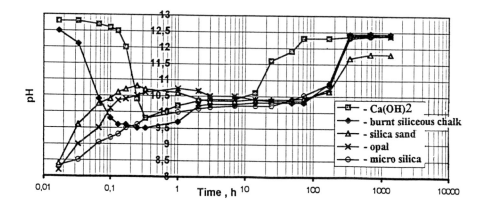

Fig. 2: Changes in pH-value during the hydration of TG-QZ (high-alumina/gypsum expansive cement) with different additives

By changing the concentration of ions and particularly the OH-content in the reaction solution it is possible to exert an active and deliberate influence on the crystallization of ettringite and the formation of a certain type of crystal / 1 /. Practically, by means of the properly adjusted pH-value during the initial hardening of expansive cements the morphological formation of ettringite and thus the formation of the microstructure of the hydrated cement may be controlled.

Our investigations have shown that different mineral additives to high-alumina/gypsum expansive cement (TG-QZ) can influence the ion-content and particularly the content of OH-ions in the reaction solution and in this way also the kinetics of ettringite formation, the morphology of ettringite and finally the properties of expansive cement. The expansive cement was produced by homogenizing a mixture of high alumina cement and ground gypsum in the stoichiometric relationship of the ettringite for 60 minutes. Mineral additives were silica powder (SF), ground opal (O) as natural puzzolanic component, siliceous chalk (K) burnt at 1000°C and ground, pure calcium hydroxide (CH) and ground silica sand.

Our investigations have shown that depending on the type of mineral additives the pH-value of the pore solution of high alumina/gypsum expansive cement (TG-QZ) differed greatly in the first 24 hours after mixing had begun (addition of mixing water) (fig. 2).

Particularly interesting are the properties of TG-QZ with calcium hydroxide (CH) and siliceous chalk (K). These additives showed an extremely great difference between amount of expansion and compressive strength (fig. 3 and 4). Although just after mixing the pH-value was nearly the same, it decreased from 12.5 to 9.7 within the first six minutes when burnt siliceous chalk was used and then remained constant for up to 72 hours. The analysis of the products of hydration showed that ettringite was formed at the same speed and after three days amounted to about 85%. In TG-QZ with calcium hydroxide the pH-value remained practically constant for the first ten minutes and decreased to 9.6 in the following 10 minutes. After a period of stabilisation on this level in the following 7 hours the pH-value increased to 12.3 and remained constant for 28 days. The formation of ettringite amounted to 70% after 24 hours and after 48 hours had achieved 90% of the final value.

This phenomenon may be explained by the formation of separating layers on the surface of the cement particles which impede the further course of hydration and thus cause a decrease in the concentration of OH-ions in the pore solution. In the further process of hydration ettringite is formed in the interspace between the unhydrated clinker minerals and the ettringite shell. The pressure of crystallization acts on the inner side of this shell. This load is increased by the difference in the concentration of hydration products on the outer and inner sides of the shell which is filled with products of hydration and, first of all, with ettringite.

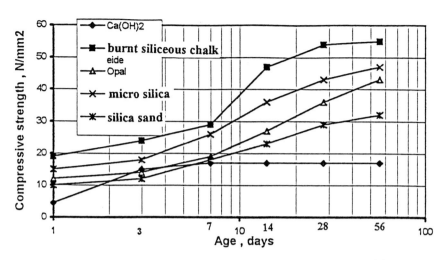

Fig. 3: Compressive strength of TG-QZ with different additives

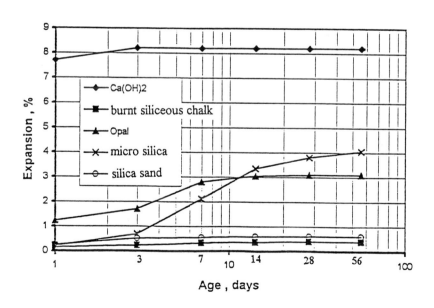

Fig. 4: Amount of expansion of TG-QZ with different mineral additives

The interspace between the separating layer and the unhydrated clinker minerals becomes too small and the ettringite shell is destroyed. After the destruction of the ettringite shell around the clinker mineral particles the process of hydration is continued as can be seen from the strong increase of the pH-value (fig. 2). After this period a highly active expansion process with an increase in volume by more than 8% can be observed (fig. 4).

The hydration process of TG-QZ with burnt siliceous chalk (K) as an additive, its kinetics of ettringite formation, process of expansion, and other properties of the hydrated cement differ greatly from the preceding descriptions, although the pH-value in the first minute was nearly the same. The phase analysis of siliceous chalk after burning at 1000°C showed that this additive contained calcium oxide as well as calcium silicate of a low basicity. Each component had a specific influence on the kinetics of the ettringite crystallization, its morphology and finally on the properties of the expansive cement. In the first two minutes the hydration process of TG-QZ is similar to that of the above mentioned expansive cement with calcium hydroxide. The higher pH-value of 12.3 is due to the calcium oxide content in the burnt siliceous chalk. It may be considered to be the cause of the increased solubility of the calcium aluminates just after mixing and of the formation of the separating layer of fine crystalline ettringite on the surface of unhydrated cement clinker particles. Burnt siliceous chalk contains only a relatively small amount of calcium oxide or calcium hydroxide after mixing with water and these amounts are quickly neutralised by calcium aluminate. Therefore, only a rather thin layer of ettringite crystals is developed whose strength and density are lower than in the case of pure calcium hydroxide. After a reaction time of about 60 minutes the pH-value increases from 9.4 to 10.3 which may be considered a consequence of the destruction of the ettringite layers. The phase analysis in this period showed an increase in ettringite content and proportionally a decrease in gypsum content in the hydrated cement. . Due to the difference in concentration of hydration products below the separating layer and beyond it ettringite nuclei "dart" at a high speed deeply into the capillary pore space near the cement particles where they crystallize. First studies into the formation of ettringite under an environmental scanning electron microscope confirm that in the presence of reacting agents ettringite forms suddenly and not as previously assumed by slow growth of the crystals.

The ettringite nuclei play the role of a crystallization centre, i.e. as the nucleus of the crystallization of the primary ettringite. Microscopic examinations showed that in the further course of hydration single clusters of ettringite developed independently and without any connection to each other. It may be concluded that in this case the crystallization pressure of the hair-shaped ettringite does not contribute to the increase in volume of the cement paste and rather plays the role of a fibre reinforcement for the gel-like hydrates.

This will also explain that after 28 days' storage the prisms, which had been coated with paraffin, showed an amount of expansion of only 0.25%, but a 1.5 to 2.2 times greater compressive strength compared to a TG-QZ with another mineral additive. Although there is no direct contact between the single ettringite clusters and the single ettringite crystals the expansive effect is also a consequence of the formation of ettringite. But the mechanism of the expansion process is different from the formation of the classical needle-shaped ettringite crystals which cause the expansion pressure in the form of crystallization pressure. From the very beginning of the hydration of the cement paste a capillary pore space is formed whose total volume depends on the water content of the system. The interconnected capillaries and pores are statistically distributed over the entire volume. In the course of hydration the microstructure changes. At a certain stage of hydration the products of hydration cause a change in this system of interconnected capillaries and pores to a microstructure of closed pores, continuous pores and dead-end pores. At a low pH-value hair-shaped ettringite in the pores can only form from an aqueous solution with an ion-content required for the formation of ettringite. As ettringite crystallizes with a considerable increase in volume this leads to a higher hydrostatic pressure in the closed and dead-end pores (comparable to the formation of ice below freezing point) / 5;6 /. Under this pressure the water is pressed out of the pores. The separating walls between the closed and dead-end pores where ettringite has crystallised have a high impermeability which may be considered the main cause of the great stress in the microstructure of the cement paste. The consequence of these stresses is an increase in volume of up to 0.25% during the hardening of the expansive cements in the first three days, the formation of fine crystalline hair-shaped ettringite which crystallises under the influence of the low pH-value of 10.2, and at the same time the development of pressure. These results confirm that the amount of expansion is not only determined by the amount of ettringite formed but that the morphology of ettringite, which depends on the pH-value, also plays an important role.

DURABILITY OF EXPANSIVE CEMENT CONCRETE

All kinds and causes of shrinkage in volume which cause cracking due to inhibited deformation pose a special danger to the microstructure of the hydrated cement and the durability of concrete and reinforced concrete structures. Problems caused by shrinkage are often aggravated by corrosion and this will lead to difficulties in the use of buildings and even to heavy damage to the structures. By the use of expansive additives the change in volume during the hardening of the concrete can be controlled in time and space so as to ensure a desired increase in volume of the hardening concrete after shrinkage and to improve the durability of concrete and reinforced concrete structures.

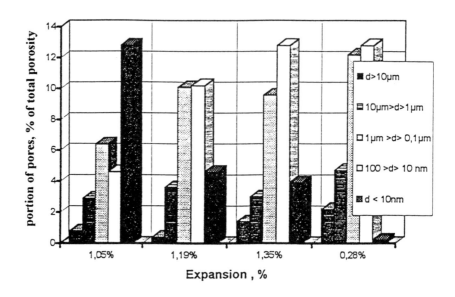

Fig. 5:   Pore distribution in expansive cement paste with different amount of expansion

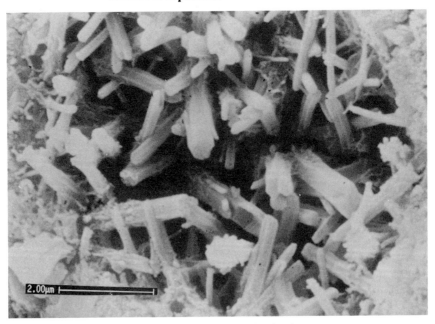

Fig. 6: The formation of ettringite in the pore space

Studies into the porosity of hydrated expansive cement with different expansive potentials (expansive potential is the maximum achievable amount of expansion when the hardening of the expansive cement is not impeded) showed that the portion of pores with a diameter < 100nm is considerably greater than in the hydrated Portland cement which forms the matrix for the expansive cement (fig. 5). The formation of ettringite is evenly distributed over the whole pore space. Thus the whole pore microstructure is made impermeable (fig. 6). This explains the great impermeability and durability of concrete made of expansive cement.

If you take the increase in volume under freezing as an example of the durability of concrete, it may be stated that in expansive cement the damage to the microstructure is remarkably smaller / 7 /. Investigations into the sulphate resistance / 8 / and the freeze-deicing salt resistance / 9 / show that the durability of properly produced expansive cement concrete is generally higher than that of portland cement concrete.

ACKNOWLEDGEMENTS

The autors gratefully acknowledge the help and financial support by the Alexander von Humboldt Foundation.

REFERENCES

1. Cohen, M.D. & Richartds, C.W. (1982): Effects of the Particle Size of Expansive Clinker on Strength expansion characteristics of Type K Expansive Cement. - Cem. Concr. Res., V.12: 717-725

2. Kusnezova, T.V. (1986): Physikalisch-chemische Grundlagen zur Technologie von Quellzementen.- NIIZement, 38: 77-84; Moskau

3. Clastres, P.; Murat, M.; Bachiorini, A. (1984): Hydratation of expansive cement correlation between expansion and formation of hydrates.- Cem.Concr.Res., V.14: 199-206

4. Deng Min & Tang Migshu (1994) : Formation and Expansion of Ettringite Crystals.- Cem. Concr. Res., V.24 : 119 - 126

5. Ludvig, U. (1991): Probleme der Ettringitrückbildung bei wärmebehandelten Mörteln
und Betonen. - 11. Intern. baustoff- und Silikattagung, 1 : 164 - 175 ; Weimar

6. Chartschenko, I.; Gathemann, B.; Stark, J. (1994) : Injektionsmörtel mit Quelleffekt zur Sanierung von historischen bauwerken. - 12. Intern. Baustofftagung, ibausil , 22-24. Sept. 1994, 1. : 12 - 17 , Weimar

7. Chartschenko, I. & Stark, J. (1993): Steuerung des Strukturbildungsprozesses bei Quellzementen sowie Betonen auf deren Basis. - Wiss. Zeitschr. Hochsch. Arch. Bauwesen Weimar, 39/3 : 163-171 ; Weimar

8. Chartschenko , I. & Stark, J. (1993) : Untersuchungen über den Einfluß des pH-Wertes auf die Ettringitbildung. - Wiss. Z. Hochsch. Arch. Bauwes., 39: 171 - 177; Weimar

9. Chartschenko , I. (1982) : Technologie und Eigenschaften von Keramsitbeton auf der Basis von Spannzement für die Außenwandplatten. - Dissertation : 184 ; Moskau

10. Michailov, W.W. & Litver, S.L. (1974): Quellzemente, Spannzemente und selbstgepannte Stahlbetonkonstruktionen.- 311 ; Moskau (Strojizdat)

STUDY AND APPLICATION OF AUTOCLAVED HIGH STRENGTH CONCRETE WITH GROUND SAND

H Shi-Yuan

Y Quan-Bing

W Zhan-De

Shanghai Institute of Building Materials

C Shi-Biao

H Yiao-Hui

Hongji Concrete Pipe Piles Company

China

ABSTRACT. This study was done for the company manufacturing prestressed high strength concrete pipe piles. Strength tests were made for autoclaved cement-ground sand pastes and concrete with ground sand. XRD and SEM analyses were also made to clarify the effect mechanism of the ground sand on cement hydration while autoclaved. Manufacture experience of prestressed high strength concrete pipe piles with the strength of $80N/mm^2$ was briefly introduced using ground sand to equally replace 30% cement in a factory.

Key Words: Autoclaving, High-strength concrete, Ground sand

Prof. Huang Shi-Yuan is honorary director of Research Lab of Material Engineering, Shanghai Institute of Building Materials, China. He is specialised in the durability of concrete; frost resistance, salt scaling and corrosion, the technology of pumping concrete and winter concreting. He is now a member of Science and Technical Committee of China National Admministration of Building Materials Industry.

Mr Yang Quan-Bing is a lecturer in Cement and Concrete Technology, Shanghai Institute of Building Materials. His main research aspects include salt scaling of concrete, chloride diffusion through concrete and repairing materials.

Mr Wang Zhan-De is an associate professor, SIBM.

He Yiao-Hui is a manager and also an engineer of Hongji Concrete Pipe Piles Company.

Radical Concrete Technology. Edited by R K Dhir and P C Hewlett. Published in 1996 by E & FN Spon, 2–6 Boundary Row, London SE1 8HN, UK. ISBN 0 419 21480 1.

INTRODUCTION

It is well known that the $Ca(OH)_2$ released by hydration of C_3S in cement can react with quartz powder to form more or less crystallized CSH at elevated temperatures, for example 170℃[1]. Therefore a good concrete can be made by replacing part of cement with ground quartz sand by autoclaving. At present, in China a great number of plants are producing prestressed concrete pipe piles, among which most are prestressed high-strength concrete (PHC) ones (concrete strength > 80 N/mm^2) produced by autoclaving. However, they were made with ordinary portland cement concrete without addition of any blended materials until the present study was carried out. Since those plants have been furnished with autoclave, why don't we try to make concrete with ground quartz sand or ordinary river sand to save cement, energy comsumption and cost of the pipe piles ?

In the present paper, tests of cement-ground sand pastes were made to determine the proper addition amount of sand, the autoclaved PHC was prepared and tested in laboratory and in production as well, and XRD and SEM analyses were also made to observe changes in hydration products and in the microstructure of hardened autoclaved pastes.

PASTE TESTS

Before concrete tests, autoclaved cement-ground sand pastes were tested in comparison with OPC pastes.

Experimental Conditions

The following materials were used in the test:
OPC of grade 525 (according to Chinese Standard) made in Ningguo Cement Plant; quartz sand with fineness of 3000 cm^2/g and SiO_2 content of 99%; ordinary river sand with fineness 3500 cm^2/g and SiO_2 content of 91%.

Sand:(sand+cement) ratios were 0, 10, 20, 30 and 40%. Water-solid ratio was 0.26 for all.

Specimen size was $2\times2\times2$ cm.

Three curing conditions were used. They were: (1) normal curing at 20 ℃; (2) steam curing comprising 2 hours in air, then temperature rising to 65℃ for 2 hours, then keeping constant temperature 65℃ for 2.5 hours and temperature dropping to room temperature for 1.5 hours; (3) autoclaving——steam-cured and demoulded specimens being autoclaved comprising temperature rising—keeping constant temperature—temperature dropping for 4-5-2.5 hours with the constant temperature of 180℃ at pressure of $1N/mm^2$.

Test Results

The strength of specimens of different sand addition amounts after different curing conditions was measured and the results are shown in Tab. 1 and Fig. 1.

Discussion

From the above results some follwing conclusions may be obtained:

(1) The strength of autoclaved paste specimens with addition of 20~30% ground sand (either quartz sand or river sand) is higher than that of OPC paste specimens normally cured for 28 days and is much higher than that of autoclaved OPC ones.

(2) For autoclaved OPC pastes strength regression occurs during consequent normal curing as expected, whereas for autoclaved cement-ground sand pastes no regression has been found.

(3) The efficiency of quartz sand is better than that of river sand with relation to the strength due to the larger SiO_2 content of quartz sand. However, from the practical view, ordinary river sand with high content of SiO_2 is also sufficient for partial substitution for cememt. The suitable replaced amount of cement by ground sand seems to be about 30%.

CONCRETE TESTS

Experimental Conditions

The same cement and ground river sand were used as those used in paste tests. Crushed rock in 5~25mm size and river sand with fineness modulus of 2.3~3.0 were used as coarse and fine aggregates with soil content < 1.5%.

Table 1   Compressive strength of cement-ground
sand paste specimens (N/mm²)

CURING CONDITIONS	SAND ADDITION AMOUNT (%)								
	Cement-quartz sand					Cement-river sand			
	0	10	20	30	40	10	20	30	40
A	121	118	116	101	78.3	120	106	89.2	64.6
B	70.9	60.9	56.6	47.8	35.3	67.2	53.8	42.7	31.9
C	105	128	126	142	112	115	122	122	114
D	91.3	109	144	137	135	124	133	118	126

* A —— normal curing for 28 days   B —— steam curing
C —— steam curing + autoclaving
D —— steam curing + autoclaving + normal curing for 28 days

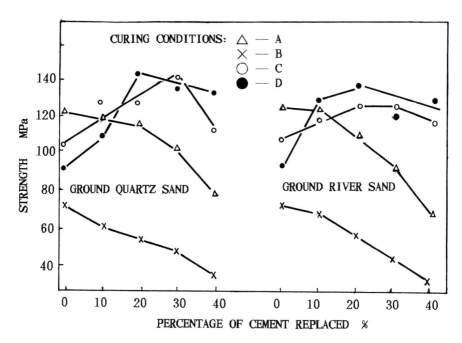

Fig.1   Effect of composition of cement-ground
sand pastes on compressive strength

Specimens size was $10\times10\times10$cm.

Three curing conditions were used. They were: (1) normal curing at 20℃; (2) steam curing 2-4-2 hours, 70℃; (3) autoclaving 4-5 -3 hours, 180℃.

The mix proportion tested are shown in Table 2.

For all, superplasticizer amount is 0.25 % of (cement + ground sand); slump 7~8cm.

Test Results and Disussion

Strength of concrete with proportions mentioned in Tab 2 are shown in Table 3.

It can be seen that the concrete test results very well conform to the paste test. The conclusions obtained from paste tests

Table 2   Concrete mixture tested

NO.	FINENESS OF SAND cm^2/g	SAND AMOUNT REPLACING %	CEMENT Kg/m^3	GROUND SAND Kg/m^3	WATER Kg/m^3	FINE AGG. Kg/m^3	COARSE AGG. Kg/m^3
1		0	520	0	148	624	1158
2	3400	30	364	156	148	624	1158
3	3400	40	312	208	148	624	1158
4	5400	40	312	208	148	624	1158

Table 3   Compressive strength of autoclaved concrete with addition of ground sand (N/mm^2)

No.	CURING CONDITIONS			
	A	B	C	D
1	86.5	82.0	91.0	83.1
2	78.5	70.3	97.3	96.1
3	61.0	59.3	84.3	91.4
4	66.2	60.3	100.0	111.6

are also available to the concrete. Ground sand may be used to replace 30% cement to make autoclaved concrete with equal strength contribution of the ground sand to concrete seems to be less than that to pastes. It is because the high-strength concrete losses its bearing capacity in some cases due to crashed rock fracture. The stregth of concrete with ground sand after steam curing is reasonablly lower than that without ground sand, but is much higher than required destressing value 35 N/mm^2, specified by Chinese Standard GB 13476-92.

XRD ANALYSES AND SEM OBSERVATION

XRD analyses were done for autoclaved OPC and cement-ground sand pastes to clarify the effect of ground sand on hydration products. The XRD diagrams of both are shown in Fig.2.

Very intensive peaks of 18.1° and 34.1° certified the presence of a large amount of Ca(OH)$_2$ formed by hydration of C$_3$S and unreacted SiO$_2$ crystals in sand respectively in the autoclaved OPC paste (Fig.2a), whereas the intensity of those peaks greatly weakens in the autoclaved cement-ground sand paste (Fig 2b) which illustrates that in paste with ground sand Ca(OH)$_2$ released by hydration of C$_3$S reacts with SiO$_2$ crystals and water to form poorly crystallized CSH (tobermorite, 14.1Å, see Fig.2b).

By SEM observation a large amount of stacks of CH lamellae can be found in the paste without ground sand (Fig.3a), but few CH lamellae have been found in the paste with ground sand and instead, there more or less crystallized CSH, or say tobermorite type CSH crystals can be obviously observed (Fig.3b), which also proves the reaction of CH with SiO$_2$ crystals to form crystallized CSH in the autoclaved paste with ground sand.

APPLICATION OF GROUND SAND IN
MANUFACTURE OF PHC

Based on the laboratory study, ground sand is being used to partially replace cement in manufacture of autoclaved PHC piles in some plants in Guangdong Province. The production experience in Hongji Pipe Piles Company is here briefly introduced.

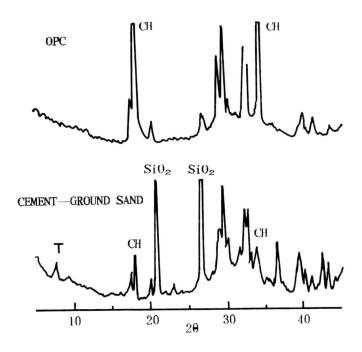

Fig.2 XRD diagrams of autoclaved pastes

CH LAMELLAE IN OPC PASTE

TOBERMORITE TYPE CSH CRYSTALS
IN CEMENT—GROUND SAND PASTE

Fig.3 SEM photos

The products are PHC pipe piles with the external diameter of 50cm and the length of 8~12m. Concrete srength grade is 80N/mm².

Portland cement of grade 525 made in Zhujiang Cement Plants is used. The quartz sand is used with the fineness of 2000 cm²/g. Crushed rock in 5~20mm and medium sand are used as aggregates. Superplasticizer used is of type sulphonated melamine formaldehyde.

The piles are shaped by centrifugal process. The centrifugal technology is unchanged while using ground sand. After centrifuging piles are cured by two-stage curing process. In the first stage, the time for rising-constant-dropping temperatures is 2-2.5-4 hours and the maximum temperature is 90 ℃. Piles after first-stage steam-curing are destressed and demoulded. The demoulded piles are transported into autoclaves and then autoclaved (second stage). The autoclaving proess is also kept the same as usual.

The concrete mix proportions are as follows (Tab 4).

The strength of the concrete tested is shown in Table 5.

Table 4 Mix proportions in production

MIX	GROUND SAND REPLACEMENT %	CEMENT Kg/m³	GROUND SAND Kg/m³	WATER Kg/m³	FINE AGG. Kg/m³	COARSE AGG. Kg/m³	ADMIXTURE Kg/m³
0	0	520	0	140	619	1144	4.16
A	30	364	156	140	619	1144	4.16
B	35	338	182	140	619	1144	4.16

0 —— routine proportion without ground sand

Table 5 Strength of concrete in production (N/mm²)

MIX	AFTER FIRST STAGE CURING	AFTER TWO-STAGE CURING
0*	60.0	90.0
A	48.9	89.3
B	46.1	88.1

* —— average strength of routine products

In routine production without addition of ground sand, some fine cracks often appear on the surface of products, but such fine cracks diappear while ground sand is added. Moreover, the field piling practice of the pipe piles with ground sand shows that these piles have high impact bearing capacity and no pile broken incident has occurred. It can reasonablly be expected because less cement is used and large crystals of $Ca(OH)_2$ are transferred to CSH gel.

The cost of products is reduced very much while ground sand is used to substitute cement. In Guangdong Province, the cost of cement and ground sand is 550 RMB/T and 87 RMB/T respectively, so the cost of raw materials of concrete is decreased about 21% (72 RMB/M^3 concrete, about 8.7 $) while mix A in Table 4 is taken for an example.

The practical production experience shows that with few changes in technology, ground sand can be successfully used to manufacture autoclaved high-strength concrete products.

CONCLUSIONS

The laboratory study as well as the production experience certified that a good autoclaved high-strength concrete can be made by addition of ground sand with high SiO_2 content to replace part of cement. The suitable addition amount of ground sand is about 30%.

By XRD analyses and SEM observation it can be clearly found that the $Ca(OH)_2$ released by hydration of C_3S reacts with SiO_2 crystals and water to form more or less crystallized C-S-H, which contibutes to the concrete strength.

A well known concrete knowlege has been transferred to great economical benefit through the application study.

REFERENCE

1.   J.J. Beaudoin and R.F. Feldman, Cement and Concrete Research, No.1, Vol.5, 1975. pp 103.

HIGH PERFORMANCE CONCRETE PREPARED WITH SULPHO/FERRO-ALUMINATE CEMENT

L Zhang

M Su

China Building Materials Academy

S Wang

Yuan Da Engineering

China

ABSTRACT. The study had been carried out on the high performance concrete with the Third Series Cement, sulpho- and ferro-aluminate cement. The experimental results showed that the setting and the workable time of the fresh concrete could be retarded and the physico-mechanical properties and the durability of the hardened concrete was improved when the special admixtures were added. The applications of the high performance concrete with this kind of cement in high-rise buildings, prestressed concrete products, ready-mixed concrete plant and mass concrete engineering etc. were also described in the paper.

Keywords: Sulpho-aluminate cement(SAC), Ferro-aluminate cement(FAC), High-strength concrete(HSC), High performance concrete(HPC), Admixture.

Mr Liang ZHANG, Master of Science, is an engineer in House Building Materials And Concrete Research Institute, China Building Materials Academy. He specializes in the properties and the application research of the third series cement concrete. Mr. Zhang has published tens of papers in this field.

Professor Muzhen SU is vice-president of China Building Materials Academy. She specializes in the production technology of cement and special cement. She is one of the inventors of the Third Series Cement. Professor Su has published widely and serves on many Technical Committees and is the vice-chairman of the Cement Society, China.

Mr Shaojiang Wang is an assistant engineer in Yuan Da Engineering Materials Development Co. His main research interests include the application of the Third Series Cement and special admixtures.

Note: The project is aided financially by the National Natural Science Fund.

Radical Concrete Technology. Edited by R K Dhir and P C Hewlett. Published in 1996 by E & FN Spon, 2–6 Boundary Row, London SE1 8HN, UK. ISBN 0 419 21480 1.

INTRODUCTION

Sulpho-aluminate and ferro-aluminate cement, which were invented in 1970's and 1980's by China Building Materials Academy(CBMA) and made the Chinese Standards in 1987 and 1991 respectively, are new type of cement series. In the 3rd Beijing International Symposium on Cement and Concrete held in October, 1993, the inventors, Prof. Wang Yanmou and Prof. Su Muzhen named sulpho- and ferro-aluminate cement (the major mineral component is C_4A_3S) as the Third Series Cement to make difference from the First Series Cement - Portland Cement (the major mineral component is C_3S) and the Second Series Cement - Alumina Cement (the major mineral is CA).

The Third Series Cement is manufactured by calcining alumina (or ferro-alumina) and gypsum together at lower temperature(1300-1350°C) into clinker with major mineral C_4A_3S, C_2S and C_4AF. The clinker is ground with appropriate amount of gypsum to produce finish product. Hydration products of the cement are mainly composed of AFt, AFm, $Al(OH)_3$ and $Fe(OH)_3$ gel et al.. The Third Series Cement bestows high strength, including early strength, impermeability, frost and corrosion resistance and low alkalinity etc. The research into preparing the high performance concrete with the Third Series Cement has been carried out in CBMA for several years. Now, we have successfully used this kind of concrete in practical civil engineerings.

SPECIAL ADMIXTURES FOR THE THIRD SERIES CEMENT

Aim of Developing Special Admixtures

In order to produce high performance concrete, we have developed a series of special admixtures for the cement to solve the following three problems emphatically: firstly to control the setting time and the workable time of the concrete; secondly to reduce water to cement ratio and ensure the consistency of the concrete required by construction; thirdly to improve the strengths and the durability of the concrete. The special admixtures included ZB-1 type of high range water-reducing and retarding admixtures for producing high strength concrete, ZB-3 type of normal retarding and water-reducing admixture for producing normal strength concrete, ZB-5 type of special pumping aid for producing commercial concrete transported from ready-mixed plant to construction site, ZB-2 type of anti-freeze admixture for making concrete at low negative temperature in winter.

Controlling Setting Time and Workable Time of Concrete

With the special admixtures, the setting time of the concrete can be controlled between 2 and 8 hours and the workable time to meet the requirements of construction on-the-spot (showed in Table 1). Table 2 shows that the slump of the fresh concrete mixture did not lose in 3 hours with the special pumping aid.

Table 1 Slump loss of the concrete with the special admixture

W/C	Admixture	Slump, cm						
		Initial	15min	30min	45min	60min	75min	90min
0.37	ZB-1	20	16	14	12	12.5	10.5	8

Table 2 Slump loss of the concrete with the special pumping aid

W/C	Admixture	Slump, cm						
		Initial	15min	45min	75min	100min	140min	180min
0.41	ZB-5	22	22	22	21	20	20	20

Therefore, we can prepare the high performance concrete of the Third Series Cement vibrocast easily with good workable time by making use of the special admixtures to construct concrete products and buildings.

Influence on the Hydration Heat of the Cement

High-early strength is one of the important characteristics of the cement which is different from OPC. From the hydration heat evolution curves of the cement with and without the special admixture as shown in Fig.1, the time of appearance of the peak of hydration heat was only postponed but the characteristic of that is not changed, that made the concrete retain high early-strength and high-strength.

PROPERTIES OF THE CONCRETE PREPARED WITH SAC/FAC

Strength Development

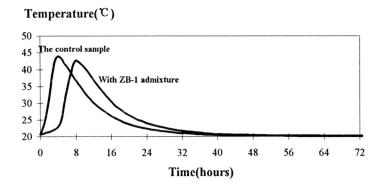

Figure 1 Hydration heat evo-lution curves of the cement with and without the special admixture

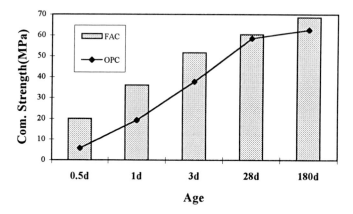

Figure 2   Comparison of compressive strength development of
FAC and OPC concrete

The strength development of FAC concrete and OPC concrete is compared in Fig.2. Obviously, the early strength of FAC concrete develops much more quickly than that of OPC concrete and the former also continued strengthening later. At the same time, the tensile splitting strength, bending strength and elastic modulus co-increase with its compressive strength as shown in Table 3 and 4. Therefore, the benefit of advancing prestressing time for prestressed concrete, quickening the circulation of moulds and shortening time limit for a project.

Strengths at Negative Temperature

Strength development of the concrete in winter could be achieved by self-curing making use of the hydration heat released by the Third Series Cement itself concentratedly. The aim to use the special anti-freezing admixture was to keep fresh concrete good workability and refrain it from freeze injury before the hydration of the cement so as to make concrete mixed normally only with simple curing measures at

Table 3   Strength development of the concrete(MPa)

Age	1 day	3 days	28 days	1 year	2 years
compressive	44.6	61.9	69.2	81.0	90.7
Splitting tensile	2.85	4.24	5.38	5.74	
Flexural		7.23	8.70	8.81	

* Cement content is 480kg/m3; W/C is 0.335, Slump is 9cm.

Table 4   Elastic modulus of the concrete at different ages

W/C	Slump, cm	Admixture	Axial Compressive Strength, MPa			Elastic Modulus, GPa		
			1 d	3 d	28d	1 d	3 d	28 d
0.37	18.8	ZB-1	40.9	43.7	50.0	34.2	35.6	37.8

Table 5   Strength development at negative temperature

Temperature	Slump, cm	Compressive strength, MPa				
		-7d	-7+3d	-7+7d	-7+28d	-7+56d
-15 °C	9.0	3.6	43.7	54.2	58.5	70.4

* Cement content is 500kg/m3; W/C is 0.31; -7+3d means that concrete specimens are cured at negative temperature for 7 days and then at positive temperature for 3 days.

the negative temperature and ensure strengths and durability of hardened concrete. So the concrete with this cement can beyond the limit of normal critical-freezing strength when used. Some experimental results is shown in Table 5.

Creep and Dry Shrinkage

Under the unsealed condition, the creep($C(t,\tau)$) and the shrinkage(ε_s) of ferro-aluminate cement concretes graded C50 and C80 tested are shown in Table 6. It is advantageous to manufacture high-strength prestressed concrete structures with the Third Series Cement to avoid the stress loss due to its quite smaller creep and shrinkage.

Expansion-Shrinkage Properties of Expansive Concrete

The shrinkage-compensating ability of the concrete is judged mainly according to the deformation under the condition of long term dry-wet circulation.

Fig.3 gives the expansion-shrinkage curve of FAC concrete which was cured in water for a month, then put into dry air room(20±3°C, 60% of relative humidity) to undergo a full dry shrinkage (one year), then put back into water to be recured (one year), and returned to dry air room once again after one year. As shown in the figure, after cured in water, the mild expansive FAC concrete still shrinked when it was put into air to dry and the shrinkage was smaller. After curing in water for second time, the specimen expansion was restored quite rapidly. Although the expansion did not reach the former level at last, but it had an obvious characteristic, when shrinked again, i.e. the shrinkage was small for the second dry shrinkage. The specimen did not shrink again and its size tended to be stable after one year of dry shrinkage. The stable-tending state was at the positive strain state, which is higher than that at first time. When used as the shrinkage compensating structural material in engineering practice, the FAC concrete is in positive strain state and the struture will not crack because the exposed structure is under the condition of dry-wet

Table 6   Creep and shrinkage

Concrete grade	W/C	Axial compressive strength, MPa		ε_s $\times 10^{-6}$	$C(t,\tau)$ $\times 10^{-6}$
		3 days	28days		
C50	0.40	38.0		510	28.4
C80	0.30	69.6		368	23.7
C80	0.30		91.1	368	17.8

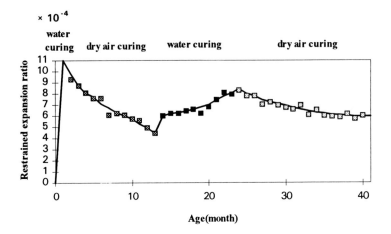

Figure 3   Expansion-shrinkage curve of FAC concrete

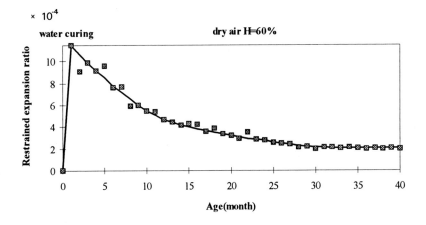

Figure 4   Dry shrinkage of FAC concrete specimen cured in water for one year

alternation. In addition, in the engineering underground with mild changes in temperature and humidity, the texture of the concrete is more stable. Because of the presence of self-stress produced by the expansion of the FAC and the crisscross growth of the hydrates, the texture is more compact.

Fig.4 shows the drying shrinkage deformation curve of the FAC concrete. After cured in water for 28 days, the specimen underwent two years of drying shrinkage. After the shrinkage, the concrete was still at positive strain state.

Durability

The excellent impermeability of the Third Series Cement concrete was attributed to its high density. The testing results made by CBMA indicate that its impermeability at 3-7 days was almost the same as that of Portland cement concrete at 28 days.

High-strength concrete with the Third Series Cement also has good carbonation resistance and freeze-thaw resistance. The depth of carbonation at 28 days under the standard condition was less than 1 mm and the compressive strength of concrete is almost without loss. The loss of the compressive strength of the concrete with ZB-1 special admixture was only 1.62% after 300 frozen-thaw circles.

Because of low alkalinity, long-term and accelerating test results convince that the Third Series Cement can effectively restrain alkaline-aggregate reaction.

The results of X-ray analysis on the long term hydration matrix of SAC show that the main hydrates AFt and $Al(OH)_3$ still exist steadily. The compressive strength of the concrete at the age of 6 years not only does not decrease but also increase continuously. Since 1970's, the Third SerPies Cement has been used in quite a lot of concrete structures and products but no strength failure has been reported. We will conduct special study on the durability of the Third Series Cement.

APPLICATIONS OF HIGH PERFORMANCE CONCRETE PREPARED WITH SAC/FAC

High-Strength Concrete

High-strength pretensioned C80 concrete pile was pilot-manufactured with FAC in 1992. At 7h to 9h, compressive strength of concrete was able to go up to 45MPa, and to reach C80 after curing 3 days in water. Traditional process of steam curing at atmospheric pressure and autoclaved curing at higher temperature and pressure for high-strength Portland cement concrete pile is not necessary for SAC/FAC concrete.

In 1993 and 1994, large span post-tensioned prestressed T-type beams were cast with C50 FAC concrete for updating projects of Eastern and North-western Third-ring-road in Beijing. Thousands of prestressed concrete beams have been manufactured with the largest span up to 32 meters. Products can be demoulded and transported out of plant without steam curing in one day, resulting in speeding-up circulation of moulds and cutting-down fuel consumption (Fig.8).

Shenyang Long Distance Telecommunication Hub is a high-rise building with 22 stories and 103m of total height. SAC/FAC and the special admixtures were used in the whole main building in which the concrete graded C60 was used for 1 to 7 stories. Statistic of concrete 28 days strengthes for the first to the seventh floor is M_{fcu}=64.5MPa (average strength from 69 group of concrete specimens), which meet

Figure 5   Pumping ready-mixed concrete   Figure 8   Concrete beams for the bridges
of SAC

Figure 6   Pouring C80 concrete column   Figure 7   High-rise building with C60 concrete
with steel pipe core

the design requirement. By using this kind of cement, the comprehensive economic
results have been achieved in respects of saving cement, reducing number of moulds
by two-thirds, raising circulation rate of moulds and speeding up construction, etc.
In winter construction, C60 and C50 concrete cast with this cement and the special
anti-freezing admixture at the lowest temperature up to -20°C, only heating water
and sand (steam boiler unused) and covered with straw bags on their surface (steel

moulds used for bottom of floor and column), was never frozen while the strength met the requirement(Fig. 7).

Liaoning Products Building of 27 stories is 102m high. C80 high strength concretes made from SAC with the special admixture were successfully used in 1 to 4 stories of main building. The cross section of column was reduced from 1.2×1.2m to 0.9×0.9m. The C80 concrete is used in the cast-on-the-spot high-rise building for the first time in China(Fig. 6).

Ready-Mixed Concrete

SAC was adopted to produce flowing concrete in Hangbo ready-mixed concrete plant in the end of 1994 for Jindu Building in Tianjin which is 8 stories high. The slump of the flowing concrete did not lose in about 3. Since the construction speed was accelerated obviously owing to the early-strength of the concrete, the user got good comprehensive economical benefit because the building was completed 15 months before its due date(Fig. 5).

Mass Concrete

The multi-function building on Shisan Wei road in Shenyang is 13 stories high and has 1,760m^2 underground construction area. Mild expansive sulpho-aluminate cement was used in the raft foundation of the building, the bottom floor of which was 1 m in thickness as rigid water-proofing. The concrete was continuously poured at the lowest temperature outdoor of -16 to -18°C and no cracking happened on the surface of it. In April, the cement was again used to pour the underground continuous wall in replace of exterior and interior flexible waterproofing layers. No leakage occurred in rainy season that year.

CONCLUSIONS

1. Cement can be controlled with the special admixtures. The setting time and workable time of the concrete prepared with the Third Series

2. The characteristics of the Third Series Cement concrete are high early strength and high early elastic modulus and smaller creep when the concrete is loaded at the early age, so this concrete is very suitable for the rapid construction of prestressed concrete.

3. Very-high strength concrete can be prepared with the Third Series Cement.

4. The Third Series Cement concrete has good impermeability, freeze-thaw resistance and carbonation resistance.

5. The results from the practical applications indicated that good comprehensive economical and social benefit can be obtained with the use of the high performance concrete prepared with the Third Series Cement which has reliable quality and achieves satisfactory use result.

ACKNOWLEDGMENTS

The authors would like to express their appreciation for the experimental work made by Mr. Wenhua SHAN and Mr. Guoyi GAO and the good cooperation during the application of the cement with vice-manager Zhihua YUE, Tangshan Kouri Special Cement Co. Ltd.

REFERENCES

1.   WANG YANMOU, SU MUZHEN. The Third Series Cement In China. Proceedings, the third Beijing International Symposium on Cement and Concrete, 1993
2.   GUAN DEJIN et al. Application of C60 Concrete Produced with Special Cement in high-rise Building. Concrete, No. 1, 1994

FURTHER DEVELOPMENTS IN PORE REDUCED CEMENT (PRC)

D E MacPhee

D Israel

University of Aberdeen

UK

ABSTRACT. Pore reduced cements are a range of cement products obtained from the mechanical pressing of immature Portland cement pastes. The pressing technique permits the partial removal of mix water which is held in the developing microstructure so that water:cement ratio is reduced from the normal mixing value of 0.3 to around 0.1. Continuing hydration of cement in the compacted paste leads to an increasing densification (porosity reduction) imparting considerably improved mechanical properties to the product. The durability of high performance cements is as important as strength. This paper therefore addresses durability, linking microstructural characteristics of PRC with its performance in aggressive exposure media.

Keywords: Pore Reduced Cements, High Performance, Durability, Porosity, Acid degradation.

Dr Donald E Macphee is a lecturer in the Department of Chemistry at the University of Aberdeen. His research interests include cement-based waste immobilisation, reinforcement corrosion, high performance cements and microstructural characterisation of cements.

Dr Dieter Israel is a Postdoctoral Research Assistant in the Department of Chemistry at the University of Aberdeen. He is a graduate of the Technological University Bergakademie Freiberg, Germany, with a Diploma in Chemistry (1989) and a PhD in gypsum-related chemistry (1994).

Radical Concrete Technology. Edited by R K Dhir and P C Hewlett. Published in 1996 by E & FN Spon, 2–6 Boundary Row, London SE1 8HN, UK. ISBN 0 419 21480 1.

INTRODUCTION

Pore reduced cements (PRC) are a class of high performance cementitious materials produced by the application of high pressure to immature pastes. The materials are compacted under an applied pressure of up to 200 MPa during which a portion of the unabsorbed mix water is expressed, reducing the effective water:cement ratio (w/c). The resulting paste is dense and contains a high percentage of unreacted cement in a matrix of low porosity hydration product. The manufacturing technique and some mechanical and microstructural properties of PRC have been described previously[1,2] and show that PRC pastes can have densities of up to 2600 $kg.m^{-3}$ (compared with around 2000 $kg.m^{-3}$ for an unpressed paste at w/c = 0.35) and strengths of up to 250 and 35 MPa in compression and tension respectively (compared with about 80 MPa and 8 MPa respectively for the unpressed equivalent.)

Microstructure and Mechanical Strength

The strength increases in PRC, relative to unpressed cements, are related to the densified microstructure. This is to be expected from a theoretical approach. In 1897, Feret[3] proposed an empirical relationship between strength and volume porosity which collectively with others subsequently proposed, approximated to an expression of the form[4]:

$$\sigma_c = \sigma_0 \exp(-bp)$$

where σ_0 is the (hypothetical) strength at zero porosity, b is a constant and p is the volume fraction of porosity. But *volume* porosity does not represent the only strength limiting factor. The Griffith's model[5] for brittle fracture relates flexural or bending strength inversely to flaw size.

$$\sigma_f = (ER/\pi c)^{\frac{1}{2}}$$

where σ_f is tensile or bending strength, E is elastic modulus, R is fracture energy and c is flaw length. In the context of cement pastes, flaws are represented by pores, as well as cracks, and so strength can be modelled in terms of volume fraction of porosity and the sizes of the pores present[6]. Cement products based on the macro-defect free (MDF) and densified system of homogeneously packed small particles (DSP) approach have also demonstrated the validity and applicability of these relationships.

Microstructure and Durability

Porosity and microstructure are also important in defining the durability of a cementitious product. Deterioration mechanisms are mostly through-solution processes and are always dependent on exposed and reacting surface area. A material of zero porosity would be exposed to the environment only on its geometrical surface, an area defined by its visual dimensions. A porous matrix has a larger exposed area through the connectivity between pores on the external surface and those in the interior of the solid. Reactive agents capable of migrating along those connected pores are then free to react over potentially a much larger surface, consequently accelerating the rate of degradation.

Essentially two types of pores are found in cement-based matrices. Interconnected pores, as discussed above, and isolated pores which are disconnected from the external surface and therefore do not influence matrix durability. The distribution of connected and isolated porosity is dependent on w/c. Pastes of high w/c generally have a high degree of pore connectivity whereas at lower w/c, the degree of isolation increases. It is expected therefore that PRC pastes have less interconnected porosity than unpressed cement pastes and this has been confirmed by nitrogen adsorption measurements[7]. Microstructurally, PRC appears very dense. Most of the area shown in electron micrographs is taken up by unreacted cement although some   reaction of these particles has taken place to bind the paste. Occasionally pores are observed but in general, the cement grains are dispersed in a dense hydration product, primarily C-S-H, whose appearance more closely resembles 'inner hydrate' than 'outer hydrate'[2]. Evidence for isolated porosity may also be derived from electron beam damage which is observed to be more severe for PRC than for unpressed cement pastes[2].

The durability of PRC products are considered in this paper. A series of experiments exposing the material to severe chemical attack is described. The performance of the PRC pastes under those conditions is discussed in relation to the microstructural considerations detailed above.

EXPERIMENTAL APPROACH

Manufacture, Curing and Exposure

Ordinary Portland cement was mixed with water at a w/c of 0.35 and cast in perspex moulds to produce cylinders of dimensions 40 mm diameter and approximately 80 mm high. After about 3 hours, the cylinders were removed from the moulds and each was placed in a hydraulic press (previously described in references 1, 2 and 7) where a uniaxial, unidirectional load equivalent to an internal pressure of up to 200 MPa was applied. Pore fluid was recovered during this operation. After pressing, each cylinder was removed from the press and its weight and dimensions were recorded. Cylinders were then stored in either moist air (over saturated copper sulphate solution (98% relative humidity)), or under water (deionised), both for 28 days at room temperature (23 ± 2°C). They were then cut horizontally in half using a masonry saw and densities of each half were determined using Archimedes method (n-hexane as the displacement fluid). Subsequently, samples were placed in either 0.203M sulphuric acid, 0.274 M hydrochloric acid or 0.05 M ethanoic acid. Acid concentrations were selected based on similar studies reported in the literature[8,9,10]. All solutions were replaced on a weekly basis.

Performance Monitoring and Microstructural Examination

Samples were removed periodically from the solutions. They were visually inspected, surface dried with absorbent tissue after which loosely bound material was lightly brushed away, and weighed.

Microstructural investigation was carried out using a scanning electron microscope (ISI-SS40) operating at 20 kV. All samples were polished sequentially using standard F600 then F1200 silicon carbide grits and with diamond paste (6 μm) and Al_2O_3 paste (0.3 μm) before carbon coating for SEM examination.

RESULTS AND DISCUSSION

Sulphuric Acid

Visual examination of these samples indicates a serious degradation. Typically, layers of degradation product, identified by X-ray diffraction as gypsum, spontaneously detach themselves from the core cylinder. The detached layers appear to have a larger volume than the parent material but this has not been rigorously measured. Preliminary results for this series of experiments were presented previously[7] and are included here with more recent data in Figure 1 which shows the variation in weight change as a function of exposure time and sample density.

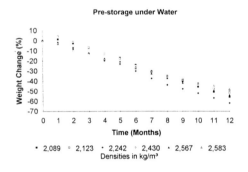

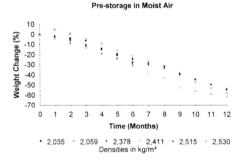

Figure 1: Effect of 0.203M sulphuric acid exposure on OPC and PRC products

It can be noted that the degradation, as defined by weight loss, increases with time as expected, except for a slight weight gain in the short term exhibited by the low density samples. The paste density does not however seem to affect the rate of degradation significantly in the longer term. This was noted from the preliminary results also[7] and the present paper shows that the trend is continuing. The extent of deterioration is considerable with up to 60% of the original weight being lost over the 12 month period and with apparently no reduction in rate of weight loss it would be expected that these samples would be completely degraded in less than two years. Any dependence on pre-storage conditions is not evident from the data.

Electron microscopy showed that even after 6 months[7] the microstructures were drastically affected by the imposed mineralogical changes and the micrographs of the 12 month samples (density = 2513 kg.m^{-3}) show further deterioration. A clear layering effect is obvious from the backscattered electron images (BSI) (Figure 2). It should be noted that the layers shown on the micrographs have underlain those which have spalled off and represent compositional changes within the core cement cylinder. Three regions can be distinguished. The unaffected cement in the interior is shown at the bottom of Figures 2(a) with a layer between this and a gypsum-based surface layer. This surface layer is primarily gypsum and contains small, darker regions corresponding to high silicon signals from energy dispersive X-ray analysis (Figure 3), probably due to decalcified cement relicts. The formation of gypsum is expansive and it is thought that gypsum crystallisation has caused mechanical stresses in the layer causing the overlying gypsum to break off. The intermediate layer, between the unaffected cement and the outer gypsum-based layer is much thinner than the outer layer and concentrates potassium. The reason for this concentration effect is not known. Sulphur is also associated with this layer but not beneath it, i.e. sulphate is excluded from the interior cement. A higher magnification image (Figure 2 (b)) of this intermediate region shows that although the gypsum morphology has not been properly established, there are clear signs of degradation within cement grains. It is not surprising that the reactivity is localised at or near the cement grains in the short-term. The calcium silicates are more reactive than the C-S-H and there is clear evidence of selective reaction within the cement grains themselves, i.e, certain regions within the cement grains are attacked preferentially. Also of interest is the growth of gypsum around grains which presumably represents the primary stage of mechanical failure in the relatively dense portion of the matrix.

Figure 2(a): Low magnification image of PRC (ρ = 2513 kg.m^{-3}) subjected to sulphuric acid attack.   Figure 2(b): Higher magnification image of the same sample showing degradation in the cement grains and formation of gypsum around them

It is also noted that these observations can be made equally for the pressed and unpressed products. This is significant since, based on the connected porosity argument presented above, intrusion of aggressive species into the PRC matrix would be expected to be reduced relative to the unpressed materials. It is suggested that it is specifically the production of gypsum which equalises the degradation rates of pressed and unpressed pastes. Expansion in the cement microstructures causes cracking and exposure of fresh surface for further attack.

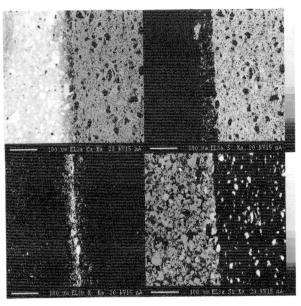

Figure 3: Electron probe microanalysis element maps from a PRC sample exposed to sulphuric acid, (left hand part of the images: cement core; right hand part, gypsum layer). Maps are: Ca (top left), S (top right), K (bottom left) and Si (bottom right).

Hydrochloric Acid

Cylinders exposed to hydrochloric acid (0.247 M) were mechanically stable (no spalling) and appeared normal with the exception of a brown surface layer. They were however apparently more porous than unexposed samples, taking longer than normal to degas when preparing for electron microscopy. Their durability performance as a function of time is shown in Figure 4.

Pressed cements appear to be more resistant to degradation in this medium than the unpressed materials and, despite the higher acid concentration, weight losses from all cements are lower than was measured for comparable cements exposed to sulphuric acid. For example, a pressed product of density 2573 kg.m^{-3} and pre-cured under water for 28 days, loses around 12% of its original weight over 12 months whereas an unpressed cement (density = 2089 kg.m^{-3}) loses about 22%. The corresponding weight losses in sulphuric acid were around 48 and 60% respectively. A slight difference in performance due to variation in pre-storage conditions was observed. Low density samples cured in moist air exhibited a higher initial weight gain than was observed for corresponding samples pre-cured under water.

The electron micrographs (Figure 5) again show a layering effect with the outermost layer being silica-rich, gel-like and cracked; the cracking may be an artefact of sample preparation. The low magnification BEI image of the pressed cement (Figure 5 (a)) shows banding in the outermost layer then a thin interfacial layer overlying the core cement. The general impression from this micrograph is that of a porous microstructure. A similar layering can be seen in the higher magnification images of the pressed (Figure 5 (b)) cement especially in the vicinity of the crack. The acid attack has again decalcified the cement grains in the outer layers although the deeper grains seem to remain largely intact despite the porous microstructure. X-ray microanalysis (Figure 6) confirms the decalcification of the outer layer and shows concentration of silicon and aluminium in these regions; some concentration of iron can also be inferred.

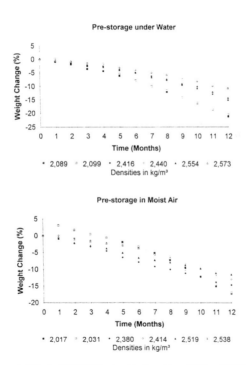

Figure 4: Effect of 0.274M hydrochloric acid on OPC and PRC products

These observations are similar to those made by Chandra[10] who studied the effect of 15% hydrochloric acid solutions on mortar blocks. He also noted an increase in the acid pH as a consequence of contact with the cement and this has been observed here also. The present study shows that the alkalising power of the cement deteriorates with time. Noting that the acid was replaced each week, pH's were raised to 12, 9.5 and 3 from 0.5 at 1 week, 7 months and 12 months respectively increases in the first week of 0.5 to 12 were reduced to 0.5 to 9.5 at 7 months and to 0.5 to 3.3 after 12 months. Such chemical gradients clearly provide conditions whereby species can be solubilised and reprecipitated elsewhere accounting for the observed formation of gel-like compounds of silicon, aluminium and iron in, and the absence of calcium from, the surface layer.

It is often proposed that chloride ions are absorbed in cements in Friedel's salt ($C_3A.CaCl_2.10H_2O$). Chandra[8] indicated from energy dispersive X-ray microanalysis data that Friedel's salt might be present but a number of calcium chloro-aluminate phases are possible[11] and indeed, the present work can not rule out the present of such phases. Alternatively, the results could be explained by the precipitation of soluble chloride during sample drying (in preparation for analysis) or by surface association with the high surface area C-S-H gel.

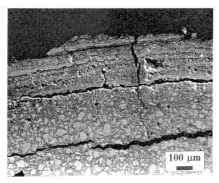

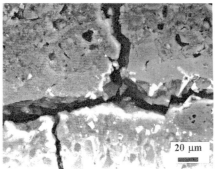

Figure 5(a): Low magnification SEM image of PRC (ρ = 2573 kg.m^{-3}) subjected to hydrochloric acid attack.

Figure 5(b): Higher magnification image of the same sample showing the transition zone between outer layer and cement core.

Ethanoic Acid

Cylinders exposed to ethanoic acid appeared unaffected visually and they were mechanically intact. The concentration of this acid (0.05M) was somewhat lower than that of the other acids but nonetheless was expected to provide a pH of 3.5 which was considered to be reasonably aggressive. Despite this, the weight losses illustrated in Figure 7 are much less severe than those for the more concentrated acids and this time, the differences between the two pre-curing conditions are more evident. All samples show an initial weight increase which in the longer term becomes a weight decrease. The more dense the sample, the less variable is the weight, i.e. the pressed cements are more stable in this environment. It is suggested that the initial weight increase is due to precipitation of degradation products in the pores. The more porous (less dense) the material, the higher the weight increase. Subsequently, the acid-induced deterioration of the solid matrix causes all this material to be lost as well as the material which is dissolving from the matrix itself.

Microstructurally, deterioration seems to be by similar mechanisms to those discussed already for samples exposed to hydrochloric acid. Figures 8 (a) and 8(b) show that although less advanced in development, three layers can again be identified and again there is attack of the cement grains in the silica-rich layer.

Figure 6: Electron probe microanalysis element maps from a PRC sample exposed to hydrochloric acid, (bottom part of the images: cement core; top part, amorphous layer). Maps are: Ca (top left), Al (top right), Si (bottom left) and Fe (bottom right)

General Observations

It has been shown that deterioration mechanisms for pressed and unpressed cement pastes exposed to hydrochloric acid and ethanoic acids are similar. Samples exposed to sulphuric acid exhibit an additional effect. Taking the hydrochloric acid case for discussion, it is anticipated that degradation is promoted by diffusion of the acid through the interconnected surface porosity. There, reaction takes place with the cementitious matrix and calcium is subsequently leached out. Such leaching causes cement pH to drop but acid pH to rise as neutralisation reactions proceed causing a number of chemical changes in the outer regions of the cement. Thus, a decalcified gel layer, rich in aluminium, silicon and iron, results at the surface whereas the interior of the cement still has significant calcium and now chloride which has penetrated to significant depth, the porosity being further increased by the deleterious reaction of the replenished acid and the cement matrix. The loss of ability of the cement to neutralise the acidity with time reflects the increasing depths to which the acid has to penetrate to dissolve the skeletal framework of the cement matrix. A similar situation can be envisaged for the samples exposed to ethanoic acid.

For sulphuric acid, a variation on this mechanism is operating. Just as the hydrochloric and ethanoic acids attacked the surfaces of the cement, so too does the sulphuric acid producing a surface layer of gypsum as the low solubility reaction product. Sulphate is excluded from below this layer and the BEI shows that the cement beneath this layer is unaffected suggesting that the acid has not penetrated this far. But physically, the most obvious degradation comes from sulphuric acid attack suggesting that while HCl and ethanoic acid can penetrate into the pore structure via connected porosity, sulphuric acid cannot do this, probably because the expansive gypsum formation causes the surfaces to break away spontaneously. Micrographs (Figure 2) show that a gypsum layer of approximately 100 μm is permitted prior to spalling but there must be some transport of acid to the underlying, intermediate layer to cause the degradation of cement grains at this interface. Such mechanical deterioration of surface layers is less likely in the hydrochloric and ethanoic acid cases. Calcium chloride, the likely product from hydrochloric acid exposure, is soluble and calcium ethanoate may have been present only in amounts insufficient to have any significant effect on surface conditioning.

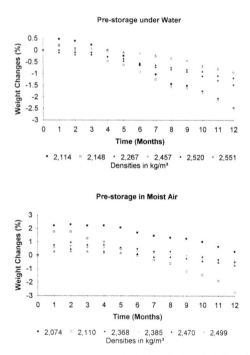

Figure 7: Effect of 0.05M ethanoic acid on OPC and PRC products

CONCLUSIONS

The durability of pressed and unpressed cements in sulphuric, hydrochloric and ethanoic acids have been studied. It has been shown that sulphuric acid has a significant deleterious effect on the stability of both pressed and unpressed cement pastes with up to 60% of the original weight being lost within 1 year, regardless of paste density. Some dependence on density was found for cements exposed to hydrochloric and ethanoic acids. Generally, high density cements were more durable. The difference between ethanoic and hydrochloric acids and sulphuric acid is the

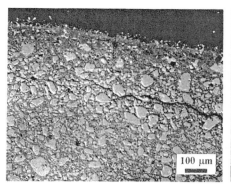

Figure 8(a): Low magnification SEM image of PRC (ρ = 2499 kg.m^{-3}) subjected to ethanoic acid attack.

Figure 8(b): Higher magnification image of the same sample showing cement degradation in the outer layer.

nature of the degradation products arising upon reaction with the cement. Sulphuric acid produces extensive gypsum precipitation on the surfaces of the samples leading ultimately to mechanical stresses in the surface layers and loss of adhesion. Material loss therefore tends to keep up with the penetrating acid front whereas in hydrochloric acid, for example, penetration through the pore structure without spalling causes porosity increases and weight loss by leaching only.

ACKNOWLEDGEMENTS

The authors gratefully acknowledge the advice and assistance of Dr E. E. Lachowski and Dr A. Coats. The financial support of the Engineering and Physical Science Research Council is also gratefully acknowledged.

REFERENCES

1 Macphee, D.E., 'PRC - Pore Reduced Cement: High Density Pastes following Fluid Extraction', *Advances in Cements Research*, 3, (12), 135, (1990).

2 Macphee, D.E., Lachowski, E.E., Taylor, A.H. and Brown, T.J., 'Microstructural Development in Pore Reduced Cement', *Proc. Mat. Res. Soc.*, 245, 303, (1992).

3 Feret, R., 'Etudes sur la constitution intime des mortiers hydrauliques' *Bull. Soc. Encour. Ind. natn., Paris II*, 1604, (1897).

4 Ryshkewitch, E., 'Compression strength of porous sintered alumina and zirconia', *J. Amer. Ceram. Soc.*, 36, 65-68, (1953).

5 Griffith, A.A., 'The phenomenon of rupture and flow in solids', *Phil. Trans. Roy. Soc. (London)*, A221, 163-98, (1920).

6 Kendall, K., Howard, A.J. and Birchall, J.D., The relation between porosity, microstructure and strength, and the approach to advanced cement-based materials', *Phil. Trans. Roy. Soc. (London)*, A310, 139-53, (1983).

7 Geslin, N., Israel, D., Lachowski, E.E. and Macphee, D.E. 'Durability and Microstructure of Pore   Reduced Cements (PRC)', *Mat. Res. Soc. Symp. Proc.*, 370, 237-44, (1995).

8 Fattuhi, N.I. and Hughes, B.P., 'The performance of cement paste and concrete subjected to sulphuric acid attack', *Cem. Concr. Res.*, 18, 545-53, (1988)

9 Pavlik, V., 'Corrosion of Hardened Cement Paste by Acetic and Nitric Acids. Part 1: Calculation of Corrosion Depth', *Cem. Concr. Res.*, 24, 551-62, (1994).

10 Chandra, S., 'Hydrochloric acid attack on cement mortar- an analytical study', *Cem. Concr. Res.*, 8, 193-203, (1988).

11 Birnin-Yauri, U., 'Chloride in Cement: Study of the System CaO-Al$_2$O$_3$-CaCl$_2$, (*PhD Thesis*, University of Aberdeen, 1993)

GEL-MODIFIED CEMENT
FOR ELECTRO-CONDUCTING CONCRETE

S Y Maxunov

M G Galuzinskiy

Scientific Research Institute on Binders and Materials

Ukraine

ABSTRACT. Electro - conducting concrete is useful alternative to the applications traditionally associated with ceramic. However, further technological progress in the area is limited with low early strength (use of admixtures which accelerate the hardening of cement is ineffective due to increasing of ionic conduction) and unstable electrical conductivity associated with covering of electro-conducting particles with hydration products. To solve these problems it has been proposed to add calcium silicate hydrate gel into the composition of electro-conducting concrete. In this paper the gel-modified cements based on ordinary Portland cement were investigated. The desiccated gel was used to substitute part of cement. It has been shown that gel admixtures in pressed cements play an important role. Thus, the use of gel results in early compressive strength increased by a factor of 1.5-2.0. Also dry compaction of gel-cement mixtures and short time water saturation lead to high strenght gain. The results obtained indicate the possibility to produce cement based composites with required electrical conductivity.

Keyword: Electro-Conducting Concrete (ECC), Gel-Modified cement (GMC), Ordinary Portland cement (OPC).

Professor Maxunov S Y is Director of the Section of Materials of Contact Hardening, Scientific Research Institute on Binders and Materials, Kiev, Ukraine.

Mr Galuzinskiy M G is a Candidate of Cathedra of Building Materials, Kiev State Technical University of Construction and Architecture, Ukraine.

Radical Concrete Technology. Edited by R K Dhir and P C Hewlett. Published in 1996 by E & FN Spon, 2–6 Boundary Row, London SE1 8HN, UK. ISBN 0 419 21480 1.

INTRODUCTION

Concrete is, in terms of its constituents and production and wide-ranging properties, useful alternative for applications traditionally associated with the ceramic products. In particular, the development of electro-conducting ceramic materials through cementitions reactions has been explored using calcium silicate cements [1].

Traditionally, Electro-conducting concrete (ECC) combines the cement mortar and particulate conductors suchas graphite and various metals powders [2-4].

Usually, ECC are intended for the production of resistors, electric heaters, ets [5]. However, the use of cementitious reactions does not provide required stability of electrical properties because of high moisture content.

The hydration products covering current-conducting particles, reduce the electrical conduction of material. One of solutions to this problem is offerred in [6]: the technology of ECC production includes the compaction of dry mixture of initial components and subsequent hydration by steam pressure feed through the holes in the punches without unloading removal of pressure. The researches on dry compaction of Portland cements [7] have shown, that this method may be of technical interest, as far as permits to obtain the very strong products in quite short time (Table 1).

Table 1 The strength of samples, obtained by dry compaction of Portland cements

COMPACTION PRESSURE, MPa	COMPRESSIVE STRENGTH, MPa AFTER COMPACTION	COMPRESSIVE STERNGTH, MPa AFTER TREATMENT IN BOILING WATER	COMPRESSIVE STRENGTH, MPa AFTER WATER STORAGE, days		
			3	28	72
40	5.8	154.5	112.2	154.5	183.3
200	18.9	176.1	177.8	188.9	197.2
1000	41.0	187.2	86.1	158.9	169.4

The obvious limitation of this method is the low strength of samples after compaction, that hinders the transport of products in technological process. Less obvious is the following advantage. As far as moulding of products does not require additional quantity of water, masture content in hardening material may be limited only of that quantity, which is necessary for the hydration of cement. This reduce to the minimum the contents of free water in product and, accordingly, to increase the stability conductivity.

In present work the results of research the new method for the production of ECC are given. Its key feature is that with the purpose of increasing the strength, cements is modified by additives of calcium silicate hydrate gel able to contact hardening. Besides, for the increasing of stability of electro conduction (due to reduction of moisture cement in composite) hardening is provided by short-time water impregnation of the pressed dry mixtures of initial components and subsequent storage of products no more the 24 hours.

EXPERIMENTAL DETAILS

Materials

The ordinary portland cement (OPC) and amorphouse calcium silicate hydrate (CSH) were used to produce modified cements.

Calcium silicate hydration were of C-S-H(I) group [8]. They possessed unstable crystalline structure and sad properties ocmplying with the TU 14-147-44-92 "Calcium silicate hydrates. Specifications". BOPC was of grade M400 complying with the GOST 10178-75. The initial dispersed components were thorougly mixed during 25 minute on different proportions using ball min with the volume of $0.8m^3$.

The samples for mechanical testswere of the samples -cylinder form with diameter and hight 1.5 cm. They were pressed at the specific load of 40 and 100 MPa. The non-pressed cubic samples of 3 cm length were used as controls.

The tile-samples of 12.1x6.5x0.6 cm size were prepared for determination of quantity of non-bound water inside gel-modified cement (GMC) measuring the electrical resistance. The samples pressed were by cold unilateral compaction under the pressure of 40 and 100 MPa. They contain incorporated wire-like electrical contacts. The samples for measuring water saturation and porosity does not have any contacts. Water saturation of samples were made by means of their short-time immersing in water.

The compositions of studied samples are given in Table 2.

Table 2 The compositions of studied samples.

CODE	COMPACTION OF PRESSURE, MPa	COMPOSITION, vol.%	
		OPC	CSH
A-1	40	0	100
A-2		30	70
A-3		80	20
A-4		100	0
B-1	100	0	100
B-2		30	70
B-3		80	20
B-4		100	0

Testis methods

The evaluation of mechanical properties were performed accordingexecuted to standard techniques.

At SHF-measurement , studied sample after short-term impregnation in water immersed in bath with oil, with purpose to exclude the evaporation of water from sample in environment, and after this bath with oil and sample placed between radiating and receiving end of measuring line. Thus the coordination of opened ends wavequides conducted at help horns. Simultaneously with measurement of easing, Determined the size of electrical resistance (R) on these tile-samples through electrical bridge P-316, which permits to measure R in range $10^{-1}...10^{6}$ Ohm, accurate to 0.2 %. Kinetik of water seturation of tiles in real time and porosity tile-samples determined by method hydrostatic the weighing in waters. The research of features hydrotation executed with use of method SHF-measurement.

The use of method permits to execute the continious check of change of physical water in hardening composite in real time scale. The use of this method is based that at frequencies 19-21 GHz the untie water inside of material has maximal factor of easing electromagnetic emission.

Results and Discussion

Features of water satiration. The degree of water saturation (DWS) of samples were determined as the ratio of weight of damp sample to weight of sample completely saturated by water.

The DWS is largely determined with the porosity of samples. The analysis of data, presented in Figure 1, shows, that the process of water saturation is enough uniform one. However, the local areas exist with sharp decrease of DWS The special observations during the saturation, have shown. That the mentioned bubbles phenomenon is connected with association of air bubbles, forming on the surface of the sample at the impregnation and their subsequent detachment. The investigation of the water saturation kinetiks has shown, that for all of the samples the changes of their weights in water one similar. At the same time, the duration of complete water saturation is mainly content determined by the ratio between OPC and CSH. Thus, lines for the composite binder are between the lines, describing water saturation of intitial components (Figure 1).

The analysis of data given in Table 3 shows that the durationof water saturation can strongly to depend not only on porosity but also, probably, on pore size distribution for the samples A-3 (9.3%) and B-2 (8.8%), as well as A-2 (11.2%) and B-1 (10.8%) having almost the same porosity duration of saturation differs, respectivly, in 4 and 2 times. First of all, it can be connected with more fine pore size in samples, pressed at 100 MPa.

However, the comparison of data for CSH samples (A-1 and B-1) shows that second probable reason of slowing down the water saturation of samples is the gel pores of CSH. As shown in [9] complete hydration of OPC need 25-28% of admixed water. Considering that the total quantity of water used is 23-27% of xamant weight, we obtaine that only 6-8% of water involved in cements hydration. Hence, water saturation of pressed products should be limited with mentioned DWS. It has not the significance that the part of water will adsorbed by the surface of CSH, as far as after hemosorbtion of water on the surface of of cement is particles, inside of material insulated from the environment (this requirement is standard to ECC) will arise the gradient of moisture, promoting the involvment of adserbed water to the hydration of cement. But it should be noted, that the part of moisture is in gel pore of CSH (the value is approximately equal to 3-5%). Then the limit of DWS should be of 9-13%. In Table 3 the values, appropriate of recommended duration of water saturation (10 s for the samples, pressed at 40 MPa, 30 s - 100 MPa) are shadowed.

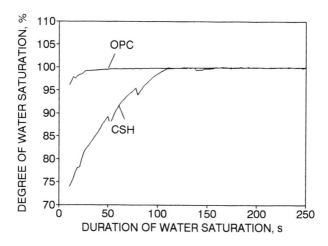

Figure 1 The kinetic of water saturation for OPC and CSH samples

Table 3 The degree of water saturation of samples depending on duratin of their saturation

CODE	POROSITY, %	DURATION OF SATURATION, s	THE DEGREE OF WATER SATURATION, % UNDER AT TIME OF SATURATION, sec		
			10	30	300
A-1	18.8	125	5.2	6.9	10.6
A-2	11.2	90	14.5	19.3	22.1
A-3	9.3	40	12.1	12.5	12.6
A-4	-	-	-	-	-
B-1	10.8	170	4.8	7.1	9.3
B-2	8.8	150	7.1	9.5	9.9
B-3	7.7	60	8.4	8.9	9.0
B-4	11.7	50	6.8	7.1	7.7

Hydration. The features of hydration were studied by means of continious control over unbaunded water change during hardening. Firstly, the features of hardening of CSH gel and OPC were investigated. For samples CSH (Figure 2) characteristically follow oneing. In the samples of CSH the change of free water, content practically does not happen. It testifie the absence of hydration processes in CSH. The small reduction of quantity of physical water in starting period (about 10 minutes) is connected, in our opinion, with the fact that some of water (0.2 - 0.6%) becomes interlayer, thus the thickness of water layer reaches 1-4 monolayer of the water molecules. That is below the sensitivity of, i.e. water becomes "invisible" for device. Thus, interlayer water provides insignificant joint of CSH layers with weak hydrogenous bounds i.e. the quasi-bounding of water in CSH happens. Obviously, it has not essential influence on value development of further processes, and significant of free water is possible to be considered as the reserve, which can be used at hydration of cement. The character of dependences "the duration of hardening - moisture content" for cement samples are entailly different (Figure 3).

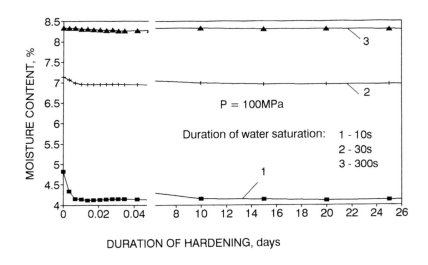

Figure 2 Free water content in CSH samples versus the duration of water saturation

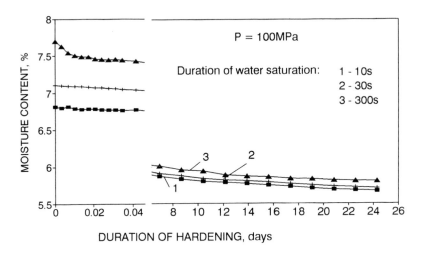

Figure 3 Free water content in OPC samples versus the duration of water saturation

Starting first minutes of hydration the intensive preaction of water in connected condition is observed. Already after 7 days the quantity of free water decreases more than 2 times. At the same time the free water content remains quite significont. This is determines the instability of electrical conduction of on OPC base.

For gel-modified cements, the following features (Figure 4) are characteristic.

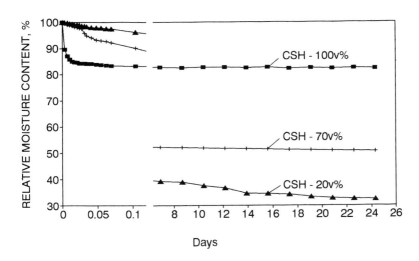

Figure 4 Kinetik of free water content change in samples pressed at 40 MPa and saturated
with water during 10 s

At the begining,the hydration prosesses in composites with higher CSH content are more
active. Probably, the water, adsorbed CSH, makes the additional reserve for hydration of
cement, as far as with the beginning of interaction, the formation of moisture content gradient
inside the material happens. After 1 day the interaction with water in such compositions is
slowed down. It should be noted, that the values of moisture content of gel-modified cements
are more lower, than those for unmodified cement. It makes possible the high stability of
electro conductivity. The of electro conductivity of hardening gel-modified cement, show more
features (Figures 5, 6).

The electrical resistance of samples of CSH (Figure 5) does not essentially change. The
quantity of charge carriers does not change because there are no any chemical processes inside
of materials. At the same time, the resistance GMC (Figure 6) are higher and continuously
increas in futher period. The comparison of data on changes of resistance and moisture content
shows, that the increase of resistance happens during the period, when moisture content has
been stabilized.

Obviously, it reflects the re-crystallisation processes on ordering of crystal structure of
substance. That prosesses make CSH "more dielectric", that, certainly, favorable for ECC. The
influence of gel admixtures on the strength of GMC.

The results on compressive strengths development of pressed and non-pressed samples are
presented in Table 4 and 5.

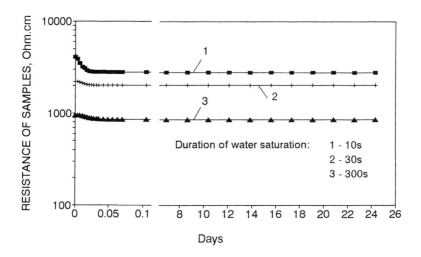

Figure 5 Resistance of CSH samples pressed at 100 MPa

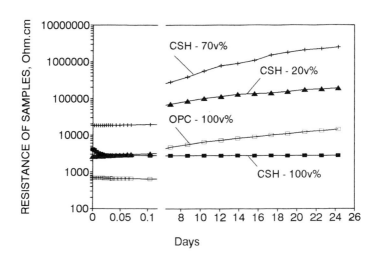

Figure 6 Resistance of samples with various CSH contents,
pressed at 100 MPa and satureted with water during 10s

The data obtained testify that the strength of gel-modified cements, sufficient for the manufacture of ESS (about 300 MPa), can be reached already on the next day after compaction. The compaction pressure of 100 MPa is, probably, optimum bacause provide formation of quite stable (mith deuse contacts) electro-condycting structure [1].

Table 4 The compressive strength of pressed of gel-modified cement

PRESSURE, MPa	CSH CONTENTS, vol.%	THE COMPRESSIVE STRENGHT, MPa AFTER STORAGE IN WATER DURING, days			
		after compaction	1	7	28
40	0	3.4	7.9	21.0	84.1
	20	8.5	15.0	31.1	80.4
	70	19.3	19.8	19.3	26.0
	100	21.0	18.7	18.1	19.3
100	0	6.1	33.4	100.2	152.3
	20	15.3	35.1	87.8	121.2
	70	31.7	36.2	54.4	57.8
	100	39.1	27.7	30.0	34.5

The time of preliminary water saturation does not influence essentially on hydration, that is confirmed with mechanical characteristics (Table 6).

Table 5 The compressive strength of non-pressed gel-modified cement

CSH CONTENTS, vol.%	THE COMPRESSIVE STRENGHT AT COMPRESSION, MPa			
	After boiling in water	After steam curing	After storage in water, days	
			1	7
0	8.0	15.0	0.9	16.7
20	7.8	12.0	0.9	13.3
70	0.0	0.0	0.4	2.2
100	0.0	0.0	0.0	0.0

Table 6 The strength of pressed gel-modified concrete

PRESSURE, MPa	DURATION OF WATER SATURATION, s	CSH CONTENTS, vol.%	THE COMPRESSIVE STRENGTH AT COMPRESSION, MPa AFTER STORAGE IN NORMAL CONDITIONS, days	
			1	28
40	10	0	21.5	23.8
		20	23.8	21.5
		70	21.5	19.2
		100	23.8	21.5
	3600	0	40.2	43.6
		20	22.6	25.8
		70	23.8	27.7
		100	20.9	20.7
100	10	0	37.4	66.8
		20	36.2	60.0
		70	31.1	48.7
		100	44.2	49.8
	3600	0	36.2	80.7
		20	37.9	69.4
		70	40.2	61.7
		100	36.2	49.3

CONCLUSIONS

1. In the presented work the gel-modified cements on OPC base intended for the manufacture of ECC were investigated.
2. Is has been shown, that the CSH gel admixtures promote high initial strength of composite and low moisture content without significant influence on mechanical properties. Besides, the hydration of GMC results in formation more "dielectric" materials. All that creates the conditions for the production of material with more stable electrical conductivity.
3. The results obtained may possible to recommend the technology of short-time-hydration the manufacture of ECC. The method include compaction of dry mixtures of initial components, their short-term saturation with water (to moisture content of 9-13% of cement's weights) and insulating the product from environment (for example, electro insulating the varnish).
4. ECC based on gel-modified cement may be used fo the production of resistors and electric heaters.

REFERENCES

1. Composite Resistors for Energy Plant's Construction /Горелов В.П., Пугачев Г.А. - Новосибирск: Наука, 1989.- 216 с.
2. 509892 СССР МКИ2 H 01 B1/04; H01 C7/02. Current-Condutiong composite material / Добжинский М.С., Логачева Г.М., Горелов В.П., Врублевский Л.Е.- N1914638/24-7, Заявл. 7.05.73., Опубл. 5.04.76, Бюл. N13.
3. Берней И.И., Автономов И.В. Electro-Conducting Concrete on Slag-Alkaline Binder // Шлакощелочные цементы, бетоныи конструкции., Тез. докл. // Всесоюз. науч.-практич. конф., Киев, октябрь 1984г.- Киев, КИСИ, 1984.- С. 300-301.
4. Petersen P.H. Metallic waterprofing // Builder.- 1967.- vol.44, N30.- p. 16-19.
5. Power resistors /Л.Е. Врублевский, Ю.В. Зайцев, А.И.Тихонов.- М.: Энергоатомиздат, 1991.- 256 с.
6. Мурадов Э.Г., Сканави Н.А., Закиров М.Х. The impovement of stability of production products from ECC // Изв. вузов. Строительсво и архитектура.- 1982.- N5.- С.72-75.
7. Глуховский В.Д., Рунова Р.Ф., Максунов С.Е. Ролькоптактно-конденсационных процессов в синтезе прочности цементного камня//Цемент. 1989.N 10. С. 7-8.
8. Taylor H.F.W. Cement Chemistry.- London: Academic Press Ltd, 1990.- 470p.
9. Binding materials / А.А.Пащенко, В.П.Сербин, Е.А.Старчевская. 2-е изд.- К.: Вища шк. Головное изд-во, 1985.- 440с.

LATE PAPERS

A CRITICAL REVIEW OF SERVICE LIFE MODELLING OF CONCRETES EXPOSED TO CHLORIDES

M D A Thomas

University of Toronto

Canada

M R Jones

University of Dundee

UK

ABSTRACT. Test methods for quantifying chloride transport rates in concrete are reviewed and results from studies on fly ash concrete are presented. Fly ash imparts a significant improvement to the resistance of concrete to chloride ingress. However, much of this benefit is manifested at later ages, with significant reductions in diffusion occurring beyond the first year. Consequently, accelerated early-age tests are unlikely to provide a true measure of the service-life extension afforded by ash utilization. It is possible that the time-dependent diffusion effects could be accounted for by using suitable accelerated curing techniques for concrete containing fly ash (and similar materials). Most of the recent tests are further limited either by failure to provide a direct measure of chloride movement (ASTM C1202 and related tests) or by the rigorous application of mathematical relationships developed for ideal conditions not representative of chloride transport in concrete. However, such tests likely provide an index of the relative ease of chloride ingress and represent a considerable advance in performance testing for durability design.

Keywords: Chloride, diffusion, fly ash, test methods

Dr Michael Thomas is an Assistant Professor in Civil Engineering at the University of Toronto, Canada. His main research interests are related to concrete durability, the use of supplementary cementing materials and service life modelling. He is a member of a number of CSA, ACI and ASTM committees dealing with these issues.

Dr Roderick Jones is a lecturer in construction materials in the Department of Civil Engineering and a member of staff of the Concrete Technology Unit at the University of Dundee, Scotland. His research interests include in the development of pozzolanic binders, particularly to enhance the durability of concrete. His current research work concerns chloride transportation processes and penetration rate estimation in concrete.

Radical Concrete Technology. Edited by R K Dhir and P C Hewlett. Published in 1996 by E & FN Spon, 2–6 Boundary Row, London SE1 8HN, UK. ISBN 0 419 21480 1.

INTRODUCTION

Corrosion of steel in reinforced concrete, due to the penetration of chlorides from de-icing and marine salts, is one of the most prevalent forms of deterioration of concrete structures worldwide. Considerable research effort has been directed towards the elucidation of chloride transport and steel corrosion mechanisms in reinforced concrete, and in the development of models to predict the service life of structures exposed to chloride environments. The service life is dependent on a great many factors, bit chief among the material considerations is the ability of the concrete cover to resist the penetration of chloride ions and, thus, provide lasting protection to the steel reinforcement. As a result, the last decade has seen the emergence of a multitude of different test methods aimed at characterizing the chloride transport properties of concrete and, indeed, some of these have recently found their way into concrete specifications for prestigious projects [1, 2].

Some test methods actually determine the "chloride diffusion coefficient" of the concrete and this parameter may be used as a direct input for a service life model. Other methods provide an indirect measure of the transport properties and a diffusion coefficient can only be obtained from previously derived empirical relationships. Many of the theories that are used in calculating diffusion coefficients were derived under specific ideal conditions, e.g. for a dilute, single species electrolyte in free solution with no chemical interactions. These conditions will not exist in concrete which is composed of a relatively high concentration electrolyte and relatively "unstable" hydrates which will interact both physically and chemically with the diffusing species. Furthermore, the properties of concrete change with time due to continued hydration and interaction with its environment. These changes are particularly marked for concrete containing slowly reacting pozzolans (e.g. fly ash) or latent hydraulic materials (e.g. slag), as significant reaction of these materials occurs at later ages. However, researchers and specifiers have tended to favour rapid tests that produce data in hours or days and which, especially for quality control purposes, may often be used at early ages (e.g. 28 days).

This paper presents a variety of data from a wide range of test methods used to characterize the performance of fly ash concrete in chloride environments. The data originate from a number separate research programmes carried out in recent years at the Building Research Establishment and Dundee University in the U.K. and the University of Toronto in Canada.

TEST METHODS

A number of methods are available for determining the diffusion of chlorides in concrete; however, most of these tests fit into one of three categories. These are:

- steady-state diffusion tests
- penetration (or bulk diffusion) tests
- electrical migration diffusion tests.

Details of each test method are given below.

Steady-State Diffusion Tests

Steady-state chloride diffusion through hardened cement paste samples has been measured by a number of workers [3-5] using diffusion cells similar to that reported by Page et al [3]. In this cell, the concentration difference between the two compartments is the only driving force and flow is assumed to obey Fick's first law for steady-state diffusion, i.e. rate of flux of chloride ions, J, through a unit area of a section of the sample is proportional to the concentration gradient normal to the section:

$$J = -D.\frac{dC}{dx}$$
(1)

where D is the diffusion coefficient and C is the concentration of chloride ions at depth x.

As the process is slow, this method has, traditionally, been used to measure flow through thin discs of cement paste. However, similar methodology has been used to measure diffusion through concrete discs [6-8]. Dhir and Byars [6] used a 25 mm thick concrete disc with a very high chloride concentration (5M NaCl) on one face. Despite, the high concentration gradient, long time periods (typically 3-6 months) are necessary to attain steady-state conditions [9] even in concrete with a moderate water/cement ratio. Other workers have used thinner concrete discs [7,8] to expedite the test; however, this may introduce problems if normal size coarse aggregate is used.

This method would probably take many years to perform using high performance concrete of reasonable sample thickness (>25 mm) and is not, therefore, suitable as a performance test.

Penetration (or Bulk Diffusion) Tests

Diffusion coefficients for concrete have been calculated from the chloride concentration profiles determined for specimens immersed in solutions of known concentration in the laboratory [10-15] exposed to salt spray in the field [7] or exposed to seawater [15-17]. The chloride profile is established by taking samples at incremental depths from the exposed surface and analyzing the sample for chloride content. The apparent diffusion coefficient can be calculated using Fick's second law [18] which states that, for a semi-infinite system:

$$\frac{dC}{dt} = D.\frac{d^2C}{dx^2}$$
(2)

Using the initial condition, $C_x = C_s, x = 0, t > 0$ and the boundary condition, $C_x = 0, x > 0, t = 0$, the diffusion coefficient can be found using Crank's solution [18]:

$$\frac{C_{x,t}}{C_s} = 1 - erf\left(\frac{x}{2\sqrt{D_a \cdot t}}\right) \qquad\qquad (3)$$

where C_s = chloride concentration at the surface, C_x = chloride concentration at distance x and time t, x = depth from the surface, D_a = apparent diffusion coefficient.

Generally, the best fit values of C_s and D_a are found by iteration using least squares. The solution produces an "apparent" diffusion coefficient, D_a, since values of total chloride are usually used whereas Eqns. 1 and 2 describe diffusion of free ions in solution. The use of Eqn. 3 assumes a linear relationship between "bound" and "free" chlorides, i.e. that D_a is constant with depth (or chloride concentration). However, binding relationships have been shown to be non-linear [14,19], and the use of linear relationships may lead to significant error in service-life prediction [20].

The value obtained from Eqn.3 gives the "lifetime average" for the concrete over the time period t. Thomas [16] and Dhir *et al* [13] found that the chloride diffusion coefficient determined from chloride profiles reduces with the length of exposure, at least up to 4 years (and possibly beyond); the reduction is particularly pronounced in concrete containing fly ash.

Electro-Chemical (Migration) Tests

Due to the time-consuming nature of chloride diffusion under concentration gradients, a number of workers have developed techniques to increase the rate of flow of chlorides using an electrical field [6,7,21-26].

The accelerated test developed by Whiting [21] has been adopted as a standard method by both AASHTO (Test T227-83) and ASTM (Test C1202). In this test, the total charge passed through the specimen in six hours is measured and reported in coulombs. The test is semi-quantitative and classifies the permeability of the concrete as high, average, low, very low or negligible. The test does not give direct information about the diffusion of chloride ions; it is assumed that the diffusion will be related to the total charge passed. However, a number of concerns have been expressed about this method. The main problems with using this test method for assessing chloride flow are: (i) the total current measured is the result of all ionic migration and not just chloride ions [27]; (ii) the measurements are made before steady-state conditions are attained [27,28]; (iii) the high voltage (60V) can lead to high temperature rises (especially in low-quality, permeable concrete) which can increase the flow of ions [27,29-31].
Other workers have produced tests that operate at lower potential differences thereby reducing the effect of heat development [6,7,23,25,26]. In addition, in these tests the migration of chlorides is actually determined directly by measuring the change in concentration in the anodic cell [6,7,23], the time to breakthrough in the anodic cell [26] or the depth of penetration in the actual sample [25].

A steady change in concentration in the anodic cell indicates that steady-state conditions have been attained and the diffusion coefficient can be calculated from the Nernst-Plank equation [25,27]:

$$J = -D \cdot \frac{dC}{dx} + DC \cdot \frac{zFE}{RT} \qquad (4)$$

where: J = flux of ions, D = diffusion coefficient, C = chloride concentration in the cathodic cell, z = valency, F = Faraday constant, R = gas constant, T = Kelvin temperature, and E = potential gradient. The solution of the diffusion coefficient may be simplified by assuming the flow contribution due to concentration gradients $(-D \cdot dC/dx)$ is negligible at voltages in excess of 10V.

Application of Eqn. 4 to concrete migration tests assumes that: (i) D is constant with time, (ii) the charge is carried by chloride ions only, (iii) the potential decreases linearly with depth ($d^2 E/dx^2 = 0$), and (iv) binding is proportional to concentration. None of these assumptions really apply to concrete. Furthermore, the relationship was developed for solutions of relatively low concentration (compared to concrete). Determining the chloride concentration in the anodic cell may be complicated by chlorine gas evolution at the anode and reaction of chloride ions with the anodic electrode [22,27], however, these effects can be reduced by selection of suitable material for the anode and avoidance of excessively high driving voltages. Using the time to chloride breakthrough [26] or the depth of chloride penetration [25] negates the effect of chloride interaction and evolution at the anode. However, use of these parameters leads to a more complex solution for D, based on a modified Fick's second law and Nernst-Plank equation. In addition, errors may be introduced due to non-steady state conditions, chloride binding effects and inaccuracies in the determination of the depth of penetration or time of chloride breakthrough.

RESULTS FOR FLY ASH CONCRETE

The results given here have been selected from comprehensive research programmes on fly ash concrete performance being carried out at the Building Research Establishment (BRE) and Dundee University in the U.K., and the University of Toronto in Canada. Full details of concrete mixes, test methods and resulting data have been presented elsewhere [6,9,13,16,17].

Selected results from chloride penetration studies at Dundee and the BRE (BRE) are shown in Figure 1. In the Dundee tests, specimens were immersion in 5M NaCl at 20°C for various periods of time up to 1 year prior to establishing the chloride profile. BRE specimens were placed in the tidal zone of a marine exposure site for periods of up to 4 years. Both sets of data show considerable improvement due to the use of fly ash and also demonstrate the effect of exposure time (up to 4 years) on the calculated diffusion coefficient. Calculating coefficients from short-term tests (e.g. 3 months) ma lead to considerable overestimation of the diffusivity over the life of the concrete and this is more marked for concrete containing fly ash. For concrete with 50% ash, the value of D_a decreases by one order of magnitude between 1 and 4 years.

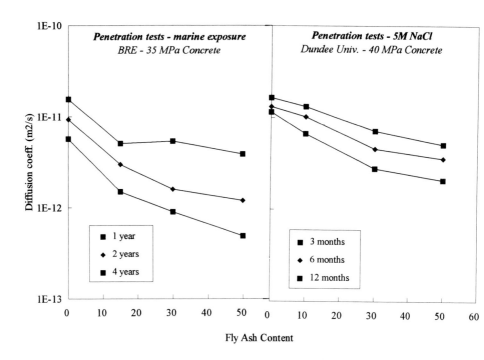

Figure 1. Comparison of chloride ingress into similar concrete systems under marine and under accerated conditions

As mentioned above these diffusion coefficients are average values for the exposure period. A numerical model was developed to calculate the diffusion coefficient that best described the change in profile between 2 ages, i.e. a "moving average", and has been described elsewhere [33]. The average coefficients between 2 and 4 years for 35 MPa concrete with 0 and 50% fly ash were 50×10^{-13} m^2/s and 2×10^{-13} m^2/s, respectively. This represents a difference of 25 times.

The increased ability of fly ash to resist the penetration of chlorides has been corroborated by field studies of structures exposed to de-icing and marine structures [34]. Figure 2 shows concentration profiles for two bridges exposed to de-icing salts (samples taken from roadside pier, 1m above ground) and a sea wall exposed to marine salts (samples taken ~ 1m above high tide). Both cases allowed the comparison between plain OPC and fly ash concrete. Diffusion curves were fitted to the concentration profiles although it is recognized that diffusion is not the only mechanism contributing to chloride ingress in these structures. Calculated diffusion coefficients were an order of magnitude lower in the ash concrete. However, of greater significance is the lower chloride content at depth in the fly ash concrete.

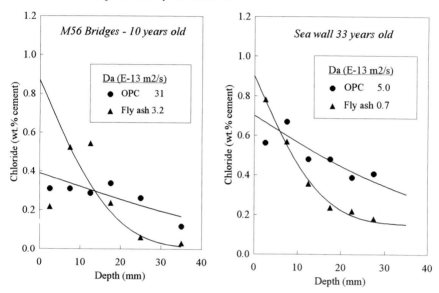

Figure 2   Concentration profiles from structures exposed to chlorides

Figure 3 shows typical concentration profiles for laboratory immersion tests with diffusion curves fitted according to Eqn. 3. The data demonstrate another potential source of error when using penetration tests to determine diffusion coefficients and

select (design) concrete for chloride resistance. There is significant variation in the chloride content close to the surface between different concrete mixes (with the same W/CM) resulting in different calculated surface concentrations, C_s, despite the use of the same external salt solution (3M NaCl). This obviously has an impact on the calculated value of D_a. For example, the concrete with 25% ash has much higher concentrations at the surface than the concrete with 56% ash, but the concretes have comparable concentrations at depth (20mm). This results in a higher concentration gradient dC/dx in the 25% ash concrete and, thus, lower value of D_a. In practice it is the concentration at depth that is important, and both concretes show equal performance in this respect. Despite this, the 25% ash concrete would be selected on the basis of the penetration test.

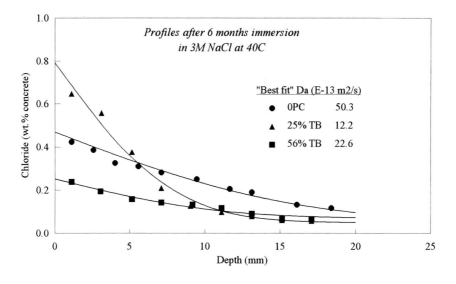

Figure 3   Concentration profiles from laboratory "Bulk Diffusion Tests"

Figure 4 shows results for three series of concrete mixes (20, 40 and 60 MPa) using different test methods. The steady-state and penetration tests both show the positive effect of fly ash, 30 to 50% ash generally reducing the diffusivity by 5 to 10 times. However, fly ash has minimal effect on the results from migration tests. This is probably related to the effect of maturity. Although all tests commenced at the same age (28 days), migration involves a relatively short test periods and the benefits of long-term curing are not manifested in the ash concrete.so shown in Figure 4 is the relationship between the diffusion coefficient and the "effective" water-cement ratio of the concrete, $W/(C + k \cdot F)$. The value k is the "efficiency" factor which can be defined as the mass of OPC which makes the same contribution to a particular property of the concrete as a unit mass of fly ash. For steady state tests the value of k is found to be 2, which implies that 1 kg of ash makes the same contribution to chloride resistance as 2 kg of OPC. A similar relationship was found for the data from BRE marine studies.

Figure 5 shows the results of migration tests for a series of concrete mixes including the variables, water-cement, fly ash level and fly ash source. Studies at Dundee have demonstrated that the quality of Class F ashes has little impact on the diffusion properties of concrete. However, a wide range of fly ashes (Class F and C) are available in North America and the chemistry (and other properties) of ashes can vary considerably between sources. Four fly ashes, selected to provide a range of CaO contents (2.0 to 27.3% CaO), were used in this study. Concrete specimens were cured for a minimum of 4 months prior to testing which included ASTM C1202 and electrical migration tests. The Class F fly ash (Lingan) was effective in reducing the diffusion coefficient by up to an order of magnitude, the effect being more marked at higher water-cement ratios.

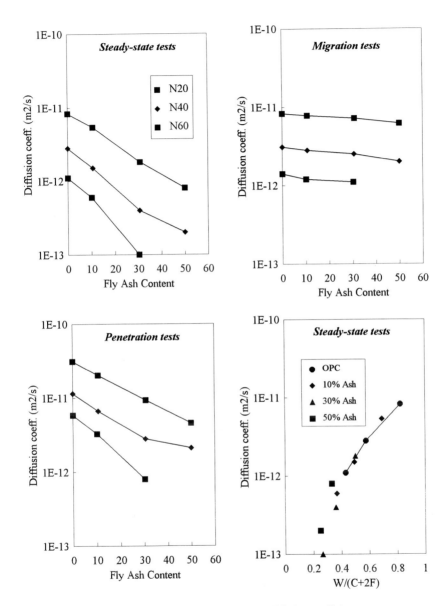

Figure 4   Effect of fly ash on chloride diffusion coefficient
as measured by various test methods

Concrete cast with two of the other ashes, with higher CaO contents (15.0 and 21.5% CaO), gave similar results; the diffusion coefficients were generally less than 10^{-12} m^2/s regardless of the water-cement ratio or level of replacement. However, the concrete with the highest-calcium ash (27.3% CaO) did not perform as well and in all cases had measured diffusion coefficients in excess of 10^{-12} m^2/s.

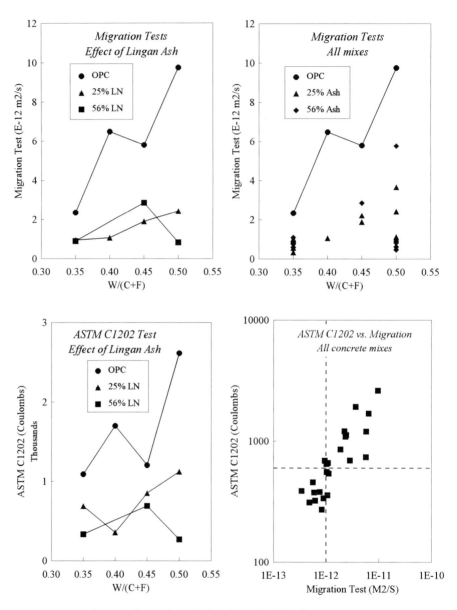

Figure 5   Comparison of migration and ASTM C1202 test results
for fly ash concrete

Results from ASTM C1202 tests showed similar trends and a fairly reliable correlation was established between the charge passed, C, in this test and the coefficient, D, calculated from the migration test (Figure 5); the relationship being:

$$D = 3C \cdot 10^{-15} m^2 s^{-1} \tag{5}$$

It is not suggested that the ASTM C1202 test result be used to provide absolute values for the diffusion coefficient. However, the test is undoubtedly useful for quality controll purposes and can be expected to provide a better indication of the transport properties of the concrete than the compressive strength! The test has the advantage that it can be completed in 6 hours; results indicate that the charge passed in 10 minutes may be equally reliable.

SUMMARY AND CONCLUSIONS

A number of tests have been developed to determine the diffusion coefficient or related properties of concrete. All of these tests have drawbacks related either to the duration of test or the rigorous application of relationships developed for ideal conditions (and certainly not intended to describe the processes in concrete). It is likely that none of these tests yields a "true" diffusion coefficient for concrete, if such a parameter exists. However, such tests represent a significant advance in an industry that places almost total reliance on compressive strength and water-cement ratio for specifying concrete for aggressive environments. The development of suitable performance tests for concrete exposed to chlorides will assist in the design of durable mixes and the development of appropriate specifications.

It is clear from the studies presented here, and numerous others, that the incorporation of fly ash significantly enhances the resistance of concrete to chloride ingress. However, much of the benefit is not manifested at early age and may not be revealed in accelerated tests. Designers and specifiers should be cognizant of the time-dependent nature of diffusivity and suitable approaches should be developed to permit the diffusion coefficient to be extrapolated to the end of the service life of the structure. It is possible that accelerated curing of fly ash concrete (and similar materials) may be allow its long-term properties to be evaluated within a reasonable time frame. Further work is needed in this area.

REFERENCES

1. Hooton, R.D. "Issues related to recent developments in service life specifications for concrete structures." *Second CANMET/ACI International Symposium on Advances in Concrete Technology,* Supplementary Papers, pp. 85-97.

2. Vincentsen, L.J. and Henricksen, K.J. "The Great Belt Link - Built to last." *Concrete International,* Vol. 14, No. 7, 1992.

3. Page, C.L., Short, N.R. and El Tarras, A. "Diffusion of chloride ions in hardened cement pastes." *Cement and Concrete Research*, Vol. 11, No. 3, 1981, pp. 395-406.

4. Goto, S. and Roy, D.M. "The effect of w/c ratio and curing temperature on the permeability of hardened cement paste." *Cement and Concrete Research*, Vol. 11, 1981, pp. 575-580.

5. Zhang, M-H. and Gjorv, O.E. "Effect of silica fume on pore structure and chloride diffusivity of low porosity cement pastes." *Cement and Concrete Research*, Vol.21, No. 6, 1991, pp.1006-1014.

6. Dhir, R.K. and Byars, E.A. "PFA concrete: chloride diffusion rates." *Magazine of Concrete Research*, Vol. 45, No. 162, 1993, pp. 1-9.

7. Bamforth, P.B. and Pocock, D.C. "Minimising the risk of chloride induced corrosion by selection of concreting materials." In *Corrosion of Reinforcement in Concrete*, (Ed. C.L. Page, K.W.J. Treadaway and P.B. Bamforth), Elsevier Applied Science, London, 1990, pp. 119-131.

8. Jackson, P.J. and Brookbanks, P. "Chloride diffusion in concretes having different degrees of curing and made using Portland cements and blended cements containing Portland cement, pulverized-fuel ash and ground granulated blastfurnace slag." *Proc. 3rd CANMET/ACI Int. Conf. on Fly Ash, Slag & Natural Pozzolans in Concrete*, Supplementary Papers (Compiled by M. Alasali), Trondheim, Norway, 1989, pp. 641-655.

9. Dhir, R.K., Jones, M.R., Ahmed, H.E.H. and Seneviratne, A.M.G. "Rapid estimation of chloride diffusion coefficient in concrete." *Magazine of Concrete Research*, Vol. 42, No. 152, 1990, pp.177-185.

10. Midgley, H.H. and Illston, J.M. "The penetration of chlorides into hardened cement pastes." *Cement and Concrete Research*, Vol. 14, No. 4, 1984, pp. 546-558.

11. Collepardi, M., Marcialis, A. and Turriziani,R. "Penetration of chloride ions into cement pastes and concretes." *Journal of the American Ceramic Society*, Vol. 55, No. 534, 1972, pp. 534-535.

12. Wood, J.G.M., Wilson, J.R. and Leek, D.S. "Improved testing for chloride ingress resistance of concretes and relation of results to calculated behaviour." *Proc. 3rd Int. Conf. on Deterioration and Repair of Concrete in the Arabian Gulf*, Bahrain, 1989.

13. Dhir, R.K., Jones, M.R. and Ahmed, H.E.H. "Concrete durability: estimation of chloride concentration during design life." *Magazine of Concrete Research*, Vol. 43, No. 154, 1991, pp. 37-44.

14. Sergi, G. Yu, S.W. and Page, C.L. "Diffusion of chloride and hydroxyl ions in cementitious materials exposed to a saline environment." *Magazine of Concrete Research*, Vol. 44, No. 158, 1992, pp. 63-69.

15. Babu, K.G. and Rao, K.V. "Chloride diffusion characteristics of concrete." In *Concrete 2000 (Ed. R.K. Dhir and M.R. Jones)*, E&FN Spon, London, 1993, pp. 1445-1452.

16. Thomas, M.D.A. "Marine performance of pfa concrete." *Magazine of Concrete Research*, Vol. 43, No. 156, 1991, pp. 171-185.

17. Thomas, M.D.A., Matthews, J.D., Haynes, C.A. "Chloride diffusion and reinforcement corrosion in marine exposed concretes containing pulverized-fuel ash." In *Corrosion of Reinforcement in Concrete*, (Ed. C.L. Page, K.W.J. Treadaway and P.B. Bamforth), Elsevier Applied Science, London, 1990, pp. 198-212.

18. Crank, J. *The Mathematics of diffusion.* Oxford University Press, 1975.

19. Pereira, C.J. and Hegedus, L.L. "Diffusion and reaction of chloride ions in porous concrete." *Proc. 8th International Symposium on Chemical Reaction Engineering,* 1984.

20. Nilsson, L.O., Massat M. and Tang, L. "The effect of non-linear binding on the prediction of chloride penetration into concrete structures." *Third International Conference on the Durability of Concrete,* (Ed. V.M.Malhotra), ACI SP-145, American Concrete Institute, Detroit, pp. 469-486.

21. Whiting, D. "Rapid measurement of the chloride permeability of concrete." *Public Roads*, Vol. 45, No. 3, 1981, pp. 101-112.

22. Li, S. and Roy, D.M. "Investigations of relations between porosity, pore structure and Cl diffusion of fly ash and blended cement pastes." *Cement and Concrete Research*, Vol. 16, No. 5, 1986, pp. 749-759.

23. El-Belbol, S.M. and Buenfeld, N.R. "Accelerated chloride ion diffusion test." *MRS Symposium*, Vol. 137, Materials Research Society, Pittsburgh, 1989, pp. 203-214.

24. Cabrera, J.G. and Claisse, P.A. "Measurement of chloride penetration into silica fume concrete." *Cement and Concrete Composites*, Vol. 12, No. 3, 1990, pp. 157-161.

25. Tang Luping and Nilsson, L-O. "Rapid determination of the chloride diffusivity in concrete by applying an electrical field." *ACI Materials Journal*, Vol. 89, No. 1, 1992, pp. 49-53.

26. Halamickova, P. *"The influence of sand content on the microstructure development and transport properties of mortars."* MASc Thesis, University of Toronto, 1993, pp. 24-54.

27. Andrade, C. "Calculation of chloride diffusion coefficients in concrete from ionic migration measurements." *Cement and Concrete Research*, Vol. 23, No. 3, 1993, pp. 724-742.

28. Zhang, M.H. and Gjorv, O.E. "Permeability of high-strength lightweight concrete." *ACI Materials Journal*, Vol. 88, 1991.

29. Malek, R.I.A. and Roy, D.M. "The permeability of chloride ions in fly ash-cement pastes, mortars and concrete." *MRS Syposium,* Vol. 113, Materials Research Society, Pittsburgh, 1988, pp. 291-300.

30. Roy, D.M. "Hydration, microstructure and chloride diffusion of slag cement pastes and mortars." *ACI SP-114,* Vol. 2, American Concrete Institute, Detroit, 1989, pp. 1265-1281.

31. Geiker, M., Thaulow, N. and Andersen, P.J. "Assessment of rapid chloride permeability test of concrete with and without mineral admixtures." In *Durability of Building Materials*, (Ed. J.M. Baker, P.J. Nixon, A.J. Majumdar, H. Davis), E&FN Spon, London, 1990, pp.493-502.

32. Buenfeld, N.R. and El-Belbol, S. Discussion of paper by Dhir et al., *Magazine of Concrete Research,* Vol. 43, No. 155, 1991, pp. 135-139.

33. Thomas, M.D.A., Evans, C.M. and Bentz, E.C. "Chloride diffusion modelling for marine exposed concretes." *In Corrosion of Reinforcement in Construction,* Society for Chemical Industry, London, 1996.

34. Thomas, M.D.A. and Matthews, J.D. "Performance of fly ash concrete in U.K. structures." *ACI Materials Journal,* Vol. 90, No. 6, 1993, pp. 586-593.

35. Dhir, R.K., Jones, M.R. and Seneviratne, A.M.G. "Diffusion of chlorides into concrete - Influence of PFA quality." *Cement and Concrete Research,* Vol. 21, 1991, pp. 1092-1102.

CONSTRUCTION OF SEGMENTAL CANTILEVERED CONCRETE BOX GIRDER BRIDGES IN AREAS OF HIGH SEISMICITY

Y-N Shih

TANFB

Taiwan

J C Tai

T.Y.Lin International

USA

ABSTRACT. Cast-in-place concrete segmental bridges constructed by form travelers using the balanced cantilever method were designed for significant portions of the Sun Yat-Sen Freeway Widening Project, one of the projects in Taiwan's Six-Year Plan. Bridges within the Project were constructed as integrated concrete box girders cast monolithically with the piers and made continuous over the mid-spans where the two cantilevers meet, thus forming a rigid framing system in resisting seismic forces. Material creep, shrinkage and steel relaxation factors were taken into consideration in the computer program for analysis. Actual construction camber at different loading stages will be measured during construction and the analysis model be adjusted to verify results. Seismic design in dynamic analysis was in accordance with the Taipei Basin Spectrum and California earthquake engineering details were applied. Earthquake in vertical direction combined with horizontal earthquake components were also considered in the project. Bridge elements of segmental design bridges subject to earthquake excitation were analyzed using different coherencies to compare and evaluate the results for seismic design of the bridges.

Keywords: Continuous frame system, Cantilever segmental construction, Spectrum dynamic analysis, Multiple seismic ground motion, Time history analysis.

Mr Yen-Nan Shih is Deputy Chief Engineer of the Taiwan Area National Freeway Bureau, Taiwan, R.O.C. He has hands-on experience on managing local projects and specializes in integrating advance techniques with local conditions, especially local construction practices.

Mr James C. Tai is Senior Vice President of T.Y. Lin International and is also Chairman of T.Y. Lin Taiwan. He specializes in the design of cable-supported, steel truss, steel plate girder and arch bridge structures, as well as seismic resistance evaluation and retrofit design.

Radical Concrete Technology. Edited by R K Dhir and P C Hewlett. Published in 1996 by E & FN Spon, 2–6 Boundary Row, London SE1 8HN, UK. ISBN 0 419 21480 1.

INTRODUCTION

The Taiwan First Freeway system, opened to service only 18 years ago, is facing severe congestion problems for many years, especially on the four-lanes in each direction near Taipei City. Expansion of the freeway lanes to increase the traffic flow has become a priority task for transportation planning and administration. Local rural traffic which uses the freeway system in between major interchanges impedes the flow of longer-distance traffic near metropolitan/urban Taipei. Relieving the traffic jams at the portion of the freeway between Hsichih and Wuku as quickly as possible was identified as the priority task in the First Freeway Expansion Program. The length of this portion is 22 kilometers.

Innovative structural and construction solutions were developed to avoid the expensive acquisition of new land right-of-ways. An elevated roadway was designed to provide a three-lane expansion along each side of the existing freeway to be constructed on the existing right-of-way and shoulder lane properties. Packages have already been awarded for construction. The bridge structural system used for most of the elevated structures were designed using single piers founded on drilled Caisson foundations. Span designed as steel box girders or precast AASHTO girders vary in length from 25m to 45m. Most of the steel box girders are simply supported on pier heads without continuity. Precast AASHTO girders provided with hinges are also simply supported on pier heads without continuity. On the project where crossing Keelung River, concrete box girder system were adopted. Segmental concrete box girders were constructed monolithically with piers and made continuous over several pier supports to form a continuous frame system. Span length varied from 75m to 175m. The continuous system provides greater redundancies, better ridability, and better capabilities of resisting seismic forces.

Only the seismic analysis and design of the cantilever constructed segmental concrete box girders are presented in this article because of the limitations on length.

CONTINUOUS FRAME SYSTEM

Starting from Wuku, the existing freeway with four lanes in each direction extends northward to Hsichih, thus covering most of the metropolitan area of greater Taipei. In 1990, a proposal was prepared to improve the efficiency through traffic by constructing a new elevated structure which would carry only the freeway's South-North through traffic. The new elevated structure would be constructed along the side of the existing freeway on existing right-of-ways and shoulder properties. The detailed design contract for this portion of the freeway expansion started in mid-1990 and was completed in mid-1992.

The pier locations of the existing freeway Yuan-Shan Bridge were matched in the design of the new parallel bridge crossings in order to minimize flood blockage and back flow which backup the water level during the typhoon season and cause upper stream bank elevations to reach critical safety margins. To provide aesthetic integration of the new and existing bridges, the piers of the new bridge structures are of the same shape as those of the existing Yuan-Shan Bridge, i.e. as a cylindrical cone (Figure 1) with an interior void. The hollow piers are tapered 1:20m. The pile caps are 4.0m in depth and rest on Caissons of 2m in diameter. Piers vary in height from 8m to 30m.

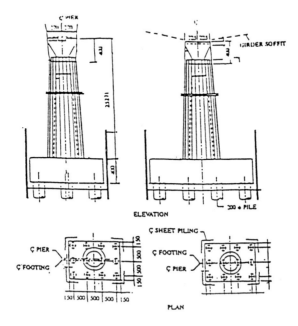

Figure 1 Typical pier

The existing Yuan-Shan Bridge was designed as a cantilevered box girder superstructure with expansion joints at the cantilever ends, thus making all piers integrated with the box girder at each side to form a self-sustained unit. The new bridges (Figure 2) are designed as a system that links several continuous girders together. This continuous framing system provides redundancy in resisting seismic and lateral or longitudinal forces. However, shrinkage, temperature and creep induced forces in the girders after the structures are made continuous complicate the design.

Computer programs "SFRAME" and "PTFRAME" developed in-house were used to analyze and verify stresses step-by-step during construction and upon completion. Creep and shrinkage model were in compliance with F.I.P. model code. Construction camber and stage stress variations were also listed for the engineer. Bridge designs were also checked for longitudinal stresses after the completion connection. Frames were checked for both longitudinal and transversal response to seismic disturbances.

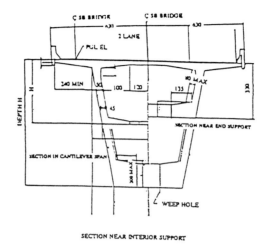

Figure 2 Typical cross section

CANTILEVER SEGMENTAL CONSTRUCTION

The segmental construction consists of North bound 7 bridge unit of total length 2611m and South bound 6 bridge unit of 2594m. The bridges twice across Keelung River and existing freeway. Bridge span varies from 75m to 175m with 2 traffic lanes and shoulder of 1m and 3m at each side totaling 12.6m. Box girder depth varies from 8.75m at pier to 2.6m at mid-span.

The pier head section was designed to have length of 14.5m box girder which is adequately to house two traveling form and then 4.5m segment for typical length. However, the contract modify the pier head length to 8.1m to house one traveling form at first step to save shoring cost, and also modify the subsequent segmental length to 5.0m to maximize the benefit of traveling form and save construction time.

Concrete pier head segments were cast-in-place monolithically with the pier shafts following placement of the reinforcement and formwork. When the formwork was removed, form travelers were hoisted and placed on each side of the pier head segment. The box girder was constructed in cast-in-place concrete segments cantilevering outward from each pier head segment. When cantilevers from adjoining piers met at mid-span, continuity reinforcing steel and stressing tendons were placed and the closure pours at the cantilever ends also varied from 2.50m to 4.00m.

The girders were designed using conventional tendon layouts (Figure 3) for segmental cantilever construction. In order to obtain more structural redundancy in the girder than provided by short straight positive tendons at the bottom of the girder, the positive tendons were designed with a parabolic shape from pier to pier. Low-relaxation tendons were used to minimize losses. The parabolic shape also provides the possibility for future repair/replacement of the positive tendons in case of future transportation overloading.

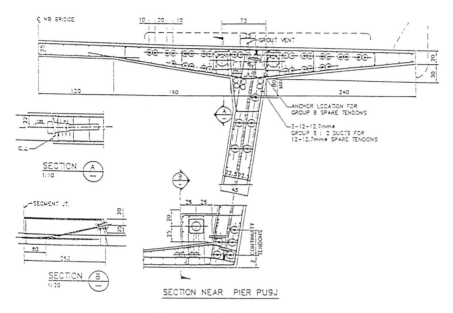

Figure 3   Tendon layout

At the span where end support will be expansion type, the construction of last quarter span near the support were modified from continuous shored pour-in-place method to a propped segmental cantilevered construction. The prop of steel frame with temporary support on footings was used in place of continuous shoring and expansive bridge form called for by P.I.P. construction.

Jacking technique were used to reduce long-term creep/shrinkage-induced shear and moments of piers and foundations and thereby minimizing the amount of reinforcements required in them. Box girders were jacked to create reverse forces equal to the built-in shear and moments at the opposite side due to predicted creep shrinkage effects.

SEISMIC DESIGN CRITERIA AND SPECTRUM DYNAMIC ANALYSIS

Seismic design was based on Caltrans' bridge design practice and design philosophy [1, 2, 3]. Following the 1989 failure of the Bay Bridge and the collapse of the Cypress Structure in the Loma Prieta earthquake, Caltrans revised its bridge design philosophy to require major bridges to remain repairable after a major seismic event. In effect, the standards for seismic design were raised to higher levels then was standard practice before that earthquake.

Seismic design was performed using a modified reduction and site-specific response spectrum developed by Prof. Penzien [4]. The spectrum curves are shown in Figure 4. The basic difference between the Caltrans' specific response spectrum for soft soils and the response spectrum of the Taipei Basin is the long plateau of the response curves reflecting the deep soft-soil character of the Taipei Basin where the response plateau extend to 1.65 second before response spectrum curve turn downwards.

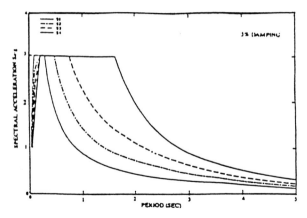

Figure 4 Normalized response spectra for site conditions S1, S2, S3 and S4

Modifications were made in the design criteria to the following: Spectrum Shape, Reduction Factor, Shear Requirement, and Confinement Reinforcement.

Elements with seismic reduction/risk factors were modified in design from Caltrans' previous 6.0 to the current 2.0 for these portions of the structures. Dynamic time history analysis demonstrated that the pier element meets a ductility requirement of 2.0 to 2.5. Time history dynamic in both elastic analysis and non-linear analysis were performed on bridge structures and compared. Column moment and load capacity interaction revealed that the ductility demand closely matches the designed ductility. The non-linear analysis response diagram also indicated that the moment capacity is well within the interaction elastic curve (Figures 5 and 6).

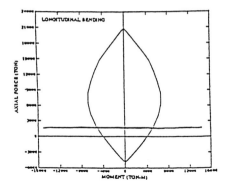 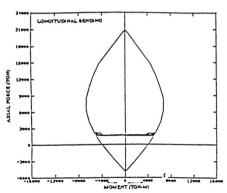

Figure 5 P-M interaction diagrams and
time-history traces obtained
by linear analysis for Pier
PU10M: Longitudinal bending

Figure 6 P-M interaction diagrams and
time-history traces obtained
by nonlinear analysis for Pier
PU10M: Longitudinal bending

Structural seismic detail criteria are similar to those used by Caltrans. Modifications
were made for local soil conditions and construction practices. Shear and
confinement reinforcement (Figure 7) were designed with special consideration for
local construction practice with circular ties used in lieu of spiral ties.

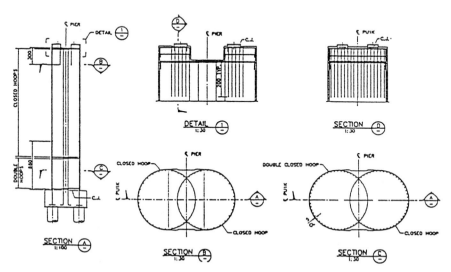

Figure 7   Typical column confinement

MULTIPLE SEISMIC GROUND MOTION - TIME HISTORY ANALYSIS
ON LONG SPAN STRUCTURE - COHERENCE CONSIDERATION

Multiple dynamic time history analysis was performed for the Yuan-Shan Bridge using seismic ground motion input which are response spectrum compatible and properly spatially correlated. One of the bridges in the expansion program was studied. This bridge has a 6-span continuous span arrangement 75-120-140-125-150-90m in length supported at two end spans on slide joints located on the two first interior pier supports to permit movement in the longitudinal direction. The other pier supports for each span are integrated (constructed monolithically) with the box girders. The cross-section is as shown in Figure 2 with a width at the top of 12.6m to carry three lanes of traffic.

Peak horizontal free-field ground surface acceleration produced by a maximum credible earthquake was taken to be equal to 0.20g. Because the bridge is continuous over six spans, covering a distance of 750m, the free-field ground motions have significant spatial variations from end pier to end pier. The coherency amplitude functions were developed from Accelerograms recorded in SMART-1 array located in Lotung, Taiwan [5]. Coherency amplitude functions are derived from array record. Three components of acceleration were used as the input to structure. Passage functions associated with these functions were correlated with the response spectrum in the investigation of multiple seismic ground motion inputs for the system.

A time history response analysis of the complete bridge model was performed for 5% Raleigh damping using longitudinal and transversal free-field ground displacement and acceleration time histories as inputs corresponding to the discrete translational foundation springs placed at the base of each pier. Notice that from ground acceleration at Pier 1 through Pier 7, the peak acceleration shifting with time from 23.5 seconds for Pier 1 to 24.5 seconds for Pier 7. The analysis was performed using computer program NEABS. The results were compared to those from uniform inputs.

Although the moment results for piers differ by 11%, the multiple input moments are judged as being more practical for the evaluation of seismic performance.

CONCLUSIONS

Rigid frame bridge structures can be designed and constructed for areas of high seismic activity. Segmental concrete box girders constructed by balanced cantilever methods are viable for crossings of rivers or areas of busy local traffic where ground activities require minimum disruption. Seismic continuous frame systems of long-span structures can be analyzed for seismic response using multiple ground motion input time history analysis techniques to verify the adequacy of the design.

Frequent dependent coherency amplitude functions can be incorporated in the time history analysis and the results obtained are more realistic for evaluating seismic performance.

Confinement reinforcement should be designed in elements to ensure ductility. New materials such as spiral coils and high-strength reinforcements may be used for construction in earthquake zones. Proper seismic reduction factors should be assigned to important bridge structures such as freeway system structures to ensure that they remain repairable after a major earthquake.

ACKNOWLEDGMENTS

Authors wish to express their appreciation to Prof. Joseph Penzien and Dr. Wen Tseng of ICEC, Inc. for the seismic design recommendations and works on multiple seismic ground motion input analysis. Also, debt are due to staff and colleagues at TANFB and T.Y. Lin International.

REFERENCES

1. Ministry of Transportation and Communications: Highway Bridge Design Code, 1976.

2. American Association of State Highway and Transportation: Standard Specification for Highway Bridges, 1983.

3. AASHTO: Guide Specification for Seismic Design of Bridges, 1983.

4. International Civil Engineering Consultant, Inc.: Task Report - Seismic Performance Evaluation of the Yuan-Shan Bridge, Taiwan Freeway Expansion Program, 1992.

5. ICEC: Task Report - Seismic Performance Evaluation of Representative Structural Systems, Taiwan Freeway Expansion Program, 1992.

INDEX OF AUTHORS

SUBJECT INDEX

This index has been compiled from the keywords assigned to the papers, edited and extended as appropriate. The page references are to the first page of the relevant paper.

ORGANISING COMMITTEE
Concrete Technology Unit

Professor Ravindra K Dhir (Chairman)

Dr Michael J McCarthy (Joint Secretary)

Mr Neil A Henderson (Joint Secretary)

Professor Peter C Hewlett
Director, British Board of Agrément

Professor Vasilia K Rigopoulou
National Technical University of Athens

Professor Sammy Y N Chan
Hong Kong Polytechnic University

Dr Nyok Y Ho
Director, L&M Structural Systems, Sinapore

Dr Frederick H Hubbard

Dr M Roderick Jones

Dr Mukesh C Limbachiya

Dr S L Daniel Ng

Dr Wenzhong Zhu

Mr Thomas D Dyer

Mr Steven R Scott (Unit Assistant)

Miss Diane H Sherriff (Unit Secretary)

INTERNATIONAL ADVISORY COMMITTEE

Dr H Z Al-Abideen
Assistant Deputy Minister
Ministry of Public Works and Housing, Saudi Arabia

Professor M S Akman
Professor of Civil Engineering
Istanbul Technical University, Turkey

Professor M G Alexanader
Professor of Civil Engineering
University of Cape Town, South Africa

Professor Carmen Andradé
Research Professor
Instituto Eduardo Tonoja of Construction Sciences, Spain

Mr J A Bickley
Implementation Manager
Concrete Canada, Canada

Professor J M J M Bijen
General Manager
Institute for Materials and Environmental Research, Netherlands

Professor A M Brandt
Head of Section
Institute of Fundamental Technological Research, Poland

Mr J C Caballero
General Director
Instituto Del Cemento Portland, Argentina

Professor A Ceccotti
Fornitek Laboratory, Canada

Professor O E Gjørv
Professor, Division of Building Materials
Norwegian Institute of Technology, Norway

Professor T C Hansen
Professor of Building Materials
Technical University of Denmark

Dr G C Hoff
Senior Associate Engineer
Mobil Research & Development Corporation, USA

Professor S Y Huang
Director
Shanghai Institute of Building Materials, China

Professor C Jaegermann
Professor Emeritus
National Building Research Institute, Israel

TECHNICAL COMMITTEE
(from United Kingdom)

Professor A W Beeby
Professor of Structural Design, University of Leeds

Dr R D Browne
Consultant, Roger Browne Consultancy

Professor J H Bungey
Professor of Civil Engineering, University of Liverpool

Dr P Chana
Divisional Director, CRIC, Imperial College

Mr P M Deason
Managing Director, Trafalgar House Technology Ltd

Professor R K Dhir (Chairman)
Director, Concrete Technology Unit, University of Dundee

Mr C R Ecob
Divisional Director, Mott MacDonald Ltd

Professor F P Glasser
University of Aberdeen

Dr T A Harrison
Technical Director, British Ready Mixed Concrete Association

Professor P C Hewlett
Director, British Board of Agrément

Mr J Innes
Chief Bridge Engineer, Scottish Office

Mr K A L Johnson
Director, AMEC Civil Engineering Ltd

TECHNICAL COMMITTEE
(from United Kingdom)
(Continued)

Professor A E Long
Director, School of the Built Environment, Queen's University of Belfast

Mr G G T Masterton
Director, Babtie Group

Professor G C Mays
Head, Civil Engineering Group, Cranfield University

Mr L H McCurrich
Technical Director, Fosroc International Ltd

Dr J B Menzies
Engineering Consultant, John B Menzies

Professor R S Narayanan
Senior Partner, S B Tietz & Partners

Dr P J Nixon
Head, Inorganic Materials Division, Building Research Establishment

Dr D J Pollock
Director, Sir William Halcrow & Partners Ltd

Professor D C Spooner
Director, Materials and Standards Division, British Cement Association

Dr H P J Taylor
Director, Costain Building Products Ltd

Professor P Waldron
Director, Centre for Cement & Concrete, University of Sheffield

Mr K R Wilson
Technical Director, G Maunsell & Partners

SUPPORTING BODIES

American Concrete Institute

Concrete Association of Finland

Concrete Institute of Australia

Concrete Society of Southern Africa

Concrete Society, UK

German Concrete Association (DBV)

Indian Concrete Institute

Institute of Concrete Technology, UK

Institution of Civil Engineers, UK

Instituto Brasileiro Do Concreto, Brazil

Japan Concrete Institute

Netherlands Concrete Society

Norwegian Concrete Association (NB)

RILEM, France

SPONSORING ORGANISATIONS

AMEC Civil Engineering

Ash Resources Ltd

Blue Circle Cement

Boral Pozzolan Ltd

British Board of Agrément

British Cement Association

Building Research Establishment

Castle Cement Ltd

City of Dundee District Council

Du Pont de Nemours International SA

ECC International (Europe) Ltd

Elkem Materials

Fosroc International Ltd

Grace-Cormix

Lafarge Aluminates

Mott MacDonald Special Services

National Ash, National Power plc

PowerGen plc

Ready Mixed Concrete (United Kingdom) Ltd

Rugby Cement

Scottish Enterprise Tayside

Scottish Power plc - Ash Sales

Transport Research Laboratory

EXHIBITORS

Advantage Precast

Allied Bar Coaters

AMEC Civil Engineering

Ash Resources Ltd

Babtie Group

Blue Circle Cement

Boral Pozzolan Ltd

British Board of Agrément

British Cement Association

Building Research Establishment

Capco Test Equipment

Castle Cement Ltd

Cementitious Slag Maker's Association

Cem-FIL International Ltd

CFPI

City of Dundee District Council

Colebrand Ltd

Du Pont de Nemours Internatioanl SA

E & FN Spon

ECC International (Europe) Ltd

Elkem Materials

Fibraflex

Fosroc International Ltd

EXHIBITORS

(Continued)

Germann Instruments A/S

GOMACO International Ltd

Grace-Cormix

Harris Speciality Chemicals

I W Farmer & Partners Ltd

John Fyfe Ltd

Kerner Greenwood (UK) Ltd

Lafarge Aluminates

Minelco Ltd

Mott MacDonald Special Services

National Ash, National Power plc

PowerGen plc

Protovale (Oxford) Ltd

Ready Mixed Concrete (United Kingdom) Ltd

Rugby Cement

Scottish Enterprise Tayside

Scottish Power plc - Ash Sales

Tarmac Topmix Ltd

TBV Stanger

Thomas Telford *Publishing*

Transport Research Laboratory

VNC Association of the Netherlands Cement Industry

Wexham Developments